Food and Famine in the 21st Century

Volume 1: Topics and Issues
Volume 2: Classic Famines

Food and Famine in the 21st Century

VOLUME 1: TOPICS AND ISSUES

William A. Dando, Editor

Santa Barbara, California • Denver, Colorado • Oxford, England

Library of Congress Cataloging-in-Publication Data

Food and famine in the 21st century / William A. Dando, editor.
v. cm.
Includes bibliographical references and index.
Contents: v. 1. Topics and issues — v. 2. Classic famines.
ISBN 978–1–59884–730–7 (hbk. : alk. paper) — ISBN 978–1–59884–731–4 (ebook)
1. Famines. 2. Food-supply. 3. Agriculture. I. Dando, William A., 1934–
HC79.F3F66 2012
363.8—dc23 2011033971

ISBN: 978–1–59884–730–7
EISBN: 978–1–59884–731–4

16 15 14 13 12 1 2 3 4 5

This book is also available on the World Wide Web as an eBook.
Visit www.abc-clio.com for details.

ABC-CLIO, LLC
130 Cremona Drive, P.O. Box 1911
Santa Barbara, California 93116-1911

This book is printed on acid-free paper ∞

Manufactured in the United States of America

To Emmaline, Anna Marie, Estella, and John and all the future decision makers of this beautiful world

Contents

Preface

Humans on Earth, although constantly in a struggle to find food, have been able to survive, multiply, overcome changing physical and cultural challenges, and flourish. Those life forms that were not able to adjust to changing cultural or physical parameters have faded into oblivion. In contrast, humans—hominids with a mental capacity sufficiently developed to think, create, and modify physical environments—spread from their centers of origin and now inhabit even the most remote lands of the Earth. They also developed value systems, traditions, and social organizations (i.e., culture). Culture ensured human survival, and cultural unity provided social stability and mutual assistance in times of acute food shortage, life-impacting hunger, and loss of life from famine. Humans, bolstered and supported by other humans, expanded their creative abilities and inventive skills along with their numbers, and the Earth began to be modified by their actions. The role of humans in changing the face of the Earth in the 21st century will undoubtedly exceed current human imagination.

It is possible—though not probable—to visualize the world in the 21st century without hunger, little seasonal or chronic undernutrition, no nutrient deficiencies, and little nutrition-related illnesses. Many political leaders, decision makers, and public opinion molders do not accept the concept of food as a basic human right. Many world citizens do not believe that food availability is a matter of facilitating food distribution. Many national financial planners do not support the inclusion of a pervasive safety net of emergency assistance, entitlements, and special need programs as a permanent component of national agendas. The world, as a result of human inventiveness and creativity, has shrunk. Places and those who inhabit places have been brought together, and what calamity befalls one nation has an impact on all nations of the Earth. A new more compassionate world society must be willing and ready to cope with the unexpected in the future. Change in the past was slow and rather steady; change in the future will be quick, affect a greater number of human lives, encompass more aspects of life, and place the environment of the world and the social institutions of billions under considerable strain.

Quotes About Hunger

There have been countless statements made by respected, knowledgeable, and concerned men and women about the suffering and pain of those who lack food or are starving. Many of these statements have been recorded, and some have been repeated in oral histories or in published works. All quotes that have passed the test of time have value to new generations. Examples of memorable quotes on hunger by Jeremiah, Chaucer, Wilson, LaGuardia, and King give authority to those who fight hunger.

Arise, cry out in the night
At the beginning of the watches,
Pour out your heart like water
In the presence of the Lord!
Lift up your hands to Him
For the life of your infants,
Who faint for hunger
At every street corner. . . .

—Lamentations 2: 19

His yonge sone, that thre
yeer was of age, unto hym seyde,
"Fader, why do ye weep? . . . Is
ther no morsel breed that ye do
kepe? I am so hungry that I
may nat slepe." Thus day by day
this child bigan to crye. . . .
"Farewel, fader. I moot dye!"
And kiste his fader, and dyde . . .

—Geoffrey Chaucer, "The Monks Tale," Canterbury Tales

"Hunger does not breed reform; it breeds madness, and all the ugly distempers that make an ordered life impossible."

—President Woodrow Wilson

"We can plant wheat every year, but the people who are starving die only once."

—New York Mayor Firorello H. LaGuardia

"When I die, don't build a monument to me. Don't bestow me degrees from great universities. Just clothe the naked. Say that I tried to house the homeless. Let people say that I tried to feed the hungry."

—Dr. Martin Luther King, Jr.

—William A. Dando

Hunger and the threat of famine exist in a world of plenty. In the first decade of the 21st century, 24,000 people died of starvation every day, and nearly 9 million each year. Between 1 billion and 4 billion people lived in grades of poverty so severe that they were unable to obtain enough food to meet their daily needs. Nearly 1 billion of the Earth's inhabitants suffered from malnourishment and a lack of clean drinking water. A silent holocaust of immense scope continues to claim lives day after day, month after month. The loss of a single life as the result of hunger or because of a hunger-related disease is tragic, because it could have been prevented. The world does produce enough food for all who live here, and, if we act wisely and prudently, it will continue to do so for future generations. The world community possesses both the financial and technical resources needed to end hunger, yet 925 million world citizens experienced issues of chronic hunger in 2010, and 50 million Americans lived in food-insecure households. In 2011, the world's population growth rate was 1.2 percent, increasing the world's population by approximately 83 million each year. The world's population in 2011 was 7 billion. Virtually all population growth is now taking place in the developing countries of Africa, Asia, Latin America, the Caribbean, and Oceania. The projected "high" world population in 2050, according to the United Nations, will be 10 billion.

In the 1960s and 1970s, the world food problem was perceived as a food-population problem. Authors such as Garrett Harden wrote on the medical aspects of this overpopulation problem, exploring new ethics for survival, the tragedy of the commons, lifeboat ethics and the case against helping the poor, and the challenges of living with death and triage (i.e., determining who will starve or who will live). Others wrote on critical issues in human ecology, ethics and health care, biomedical ethics in a starving society, impending famines in the world, and the question of who America should feed. Some concerned authors contended that if humans were to avert famines, particularly in developing countries with high population growth rates, world food production had to double—a feat that was thought unlikely. To the dismay of many human ecologists and demographers, growth in world population actually exceeded the most pessimistic expectations in the 1980s and 1990s. In the developed and developing countries of the world, however, food production rose by more than 3 percent per year, even as population increased each year by less than half that amount. Gross world food production kept pace with population growth.

In the first decade of the 21st century, the world food problem was perceived as a problem of food distribution, entitlements, poverty, and affluence. Compounding the concerns of agricultural specialists and world food planners was the uncertainty of food production in a world entering a period of rapid climatic change, the pressure on increased food production exerted by world affluence and revolutionary dietary expectations in developing countries, and the competition for food grains by those who wish to convert grain into fuel and feed for livestock and poultry. National priorities and diets have changed. Diets once composed basically of wheat, rice, potatoes, or cassava are being replaced by diets dominated by meat, fats, oils, sugar, vegetables, and dairy products. Not even the most totalitarian governments would now attempt to reverse this trend in dietary improvement. There exists a world revolution in consumer expectations, and failure to meet these expectations in food and nutrition is a more explosive topic than failure to reach nonfood consumption goals. Deprivation relative to expectations leads to social change; persistent deprivation relative to expectations leads to revolution and strife, especially in large urban centers of developing countries.

Ending hunger in the world and ensuring that famines will not occur again on Earth is a human issue. It is central to resolving a myriad of social issues, such as population growth, civil unrest and revolutions, and environmental destruction—issues that will threaten the quality of life for all earthly beings in the 21st century. Ending hunger and eliminating famine from the human vocabulary pose a highly complex challenge.

In framing this volume, 50 topics were considered essential to provide a comprehensive view of hunger and famine and to pinpoint interconnected responses necessary to create a strategy and action plan for a future free from hunger and famine. The paradigm for success must be based on people's own creativity, skills, resources, and needs. Simply feeding those in need of food will not end hunger; in contrast, creating an enabling environment in which people are provided with guidance, support, technologies, and models to grow more food will increase self-esteem and elevate self-reliance. Women and girls are the population first and most seriously affected by poverty, hunger, or famine. They must be empowered by knowledge and by assistance in food production, basic nutrition, family planning, basic health care, and education—key areas that affect both their lives and the lives of their families. Old methods of hunger relief and famine aid must be replaced by new ways of thinking and new approaches. Leadership at all levels is of great importance, and individuals with deep commitment, demonstrated integrity, and openness to all within a society and all suggestions must be mobilized. Leaders, once identified and their responsibilities defined, must bring together decision makers and activists from all critical sectors, government agencies, nongovernmental organizations (NGOs), academia, business and religions, working in equal co-partnership. They must gain insights into and a shared understanding of the problem, identify the priority areas where action is required, and develop clear strategic objectives. Action then must be taken where it can succeed and produce desired results. A momentum of accomplishments will lead to more accomplishments, more success, less hunger, and no famine.

Volume 1, *Topics and Issues,* of the two-volume *Food and Famine in the 21st Century* book set covers 50 topics, carefully selected and linked together to provide a reader with an understanding of the basic components and central issues that create or solve hunger problems and initiate or mitigate famines. Topics were determined from careful analysis of what was published in the past, what is being discussed in professional meetings today, and what were identified as critical components of the hunger–famine problem in the future. Every author who was invited to contribute an entry in Volume 1 either had published on the topic, wrote a dissertation on the topic, gave a professional paper at national or international scholarly gatherings, or was recognized by his or her peers to be a specialist in the topic. Each entry was carefully reviewed and edited by professionals who had backgrounds in food, hunger, and famine issues. To our knowledge, there is no other publication available that is as comprehensive and includes the results of so much research and intense investigation as are found in Volume 1, *Topics and Issues,* and Volume 2, *Classic Famines.* I am especially grateful for the time, effort, and commitment to this writing project by my wife and partner, Caroline Zaporowsky Dando, and to those authors who wrote entries for Volume 1 and chapters for Volume 2.

This writing project was not initiated by me. It was a topic that the editors of ABC-CLIO determined needed to be addressed, where their survey of the literature found a void. Much had been written about hunger and famine in the 1960s and 1970s, but slowly the interest in the topic declined even though the problem persisted. I was selected to be the author/editor of this project by ABC-CLIO and offered the opportunity to frame a two-volume book set, determine topics, select contributors, and approve or disapprove each entry or topic. Kaitlin Ciarmiello, Acquisitions Editor with ABC-CLIO, approached me officially on December 17, 2009, and asked if I would be interested in taking on this project as the author/editor. After much thought and after conducting an international survey of potential entry and chapter articles, I signed a publishing agreement on February 8, 2010. I would like to thank Kaitlin Ciarmiello for the invitation to author and edit Volume1 and Volume 2 of the two-volume book set, *Food and Famine in the 21st Century.* Also, I would like to thank Cathleen Casey, an ABC-CLIO Editorial Assistant; Barbara Patterson, Project Coordinator, Books; Kim Kennedy-White, Development Editor; and Christian Green, Development Editor. Although I have had wonderful authors to work with and the fantastic editorial skills of my wife, Caroline, I assume full responsibility for any errors found in the final copy.

—William A. Dando

Introduction

Topics and Issues: Scope and Application

The Earth is the only known home of humankind. People live on the surface of the Earth in physical and social environments that are extraordinarily complex, extremely diverse, infinitely renewing, and very fragile. Humans have evolved to conform with their environments and, for the greater part of human history, have constantly been molded and supported by their physical and cultural environments. They have learned that the key to survival is not resistance to change, but rather meeting change with change. In the early years of human history, surviving in a hunting and food-gathering culture required enormous areas of territory. Climate and cultural changes prompted people to migrate from the place where they were born to places where food was available. Hunger and famine plagued those who could not find adequate foodstuffs to maintain their bodies.

Substances in food (i.e., nutrients) are required by the human body for energy, growth, maintenance, and repair. Also, the effects of a balanced diet on other health dimensions are far-reaching. Physical health is extremely dependent on the quantity and quality of food-supplied nutrients to the body. Intellectual health relies on a well-functioning brain and central nervous system. Cognitive abilities and learning problems are, in turn, influenced by nutrient levels—and detrimentally affected by deficiencies. Emotional health is damaged by undernutrition and malnutrition, manifested by feelings of anxiety, rage, confusion, fear, and trembling. Social health centers on food-related occasions, and it is sometimes affected by the quality of a person's relationships with family and friends. Meals anywhere can be an enjoyable experience or a tense ordeal. Finally, spiritual health has ties to food. Some religions prohibit the consumption of specific foods. For example, both Islam and Judaism forbid consumption of pork; Seventh-Day Adventists consume only plant foods and dairy products. In India, cows are sacred animals and are not eaten. Humans, through time, have learned to adapt, secure foodstuffs, improve the quality of their diets, and survive. Even so, the world's population has always been vulnerable to hunger, starvation, and famine.

Famine: A Life-Taking Cultural Hazard

Natural and cultural hazards were once considered to be the consequences of an extreme physical event (acts of God), technological failures (acts of machines or operators of machines), or cultural failures (acts of humans and results of human decision making). Today, however, hazards are no longer viewed as singular events, but rather as complex interactions between natural, social, and technological systems. Hunger, starvation, and famine were, at one time, considered natural hazards and a means by which nature controlled population growth. Today, they are acknowledged as cultural hazards, interrelated with natural hazards, increasingly complex, and difficult to resolve. Natural hazards may cause crop failures or reduced yields; humans, by denying foods they have available to those in need, may cause famines. In this way, hunger and famine are embedded in larger political, social, economic, and technological systems. It is impossible to separate these influences from the problems of hunger and the loss of life in a famine. Reducing hunger on Earth and eliminating famines are very complex challenges, requiring new approaches at the local, regional, and global scales. In a world linked together by telephones, television, radio, cell phones, and international air carriers, hunger and famine issues become increasingly politicized, and life-saving food-sharing options are debated in the court of public opinion.

Humans' current increased vulnerability to cultural hazards, such as food shortages, hunger, starvation, and famine, is a function of both socioeconomic and political forces. In the 21st century, societal processes and decision making will determine the extent and occurrence interval of food deficiency-related cultural hazards. Hunger and famine will be intensified and populations made more vulnerable by social, political, and economic constraints on their responses and food aid. Tempering and resolution of food-deficiency events will require an understanding of physical settings, population distributions and density, socioeconomic stages of development, political contexts, and the role of individuals and impacted-area leaders if the world is to effect positive change.

Food-Famine Resolution Paradigm

A basic food-famine solution paradigm evolved as more data became available, as new methods of spatial analysis were developed, and as knowledge of human food-nutrition needs expanded. This paradigm includes the following elements:

1. Determining the food-producing and food surplus regions of the Earth
2. Creating world, national, and civil division maps of actual and potential hunger and famine regions
3. Calculating the food-population imbalances and the number of people affected by infrequent and frequent food shortages, hunger, and potential famines
4. Studying the spatial and temporal natural and cultural factors that led to acute food shortages, hunger, starvation, and famine in regions or zones of food-deficiency disasters in the past
5. Listing and describing the mitigation measures taken, economic development needs, and the positive and negative responses to requests for food aid, local actions, and "famine foods" consumed
6. Planning for future acute food shortage events that might be triggered by natural events and regional or national cultural–political decisions, then potentially magnified by political callousness and political expediency, in both rural and urban megacities in the developing world
7. Identifying and suggesting methods to increase local, regional, national, and international social discourse and social activism and to reduce resistance to end temporary food shortages, hunger, starvation, and famine
8. Formulating an all-encompassing international educational thrust that changes the way many world citizens perceive poverty, hunger, and life-impacting food, hunger-to-famine continuum problems

That All May Eat

The first decade of the 21st century has provided a foretaste of what humans can expect to occur in the next 90 or so years. Large area-encompassing droughts in major grain-producing regions, massive flooding of productive farmland, and devastating earthquakes and tsunamis have negatively affected food production and processing, and life-claiming tornadoes have destroyed food storage capabilities and food transportation-to-market systems. The world also is experiencing costly military conflicts, life-taking religious slaughters, and an international recession. Despite these challenges, however, a major food shortage or famine has not been reported anywhere on Earth.

Basically, there always has been food available to feed the world's people. Enough food is available today to feed more than 7.3 billion humans. Unfortunately, attempts to elevate the quality of life of all humans and to deal with the food–poverty–population problems of the future have been hampered by the passivity of many world citizens and the complexity of the issue. Millions of impoverished, undernourished people in remote areas of economically developing agrarian countries have been forgotten by their own governments and urban dwellers, and they have never been made known to the leaders of affluent, technically advanced nations. Also, the complexity of the current problem is greatly underestimated, because few have given thought to how to provide water and food for 9 to 10 billion people who will populate the Earth in 2050.

The solution is not simply to produce more food; rather, the catalyst for action rests with the governments of developing countries. Burgeoning food deficits in the poorer countries of the world represent ominous warnings. This life-claiming threat must not be overlooked during periods of food surpluses and on basic developed world nations' cooperative agendas. It is politically important that all nations of the world be self-sufficient in food staples. International natural and cultural hazards' uncertainties must be expected. Each government, by its own action or inaction, will determine its country's ability to feed its citizens. Short-term political considerations by the governments of both developed and developing countries of the world must be replaced by long-term planning action programs—so that all may eat.

Framing Volume I

Determining what should be included in Volume 1, *Topics and Issues*, of this two-volume book set focused upon food and famine in the 21st century, was a daunting task. The topic is extraordinarily complex. The book design had to include or cover aspects of the major topics and issues to which readers should be exposed if they want to understand the significance and ramifications of the world's agricultural systems, food production, food availability, food shortages, hunger, starvation, and famine. The intent was not to write or edit an exhaustive treatise—indeed, any article or chapter easily could have been expanded to fill a volume on its own. Instead, the goal was to provide a general introduction to the issues of food and famine in the 21st century that presupposes no background in the sciences and requires only an inquisitive mind to grasp. It is framed so that any college student, interested adult reader, or high school honors or AP student could read the text with understanding and profit from it. A "Further Reading" list is supplied at the end of each entry; a comprehensive bibliography can be found at the end of the volume; and an index assists readers in finding specific topics or locating where specific issues are discussed. The book begins with a preface, then a brief introduction, followed by 50 topic entries. These 50 topic entries provide critical insights into what has become "A Food and Famine System" (Table 1). Entries can be read individually

Table 1. A Food and Famine System (a book-use framework by entry title)

or in a clustered theme. Thus Volume 1, *Topics and Issues*, can be used as a reference work or as a classroom textbook in a 10-week quarter system or a 15-week semester system. The theme statement to Volume 1 is entitled "That All May Eat" and includes short discussions and comments on five topics.

To introduce the topics and issues involved in producing food, six topic entries were selected from the 50 included in Volume 1. These six topics focus upon the sources of food, world agricultural systems, and the Green Revolution. Their theme can be summarized as **"Experiment, Adapt, or Starve: The Agricultural Revolution."** This collection of entries provides the reader with an understanding of the significance of the Agricultural Revolution and agriculture, the great diversity of plant and animal food sources, the importance of animal husbandry, the impact of agriculture and food economics in the daily lives of humans, the evolution of a complex world agricultural system, and the impact of the Green Revolution.

Eight topic entries were grouped into the theme **"By the Sweat of Their Brow: The Basics of Food Production."** This collection offers insights into the importance of soil and arable land; discusses the term "carrying capacity"; notes the importance of climate in agriculture; and cites factors involved in climate change, the significance of conservation and sustainable agriculture, trends in agriculture, including organic agriculture, and means to sustain a food-producing physical environment in the future.

Four topic entries were grouped into the theme entitled **"The Quiet Revolution: Enriching the Human Diet."** These entries pinpoint food sources, list the basic foods of the world in the 21st century, describe the diffusion of foods from one part of the world to another, and analyze the role of food in popular culture and the perception of hunger and famine.

Eight topic entries were clustered into the theme defined as **"Natural and Cultural Hazards: Current Problems in World Food Provisioning."** Today humans are faced with undeniable evidence of climate change. The ice caps of Antarctica, Greenland, and the Arctic are melting at a rapid rate, sea level is rising, storm patterns have changed and storms intensified, and droughts and flooding have reduced food production in many parts of the world. This group of entries provides evidence of future climate change impacts, outlines the problem of desertification and expanding deserts, explores the role of drought and water shortages on crop yields, provides maps of changing North American agroclimatic regions, and stresses the ramifications of food poisoning and the need for food safety.

Seven topic entries were grouped into the theme of **"Silent Deaths: Hunger and Famine."** In recent years, there has been an upsurge in historical studies of local food shortages, hunger incidents, and famines. Included in this selection is a historiography of food, hunger, and famine; a critical examination of Thomas Robert Malthus's impact on food aid and famine relief in the period between 1800 and 2000 CE; and discussions of the horrors of hunger and starvation, the suffering during a famine, and the deterrents to hunger and famine posed by unbounded foodstuff trade.

Eight topic entries were collected under the theme of **"Human Resourcefulness: Responding to Food and Famine Issues in the 21st Century."** From the beginning of time, their ingenuity and resourcefulness have enabled humans to overcome adversities and life-taking problems. Today, in the early years of the 21st century, humans have the tools and the capability to monitor worldwide food production, land-use change and conversion, and environmental destruction. This group of entries illustrates the capabilities of satellite remote sensing and an amazing array of geotechniques that provide very accurate data on food production and areas of food production problems, comments on the significance and future of genetically modified foods, offers an example of farm adjustments to climate change, and provides insights into new energy resources, the growth of community-supported agriculture, and innovations occurring within the U.S. farm machinery industry.

Five topic entries were grouped into the theme of **"Saving Lives: Food Aid and Famine Response."** For political reasons, until 1969, the United States' foreign assistance agencies were reluctant to support overseas development programs to increase productivity of basic food crops. At the same time, few decision makers understood the complexity of hunger and famine issues, which were primarily biological and social in nature. This cluster of entries begins with an overview of food aid policy debates and then explores food policies of the United States, food assistance programs and landscapes within the United States, and mapping hunger in the United States.

Four topic entries were grouped into the theme of **"Educating the Decision Makers of the Future."** For humans to reduce poverty and alleviate the possibility of famines, they must have the will and wisdom to identify problems, determine critical issues, devise plans to overcome the problems, and implement these plans quickly and effectively. People with vision must be nurtured, educated, and supported. Education is the basis of human progress and a conflict-free world of the future. The four topic entries included in this group begin with teaching concepts, teaching data sources, and teaching definitions. They conclude with an in-depth study of World Food Day as a teaching-action forum.

The final theme of **"Feeding 10 Billion More Humans"** is covered by two topic entries. These entries comments on world population trends and projections to 2050, the second on climate change. The world's population is increasing, more than half of the world's population now lives in developing countries, and these low-income countries have the most acute food problems. Unfortunately, food production gains have been offset by population increases. Unless significant changes occur, the world's food–poverty population situation will become significantly worse. Governments that fail to respond to their citizens' demands for food will be replaced.

Regionalizing the Factors That Create and Sustain Famine

Hunger and famine regions are defined and delimited on the basis of similar, associated, or combined factors that provide internal unity or homogeneity and distinguish famine regions from surrounding areas. The 10 famine regions defined and delimited in Volume 2 are based on multiple factors or a combination of complex factors. Most of these factors are described in the 50 entries found in Volume1. Regionalizing famine factors to determine the extent of death-producing starvation and disease may create either a small famine region, such as a city or place, or a larger region, such as a country. In the past, world maps of famine events were based on hundreds of famines and were, in essence, multifactored physical and cultural combination famine zones. After delimiting and defining a famine region, a great deal of time and effort was devoted to determining the single, multiple, or interacting factors that reduced loss of life or ended the famine. Volume 1, *Topics and Issues*, provides readers with insight into those factors or associated factors that, when imbalanced, produce a famine. Volume 2, *Classic Famines*, synthesizes, regionalizes, and analyzes factors to produce a descriptive narrative that gives the reader an understanding of the famine event that could not be gained otherwise. Each famine was critically researched and the triggering or contributing factors examined carefully and in detail to determine why it happened and, if the same combination of triggering or contributing factors were to emerge in the future, why famine might be expected to happen again. All famine regions have distinct characteristics, and they are bonded primarily by one factor—many people have suffered horribly and died.

The 10 classic famines included in Volume 2 were identified and selected from a data bank of thousands of famines. They are models for potential future famines.

1. **Russian and Soviet Famines: 971–1947.** Repressive social systems controlling food production and utilizing funds secured from the sales of agricultural products to support the elite, the government, or the purchase of equipment for industrial development led to famines. See Volume 1 topic entries beginning on pages 1, 11, 23, 71, 120, 139, 269, and 279.
2. **Indian Famines: 1707–1943.** Famines were nurtured by poverty, entitlement conflicts, and colonialism, which collectively produced an environment in

which food was used to fund governmental and military expenditures, support a ruling class, and enrich colonial masters. See Volume 1 topic entries beginning on pages 1, 39, 66, 71, 120, 139, 225, 269, 279, 319, and 415.

3. **Chinese Famines: 108 BCE–1961 CE.** Population pressure on a limited amount of good arable land, internal strife and wars, expanded settlement in drought and flood-prone marginal food-producing land, and unwise political decisions resulted in famines. See Volume 1 topic entries beginning on pages 1, 11, 66, 71, 79, 120, 139, 225, 269, 279, 319, 367, and 429.
4. **Great Irish Famine: 1845–1850.** An exploitive political system and population pressure on arable land provided the setting for a potato blight that caused crop failures and mass starvation, unleashed cultural prejudice that subsequently restrained famine aid, and resulted in the teachings of Malthus being applied to a real-life situation. See Volume 1 topic entries beginning on pages 1, 39, 66, 103, 139, 162, 181, 208, 225, 259, 269, 279, 319, and 445.
5. **Bengal Famine: 1943–1944.** The Japanese invasion of South Asian countries during World War II and the threat of the invasion of India led to an Indian government/British military "food denial" policy and the famine that claimed the lives of millions of noncombatants in a food-surplus, densely populated region of India. See Volume 1 topic entries beginning on pages 1, 11, 66, 71, 103, 139, 208, 225, 269, 279, 319, 429, and 445.
6. **Dutch Famine: 1944–1945.** The German invasion and occupation of traditionally neutral Holland in the early years of World War II led to much hardship and deprivation. Near the end of the war, a food embargo by the Germans led to an unwarranted, primarily urban, famine. See Volume 1 topic entries beginning on pages 139, 191, 208, 225, 269, 279, 319, 339, and 415.
7. **Mao's Famine: 1959–1961.** The decision by Mao to rapidly industrialize China in 1959, using rural food producers' manpower and funds secured from internal and export sales of foodstuffs for industrial machinery, created the largest and most life-taking famine in history. Indifference to peasant loss of life extended this famine. See Volume 1 topic entries beginning on pages 1, 66, 139, 208, 225, 269, 319, 339, and 429.
8. **Ethiopian Famine: 1984–1985.** A Marxist military dictatorship's demand that farmers produce and sell, at less than production costs, food to feed urban dwellers; a catastrophic drought; desertification; and a civil war collectively produced a devastating famine. See Volume 1 topic entries beginning on pages 23, 32, 39, 66, 71, 79, 110, 120, 139, 191, 208, 225, 259, 269, 319, and 339.
9. **North Korean Famine: 1995–2000**. A Soviet Union–implanted military dictatorship in North Korea adopted a Soviet-style agricultural organization

system that failed. These factors, combined with a series of natural disasters and a political decision not to inform the world or seek food aid, engendered a brutal, multiple-year famine. See Volume 1 topic entries beginning on pages 1, 39, 66, 71, 79, 103, 120, 139, 162, 208, 225, 269, 319, 339, 347, 415, and 429.

10. **Famine in the United States: A Future of Uncertainty.** This study concluded that only if a "perfect storm" of circumstances occurred or if Americans considered themselves as a favored segregated entity, detached from the citizens of the world, would the United States experience a famine. See Volume 1 topic entries beginning on pages 1, 11, 39, 66, 79, 86, 94, 129, 154, 162, 171, 208, 235, 242, 288, 299, 308, 329, 347, 375, 415, 429, and 445

Readers are encouraged to not only read the Introduction and the 50 topic entries and issues included in Volume 1, but also the Introduction, the 10 classic famines, and the Postscript in Volume 2. Both volumes are designed to be standalone entities—that is, information sources for specific reader needs. Volume 2, *Classic Famines*, also can be used simply as a reference work, as the second book in a two-volume textbook set for a year-long class or seminar on food, hunger, and famine; or as a one-book text for a 10-week quarter or 15-week semester class devoted to famine.

Time for Renewed Commitment and Positive Thinking for a World Free of Hunger and Famine

Despite the abundance of books, articles, and pamphlets about the factors that contribute to nutrition, food, hunger, and famine, and the differences in opinions found in these publications, there are many points of agreement on these subjects. First, the population-food issue or Earth's carrying capacity was first vigorously debated in Europe in the mid-19th century. Many figures were proposed, but the general conclusion in the 1890s was that the world could support a population of not more than 6 billion people. This figure was projected to be reached in the 2070s. Many additional variables and factors were then introduced into the conjectures—specifically, world climate and weather data, which were then becoming more comprehensive. The "highest conceivable" population of the Earth, in 1925, was estimated at 16 billion, with 1 billion people in the United States alone. Projecting improvements in seed and livestock and combining these factors with changing technological levels, one group of demographers-geographers contended, in 1925, that the Earth could support, at most, 6.5 billion at the living standards of the French. In later years, numerous projections of the Earth's carrying capacity were developed, based on the extension of irrigation and land reclamation, yield intensification, exports of farm products, and shifts in dietary emphasis. The wide variation in the population estimates used and the inability to compare the results because of the different research methods used and factors considered, however, limited the usefulness of their findings. Also, some of the absolute maximum estimates were quickly exceeded. The optimum carrying capacity of the Earth is now estimated at 10 to 15 billion people, dependent upon a favorable complex of economic, social, political, and natural conditions or hazards rather than on merely the productive capacity of the Earth's land and oceans.

Ever since the clergyman-economist Thomas Malthus published his thoughts in the *Essay on the Principle of Population* in 1798, most researchers and authors have defined or understood the problem of food and famine largely in terms of food production and population growth. The relevant questions to gain insights into hunger and famine in the 21st century are whether the Earth's ecosystem can withstand the stress placed upon it by human exploitation of the Earth's resources and whether there is a commitment by the decision makers of the world to reduce poverty, eliminate hunger, and confront any potential of famine through the concerted actions of all food-surplus nations. The tools and the ability to eradicate hunger and famine are in place. But does the commitment necessary to achieve these goals exist?

—William A. Dando

Agriculture

Humans have always realized that they must eat to live. Today, as in the past, they spend much time, effort, and thought on food. Food nourishes our bodies; contributes to our physical, mental, and emotional health; and provides us with pleasure. Beginning approximately 10,000 years ago, humans made great strides in developing the means to improve the basic foods they were consuming, to increase food production at a place, and to feed more people by exchanging food products. At the beginning of the 21st century, the people of the world are surviving primarily on the yields of cropland equivalent to one-half acre per person. Only one-fourth of the total area of the world is used for agricultural purposes, and only one-tenth of this area is cultivated. Even so, the world's food producers provide enough food to supply the current needs of over 7 billion people. This is a remarkable achievement and a tribute to the skills, efforts, and ingenuity of a diminishing number of agriculturalists.

Agriculture is the art of understanding the opportunities and potential for food production at a place, determining which crops or food-producing activities are best suited to the location, and making the site or place more productive by preparing the land, planting and raising the crop, harvesting it, storing it, and getting the surplus to markets. In some parts of the world, agricultural food-producing methods and yields are very similar to those employed hundreds of years ago. However, in most parts of the world, with the aid of science, new and improved technologies, better seeds and hybrids, quality-enhanced livestock and poultry, mechanization, and fertilizers, pesticides, and herbicides, more food is produced today than ever before. Modern agriculture is one of the great success stories of human society. In almost every case of acute food shortages or famine, the cause of human suffering and death was not the lack of food somewhere in the world, but rather the will and commitment to supply food to the stricken area or place from food-surplus areas or from storage facilities. Challenges to food producers in the 21st century are related to the projected increase in world population, changing dietary expectations, and the need to supply food to 9 billion or more people from relatively limited cropland with unequal distribution of water in a century of rapid climate change and destructive weather variations.

Origins of Agriculture

Evidence strongly suggests that agriculture was originated by societies who had inadvertently adopted a rudimental sedentary survival system at places they found and occupied that had an intensive assortment of wild food plants and wild food animals. Four major centers of food plant domestication have been posited: the Middle East, the Sahel zone in northern Africa, Southeastern Asia, and sites in Central and South America. The Middle East center also is believed to be where food animal domestication began. Here, plant and animal husbandry became closely related, whereas in the Central and South American center, animal domestication and husbandry were less important. In the Fertile Crescent of the Middle East, goats, pigs, sheep, and cattle (the major food animals in modern agriculture) were first hunted and eaten; later, others were domesticated. Here, wheat and barley seeds were first gathered, then planted. Some sources contend that by 6000 and 5000 BCE, farming villages extended from the Nile River Valley in Egypt to the Tigris and Euphrates valleys of Mesopotamia.

Successive migrations of farmers and herders introduced these basic crops and animals north and westward into Europe by 4000 BCE and then westward into the Sahel zone of Africa. Food crops and animals also diffused eastward into Iran from the Fertile Crescent, then into the Indus Valley and India in approximately 3500 BCE. Chickens are descendants of jungle fowl of India and neighboring countries. Millet, rice, and soybeans were domesticated in China in this time period, and slowly cattle, pigs, sheep, and goats began to be raised for food and skins.

Corn and potatoes were the most important food plants in Central and South America and may have been domesticated at least by 2000 BCE, as well as beans. Two thousand years after the domestication of corn, village farming appeared in the North American southwest and, and eventually beans, squash, potatoes, and corn became the primary food crops. Nowhere in forested eastern North America did crop agriculture—which was primarily a part-time occupation for women—play as decisive a role as in the Middle East, the Indus Valley, China, Central and South America, and the Sahel zone of Africa.

The domestication of livestock and the use of livestock products for food and clothing were initially a survival adaptation by nomadic peoples. Exchange of livestock and livestock products for grain, fruit, and vegetables became common. Eventually in many parts of the world, the distinction between herdsman and cultivator disappeared, and a mixed farming system evolved to take its place. Diffusion of new agrotechniques, the realization that animal manure would restore soil fertility, the introduction of new crops, and improved farm implements gave stimulus to intensification of agriculture. A system of modified pastoral nomadism has survived in areas of mountains and valleys with lowland areas that can grow planted food crops. Transhumance—the spring migration of herds, herdsmen, and, at times, families from lowland valleys up to hill and mountain pastures and then back down to the valley lowlands in the fall—is a proven and successful human adaptation to the opportunities available to increase food availability at places. Over time, selective breeding of livestock led to the creation of the animal stock currently found in the world. Products from ranches and farms now provide more than 90 percent of the food and beverages consumed on Earth, and more than 50 percent of the world's population is employed in agriculture.

The Modern Agricultural Revolution

The growth of industry in the 18th and 19th centuries and the demand for food by the new urban industrial centers led to a modern Agricultural Revolution. From approximately 1750 onward, an expansion of food needs occurred in the industrial nations of Europe. Vast areas of very productive new agricultural land became available in the New World, as did new crops, wealth, and a surplus of hard-working and talented people willing to leave Europe and exploit the opportunities found in the New World. When settlers left their European homelands seeking a new life in the Americas, they brought with them proven agrotechniques, horses, implements, cattle, sheep, pigs, fowl, seeds, and plants. They found the weather and climate suitable for European crops and animals, and the soils rich enough to produce large amounts of food for export. In segments of the New World, plantation and estate farming became important. Food and beverage crops, such as coffee, tea, cacao, coconut, pineapple, bananas, and rice, were introduced into the countries and colonies of the New World and other regions of the world influenced by European commercial expansion; all were linked with Europe. A blend of Old World and New World crops, livestock, and agrotechniques emerged in many parts of a more commercially interconnected world.

Arable land and land planted to food crops are unequally distributed in the world. Asia, with more than half of the world's population, has only 31 percent of the world's arable land; North America and Central America, with less than 10 percent of the world's population, have almost 20 percent of the world's arable land; Europe, with 14 percent of the world's population, has slightly more than 4 percent of the world's arable land; and Africa, with more than 9 percent of the world's population, has 18 percent of the world's arable land. The crops planted, agrotechniques and agrochemicals used, and crop yields vary from continent to continent and from country to country (see Table 2). Wheat, rice, and corn are primary food grains. Wheat is produced mainly in North America, Western Europe, Russia, Ukraine, China, India, Argentina, and Australia. Rice is produced mainly

Table 2. World Wheat, Rice, and Corn Production (in million tons)

	1935–1939 Average	1956–1958 Average	1965	1999
1. Wheat	166	209	274	536
2. Rice	162	203	256	512
3. Corn	121	173	226	550

Sources: FAO. *1965 Production Yearbook*, 1966, 19; *2001 World Almanac and Book of Facts 2001*. Mahwah, NJ: World Almanac Books, 2001, 160.

in India, China, Japan, Indonesia, Vietnam, and Thailand. Corn is produced primarily in North and Central America, South America, Europe, China, and India. The major wheat-exporting country in the world is the United States, the major rice-exporting country is Thailand, and the major corn-exporting country is the United States. The major wheat-importing country in the world is Egypt, the major rice-importing country is Bangladesh, and the major corn-importing country is Japan (see Table 3).

Fruit and vegetable production are intensive agricultural activities that are concentrated in select portions of many countries of the world where weather, climate,

Table 3. World's Leading Wheat, Rice, and Corn Exporters and Importers: 1998 (in thousands of tons)

Exporting Country	Amount	Importing Country	Amount
Wheat			
1. United States	27,004	1. Egypt	7,340
2. Canada	17,702	2. Italy	6,916
3. Australia	15,231	3. Brazil	6,395
4. France	13,733	4. Japan	5,758
5. Argentina	10,371	5. South Korea	4,695
Rice			
1. Thailand	6,356	1. Bangladesh	2,635
2. India	4,800	2. Philippines	2,200
3. Vietnam	3,800	3. Iran	2,000
4. China	3,792	4. Indonesia	1,895
5. United States	3,113	5. Brazil	1,305
Corn			
1. United States	42,125	1. Japan	16,049
2. Argentina	12,442	2. South Korea	7,111
3. France	7,979	3. Mexico	5,212
4. China	4,687	4. China	5,009
5. Hungary	2,109	5. Egypt	3,043

Source: World Almanac and Book of Facts 2001. Mahwah, NJ: World Almanac Books, 2001, 160.

Corn-Fed Fuel

Ethanol has been proclaimed by many as the United States' economic and environmental salvation. Others, however, have expressed a genuine concern about diverting cropland from food production to fuel conversion. A third group of Americans is concerned about the huge subsidies which are stimulating the ethanol boom.

Whatever the motives of supporters or critics, the U.S. government is strongly supporting this energy option, recognizing that it offers a green alternative to fossil fuels. In the United States, 95 percent of the total ethanol production is derived from corn. The Congressional Research Service (CRS) reported that, when it is blended with gasoline, ethanol can reduce emissions and extend gasoline supplies. The American Coalition for Ethanol (ACE) and others contend that ethanol reduces tailpipe–carbon monoxide emissions by approximately 30 percent, and suggest that the growth of ethanol production has caused retail gas prices to be 29 to 40 cents lower per gallon. Nevertheless, ethanol has a lower energy content and more fuel is required to travel the same distance as when petroleum products are used as the source of fuel. Ethanol-producing facilities also need huge amounts of water, and ethanol cannot be transported via gasoline pipelines.

The demand for corn as a source of ethanol has created a shortage and driven up corn prices. Demand for corn is so great that U.S. livestock and poultry producers are now often unable to find or afford the animal feed they need. The exports of corn for food are also being affected by the push to increase ethanol production. The cost of corn or cornmeal in developing countries has put stress on the poor to buy fewer corn products, and they have already had to reduce the amount of corn in their diets. In January 2011, corn stockpiles in the United States were among the lowest levels ever recorded—just 5 percent of the total corn used or exported.

—William A. Dando

and water are adequate for the specific crop. However, commercial apple and pear production is concentrated in Europe and North America; grapes, in Europe; figs, in the Near East and Europe; and bananas, in Central America and South Asia. In the modern world agricultural system, food crops are harvested in almost every month of the year somewhere in the world (see Table 4, as an example).

Modern Agriculture Types

Agricultural enterprises are classified according to crops and animals raised and the purpose of production. The basic modern agricultural types are as follows:

1. **Mixed farming** occurs when crops are cultivated for sale and partly for animal feed. The distribution of mixed farming relates to environmental

Table 4. Wheat Harvesting Months in Selected Exporting and Importing Countries

Country	Jan.	Feb.	Mar.	Apr.	May	June	July	Aug.	Sept.	Oct.	Nov.	Dec.
Argentina	*	*										
Australia	*											
Chile		*										
India			*	*	*							
United States						*	*	*	*			
Canada							*	*	*			
Russia							*	*	*			
Japan								*				
United Kingdom								*	*	*	*	
New Zealand												*

Source: C. F. Jones and G. G. Darkenwald. *Economic Geography.* New York: Macmillan, 1954, 287.

conditions, but it primarily occurs in areas characterized by rolling to flat agricultural land, deep rich soils, and adequate moisture and heat during the growing season. Mixed farming is found in Western Europe, Eastern Europe, Canada, Australia, and New Zealand. Typical food crops are cereals, corn, soybeans, potatoes, and root crops; typical animals raised for food are dairy cows, beef cattle, sheep, pigs, and chickens. Farm output and yields are very high.

2. **Dairy, fruit, and small grain farming** is located generally near, and in some cases surrounds, large population centers in the Northern Hemisphere, Australia, and New Zealand. Environmental requirements include mild winters, rainy and cool summers, and soils that will support pasture grass and nutritious hay. Milk brings a premium price in urban regions. Butter, cheese, and other dairy products have high value per weight and are transported over greater distances to market than milk. Many dairy farms resemble mixed farming enterprises except that they concentrate on milk and milk products rather than meat.
3. **Mediterranean agriculture** is found in selected regions of the world that have a dry summer, subtropical type of climate. In these areas, the weather and climate, which feature winter rains and much summer sunshine, favor the growth of winter wheat and barley, drought-resistant vines, grapes, citrus fruit, vegetables, dates, and olives. Growing grapes for wine is common. This modern agricultural type is found primarily in selected regions around the

Mediterranean Sea, California in the United States, Central Chile, Southwest Australia, and the tip of South East Africa.

4. **Wheat and small grain farming** is a specialized type of modern agriculture that came to the forefront when growing urban populations in Western Europe and the United States required vast quantities of bread grains. This crop type is found in the drier margins of the humid continental climates, the steppe climates, and the humid fringes of the semi-arid climates. Vast acreages of land planted in wheat, rye, barley, and oats are found in the northern Great Plains of the United States and Canada, in Australia, and in Argentina. The relatively low yields per acre are countered by very large farm size and the mechanization of all agricultural activities. Wheat and small grain farms are the largest of all crop-growing agricultural enterprises and the most mechanized. Weather variations and drought are ever-present dangers in most non-irrigated grain-producing regions of the world.
5. **Range livestock ranching**, like nomadic herding, requires much land and extensive pastures. Beef cattle, sheep, and goats are the primary food-producing animals found on vast ranches in the western United States and Canada, southern Russia, northern Kazakhstan, northern Chile, southern Brazil, Uruguay, Argentina, South Africa, extreme northwest and central Australia, and portions of New Zealand. Great variability in the limited amounts of precipitation from year to year in marginal agricultural areas of the world and, in many cases, on desert fringes exposes both the livestock and the humans who tend the livestock to severe risks.
6. **Nomadic herding** is an ancient and traditional agricultural type now confined primarily to remote areas of Eurasia, Africa, Canada, and the United States. It is essentially a subsistence economic system in which nomads depend chiefly on the milk, blood, meat, wool, and skins supplied by their livestock. Nomadic herders are constantly or intermittently on the move, searching for forage (animal feed) and water. Sheep, goats, cattle, alpacas, yaks, llamas, camels, reindeer, and horses are tended. Life for nomadic herders is difficult, as they roam in some of the coldest and driest regions of the world, classified by many sources as "non-agricultural areas."
7. **Plantation agriculture** is a specialized agricultural type found widely in the tropics. A very distinctive form of agriculture, it is one of the oldest of the modern large-scale agricultural types. Plantations were formed to grow and process cash crops for exports to temperate lands. Most products of plantations have little competition with crops grown where their produce is sold, mainly Europe and North America. Plantation owners recruit workers from local or nearby districts, hire administrative and technical staff from Europe

or North America, and purchase plantation equipment, implements, machinery, and even a portion of the food consumed from other parts of the world, but especially North America, Europe, and Japan. Plantation food crops include bananas, cacao, tea, coffee, sugar, pineapple, and nuts.

8. **Corn, beans, and livestock farming** is one of the most significant and efficient means of producing foods in the world. Field crops include corn, soybeans, wheat, oats, and barley. Crop surpluses are sold and much of the grain is utilized to feed cattle, hogs, and chickens. Farming methods include use of much machinery, close attention to breeding and plant selections, field rotation of crops, fertilization of the soil, and careful selection, tending, and feeding of livestock. The farmer and his or her family supply most of the farm labor. Overall, no other type of farming has more progressive farmers or a higher standard of living. This type of agriculture is practiced in the humid middle latitudes of the world of all continents, except Asia. The model and outstanding development of this farm or agricultural type is the corn belt of the United States.
9. **Specialty crop farming** is found throughout the world where unique combinations of weather, climate, soils, topography, food plants and trees, and animals are found in close association with a market or where products can be transported to distant markets with ease. A large band of specialty crop farming extends from coastal Connecticut, New Jersey, Virginia, North Carolina, South Carolina, Georgia, Florida, Alabama, Mississippi, and Texas in North America, producing tobacco, vegetables, citrus fruits, peaches, and apples. In Europe, greenhouses cover extensive areas, producing table grapes, tomatoes, and other fruits and vegetables. Most cities of the world have "truck farms" or "specialty farms" that grow and sell the specialized products demanded by consumers.

Trends in Modern Agriculture

The world's agricultural system in the late 20th century was able to supply the population of the world with high-quality, diverse, fresh, and nutritious foodstuffs at a reasonable price. For example, world food production in 1960 was approximately 40 percent greater than in 1940, representing a nearly 2 percent increase per year. Likewise, the world's population between 1940 and 1960 increased nearly 34 percent in total, or an average of 1.7 percent per year. The average annual rate of increase in agricultural production exceeded population growth by only a small amount during this period. Since then, spectacular achievements have been made in the breeding and feeding of livestock. In the United States in 1900, it was estimated that one cow supplied the daily need of dairy products for 5 persons; in 2011, it was estimated that a cow supplied the needs of 30 to 35 individuals. In 1930, beef cattle were fed to market weight in 2½ to 3 years; in 2011, only 1½ years or less of feeding was needed to reach this milestone. Chickens that reached a weight of 1½ pounds in 9 weeks in 1930 now weigh 3 to 3½ pounds in the same time span. Farmers in 2011 could choose from almost an infinite variety of improved hybrids that are designed to thrive in local soils, grown under local precipitation and temperature regimes, and produce high yields. Crops are made resistant to drought and disease, adapted to machine cultivation and harvesting, and given qualities desired by consumers. Also, the employment of modern, efficient implements in the United States has accelerated output per farmer at an astonishing rate. Unfortunately, this modern miracle in food production and the achievements in farming are not found in all countries of the world and have not affected all who grow food in the world.

In most cases, a farm undergoing a transformation from traditional practices to a modern enterprise requires large investments of capital and competent management. Farms must be mechanized and the products produced must be harvested quickly and efficiently, so as to satisfy consumer preferences. Modernization of agriculture leads to larger-sized enterprises, reduction in the number of farms, and release of farm workers for other socioeconomic needs. It also leads to the enhancement of the quality of life for those who produce food.

The demand side of world agriculture is influenced by three great pressures in the 21st century: (1) population increase and redistribution; (2) changing consumption standards and improved diets; and (3) climate change. According to the United Nations and other estimates, the world population reached 6 billion in 1999 (see Table 5). This population doubled in approximately 40 years, and the world gained 1 billion people in just 12 years. The world's population is expected to exceed 9 billion in 2050, with most of the increase occurring in countries that are less economically developed and in which citizens suffer from dietary deficiencies and food shortages.

Citizens of the world are attempting to emulate the diet patterns and the eating standards of their counterparts in North America, Europe, Australia, and New Zealand. Such a diet includes more meat, socially superior food grains, high-quality fruits and vegetables, dairy products, and condiments. An even more important factor in changing diets are the new consumption standards set by modern nutritional science. Food taste and diet changes are most striking in urban-industrial societies and accompany economic and social changes.

Agriculture is the most important of all primary economic activities, yet the most dependent upon weather and climate. In the same way weather and climate

Table 5. Population of the World, 1950–2000 (estimated in thousands)

	1900	1950	2000	Percentage of World Total in 2000
North America	106,000	372,000	481,000	8%
South America	38,000	110,000	347,000	6%
Europe	400,000	392,000	729,000	12%
Asia	932,000	1,411,000	3,688,000	61%
Africa	118,000	229,000	470,000	13%
Former USSR	—	180,000	290,000	—
Oceania and Australia	6,000	12,000	31,000	1%
World	**1,600,000**	**2,556,000**	**6,080,000**	**—**

Source: World Almanac and Book of Facts, 2001. Mahwah, NJ: World Almanac Books, 2001, 860.

help to define climax vegetation formation, they also set limits for crop and animal production. If current climate change models are correct and if the trends noted in the past few decades continue, human survival within current dry or drought-prone regions will be threatened by overpopulation in marginal areas, environmental degradation in once prime agricultural areas, changing weather patterns, acute water shortages, and agricultural region shifts due to climate region reconfigurations and alterations. The greatest negative impact of climate change in the 21st century is occurring in the non-industrial or developing countries of the world with the greatest population increases and with the fewest resources to develop new farmland or make existing farmland more productive.

To satisfy the pressure placed on the world's food producers to overcome the pressures exerted by 9 billion or more people, changing consumption standards and diets, and climate change, agriculture in the 21st century must increase production by biological and technological means, expansion of food-producing crop land by utilization of empty or under-used areas, accepting change and taking advantage of climate change opportunities, and integrating food commerce and food supply systems.

—William A. Dando and Vijay Lulla

Further Reading

Broek, Jan O., and John W. Webb. *A Geography of Mankind*. New York: McGraw Hill, 1978, 264–280.

Brown, Lester R., Christopher Flavin, Sandra Postel, and Linda Starke. *State of the World: 1990*. New York: W. W. Norton & Company, 1990, 59–78.

Dando, William A. "Food." In A. S. Goudie and D. J. Cuff, eds., *Encyclopedia of Global Change*. Oxford, UK: Oxford University Press, 2002, 455–461.

Dando, William A., and Vijay Lulla. "World Agriculture and Food Provisioning, A.D. 2100." In *Climate Change and Variation: A Primer for Teachers*. Washington, DC: National Council for Geographic Education, 2007, 157–165.

Jones, Clarence F., and Gordan G. Darkenwald. *Economic Geography*. New York: Macmillan, 1954, 145–354, 287.

Phoenix, Laurel E. *Critical Food Issues: Problems and State-of-the-Art Solutions Worldwide*. Santa Barbara, CA: ABC-CLIO, 1 (2009): xiv–xx.

Statistical Abstract of the United States: 2002. Washington, DC: U.S. Government Printing Office, 2002, 515–538.

World Almanac and Book of Facts 2001. Mahwah, NJ: World Almanac Books, 2001, 160, 860.

Agriculture, Economics, and Milk

The cost of food in a market economy is often a concern for people who live on limited means and attempt to stretch small incomes to meet family food needs. Prices for food have been on the rise, even in developed nations. In the United States, it is becoming more common for limited supplies of grains to be used for the production of biofuels, such as ethanol, instead of for human consumption. This phenomenon can be termed the "food-to-fuel economy." The prices of common grain crops, such as wheat, corn, and soybeans, have been on the rise over the last few years (see Figure 1). Corn prices are highly cyclical and are closely tied to the price of oil on the world market. As the price of oil goes up, more corn is used for ethanol production, and the price of corn goes up accordingly. This trend, in turn, increases the costs of production for farmers who need to buy not only corn, but also soybeans to feed livestock and generate livestock products such as milk and eggs. Between 2007 and 2008, average feed costs in the United States increased by 26 percent. Feed accounts for more than one-third of the costs of production for the average hog farm. In other types of livestock operations, animal feed may account for as much as 70 percent of production costs. These cost increases are eventually passed along to the consumer in the form of inflated prices at the grocery store.

One of the most basic foodstuffs in the human diet is milk. Milk prices are affected when farmers choose to sell grain crops for conversion to fuel instead of food. Dairy cattle are dependent on a specific diet, usually made up largely of grains, which is designed to generate the energy necessary to

Figure 1. The increasing price of grain crops in a food-to-fuel economy

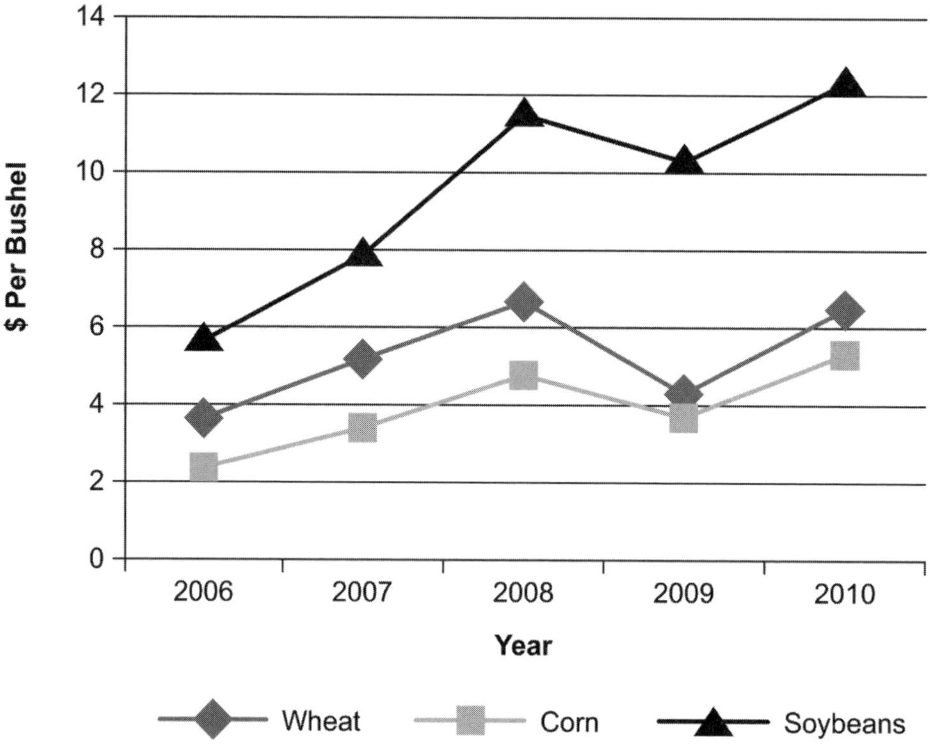

Source: Dawn M. Drake. Data from the University of Illinois, 2010.

produce milk daily. The ever-increasing cost of grains as well as other inputs has, in turn, been driving costs of production steadily upward, while the value of milk has fluctuated to a lesser extent. This pattern yields lower profits to the farmer who produces dairy products (see Figure 2). As a consequence, many farmers are seeking other opportunities and exiting the dairy industry. A mass exodus of farmers from the dairy industry will lead to scarce supplies of milk and dairy products and, in turn, increase prices to the consumer. Increased prices to the consumer will result in economic decisions for cash-strapped households and perhaps, in the long run, issues of dietary deficiencies and even hunger.

If the pursuit of a food-to-fuel economy in the United States is driving up the costs of livestock operations as well as the costs of food in the marketplace, why does it continue to happen? Farmers, like most business people, seek profit

Figure 2. Average production costs and profit in the U.S. dairy industry

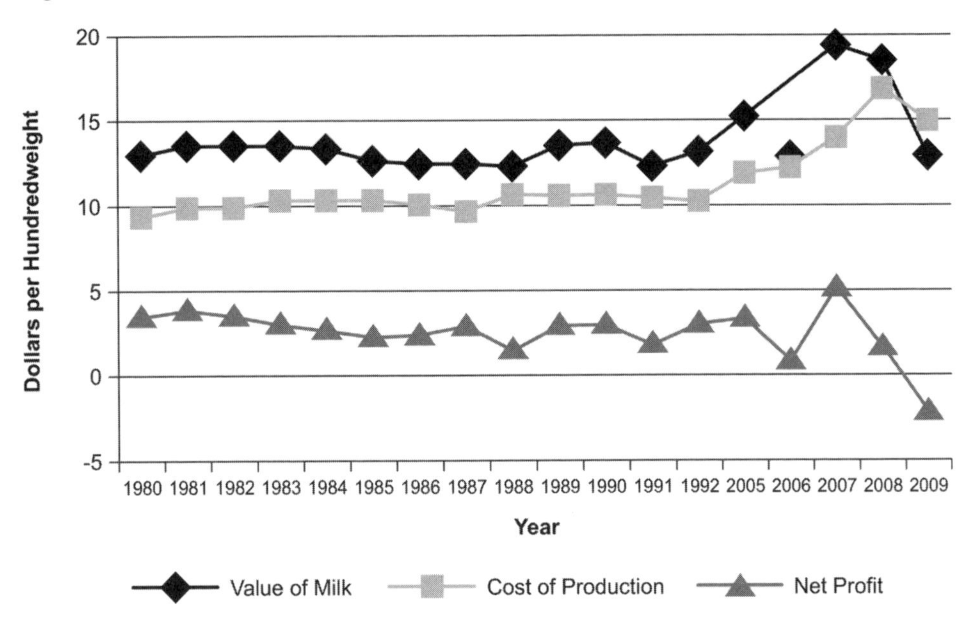

Source: Dawn M. Drake. Data from U.S. Department of Agriculture, 2010.

maximization. A U.S. government policy has made it attractive to turn grain crops, once destined for the marketplace as livestock products, into biofuels, like ethanol, which is consuming an increasingly larger portion of U.S. corn production (see Figure 3). As more of the corn production is diverted into the fuel economy, less is available for other uses, making grains scarce and continuing to drive prices and profits to the grain farmer upward. Tax credits to ethanol production facilities effectively make it cheaper to produce ethanol than gasoline and allow ethanol to be sold at a lower price point in the marketplace, thereby increasing demand for this fuel. The demand for ethanol also is artificially stimulated by government mandates that require gasoline sold in the United States to be blended with as much as 10 percent ethanol. Government-sponsored blenders' credits further reward the firms that blend gasoline and ethanol for sale to the consumer. Moreover, the introduction of vehicles that can operate on blends containing as much as 85 percent ethanol, yet are sold at no more cost to the consumer than vehicles using a regular internal combustion engine, has stimulated ethanol demand and diverted grain away from livestock feed and human grain-based food uses.

Figure 3. The increase in U.S. corn used in ethanol production

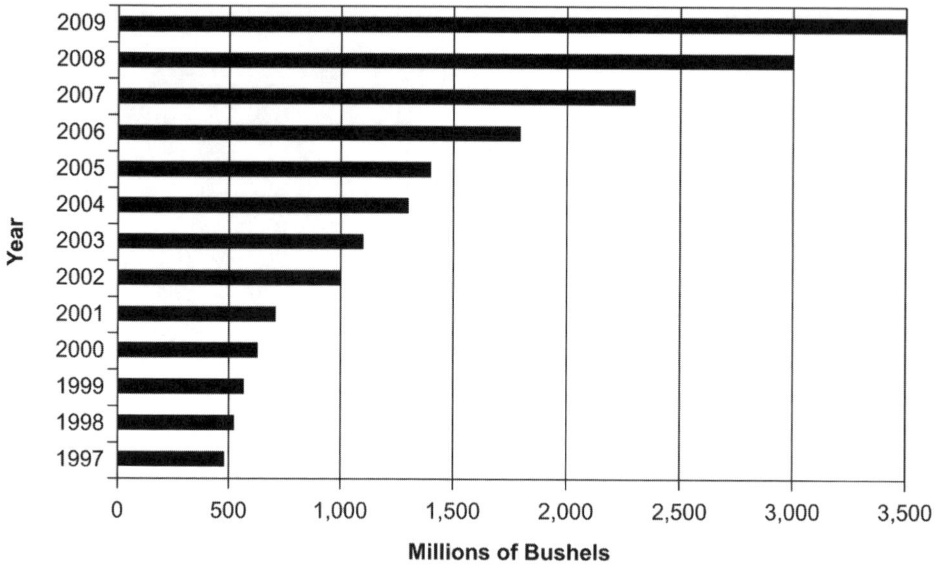

Source: Dawn M. Drake. Data from Kansas Ethanol, LLC.

One needs simply to look at recent market prices for grains to see the results of the food-to-fuel economy (see Figure 1). In the summer of 2008, average prices of grain crops hit record highs, largely due to demand for corn and soybeans for biofuels. This spike in the price of grains encouraged farmers to convert more and more acreage to grain crops that were sold to the biofuel industry. The number of acres of corn planted in the United States has been steadily on the rise over the last 20 years (see Figure 4). While the average price of grain in the marketplace did return to normal levels in 2009, facilitated largely by the expiration of some of the tax incentives to ethanol producers and blenders, high oil prices and poor growing seasons in parts of the Midwest drove the price upward again in 2010–2011. The increasing costs of production to livestock farmers leave many of them with a hard economic choice: either stay in livestock production and risk losing money or exit the industry and sell grain crops, once produced on their farms to feed the herd, to a market clamoring for more grain. Looking more closely at a few examples from the industry might provide insights into how feeding corn to SUVs instead of cows can lead to higher dairy and meat prices, and milk shortages, and have health-impacting ramifications in the United States.

Figure 4. Acres of corn planted in the United States

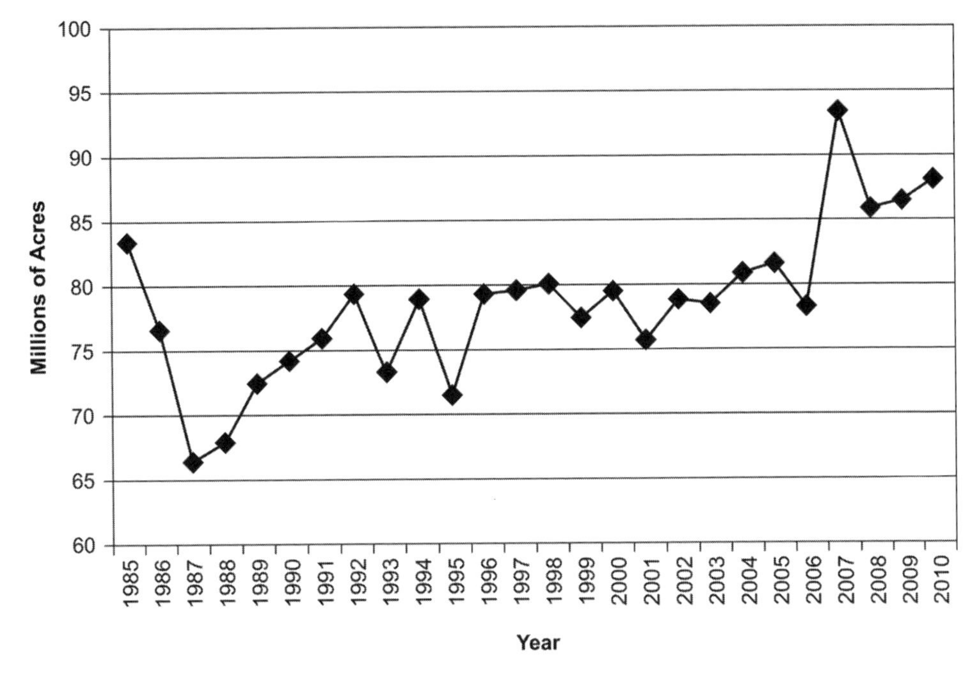

Source: Dawn M. Drake. Data from U.S. Department of Agriculture, 2010.

Case Studies from the Pennsylvania Dairy Industry

As health organizations can attest, milk and other dairy products are an essential part of a complete and nutritious diet. As such, the dairy sector of a country's agricultural economy is a vital building block for its human population's physical growth and health. Yet every day, dairy farmers are faced with economic opportunities that can entice them to modify their operations or abandon them to maximize profits. Many of these options are the direct result of government policies in a food-to-fuel economy. The number of dairy producers selling milk to the general marketplace has clearly dwindled in recent years. Farmers see their profits declining as wholesale milk buyers do not respond to market signals as rapidly as the farmers themselves must do to remain solvent. Despite increasing production costs related to higher grain and oil prices, milk buyers have lagged in the prices they pay to dairy farmers (see Figure 2). This leaves milk producers more susceptible to intervening economic opportunities that offer them a chance to earn higher profits than can be secured by the mainstream dairy farmer.

Examples of the dilemma faced by dairy farmers can be drawn from case studies of two small dairy operations in Pennsylvania, one of the largest dairy-producing states in the United States. The consequences of the food-to-fuel economy on milk production are clearly best seen on farms where dairying has been a tradition for decades.

Klein Farms

One intervening economic opportunity in the dairy industry is niche production, usually in the form of organic milk, raw milk, or gourmet products. Gourmet cheeses and other dairy products as well as organic fluid milk and milk products command a premium in the market and have larger profit margins to compensate for the increased costs associated with their production. This relationship also holds for farmers who choose to enter the market by producing raw milk or milk that has not been pasteurized or homogenized. Raw milk, in those states that permit its sale, can be marketed directly to the consumer at as much as four times the price that the dairy farmer would get by selling the milk to a wholesale buyer. Klein Farms (see Figures 5 and 6) is one of approximately two dozen dairy farms in Pennsylvania to hold Raw Cow Milk and Raw Milk Cheese licenses. Such licenses permit their

Figure 5. View of Klein Farms in Forks Township, Pennsylvania. (Photo by Dawn M. Drake)

Figure 6. Cows at Klein Farms on pasture. (Photo by Dawn M. Drake)

holders to sell fluid raw milk and cheese made from raw milk directly to the consumer, often at much higher prices than if they sold to a wholesaler.

Raw milk can be a very profitable niche market for dairy farmers who are struggling to survive in an environment characterized by rising production costs. However, it is a very narrow market. Currently, only 10 states allow direct retail sale of raw milk. Laws regarding the marketing and sale of raw milk often are complicated and convoluted, involving partial ownership of cows, prescriptions from a medical doctor, or purchases of milk meant for pet food uses. In states that do allow retail sales, dairy farmers remain vulnerable to shifts in the legislative climate, making raw milk sales profitable but also very risky. In 2007, a farmer selling directly to a wholesaler in Pennsylvania might expect to get a little more than a dollar a gallon for milk, whereas Klein Farms sold a gallon of raw milk direct to the consumer for approximately four dollars.

Presently, the niche production shifts have had few effects on the price of milk to the American consumer. Niche products such as organic and raw milk command higher prices, but the U.S. federal government heavily regulates the price of regular milk and milk products to the consumer through Federal Milk Marketing orders. These price-support programs not only set a ceiling for the price at

Raw Milk

There is an ongoing debate in the United States about whether the human consumption of raw milk is safe and should be legal. Various states have enacted legislation complicating the sale of raw milk, as part of an effort to discourage human consumption of this product. Yet no conclusive study has demonstrated that raw milk, when produced on a farm that meets sanitary laws, is not as safe as milk that has undergone pasteurization and homogenization. In fact, raw milk contains more vitamins and minerals, including the beneficial vitamins C, B_{12}, and B_6, than milk that has been cooked in a pasteurization process. Pasteurized milk also has been linked to increased allergies and lactose intolerance. People who cannot drink pasteurized milk often find that they can consume raw milk and products made from raw milk. Furthermore, homogenized milk, in which butterfat globules are broken down to prevent separation, has been linked to increased heart disease. On the whole, some claim that raw milk is no more dangerous, and, in fact, may be healthier than, pasteurized and homogenized milk for human consumption.

—Dawn M. Drake

which milk can be sold to consumers, but also set a minimum price at which a wholesaler can buy milk from the producer. This minimum price is often below the costs of production for the farmer, as it does not take into account rising costs of grain, fuel, and management strategies to handle disease. In the long run, as more dairy farmers exit mainstream milk production and move into niche opportunities, the supply of milk and milk products to the average consumer will decrease, forcing prices upward and leaving cash-strapped households with tough decisions to make about how to spend their limited food dollars.

There is another caveat attached to niche production, in addition to the risk involved in raw milk legislation. The niche market can support only a small number of producers. At a certain point, the niche market will reach saturation and the availability of organic milk, raw milk, and gourmet milk products will be such that they will no longer command higher prices in the marketplace. Not everyone can become a niche producer. Some dairy farmers have selected other intervening economic opportunities to combat decreasing profits from milk production in a food-to-fuel economy.

Farview Farmstead Farm

With the price of grain at record highs, another intervening economic opportunity that has become increasingly attractive to dairy farmers is total exit from the dairy industry. The number of dairy cows in the United States has decreased from a high of 11 million head in 1983 to approximately 9.3 million in 2007, with the number of farms decreasing proportionately (see Figure 7). Small family farms, especially dairy farms, are most susceptible to the influences of fluctuating markets and are most likely to choose other options when faced with an economic crisis. One just needs to look at the case of Farview Farmstead Farm (see Figures 8 and 9) to see the economic benefits of exiting the dairy industry all together. Farview Farmstead Farm is a small family farm in eastern Pennsylvania that chose to sell off its herd. The grain grown on the farm is now sold commercially rather than serving as feed for the farm's dairy cattle. Producers can obtain much larger profit margins in grain production than milk production, as can be witnessed in the changes to Farview Farmstead since the sale of the herd. Figure 10 is a photograph of Farview Farmstead just prior to the sale of the dairy herd in 2007. Figure 8 is a more recent photograph of the farm, complete with new grain bins and a grain dryer. These large capital expenditures have been made in response to increased profits, both from the rising market price of grain and the additional bushels that Farview Farmstead can sell because the grain is no longer being used to feed dairy cattle.

Total exit from the dairy industry has economic benefits for the producer, but it has potential consequences for the price of milk. As more dairy farmers exit

Figure 7. Number of dairy farms in the United States

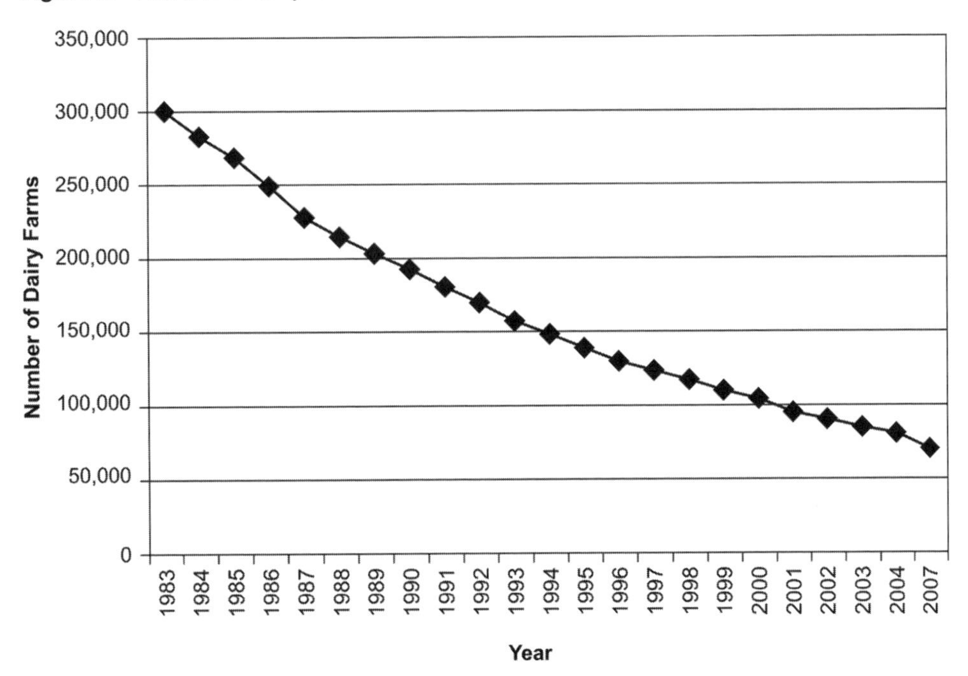

Source: Dawn M. Drake. Data from U.S. Department of Agriculture, 2010.

Figure 8. A view of Farview Farmstead Farm in Palmer Township, Pennsylvania. (Photo by Dawn M. Drake)

the industry, the supply of milk available for sale to the consumer will decline. While federal programs will control the price in the marketplace for the short run, in the long run wholesale milk buyers will need to offer economic incentives to entice dairy farmers to remain in the industry or new dairy farmers to get into the industry. Increased costs to wholesalers will be passed on to consumers in the form of inflated milk and milk product prices. Families with limited food budgets will be forced to make hard choices when selecting food or planning family food menus as costs increase and family incomes either remain stationary or decline.

Prospectus of the Dairy Industry

It is highly unlikely that the U.S. dairy industry would ever become so crippled by diminished supply that average Americans could not afford milk. This may not be the case worldwide, however. Dairy farmers in other countries are experiencing the same increased feed costs due to rising grain prices as dairy farmers in the United States. If U.S. government policies continue to encourage conversion of grain into fuel and subsidize a food-to-fuel economy, many farmers throughout the world will face the reality of not being able to afford to continue in dairy

Figure 9. Cows in pasture at Farview Farmstead. (Photo by Dawn M. Drake)

production. Whether they choose one of the aforementioned intervening economic opportunities or simply are forced to cease operations, a worldwide decline in dairy production could lead families to limit their intake of dairy products. Dairy products are important components of a healthy and balanced diet in many parts of the world.

Increasing the amount of farmland to grow food grain crops is theoretically possible in developing countries, but remains improbable in the short run. Governments worldwide need to seek solutions that limit the effects of the evolving food-to-fuel economy on dairy farmers and put in place the means to control the supply and price of milk and milk products. Lapses of ethanol and blenders' tax credits briefly helped the situation, but recent approval of increasing gasoline/ethanol blends from 10 percent ethanol to 15 percent ethanol most likely will again result in grain prices increasing. At the same time, rising demands for high-quality food products and changing dietary compositions in countries such as India and China have resulted in those countries increasing selected food imports. Their demands for beef and dairy products have stimulated agriculture in grain and dairy export countries. Grain prices in the United States, for example, will continue to rise, as will the feed costs to dairy farmers. The price consumers pay for milk and milk products in the future will increase if more and more grain is used for the production of bio-fuels.

—Dawn M. Drake

Figure 10. Farview Farmstead in the winter of 2007. (Photo by Dawn M. Drake)

Further Reading

Becker, Geoffrey S. *CRS Report for Congress—Livestock Feed Costs: Concerns and Options*. Washington, DC: Congressional Research Service, 2008.

Drake, Dawn M. *Connections Between Mastitis and Climate: A Study of Holsteins of Pasture in Northampton, Pennsylvania*. Master's thesis, University of Delaware, 2008.

Economist staff. "Cheap No More—Food Prices." *The Economist*, December 8, 2007.

Giampietro, Mario, and Kozo Mayumi. *The Biofuel Delusion: The Fallacy of Large-Scale Agro-biofuel Production*. London: Earthscan, 2009.

"Growth Energy. In Landmark Move, EPA Approves Higher Ethanol Blend for Vehicles Built in Last Decade." Accessed January 22, 2011. www.renewableenergyworld.com/rea/partner/growth-energy/news/article/2011/01/in-landmark-move-epa-approves-higher-ethanol-blend-for-vehicles-built-in-last-decade.

"Kansas Ethanol. US Ethanol Facts." Accessed October 26, 2010. www.ksgrains.com/ethanol/useth.html.

Lowe, Marcy, and Gary Gereffi. *A Value Chain Analysis of the U.S. Pork Industry.* Durham, NC: Center on Globalization Governance and Competitiveness, 2008.

United States Department of Agriculture. "Dairy Policy." Accessed November 30, 2010. www.ers.usda.gov/Briefing/Dairy/Policy.htm.

University of Illinois. "Farm Decision Outreach Central, University of Illinois at Urbana–Champaign." Accessed October 25, 2010. www.farmdoc.illinois.edu.

Westhoff, Patrick. *The Economics of Food: How Feeding and Fueling the Planet Affects Food Prices.* Upper Saddle River, NJ: Pearson Education, 2010.

Weston A. "Price Foundation: Campaign for Real Milk." Accessed January 16, 2011. www.realmilk.com.

Animal Husbandry

Modern farm livestock are descendants of wild animals that were domesticated thousands of years ago. In some cases, they were raised for their ability to carry items or to pull them; in most cases, they were raised as a source of food and animal products. Farm animals are a renewable food source. To survive and to maintain their health, they consume plants or plant products—another renewable resource. Almost every part of a farm animal has been used by humans over the course of history, including animal waste/manure. Notably, manure was discovered to renew soil fertility and to aid in the growth of plants.

Food is the most important contribution animals make to human survival, although they rank well behind plants in total quantity of food products supplied. It is estimated that plants supply 80 percent of the total calories consumed by humans. Farm animals and their products are a more important source of protein than of calories. Meat supplies 35 percent and milk 34 percent of the world's total supply of protein. Farm animals are not the most efficient converters of raw plants into human food, yielding only 1 pound of human food for each 3 pounds of plant mass eaten. Nevertheless, two-thirds of what farm animals eat consists of substances considered either undesirable or completely unsuited for human consumption. Their ability to convert inedible plant materials into edible human foods aids greatly in improving the quality and quantity of human diets. Thus farm animals do not compete with humans for food, and they do contribute substantially to the world's food supply.

Only 35 percent of the world's land area is classified as agricultural land; only 10 percent or so is classified as arable or cultivated land that can produce plant products suitable for humans' direct consumption. Grass and shrubs cover approximately 25 percent of the world's agricultural land, and these forms of vegetation can be digested only by ruminant or farm animals (see Table 6).

Table 6. World Agricultural Land: Cultivated and in Permanent Pastures

Geographic Region	Total Land Area (1,000 sq.mi.)	Percentage of Agricultural Land		
		Total	Cultivated	Pastures
World	50,495	35	31	67
Developed Countries	21,176	36	33	66
Developing Counties	29,319	34	29	69
Africa	8,994	37	19	79
Asia	10,334	38	45	53
Europe	1,826	49	55	38
Oceania	3,254	61	9	91
North America	7,084	27	46	53
South America	6,771	31	15	81
U.S.A.	3,524	47	43	56

Sources: FAO Production Yearbook. 2008; U. Desilva and J. Fitch, *Breeds of Livestock*. Stillwater: Oklahoma State University, 2010, pp. 1–3.

Origins of Animal Husbandry

Domestication of food animals had its origin in the transition period, whereas humans evolved from roving hunter-gatherers into settled agriculturalists living in communities. Wild animals were domesticated when their breeding, living conditions, and life cycles were controlled or interfered with by humans. The most significant food animals are believed to have originated in the Middle East food crop center. Evidence indicates that plant cultivation and wild animal domestication began at approximately the same time. Of the 50 or so wild animals domesticated, only about a dozen are considered as significant food sources on a worldwide basis (see Table 7). A domesticated farm animal is not only a source of protein; rather, it is, in a sense, stored food on the hoof. Even-toed ungulates (hoofed animals) have been the prime domesticates, and cattle, sheep, and goats rank highest as food sources. These ruminants (cud-chewing animals) are able to covert second-rate protein sources into high-quality, first-rate protein usable by humans.

Domestication of most animals used as food sources was facilitated by the social rather than the solitary habits of these animals. Sheep, it is believed, were the first animals to be domesticated. Archaeologists have found sheep remains dating back to 10,000–9000 BCE. Sheep are capable of surviving in harsh environments that are unfavorable for many other food animal species. Their panting

Table 7. Food Animal Ancestors, Place of Origin, and Current Use

Animal	Wild Ancestor	Date of Domestication	Place of Origin	Current Use
1. Camel	Wild camels	c. 4000–1400 BCE	Asia	Meat/Dairy
2. Cattle	Wild Aurochs	c. 7000–6000 BCE	Middle East	Meat/Dairy
3. Goat	Wild goats	c. 8000 BCE	Middle East	Meat/Dairy
4. Llama	Guanaco	c. 3500 BCE	Andes	Meat
5. Pig	Wild boar	c. 7000 BCE	Middle East	Meat
6. Reindeer	Reindeer	c. 3000 BCE	Northern Russia	Meat/Dairy
7. Sheep	Mouflon sheep	c. 10000–9000 BCE	Middle East	Meat/Dairy
8. Water buffalo	Wild Arni buffalo	c. 4000 BCE	East South Asia	Meat/Dairy
9. Yak	Wild yak	Unknown	Tibet/Nepal	Meat/Dairy
10. Rabbit	Wild rabbit	c. CE 400–900	France	Meat

Sources: C. Sauer. *Agricultural Origins and Dispersals*. New York: American Geographical Society, 1952; J. Harlan, "Agricultural Origins' Centers and Noncenters," *Science* 174 (1971): 468–474.

mechanism for dissipating heat enables them to tolerate hot climates, and their coats of wool help them survive in colder climates. Dated remains of sheep are at least 2,000 years older than goats, though both animals are believed to have been domesticated in the Middle East. Goats can survive in areas where the natural food supply is minimal because they are browsers and grazers (nibblers on grass, twigs, and leaves). Goats are sometimes called "poor men's cows," reflecting the fact that their meat is not relished by humans as much as cattle or sheep meat. In addition, relative to goats, sheep are also superior fiber producers; cattle are superior milk producers.

The earliest archaeological evidence of cattle remains date to 7000–6000 BCE. Cattle (wild Aurochs) were hunted, trapped, and worshiped long before their domestication. There is some disagreement over the food center in which cattle were first domesticated, but most scholars believe that it was probably the Middle East. Early domesticated cattle were very large, strong, and temperamental. Humans selectively bred cattle to reduce both their size and their ferociousness. Cattle provided humans with milk, a very fine protein source, along with meat, hides, bones for tools, power for field work, and manure for fertilizer, fuel, and plaster. Water buffalo were first domesticated in South Asia (possibly India) in about 3000 BCE. They are adapted to hot, humid, tropical lowlands, and they consume aquatic and semi-aquatic grasses. Water buffalo live on vegetation that will not support traditional cattle. From the water buffalo, both meat and milk are secured.

The earliest domestication of pigs probably occurred in the Middle East. Wild pigs were native to both the Middle East and Europe. Pigs are believed to have been domesticated in about 7000 BCE, thousands of years after sheep and goats.

Table 8. World Livestock, 1972 and 2008 (million head)

Animal	1972	2008	Increase
1. Cattle	1,131	1,347	19%
2. Sheep	1,043	1,078	3%
3. Goats	392	862	219%
4. Camels	14	24	71%
5. Pigs	646	941	45%

Sources: Adapted from *FAO Production Yearbook*, 1974 and 2008.

Some believe that pigs were domesticated later because they cannot survive on grass, leaves, twigs, and other plant sources, as ruminants do. Instead, pigs compete directly with humans for food. They primarily are scavengers and household or barnyard animals. Pigs produce more meat per unit weight of all the major food animals.

Camels, llamas, reindeer, yak, and rabbits are also important food sources in specific portions of the world. Camels, reindeer, and yak provide meat and milk. Llamas and rabbits provide meat. All five were domesticated later than sheep, goats, cattle, pigs, and even water buffalo.

All 10 of these major food animals are important components of modern agriculture and contribute to human diets. However, in a number of countries in the world, for cultural or religious reasons, livestock make limited contributions to the food supply. India, for example, is home to millions of cattle and water buffalo, but these animals contribute little meat or animal products to those who live in India or in other countries of the world. Nomadic tribes in portions of Africa measure a farmer's wealth by the size of his herd. In many nations of the developing world, the rate of animal growth and the yield of animal products are low because of inferior genetic stock, limited food supply, little water, diseases, parasites, and low-level herd management. Even so, the number of livestock animals in the world is impressive, and these animals do constitute a significant food source and food reserve.

The number of livestock (excluding water buffalo) found around the world and the percentage increases in their number from 1972 to 2008 are shown in Table 8. The percentage increase in the number of goats and camels was greatest, while the smallest percentage increase occurred for sheep and cattle. While the number of livestock increased on a worldwide basis in the last decades of the 20th century, a decline of livestock occurred in the United States during this period. This decline in the United States was related to herd improvement, feed costs related to weather factors, and the changing diets of American food consumers.

Texas is the state with the greatest number of cattle on farms and ranches. California has the largest number of milk cows on dairy farms, while Iowa has the largest number of pigs on farms (see Table 9). Cattle ranches are located

Table 9. Leading Cattle-, Milk Cows–, and Pig-Producing States, 2000 (in thousand head)

Cattle	Number	Milk Cows	Number	Pigs	Number
1. Texas	13,900	1. California	1,526	1. Iowa	15,100
2. Nebraska	6,650	2. Wisconsin	1,344	2. North Carolina	9,300
3. Kansas	6,600	3. New York	686	3. Minnesota	5,800
4. California	5,100	4. Pennsylvania	617	4. Illinois	4,150
5. Colorado	3,150	5. Minnesota	534	5. Indiana	3,350
U.S. Totals	98,198		9,206		59,138

Source: *Statistical Abstract of the United States*, 2002, 536–537.

primarily in the dry, semiarid, short-grass states of the American Southwest. Milk cows generally are found in states with mild climates, moderate rainfall, and lush pastures. These states contain large urban populations or are near large metropolitan centers. Pigs are raised primarily where there is relatively inexpensive feed and where there is a tradition of combining the growing of feed grain and caring for pigs. Although "pig factories" are found in many states, Iowa and North Carolina have set the standards for the world.

Total agricultural exports from the United States were valued at more than $40 billion in 1990 and approximately $51 billion in 2000, an increase of more than 25 percent. Animal and animal product exports increased by more than $8 billion from 1980 to 2000, or greater than 330 percent (see Table 10).

In the United States and other developed countries of the world, efficiency in animal production has increased substantially in the past 30 years. Improvements in animal quality and animal products have been largely the result of better animal feed, advances in disease control, and innovations in herd management. The results have been faster animal growth rates with less feed and higher levels per animal of meat and milk. In many developed nations of the world, the number of dairy cows has been reduced by more than 50 percent while milk production has remained essentially the same.

Table 10. U.S. Animals and Animal Product Exports, 1980–2000 (in billion dollars)

	1980		1990		2000	
	Value	Percent	Value	Percent	Value	Percent
Total Agricultural Exports	$41	100%	$40	100%	$51	100%
Meat and Animal Products	$4	9%	$7	17%	$12	23%

Source: *Statistical Abstract of the United States*, 2002, 529.

In 2007, the United States exported $72 billion worth of agricultural products and $9 billion in animal and animal products. Animal and animal products exported from the United States in 2007 were less than the comparable exports in 2000. This decline reflected a world economic slowdown, weakened international markets, inclement weather in the United States, and domestic dietary changes. Many of the U.S. exports of animal and animal products are sent to Asia, Western Europe, and Latin America.

There exists a great difference in meat availability and meat consumed between developed and developing countries, however. Cattle and pigs are the source of 75 percent of the world's meat. In 2007, the countries of the world imported $19,569,672 worth of beef cattle meat and $19,638,555 worth of cheese made from cow's milk.

Modern Animal Husbandry

Animal husbandry practices vary greatly worldwide between different agricultural systems, cultures, climatic regions, sources of animal feed, and types of animals. Historically, livestock rearing was a nomadic subsistence occupation, unassociated with sedentary agriculture. Animals were not kept in enclosures, but rather secured feed from natural sources and drank water from rivers, streams, ponds, springs, or wells. These animals were allowed to breed freely, and then were rounded up and driven to market. In modern society, nomadic herding exists on a minor scale; camels and reindeer are tended in this way, as well as a few other animals of local importance. On a larger scale, in the prairie and steppe vegetal regions of the world, such as the western United States and Canada, Russia, Kazakhstan, Mongolia, west Africa, and Pampas region of Argentina, cattle and sheep were once raised on the open range (no fences).

Over time, population pressure and changing world market demands led to a modification of the open-range system into a controlled range system. In modern livestock-raising range operations, animals are monitored and occasionally inspected, and they roam on massive ranches or large farms. They are restricted somewhat in movement by barbed wire or electric fences, rounded up at times to a central place by cowboys, ranch hands, stock-men, or sheep herders with the help of dogs, horses, and all-terrain vehicles, all-terrain vehicles, and helicopters. Livestock may be branded, ear-tagged, or micro-chipped to indicate ownership. Cattle and sheep are moved from one pasture to another when over-grazing is an issue or to avoid pasture damage. In some areas of the American West, livestock is moved, at times, from private land to federal land if the rancher has a permit. Animals are bred through artificial insemination or

Farm Veterinarian Shortage

American agriculture is an integral component of the global economy. In 2007, world agricultural exports exceeded $90 billion. The worldwide growth in agricultural production, in some cases stimulated by the "Green Revolution," has increased competition for export markets and for specific items. Dietary expectation changes, along with population growth, have increased the demands for livestock and livestock products worldwide.

In 2007, American farmers exported animals and animal products valued at more than $9 billion. The states of Texas, Nebraska, Kansas, Illinois, and North Carolina are major livestock and livestock commodity producers. In 1980, red meat and poultry production in the United States totaled 53,150 million pounds; in 2001, the total was 82,643 pounds.

In their attempt to help satisfy the worldwide demand for meat and animal products, American farmers increased production substantially—and created a farm veterinarian shortage. Currently, there are not enough farm veterinarians to fill the vacancies at key federal agencies responsible for protecting the nation's food supply. Nearly 30 percent of the veterinarians working for the U.S. federal government are eligible to retire in three years. At the same time, 1,300 counties in the United States do not have a farm veterinarian available. Interest in becoming a large-animal veterinarian has declined, even as the pet industry—and the number of pet-oriented veterinarians—has grown dramatically. An estimated $3.4 billion was spent on pet services in 2010. The vast majority of new veterinarians choose to care for dogs and cats—not cows, pigs, and chickens.

—William A. Dando

by controlled breeding. Cattle are sometimes transported to "feed lots" after a round-up where they are confined into a small fenced area and "fattened" for slaughter.

Housed or indoor climate-controlled animal production facilities are a new development in livestock management. "Pig factories" are an example of high-technology, large-scale, standardized-product, industrial-style agricultural facility. Factory farms are successful in producing high-quality meat-producing animals, with fewer losses, in less time, and at less cost than a traditional small, mixed grain–pig farm. Unfortunately, pig factories are so large that they have problems with the quantity of animal waste and odors, groundwater contamination, and disease. Modern animal husbandry practices are designed to produce high-quality, reasonably priced products with minimum human labor, improved animal health, sound environmental conservation, and humane animal welfare.

Animal Welfare

Animal welfare activists contend that animals under human care should be treated in a way that they do not suffer unnecessarily. Concern for farm animal welfare is reflected in recent European Union (EU) legislative efforts. The EU officially recognizes "five freedoms" for farm animals:

1. Freedom from hunger and thirst
2. Freedom from discomfort
3. Freedom from pain, injury, and disease
4. Freedom to express normal behavior
5. Freedom from fear and distress

Traditional animal husbandry practices involved commitment to and provisions for animal welfare. However, modern animal husbandry practices and factory farm complexes subordinate animal welfare to the economic pressures for increased numbers of animals and greater meat or milk output per animal. In the United States alone, more than 8 billion animals are killed for food each year, with 23 million farm animals being slaughtered every day. Americans are great meat eaters. Each year, the average American consumes 63 pounds of beef, 45 pounds of pork, 1 pound of veal, and 1 pound of lamb. Worldwide, there is an ever-increasing demand for meat and livestock products. Industrializing livestock husbandry is one modern attempt to satisfy this demand.

Future Animal Husbandry

Pastoral nomads relied, for their survival and for peer recognition, upon their livestock. Milk from cattle, sheep, goats, and camels provided the basis of their diet. Meat was eaten rarely, for herd animals were considered too valuable to kill and took so long to rear. Livestock was rarely sold because they were considered capital, a herder's wealth, and animal numbers, prestige. Nomads made no attempt to improve pastures, had limited knowledge of disease control, and did not practice selective breeding. In most instances, an animal's ability to withstand the area's weather and climate, and particularly drought, was more important than the quantity of meat or milk produced from it. Pastoral nomads relied solely upon natural vegetation for their livestock feed. The true nomads had no dwelling places and lived in tents. Their life was hard.

Extensive commercial grazing or ranching, in many ways, is similar to pastoral nomadism and may be considered an advanced form of pastoral nomadism. In the modern world, ranching has been replaced by more intensive farmland use and has been pushed to drier, less hospitable areas, where it is considered the best use of poor vegetation and scarce water. Ranching, everywhere in the world, is a large-scale activity and reflects the low productivity of the land used for grazing. In the United States, 60 percent or more of the land used for cattle or sheep grazing is federal land (ranchers pay for grazing rights). In Latin America, ranches have been large private holdings since the time of Spanish settlement. Large herds are grazed on extensive commercial agricultural units. Ranching is very specialized, generally involving beef cattle or sheep. Returns to the rancher are low and vulnerable to weather and climate, and the price per animal fluctuates. The rise of large urban markets, competition for the best grazing land, and dietary shifts have displaced extensive commercial grazing for intensified animal husbandry activities.

Modern livestock husbandry, focused on providing meat, milk, and other products for the increasing world's urban population, has evolved into a rural farm factory system. The transition began with the fattening up of lean range-fed cattle on grain farms, then to extensive feed-lots that encompassed hundreds of acres, and in the first decade of the 21st century, to farm factories that require little land but produce tens of thousands of animals for slaughter, every three or four months. The application of state-of-the-art technology to animal husbandry practices has led to an increase in meat and milk availability at reasonable prices to world consumers. The increasing demand for beef and dairy products in a world population of nearly 7 billion in 2010 and possibly 9 billion by the mid-21st century will require further intensification of world animal husbandry practices and bring greater concerns related to environmental protection and conservation.

—*William A. Dando*

Further Reading

Dando, William A. *The Geography of Famine*. New York: John Wiley & Sons, 1980, 23–27.

Desilva, Udaya, and Jerry Fitch. *Breeds of Livestock*. Stillwater, OK: Oklahoma State University, 2010, 1–3.

FAO Production Yearbook. Rome: Food and Agriculture Organization of the United Nations, 1974 and 2008.

FAOSAT. faosat.fao.org/site/339/default.aspx.

Fiala, Andrew. "Animal Welfare." In *Critical Food Issues: Problems and State-of-the-Art Solutions Worldwide. Vol. 2: Society, Culture, and Ethics*, edited by Lynn Walter. Santa Barbara, CA: ABC-CLIO, 2009, 228–238.

Grigg, D. B. *The Agricultural Systems of the World.* Cambridge, UK: Cambridge University Press, 1974, 241–255.

Harlan, J. "Agricultural Origins, Centers, and Noncenters." *Science* 174 (1971): 468–474.

Jacobson, N. L., and G. N. Jacobson. "Animals: Potentials and Limitations for Human Food." *Dimensions of World Food Problems*, edited by F. R. Duncan. Ames, IA: Iowa State University Press, 1977, 136–149.

The New York Times 2010 Almanac. New York/London: Penguin Group, 2010, 316–318.

Sauer, Carl. *Agricultural Origins and Dispersals.* New York: American Geographical Society, 1952.

Animal Proteins

Proteins are composed of long chains (usually more than 100) of amino acids held together by peptide bonds. Amino acids are molecules with a central carbon atom linked to nitrogen, containing an amine group (—NH_2), a carboxyl group (—COOH), a hydrogen atom, and a functional group (indicated by "R") that differs from molecule to molecule. The R group is what gives the amino acids their individual characteristics. The amino acid composition is designated as the primary structure of a protein. Proteins have secondary, tertiary, and quaternary structures in which parts of the molecules twist to form complex shapes that give the proteins many of their functional characteristics. Proteins provide structural attributes; for example, collagen is a component of connective tissue. In addition, proteins provide a variety of functional characteristics, such as trypsin, an enzyme that digests protein; hemoglobin, the iron-containing oxygen transporter protein found in the blood; immunoglobulin, disease-protecting molecules; and thyroid-stimulating hormone (TSH), which stimulates the thyroid to produce thyroxin (the hormone that controls metabolic rate). As one can see, numerous important activities are associated with protein molecules.

Physiologically, humans break down proteins into amino acids in their digestive system and use these amino acids to form and replenish an amino acid pool in the cells used to make the body's own proteins. Approximately 20 amino acids commonly are found in animal protein. These proteins have similar composition to the proteins found in the human body. Proteins derived from animal sources are called complete proteins because of their similarity in amino acid composition.

Animal proteins also contain all of what are called essential amino acids. Essential amino acids are those amino acids that the human body cannot make or cannot make in sufficient amounts to meet the body's needs. In the human diet, they consist of histidine, isoleucine, leucine, lysine, methionine, phenylalanine, threonine, tryptophan, and valine. Conditional essential amino acids are arginine, cysteine, glutamine, and tyrosine; they are considered conditionally essential because they are made from precursor amino acids that are essential; for example, tyrosine is made from phenylalanine.

Plant proteins usually are incomplete proteins. Because some plant materials are missing one or more of the essential amino acids, a person must eat plant proteins in combinations, such as rice and beans, to have sufficient overlap of amino acid content to supply all the amino acids needed in a single meal. Soybeans (*Glycine max*) are an exception. Genetically modified plants may be engineered to have complete proteins with all of the issues associated with these procedures (see the "Protein and Protein Deficiency" entry).

The general nutritional guidelines for the amount of protein in the diet call for proteins to make up between 10 percent and 35 percent of the total nutritional intake. People from the United States usually consume approximately two-thirds of their proteins in the form of animal proteins, including poultry, beef, pork, milk, cheese, and fish. The proportion of animal proteins in the diet varies widely throughout the world (discussed later under "Risks/Disadvantages of Animal Proteins").

Benefits/Advantages of Animal Proteins

From the preceding description of animal proteins, it is easy to understand that one of the benefits of eating animal proteins derives from the similarity in amino acid content and ratios needed by humans. In countries where animal proteins are available and relatively inexpensive, eating animal proteins (meat, milk, and cheeses) is convenient and easy. In some countries of the developed world (e.g., the United States), convenience is a definite lure for meeting protein requirements from animal sources. The current estimate of the percentage of protein consumed in the form of animal sources is 64 percent in the United States. Although this is not the highest level (Luxembourg has the highest rate, 71 percent), it is substantial. Some countries have exceptionally low levels; for example, in Burundi, this rate is 6 percent, and in the Democratic Republic of the Congo, it is 16 percent. In some countries known for raising cattle, such as Kenya, residents average only 27 percent of their protein from animal sources.

Animal Protein

Many scientists and world leaders think that the use of animal proteins will be unsustainable in the future. According to the FAO of the United Nations, almost half of the world's cereal crops is used to feed animals. The appetite for animal protein is predicted to increase approximately 40 percent by 2050. If these cereals used to feed animals were reallocated to feed humans, 3.5 billion people would have adequate calories for a year, according to the UN Environment Programme (estimate made in 2009). Is eating meat possible in the long term? What is a sustainable plan to meet the nutritional needs of all peoples of the world?

—Joyce V. Cadwallader

Many arguments are made for eating animal proteins. Proteins from different sources have different levels of biological availability (BA) and digestibility. Scientists use a variety of methods to measure protein activities in the body, including biological value (BV), net protein utilization (NPU), and protein digestibility (Protein Digestibility Corrected Amino Acids Score [PDCAAS]). The Food and Agriculture Organization (FAO) of the United Nations and the World Health Organization (WHO) use a standard for proteins called the essential amino acid requirement for preschool-aged children.

The easiest way to judge protein quality is to compare its amino acid composition to a reference protein. In the past, an egg white was used as the standard (score: 100), and other proteins were then compared to it. If the new protein had 70 percent of a limiting amino acid, it received a score of 70. This measurement did not reflect on the protein's digestibility, defined as the amount of amino acid that is absorbed. To correct for digestibility, the FAO has come up with the new measure, PDCAAS, which corrects the amino acid requirement for preschool children with digestibility. To measure digestibility or its absorption, animal testing usually is done, and much of this research is performed using rats—rats have digestive systems similar to humans'. Thus, if one compares the PDCAAS of beef, 0.92, to that of wheat gluten, 0.25, there is a substantial difference. This pattern generally holds for animal versus plant proteins. Eggs and milk, both of which are renewable animal protein sources, have a PDCAAS value of 1.00.

Sources of animal proteins such as meat also contain other nutrients. One of the interesting anthropological hypotheses for the change in human society that promoted the hunter-gatherer phase was the impact of the iron-containing animal protein, hemoglobin. The conjecture is that because female *Homo sapiens* menstruate and thereby lose hemoglobin, the males who went out, hunted, and returned with animal meat that was high in iron-containing hemoglobin were the most favored males. Although the hunters' contribution to the total food available was sometimes meager, the hypothesis is that this stage in human development persisted because it was important to maintain the iron levels of women in the group. According to L. Shlain (2003) in his research on women and human evolution, this hypothesis points to the role animal proteins may have had in influencing the formation and maintenance of certain human behavioral patterns of high animal protein consumption. Even as the hunter-gatherer societies were replaced by food-producing ones, diets have continued to contain meat from domesticated animals in many regions of the world.

Besides iron-containing hemoglobin, animal-derived foods have higher concentrations of other essential nutrients. As previously mentioned, animal meat is an excellent source of iron. Zinc and the B vitamins also are found in higher levels in animal meat. Calcium and vitamin D are found in milk products. Plant sources contain phytates, oxalates, and tannins, which have a tendency to bind minerals together and make them less absorbable, thus arguing for the inclusion of meat in the diet. In addition, some individuals have problems with gluten (wheat protein) and cannot digest it properly.

Risks/Disadvantages of Animal Proteins

There are some disadvantages to consuming one's protein from animal sources. Animal sources contain cholesterol, a substance not found in plants. Consequently, humans who eat large quantities of protein from animal sources tend to have higher cholesterol levels; higher cholesterol levels have been associated with heart disease. The higher fat content of animal protein also increases the probability of obesity, which is currently on the rise in most of the world's developed countries. Substantially less fiber is found in the diets of those who consume most of their protein from animal sources, and fiber has been shown to protect against colon cancer. Higher risks of developing other cancers, such as breast, prostate, and pancreatic cancer, are associated with increased fat in the diet. As this brief discussion suggests, health issues raised by too much or too little animal-derived proteins are complex.

Metabolically, when people do not have sufficient calories coming into their diet, the body breaks down carbohydrates and fats to maintain energy levels. If there is no fat reserve, then the body begins to break down proteins. A relatively small amount of protein is all that is required per day to maintain the nitrogen balance (10 to15 percent of incoming macronutrients or appropriately 100 grams [3 ounces]). Proteins must be eaten daily because amino acids are not stored in the body. If excess proteins are brought in, they are deaminated (have the amino group removed) and used for energy or stored as fat. However, when proteins are broken down for energy, the resulting amino acids must be deaminated before being shunted into the metabolic pathways for energy. The amino groups must be detoxified by the liver, forming urea, which in turn is excreted by the kidneys. Humans remove most of their nitrogenous waste in the form of urea, whose production requires a moderate amount of water to excrete the waste from the body. The elimination of water to rid the body of the nitrogenous wastes can interfere with fluid balance in the body if proteins are taken in excess. Thus dehydration may result, with all the attendant problems associated with dehydration. The problem of dehydration is exacerbated where the supply of clean water is limited.

Excess animal proteins also can cause excessive loss of calcium. Although a person obtains more calcium in meat, some of the amino acids that are acidic tend to cause the calcium to be removed from the bones and increase the risk of osteoporosis when taken in excess. The high animal protein intake is one reason that the required intake of calcium is higher in the United States than in other countries.

A different problem, associated with the lack of animal protein, is protein-energy malnutrition (PEM)—a condition that is prevalent in Africa, Central and South America, the Middle East, and Southeast Asia. PEM is found in the United States among the poor, elderly, homeless, and alcohol and drug abusers. In addition, infections such as HIV and tuberculosis may cause protein deficiency.

Two forms of PEM are distinguished. The first form, known as marasmus, results from chronic protein deprivation. This form of PEM is common among children, especially those from 6 to 18 months of age, in the least developed areas of the world. When children do not have adequate nutrition, their heart and muscles weaken, brain development is impaired, metabolism is slowed, and body temperature is difficult to maintain. Digestive enzymes are not produced and new tissues to replace the digestive linings are not manufactured, which causes the digestive lining to deteriorate such that the food consumed is not properly absorbed. Kwashiorkor, the second form of PEM, occurs later in childhood and is usually due to sudden deprivation of proteins. Kwashiorkor is seen more often in hot, tropical areas where mold is more common and found infecting the grains. Interruption of food supplies, caused by social unrest or war, also can contribute to kwashiorkor. Kwashiorkor's symptoms include edema caused by a lack of the blood proteins that normally tend to keep the fluid in the bloodstream. A picture of children with bloated stomachs (symptomatic of kwashiorkor) frequently signals hunger in various regions of the world.

What are the environmental and ecological impacts of eating animal protein? One of the concepts that a student learns early in ecology class is the "rule of 10": As nutrients go through different organisms, only about 10 percent of the energy is passed to the next group. What does this rule of thumb have to do with animal protein? Instead of the animals in the developed countries getting feed from pastures, these animals are placed in large factory-like farms where they are fed grains. According to Pond and his associates in their 2005 study of animal nutrition and feeding, even if the energy yield is higher than 10 percent (it does vary from 5.3 percent to 37.8 percent from beef to swine), animals are fed grains that could otherwise provide food for people. The grains grown for animal feed prevent the use of the land to grow foods that could be consumed directly by humans. This relationship explains why some scientists talk about eating "lower on the food hierarchy" in an effort to feed the hungry in the world. Even if only plant sources for protein were eaten, however, it is doubtful that the problems of hunger and starvation in the world would be eliminated. The current estimate is that there is sufficient food (grains) to feed all the people of the world and millions more. The real problem is that distribution of food parallels distribution of wealth. Food costs money, and many of the poor do not have the resources to buy sufficient food to meet their nutritional needs. Giving food to starving people in times of a food crisis or famine saves lives, and assisting those with food problems grow their own food or find employment so they can purchase food eliminates a food crisis. Unfortunately, convincing the world's wealthy countries to assist those nations with chronic food problems to develop agriculture, plant new crops, and apply new agricultural techniques is difficult. As a consequence, these underdeveloped countries continue to import agricultural products from the wealthy countries, thereby strengthening the economies of wealthy countries.

In this world, hundreds of thousands of people are malnourished. Although not all malnourished people are protein deficient and specifically not animal protein deficient, a substantial number of malnourished people suffer from protein deficiency. Protein deficiency results in the condition of PEM described earlier. Whitney and Reites (2011) attribute PEM as the most prevalent form of malnutrition, affecting 500 million children and leading to 33,000 deaths per day. According to the most recent United Nations data, compiled in 2010, there are an estimated 925 million malnourished people in the world.

Regardless of the question of advantages and disadvantages of an animal protein (meat-based) diet versus a plant-based diet, many consider the meat-based diet as unsustainable. Put simply, an animal protein diet requires too many resources (land, water, and fossil fuels) to be sustainable for the long term as currently practiced. Pimental and Pimental (2003) estimated that the grain fed to livestock in the United States alone would feed 840 million people on a plant-based diet—most of the estimated number of malnourished people in the world.

Future Importance

While the future is always difficult to predict, the consumption of animal proteins in the future is particularly problematic to forecast. If genetically modified (GM) plants were to be created with an amino acid ratio similar to the animal protein ratio, would they be accepted? Many parts of the world continue to reject GM organisms. Will the food distribution problems that lie at the heart of the hunger question be resolved? If it were easier to obtain proper nutrition from plant materials, would the people of the world change their eating behavior? In China, the most populous country in the world, residents have increased the percentage of animal protein in their diet from 24 percent to 41 percent over the last several decades as the country has become more developed. This increase in using animal proteins seems to be a general trend. Will the increased competition for animal proteins be sustainable, environmentally harmful, or a threat to peace? Will the cost of animal proteins increase so much that they will continue to be out of the financial reach of most people? There are far more questions than answers related to the role that animal proteins will play in the future. Some of these questions must be answered for the future of humanity and the environment.

—Joyce V. Cadwallader

Further Reading

Diamond, Jared. *Guns, Germs, and Steel.* New York: W. W. Norton, 1997.

Food and Agriculture Organization of the United Nations. "Food Security Statistics." February 2011. www.fao.org/economic/ess/food-security-statistics/en/

Insel, P., D. Ross, K. McMahon, and M. Bernstein. *Nutrition*, 4th ed. Sudbury, MA: Jones and Bartlett Publishing, 2011.

Pimental, D., and M. Pimental. "Sustainability of Meat-Based and Plant-Based Diets and the Environment." *American Journal of Clinical Nutrition* 78 (2003): 660S–663S.

Pond, W., D. Church, K. Pond, and P. Schoknecht. *Basic Animal Nutrition and Feeding*, 5th ed. Hoboken, NJ: Wiley, 2005.

Shlain, L. *Sex, Time and Power: How Women's Sexuality Shaped Human Evolution.* New York: Viking, 2003.

Team C002291. "An End to World Hunger: Hope for the Future." Oracle ThinkQuest, February 2011. library.thinkquest.org/C002291/high/present/who.htm.

Whitney, E., and S. Rolfes. *Understanding Nutrition*, 12th ed. Belmont, CA: Wadsworth Publishing Company, 2001.

Arable Land

The term "arable," derived from the Latin word *arare*, meaning "to plow," is an agricultural description for land that is capable of being used for growing crops. It also includes all land that possesses the soil and climate conditions suitable for growth of forests and natural grasslands. Arable land has geographical significance when it is considered in the context of an ability to grow and sustain crops that, by extension, provide a food source for a regional population of people. The most productive arable land is overlain by soils derived from glacial till and sediment. Such soils were deposited by glacial outwash, major rivers, and, within near-shore areas, shallow sea environments through evolutionary processes of weathering and erosion, extending over thousands of years of Earth history.

A number of critical factors must be examined when determining whether land is suitable for an arable classification, and by its value, in the type of crops it can yield. Physical and chemical properties of the soil, water availability, irrigation needs, climate factors, land use and deforestation practices, desertification, landfills, urban sprawl, and topography influence the quality, distribution, and ultimately the land's ability to render itself as agriculturally productive. Arable land that is considered unsuitable for farming will be affected by one or more of the following factors: no source of fresh water, extreme temperatures (too hot or too cold), rocky terrain, too mountainous (steep slopes), saline or alkaline soil, too little or too much precipitation, too snowy with a short growing season, too much pollution in the soil or ground water, and generally nutrient-poor soil.

A Growing World Population

Arable land and crop production play important roles in the ability of nations to feed their citizens. Unfortunately, while many countries possess reasonable percentages of arable land (see Table 11), they are constrained by at least one or more variables (infertile soil, unsuitable climate conditions, poor land-use practices, or a topography that is too steep, too high, too rocky, or otherwise unsuitable) that limit the total amount of area that can be planted to grow a significant volume of food.

The Earth's land surface area totals approximately 57,506,055 square miles (36,803,728,290 acres), which is about 29 percent of the total surface area of the planet. Of this area, 6,078,390 square miles (3,890,154,081 acres), or 10.6 percent of the total land area, is arable; 598,063 square miles (382,758,774 acres), or 1.04 percent, is being used to grow permanent crops. A lack of uniformity in the distribution of natural resources leads to a number of problems, such as food

Table 11. Land Utilization and Urban Growth Rates

Country	Arable Land (%)	Permanent Crops (%)	Other Land Use (%)	Urban Growth Rate
Argentina	10.03	0.36	89.61	1.4
Australia	6.15	0.04	93.81	1.4
Brazil	6.93	0.89	92.18	2.0
Canada	4.57	0.65	94.78	1.2
China	14.86	1.27	83.87	3.2
Egypt	2.92	0.5	96.58	2.1
France	33.46	2.03	64.51	0.7
Germany	33.13	0.6	66.27	0.3
Guatemala	13.22	5.6	81.18	3.4
India	48.83	2.8	48.37	2.3
Indonesia	11.03	7.04	81.93	3.9
Iran	9.78	1.29	88.93	2.3
Ireland	16.82	0.03	83.15	1.5
Japan	11.64	0.9	87.46	0.3
Mexico	12.66	1.28	86.06	1.8
Netherlands	21.96	0.77	77.27	1.3
Pakistan	24.44	0.84	74.72	3.4
Russia	7.17	0.11	92.72	–0.6
Saudi Arabia	1.67	0.09	98.24	3.5
South Africa	12.1	0.79	87.11	1.4
Sweden	5.93	0.01	94.06	0.1
Thailand	27.54	6.93	65.53	1.9
United Kingdom	23.23	0.2	76.57	0.4
United States	18.01	0.21	87.78	1.4
Vietnam	20.14	6.93	72.93	3.2

Source: Created from data available from http://www.di.net/articles/archive/2572/, January 2007.

shortages and famine throughout many regions of the world. The disproportion and quality of arable land, in association with the factors previously mentioned, play a significant role in whether a country's population is able to maintain its basic food base. This imbalance in the distribution and availability of the necessary resources has resulted in many of the world's principal survival and longevity issues.

On October 12, 1999, the planet reached a milestone population of 6 billion inhabitants, with most new births occurring in developing countries. The total population for the Earth is projected to be approximately 9.3 billion by 2050. Africa and Asia will account for 20 percent and 60 percent of the world's population, respectively. Arable land is a critical resource in providing a large percentage of food to the world through agricultural development. The availability of productive land, coupled with climate adjustments and human behavior, ultimately will affect the planet's ability to sustain human life or result in catastrophic consequences.

Soil Retention, Climate Factors, and Productivity

Soils are among the most complex and varied of the Earth's surface materials. Soil is a collection of natural bodies, consisting of weathered rock and minerals, decaying organic matter, which, when combined with sufficient quantities of air and water over time, will promote and support plant life. The thicknesses of soil horizons and mineral and organic content vary with respect to time, climate, geographic location, and topography. A productive soil is one that possesses an optimal measure of sustainable nutrients and water; its availability is, in part, governed by climate, topography, and prudent land-use practices.

The development and practice of agriculture has been a means of obtaining food for only some 10 percent of documented history. Plowing degrades the soil, causing chemical and biological losses of the three critical elements that are necessary to grow healthy crops and maximize yields: nitrogen, phosphorus (phosphate), and potassium (potash). For example, deforestation and land-use changes to agriculture in Paraguay and Brazil, through conventional (mechanical) tillage practices using a heavy disk plow and harrow, were causing a reduction in organic content in the upper layers of soil. A return to more manual tilling practices has since improved the organic content in the soils of these countries, although no defensible scientific evidence indicates that this practice is the principal cause for the increase.

Fallow arable land (i.e., land not tilled or sown) can play an important part in affecting air quality, directly influencing local climate and crop yields. For instance, in March 2007, a dust cloud, composed of very fine-grained topsoil from the lower reaches of the fertile Dnieper Valley in the southern Ukraine, traveled more than 932 miles and settled over Slovakia, Poland, the Czech Republic, and Germany. Peak concentrations of between 200 and 1,400 micrograms of PPM_{10} particulates per cubic meter were measured. The daily average limit over the European Union is typically 50 micrograms per 1.3 yards. The total load of dust was estimated at 2,952,619 long tons. These massive movements of dry topsoil had the potential of destroying large tracts of farmland, resulting in human-induced desertification and micro-climatic change. In another case, since the 1930s, fallow fields throughout the southern Ukraine have been exposed to wind erosion, placing 84,943 square miles of highly productive farmland under siege. One-fifth of the particulates distributed about the planet are thought to be caused by the cultivation of fields.

Weather and climate are important in recharging ground water reservoirs and moisture content within the overlying soil layers. Timing, amount, and duration of precipitation and runoff from snowpack are controlling factors. Water-holding capacity is critical for arable soils in arid climates where evaporation rates are normally very high. Seeds will germinate and mature into productive crops only when these basic climatic requirements are met. A deficiency or surplus of water can result in soils losing their ability to sustain plant growth, thereby becoming agriculturally nonproductive. Too much water will drive the air (oxygen) out of the soil, so that the plant drowns. Insufficient water desiccates (wilts) the plant, which causes its stomata pores to close, thereby preventing carbon dioxide from entering the plant. A deficiency in carbon dioxide will inhibit plant health and ultimately stunt its growth.

When arid regions, not climatically conducive to growing food crops, are cultivated, irrigation techniques become necessary to provide sufficient water in consistent amounts and over a sustained duration of time to ensure plant vigor and a healthy yield. Irrigation will always be a controversial issue, because it requires transporting water from a location of abundance to a location in need. Irrigation is necessary only when the amount and distribution of water do not sustain healthy plant growth, although many regions are irrigated unnecessarily, reducing available water supplies for all concerned. A problem develops when a source of abundant water is drawn down to levels at a rate that is not recharged sufficiently to replenish the source. This action results in shortages throughout both the water source region and the more distant regions to which the water is being transported.

On average, only 45 percent of irrigated water reaches crops. This value varies, however, with higher percentages of water loss occurring in hot, arid climates due to evaporation. The Yellow River (China) was dry for nearly 200 days in 1997, due to low rainfall amounts and excess use of water for industry and agriculture. In turn, China's soybean production was reduced to levels that required the country to import products to meet its current needs from Brazil. While Brazil has an abundance of water in the large tropical savanna known as the *cerrado*, encompassing an area of about 250 million acres, the land is nutrient deficient in phosphorus and potassium. The total arable land within the *cerrado* is approximately equivalent to the entire amount of arable land in the United States; however, while this land is water rich, it is nutrient-poor, requiring replenishment of large volumes of phosphorus and potassium to sustain a decent crop (soybean) yield. An average acre of recently cultivated arable land within the *cerrado* requires 14 times the amount of phosphate and three times the amount of potash of a typical American acre.

Increasing the amount of productive land for agriculture by destroying forests and other naturally vegetated regions will merely continue to destabilize the world ecosystem by destroying the wildlife habitat necessary to ensure a proper balance. Sufficient percentages of arable land (see Table 11) exist throughout the world; however, land-use practices and technological advances will be the key to producing sufficient food supplies for the future. Increasing the efficiency of nutrients beyond just allowing a natural course of action through basic recycling of water, biological processes (nitrogen fixing), and replenishment of minerals will require reliance on worldwide advances. Technology must be used in a prudent and responsible way to ensure that regions do not continue to poison the soil while increasing their crop yields. The basic question of how to create better fertilizers and pesticides to produce higher yields, while simultaneously eliminating waste products that are detrimental to the total ecosystem, must be resolved. The negative consequences of industrial waste, including heavy metals and other carcinogenic compounds, plus increased concentrations of pesticides into the river systems, have diminished the world's fresh water supply. On a positive note, the United States returned more than 308,882 square miles of farmland to wilderness over a few decades leading up to 1999. This practice must be followed throughout all developing countries.

The causes of global climate change are complex, and predicting weather patterns with any degree of certainty is nonlinear and problematic. There is little doubt that the Earth is warming, glaciers and ice sheets are melting, sea level is rising, storm activity is becoming more intense, and precipitation patterns are becoming more inconsistent in the amount and timing of events. All of these measurable changes will continue to have pronounced effects on the quality of arable land available for agriculture, its geographic distribution, the type of crops grown, yields, and eventually the places where people choose to live. While some countries are land rich, others such as Bangladesh will lose large percentages of valuable land necessary to provide an adequate food supply for their populations. Bangladesh is principally a delta formed by two rivers, with the heaviest concentrations of sediment in the world. World increases in sea level could inundate large tracts of fertile cropland in this country, further reducing an important food source. Shortages in food will prompt violence as a final means for survival. This issue is already occurring in sub-Saharan Africa, where drought is the natural catalyst for starvation, while politics and intertribal conflict have resulted in mass genocide, serving only to compound the problem of food shortages. Negative human behavior has created long-term social enmity in nations stressed by environmental constraints or calamities.

A change in the flow pattern or even the fate of the future existence of the Gulf Stream was predicted by Edwards in 1999. The impact on temperature variation and ultimately crop production in countries bordering the eastern Atlantic Ocean, such as the United Kingdom, could be devastating. Cooler temperatures and a shorter growing season could change the type and overall yield of food crops for many countries aligned with this warm ocean current. Florida and other coastal regions, including most major port cities throughout the world, will be affected economically as sea level rises and inundates major metropolitan areas. Massive flooding and salt water impingement will require major adjustments in land use.

Climate Change and Global Distribution of Arable Land

The Earth's climate is changing; an abundance of scientific data supports this conclusion. A number of global databases have been created, examining soil properties, slope, temperature, and precipitation in an attempt to simulate arable land and its agricultural suitability, from a series of low- and high-emission circulation models. Countries at higher latitudes within the Northern Hemisphere are more likely to benefit from climate change, whereas countries at middle and lower (equatorial and subtropical) latitudes may suffer various levels of loss of arable land. Expansion of potential arable land might occur in regions such as Russia, North China, and the northern United States. Arable land loss is predicted in selected countries of South America, Africa, and Europe, as well as India. The greatest potential for arable land increases and agricultural expansion also lies in a number of countries within Africa and South America, where the currently cultivated land accounts for approximately 20 percent of the net potential arable land in the world. Climate change will alter the global distribution of arable land and further influence agricultural-related socioeconomic aspects by the end of this century.

—Danny M. Vaughn

Social and Cultural Factors

In the 1990s, a United Nations Food and Agricultural Organization survey of 57 developing countries found that more than half the farms in these countries were smaller than 2.5 acres and not sufficient enough in size to generate enough food to feed a family of four to six children, while also producing a surplus for profit. More than 70 percent of India's population farm and eat all of what they grow. Three-fifths of all of their farms are smaller than 2.5 acres. By 2025, India will be the most densely populated country in the world, with more than 1.5 billion people compressed into an urban setting that will be unable to sustain even the minimum nutritional requirements. In India, sons receive equal shares of land as an inheritance; therefore, the size of the farms gradually diminishes with each generation. In this country, the concentration of arable land is being reduced to a few farmers at the cost of the more traditional smaller family farms. Sixty-five percent of the most productive land in Guatemala is controlled by 3 percent of the farmers, typical of many Latin American countries. In this case, the problem is not yield, but rather the fact that the crops produced are exported and are not feeding the host population in need.

The demand for arable land has resulted in clearing of marginal land such as hill slopes, causing erosion of fertile soil layers, loss of essential nutrients, and salt contamination in arid regions. Poor farming practices also contribute to soil nutrient degradation, while huge amounts of land are lost each year to rapid urbanization from a growing population.

Possessing large tracts of land classified as arable (see Table 11) does not ensure that a country's population will be fed adequately. China has about 7 percent of the world's arable land, but it also contains nearly 25 percent of the world's population. The rapid growth of China's economy has prompted massive construction projects that have reduced the amount of arable land by 1,185 square miles in a single year. Illegal land-use practices have also escalated by more than 17 percent, threatening the government's red line of 296,525,271 acres necessary to maintain the country's level of food safety until 2020.

The value of arable land is critical in countries at risk of not having enough to feed their population or with serious environmental issues that are damaging their natural resources. China, South Korea, the United Arab Emirates, Japan, and Saudi Arabia have acquired and now control more than 18,532,829 acres of cultivatable land outside their national territories. The fall in land prices during the early part of the 21st century and a world economic implosion have redirected investors to more stable and potentially lucrative real estate. The distribution of arable land is falling into the hands of countries outside the political boundaries of the people who reside there as its citizens. Those countries that have a strong monetary base, such as the petroleum monarchies, are grossly deficient in healthy, productive agricultural land; money buys them security in food production through the purchase of arable land in underdeveloped countries. As an example, Qatar controls land in Indonesia, Bahrain in the Philippines, and Kuwait in Myanmar (Burma).

Future Sustainability of Arable Land

The United Nations forecasts that the area of arable land will increase 13 percent by 2030. Sufficient arable land should be available to feed a population of 9.3 billion until 2050, assuming a number of factors remain constant. Arable land will have to expand by approximately 296 million acres in developing countries, mainly in sub-Saharan Africa and Latin America. Mechanisms for equitable food distribution, effective technical assistance, and available capital for infrastructure development will be necessary to ensure developing countries within these regions have a reasonable chance of sustaining their growing populations. They will be particularly critical in sub-Saharan Africa and throughout many Asian countries. Arable land currently in use in developed countries is expected to decline by some 193,051 square miles, although this trend could be reversed by the demand for biofuels.

Pesticides protect crops around the globe; without them, 70 percent of the world's harvest would be lost. Currently, more than 40 percent of food crops are lost annually to insects and fungal diseases. The chemical industry has developed pesticides that are biologically unstable so that the levels of residues detectable in fruits and vegetables are 100 to 1,000 fold below safe levels. If pesticides were not used, an increase in 1,544,408 to 21,621,721 square miles of arable land would have to be made available for agricultural cultivation in the future to equal year-2000 yields. Without the technological advances currently in place, current food production would require substantially more than 4,942,087 acres to be plowed up.

Whether one subscribes to the theory that natural or human-induced effects are to blame for climate change, the outcome of this trend is clear: Such change will be destabilizing for a world village. The physical and chemical properties of soils and climate factors no doubt will play important roles in the health and long-term ability of arable land to produce food crops. While environmental factors are not easily controlled, human stewardship, through prudent land-use practices, will be critical, affecting crop yields throughout the world.

The use of satellite imagery has reached the level of resolution that now enables scientists to measure soil moisture over large tracts of land. The launch of the European Space Agency's Soil Moisture and Ocean Salinity (SMOS) satellite will revolutionize the study of Earth's hydrologic cycle and, by extension, measure changes in land surface moisture. The Microwave Imaging Radiometer, using Aperture Synthesis (MIRAS), will enable scientists to identify fields that are stressed with water shortages, enabling irrigation practices to be more precisely directed by farmers. Identifying stressed fields can be a precursor to identifying regions susceptible to desertification. One-third of the Earth's land surface (19,150,667 square miles) is threatened by desertification, with 24 billion tons of fertile soil disappearing annually. More than 250 million people are affected directly. The human impact on the amount of usable arable land will be magnified as the Earth's population increases and land is taken up by homes and other structures. As the planet becomes more crowded, societies will be required to address the limits of arable land, climate factors that govern sustainable agriculture, and water availability.

—Danny M. Vaughn

Further Reading

Baudet, M. B., and L. Cavreul. "The Growing Lust for Agricultural Lands." *Truthout*, April 14, 2009, 1.

Design Intelligence. "Land Utilization Review." January 2007. www.di.net/articles/archive/2572/.

Dingding, X. "National Population and Planning Commission of China." *The China Daily*, 2007.

Edwards, R. "Freezing Future." *New Science* 164, no. 6 (November 27, 1999), 2214.

Goklany, I. M. "Saving Habitat and Conserving Biodiversity on a Crowded Planet." *Bioscience* 48, no. 11 (November 1998): 941–953. www.chicagomanualofstyle.org/tools_citationguide.html.

International Development Research Centre. "Population Explosion in the World's Cities." 2010. www.idrc.ca/en/ev-97406-201-1-DO_TOPIC.html.

Maley, C. "Agricultural Food Production and Arable Land and Water Trends." *The Market Oracle*. December 17, 2009. www.marketoracle.co.uk/Article15890.html.

Pimentel, D. "Pest Management in Agriculture." In *Techniques for Reducing Pesticide Use: Economic and Environmental Benefits*, edited by D. Pimentel. Chicester, UK: Wiley Press, 1997, 1–11.

Repetto, R. "The Second India Revisited: Population Growth, Poverty, and Environment Over Two Decades." In *Proceedings of the Conference on Population, Environment and Development* (March 13–14, 1996): 2–31. Washington, DC: Tata Energy Research Institute and World Resources Institute.

Smi, I. Vaclov. *Feeding the World: A Challenge for the 21st Century*: Cambridge, MA: MIT Press, 2000.

Southgate, D., and M. Basterrechea. "Population Growth, Public Policy and Resource Degradation: The Case of Guatemala." *Ambio* 21, no. 7 (November, 1992): 460–464.

United Nations Food and Agriculture Organization (FAO). *Dimensions of Need: An Atlas of Food and Agriculture*. Rome: FAO, 1995, 16–98.

Williamson, R. A. "Water and Space Systems." *Imaging Notes* 25, no. 1 (2010): 10–11.

Wolfram, B. "Arable Land Can Have a Negative Impact on Air Quality." Leibniz Institute for Tropospheric Research, May 6, 2008. www.eurekalert.org/pub_releases/2008-05/haog-alc050608.php.

The World Fact Book. Washington, DC: Central Intelligence Agency, July 24, 2008. https://www.cia.gov/library/publications/the-world-factbook/geos/xx.html.

B

Biodiversity

Human beings depend on other living organisms for food and nourishment. For 12,000 years or more, humans have been successful in domesticating a variety of plants and animals not only to satisfy their immediate needs, but also to provide them with long-term food security. Domestication, a significant innovation, would not have been possible without an understanding of and deference to the natural systems and their limits. After all, the same mechanisms that are responsible for the existence of a myriad of living beings on Earth have allowed humans to select plants and animals to fit their needs. The diversity of living organisms is critical for both plants' and humans' continued existence and evolution. It is important to understand the significance of biological diversity if humans are to aspire to sustained food security.

Biodiversity refers to the number and variety of all plants, animals, microorganisms, and ecosystems. It is expressed as the variation of living organisms within a given location or across geographical regions. The diversity of life-forms is uneven across the globe. While some areas, such as the tropical forests, abound with a variety of life-forms, others, including arid deserts or frigid poles, are virtually devoid of life-forms. Most areas on the Earth lie on a continuum between these two extremes. The goals of ecologists and biogeographers are to try to understand these variations and their geographic distribution. Biodiversity, however, is not limited to a simple count of distinct life-forms, but rather can be explained at three levels—genetic diversity, species diversity, and ecosystem diversity.

Genetic diversity refers to the variation in genetic information contained in the deoxyribonucleic acid (DNA) of living things. Genes contained in DNA can occur in nearly infinite combinations, giving rise to limitless varieties of shapes, sizes, colors, and behaviors of individuals. Depending on the given environmental conditions, these traits can be either advantageous or disadvantageous in terms of an individual's survival and its ability to pass those traits on to its descendants. The process that allows for the survival of individuals with such optimal traits and the extinction of individuals with incompatible traits is natural selection, which plays a key role in evolution. Although individuals have genetic variations that make grouping them into various categories very difficult, biologists classify individuals with morphologically similar traits based on their ability to interbreed and produce fertile offspring. Such groups are called species.

Species diversity refers to the total number of living species. It can be understood from two perspectives—species richness and species evenness. The total number of species in a given area indicates its species richness. Some areas of the Earth, such as the Amazonian rainforests, are called biodiversity hotspots because species richness in such areas is particularly high. Globally, species richness is generally greatest near the equator and declines toward the poles. Species evenness is the distribution of individuals among different species; it incorporates a determination of whether all species share equal numbers of individuals. For instance, it is likely that in an area containing several species, one species has a greater number of individuals when compared to another, which indicates low species evenness. Biological diversity within an area is indicated by the variation in the number of different species, the degree of variation among them, the frequency of their occurrence, their survival strategies, their interactions with other species in the community, and their interaction with the environment. Out of the 10 million to 100 million species estimated to inhabit this planet, biologists have identified only some 1.7 million species.

A collection of species that interact with one another and with the environment is called a community. An ecosystem is made up of different habitats that are used by a variety of communities. For instance, a forest ecosystem is different from a grassland ecosystem owing to the difference in the physical conditions (e.g., soils, climate, topography) and the variation in species composition and behavior (e.g., nutrient use, leaf area, root depth). Ecosystem diversity refers to the variety of ecosystems on Earth. The diversity of an ecosystem depends on the physical characteristics of the environment, number and types of habitats, the diversity of species present, their mutual interactions, and their interactions with the environment.

Significance of Biodiversity

Maintenance of biological diversity is important for the health and wellness of ecosystems. Due to the dynamic nature of Earth environments, an area that is favorable for the survival of a species could easily become hostile. Persistence of life through small or catastrophic environmental disturbances is made possible by biodiversity. Some species that are not resilient will perish, whereas others that possess advantageous traits may survive in the new environment. As a result, while the type of species living in that environment may change, life itself persists. The importance of biodiversity is vividly seen when a disease kills certain plants, but not others, because the surviving plants happen to be resistant to that disease.

If all plants had the same traits and susceptibility to the disease, then that disease would drive all of them to extinction. The effects of environmental disturbance can also be visualized through another example. After a violent volcanic eruption, it is likely that all plant and animal life will be destroyed in the immediate vicinity of the event. This environment becomes unfavorable for woody trees, as the soil is replaced with new rocky material. However, grasses or shrubs may quickly colonize the area to take advantage of the space and sunlight. This spread is possible because grasses and shrubs, unlike trees, do not require deep soils to survive. If these species had not possessed traits that were different from those of the trees, life could not have continued.

Historically, humans have modified the genetic makeup of organisms that provide them with food. For example, they have selectively bred out the bitterness from almonds, subdued the toxicity in potatoes, and created docile cattle from wild Aurochs. As humans began to move away from hunting and gathering to agriculture a few thousand years ago, they started modifying natural ecosystems to generate an increased amount of edible food. Largely inedible wooded ecosystems were removed and replaced with grassy ecosystems containing edible cereal grasses. Fertilizers were applied to nutrient-poor soils, pests were kept at bay or eradicated, weeds were kept in check, and the grasses themselves were harvested and replanted seasonally. Essentially, agroecosystems evolved as simplified ecosystems established to provide energy for human consumption. Whether agroecosystems have higher biodiversity or not depends on the agriculturalists and the economic pressures to which they are exposed. In subsistence systems where the agriculturalists consume most of the food they grow, a high level of agricultural diversity is required to ensure sufficient food and balanced nutrition. In systems that are integrated with the market, export motivation usually drives the cultivation of one or two principal cash crops, thereby reducing agricultural diversity. Although agroecosystems are maintained by humans, much of their success depends on natural systems that provide services, including water, nutrient, and biogeochemical cycling.

For human well-being, biodiversity is like a savings account, library, and municipal services bundled together. First, it acts as a savings account for the following reason: Humans have come to rely on a few food sources that are economically sensible. For example, food crops that generate the largest amount of money when sold are grown in large quantities. By growing one large genetically identical food source year after year, an agriculturalist takes the risk of exposing the food crop to quickly evolving pests and diseases that may cause large-scale crop failures. According to the Food and Agriculture Organization (FAO), since humans first began farming, they have cultivated approximately 7,000 species of plants. Today, however, only 30 crops provide 90 percent of the world population's dietary energy requirements. Wheat, rice, and maize alone provide 50 percent of the global dietary energy consumed. Similarly, out of the 30 to 40 species of animals and birds that were domesticated, only 14 species account for 90 percent of global livestock production today. Growing a genetically diverse variety of food sources ensures that even if one variety is attacked, the resistant varieties will survive.

Diversity also ensures a better balance in human nutrition. Oversimplification of diets is one of the major causes of malnutrition in the world. One-third of the world population suffers from diseases associated with malnutrition and inadequate access to food. Simultaneously, conditions such as diabetes, obesity, and heart disease are on the rise among the world's peoples, regardless of their economic status. A broader variety of foods grown in agroecosystems with higher diversity can improve nutrition. Even different varieties of the same crop may contain significantly different nutrient contents. For example, the protein content in rice varieties can vary 5 to 14 percent, and some maize varieties contain high protein while others contain high sugar content. Ethiopian farmers have identified at least three varieties of sorghum that contain 30 percent more protein than other varieties and 50 to 60 percent more lysine, a limiting amino acid required for the body's utilization of protein as an energy source.

Second, the multitude of species, each with specific genetic information, can be viewed as analogous to books in a library. One use of this library has been in the field of medicine. By understanding the makeup of plants, humans have been able to harness their many chemical compounds for medicinal use. For example, a medicine was developed from the rosy periwinkle for treating childhood leukemia. The foxglove plant was found to be useful in the treatment of heart disease. Plants, being immobile, defend themselves against a wide range of predators through the use of natural toxins. By elucidating the evolutionary history of the plants, the nature of these toxins, and their effects on predators, humans can identify specific uses for these compounds. Perhaps cures for many diseases are locked in the many different plant species that have not yet been examined.

Third, like a municipality, biodiversity provides humans with several environmental goods and services. Ecosystem goods include food, fiber, fuel, biochemical resources, genetic resources, and fresh water. Ecosystem services include flooding, pest control, pollination, seed dispersal, erosion regulation, water purification, and climate and disease control. Additionally, nutrient cycling, production and regulation of atmospheric gases, water cycling, and other functions provided by the ecosystem are services on which humans rely. An estimated 100,000 different species of pollinators keep food crops alive at no monetary cost. Pollinators, such as bees, transport the male sexual cells from one plant to the female parts of other plants, thereby facilitating sexual reproduction in immobile plants. Consequently, scientists warn that the decline of pollinators could have a profound negative impact on crop survival and, ultimately, food prices. Trees have a significant effect on regional temperatures and precipitation through transpiration, which is the process by which plants pump ground water up through their root systems, allowing for evaporation into the atmosphere from their leaves. Research shows that plants influence the composition of atmospheric gases such as oxygen and carbon dioxide and play an important role in the biogeochemical cycles. The availability of soils for human use is yet another important environmental service that cannot be taken for granted. Soil fertility and suitability for agriculture are functions of soil structure, composition, and chemistry that are determined by both physical and biological components. Weathering due to sun, wind, and water action breaks down rocky material into soils, which then act as natural filters to purify the groundwater. Soils also act as water and nutrient storage systems, which are essential for growing food crops. Biological organisms such as earthworms and fungi recycle nutrients for plant use by breaking down and decomposing organic material.

Current Status and Future Prospects

Natural selection favors species that possess traits that are successful; this implies the extinction of species that possess unsuccessful traits. Extinction, an integral part of evolution, occurs when no living representative of a given species exists any longer. Of all the living things that ever existed on Earth, only 1 to 5 percent exist today. Based on fossil record estimates, the average rate of extinction appears to be about one species every four years. In addition to this background extinction, sudden extinction events, known as mass extinctions, also have occurred. Fossil records indicate that five such events occurred in the past, during which most living organisms perished due to a cataclysmic event or a rapid change in the environment. Scientists suggest that such a mass-extinction event is occurring today as well, with species vanishing at a rate 1,000 times greater than the natural rate of species loss. The major driver of this event is linked to human modification of the environment.

Habitat alteration in the last five decades alone has caused major loss of biodiversity. Conversion of natural landscapes to human land uses, especially for agriculture, has fragmented continuous landscapes that once provided food, cover, and shelter for several species. Human activities have disrupted the biogeochemical cycles by speeding up the loading of nutrients, such as nitrogen and phosphorus, derived from fertilizers and farm effluent. These effects have caused drastic changes in terrestrial, freshwater, and coastal ecosystems. The amount of atmospheric carbon dioxide has been elevated by burning fossil fuels for energy generation, while simultaneously large expanses of vegetation and coral reefs that act as carbon sequesters have been destroyed. Manufacturing and industrial farming have contributed to increased methane, nitrous oxide, and other gases in the atmosphere. Consequently, shifts in global climate patterns have been observed. Numerous species, not resilient enough to adapt to the rapid environmental changes brought about by such climatic shifts, are being driven to extinction.

Additionally, human exploration and travel have consequences. Intentional and unintentional introductions of non-native species into new locations have been a significant driver of species extinction. Alien species often thrive in new locations because the predators and diseases with which they coevolved do not exist in these locations. Native species are often unprepared for such sudden and aggressive competitors, and they are rapidly edged out and driven to extinction. Humans themselves have played the role of invader and predator. They have harvested and hunted plants and animals and driven many species to extinction in the past. For example, the large marsupials of Australia were driven to extinction by the invading aboriginal humans approximately 30,000 years ago. Currently, the large human population and its insatiable demand for plant- and animal-derived products have increased the rate of harvesting to unsustainable levels worldwide. In other words, the speed at which humans are harvesting does not allow enough time for these species to repopulate.

Preserving biodiversity is not only critical to prevent a mass extinction of species, but also to protect human beings' food security. The biggest challenge that people face today is to discover ways to live in a sustainable manner. That is, at a minimum, humans need to figure out a way to feed, clothe, and shelter all people without exhausting the resources for future generations. Broadly, the following actions may be required: (1) discover clean, inexhaustible, and practical sources of energy; (2) provide incentives for agricultural biodiversity and discourage monoculture farming methods; (3) view funding of active plant and animal species conservation efforts as a long-term investment; (4) discourage excessive consumerism and encourage sustainability through education and public policy; (5) use market-based mechanisms, such as estimating the monetary value of ecosystem services, and add that value into the cost of the products, with the additional revenue generated being allocated for the preservation of biodiversity; and (6) reevaluate the fundamentals of economic practices that promote maximization of profits without regard to the environmental burden that is thrust upon future generations.

—Bharath Ganesh-Babu

Further Reading

Kauffmann, Robert, and Cutler Cleveland. *Environmental Science*. New York: McGraw Hill Press, 2008.

United Nations, Food and Agriculture Organization. Biodiversity. Accessed February 2011. www.fao.org/biodiversity/en/.

Wilson, E. O., and Peter M. Frances, eds. *Biodiversity*. Washington, DC: National Academy Press, 1988.

Carbon and Conservation Tillage

All life on Earth—and, as far as scientists on Earth can determine, all life in the universe—is based on carbon. If the carbon atom is the basis for biochemical life, then agriculture is the basis of human civilization. Indeed, "civilization" is equated with the adoption of agriculture. Key developments in agricultural technology were the tools used to "till" the soil. Tillage generally inverts the top layers of soil, making it possible for humans to concentrate on producing a particular suite of crops. The plowshare appears to date to 3500 BCE in Egypt. This initial crude tool was followed in the 11th century by the invention of the moldboard plow in Europe. The widespread adoption of this new tool and the enhanced ability of the moldboard to bring into production the heavy soils of much of Northern Europe, coupled with the warm stable climate of the Medieval Warm Period, led to significant expansion of the human population of Europe. This expansion was destined for reversal soon thereafter, however: During the Little Ice Age and the spread of the Black Death throughout Europe, between 30 percent and 60 percent of the population died.

Since that time, populations have grown steadily, and innovations in agriculture to support this expansion have centered on the plow. In 1837, innovations in metallurgy and efforts to improve farming practices came together to again expand the range of soils that could be farmed. The invention of the steel moldboard plow by John Deere in the United States enabled farmers to plow the thick prairie soils of the world. The result was revolutionary change in the agricultural industry: Now the world's "bread baskets" were no longer prairies. Even so, more than 100 years after Deere changed agriculture and less than 50 years into the mechanization of farming, a global economic panic, coupled with severe regional climate change (called the "Dust Bowl" in the United States), resulted in a crisis for global agricultural production. Edward Faulkner, among others, reflected on the American Dust Bowl and concluded that perhaps tilling the soil was not the best way to feed the world. In *Plowman's Folly* (1943), Faulkner exposed the problems connected with conventiontal tillage and tested one of the first versions of no-till agriculture.

Stored soil carbon is "burned out" by traditional tillage methods. This reduction in organic carbon in the soils dramatically reduces the inherent fertility and productivity of the soil. It has long been recognized that the oxidized organic carbon

Thomas Jefferson and the Moldboard Plow

In an 1813 letter to Charles Wilson Peale, Thomas Jefferson noted that "The plough is to the farmer what the wand is to the sorcerer. Its effect is really like sorcery." Thomas Jefferson, statesman, inventor, and farmer, was one of the premier examples of a true "renaissance man" in early America. He was passionate about improving the efficiency of the plow. In 1794, using specific, repeatable, mathematic calculations, he developed a light moldboard plow that was revolutionary in its performance. Jefferson declared that the moldboard, as he designed it, was "mathematically demonstrated to be perfect."

Jefferson's plow was designed specifically for use in the hilly ground of Monticello. His plowing methods represented early efforts at contour farming. He wrote:

> We now plough horizontally following the curvatures of the hills and hollows, on the dead level. . . . Every furrow thus acts as a reservoir to receive and retain the waters, all of which go to the benefit of the growing plant, instead of running off into streams. . . . In point of beauty nothing can exceed that of the waving lines and rows winding along the face of the hills and vallies [sic].

—Gregory Gaston and Kacey Mayes

from soil is released to the atmosphere and represents a significant source of anthropogenic atmospheric carbon dioxide. Conversion of lands to agricultural production results in a sharp decrease in organic carbon stored in the soil. In 1983, Houghton and his colleagues noted that a significant percentage of previously stored organic carbon in soil is released to the atmosphere upon conversion to agriculture. Since the invention of agriculture, the total amount of atmospheric carbon released through conversion of soils to agricultural production on a worldwide basis is estimated to be 180 Gt (gigatons), or about 6 percent of the total 3,000 Gt of carbon in the current atmosphere.

Basics of Conservation Tillage

Traditional tillage practices rely on the moldboard plow to invert the top layers of the soil, burying surface residue and exposing the middle layers of soil fully. The bare soil is further broken down through additional tillage, preparing the seedbed for the few annual crops that form most of our agricultural base. Additional tillage is used during the growing season to suppress other vegetation (weeds). The bare exposed soil between agricultural plants experiences significantly higher soil temperatures and exposure to abundant atmospheric oxygen, which releases carbon dioxide from decomposing soil organic matter. This continual disturbance breaks down soil structure, and the exposure to wind and rain produces rapid rates of erosion.

Technology has made it possible to reexamine the whole process of soil preparation in agricultural fields. Conservation tillage, reduced tillage, and no-till are all terms that are applied to agricultural management practices with a common focus, as defined by the UN-FAO: "to maximize ground cover by retention of crop residues and to reduce tillage to the absolute minimum . . . to achieve a sustainable and profitable production strategy."

Critical new tools for this form of agriculture include herbicides that suppress competing vegetation. Because conservation tillage requires minimal soil disturbance, a method of removing unwanted vegetation is required. Herbicides seem to be the most efficient way to accomplish this goal. Species-specific types such as atrazine, which affects only grasses; 2-4D, which is tailored to affect only broadleaf species; and glyphosate, which is a broad-spectrum, nonspecific herbicide; represent a significant change in agricultural technology. Chemical herbicides have provided a technological solution to the need for mechanical cultivation to suppress competing vegetation.

Advantages of Conservation Tillage

While sequestration of carbon in the soil, rather than allowing its release to the atmosphere, is a fairly recent concern, a much more significant problem addressed through conservation tillage is the almost universal reduction of soil erosion. When a layer of crop residue is maintained on the soil surface, it provides a mechanical barrier to wind and water erosion. Both wind and water erosion are sharply reduced using conservation tillage.

Dramatically reduced rates of soil erosion are possible with conversion to conservation tillage practices. As much as a 98 percent reduction in soil erosion has been observed in fields where no-till agriculture was implemented. Just this benefit alone would convince many farmers to investigate conservation tillage. After all, the primary source of capital for agriculture is the fertility of the soil. If it is practical to preserve that capital through better management processes, very few farmers will refuse to adopt new practices. It has been estimated that 50 percent of the carbon transported from agricultural soils by erosion ends up in the atmosphere. While several management options affect the storage of organic carbon in soils, only no-till agriculture results in significantly increased carbon sequestration.

In modern agricultural practices, the line between profit and loss is very narrow indeed. Seemingly small reductions in production costs could very easily mean the difference for farmers between making the payments and having a little left over and defaulting on the loans required to purchase seed, fertilizer, equipment, and fuel. Conservation tillage can dramatically reduce the number of passes over the field required to produce a crop. Specialized seeders are able to push back the residue from the previous crop, prepare a narrow strip of seedbed, plant the seeds, and apply both fertilizer and herbicides in a single pass. While herbicide inputs are necessarily higher and the cost of specialized equipment is high, total production costs are lower.

When a layer of decomposing organic material is kept on the majority of the soil surface, overall temperatures in the soil are reduced. Lower temperatures produce much slower rates of oxidation. The lack of continual mechanical disturbance keeps the soil peds (the basic units of soil structure) intact and reduces the mixing of atmospheric oxygen into the soil column. Conservation tillage also produces a shift in distribution of carbon throughout the soil profile, with a significantly higher concentration of carbon being found at the very top layers of the soil. This shift in the distribution of carbon through the soil profile produces measurable changes in rooting, nutrient cycling, and distribution of soil microbes. Cooler soils retain more organic carbon and act as a storehouse of organic carbon that prevents the release of a high percentage of carbon into the atmosphere. Organic matter is stored carbon that is not available to contribute to global warming.

Adoption of Conservation Tillage

Worldwide conservation tillage was estimated to exceed 247 million acres in 2008, increasing from 111 million acres in 1999. While conservation tillage is expanding globally, however, less than 7 percent of global agricultural lands is managed with conservation tillage. By far, the highest percentages of agricultural lands converted to conservation tillage (85 percent) are concentrated in North and South America. For a variety of reasons—economic, social, and climatic—farmers in Africa and Asia have been much slower to adopt conservation tillage.

Although a number of press releases have discussed the importance and widespread adoption of conservation tillage in Russia, the FAO analysis of Russia notes that "it was not possible to get realistic numbers." While Derpsch and Freidrich cite anecdotal evidence of the importation of conservation tillage equipment and the issuance of press releases, they are limited to noting that there should be a considerable area under no-tillage practices in Russia. The uncertainty of actual adoption of conservation tillage in the agricultural soils of Russia is significant, but it should be possible to assess large-scale patterns and potential for conservation tillage.

A Case Study: Soil Carbon in Russian Agricultural Soils

According to the U.S. Department of Agriculture (1990), the cultivated area of Russia and associated states totaled 522 million acres in 1988. This huge area was almost twice the cultivated area of the United States (296 million acres) and more than four times the area of Canadian agricultural lands (114 million acres). Agricultural lands of Russia were located using a digitized version of the map produced by Cherdantsev. Various soil types within the larger agricultural region were established using the soil/vegetation association map of Ryabchikov.

Estimating the potential for conservation tillage to sequester carbon in agricultural soils first requires that the current carbon content of the soil be established. Kobak compiled a database of soil carbon parameters from 70 Soviet and international sources. Carbon profiles were estimated for each major soil association in Russia. Carbon contents were presumed to be representative of an initial or pre-cultivation condition.

Conversion of soil to agricultural production releases carbon to the atmosphere. The rate of carbon release is relatively high after initial cultivation, but decreases over time. The greatest rates of carbon storage occur in the first 20 years of cultivation. Mann used published data from 625 paired soil samples to develop a generalized regression equation that predicts the current carbon content as a function of initial (pre-cultivation) carbon content. He found that most soils lose at least 20 percent of their carbon content after cultivation. The loss of soil carbon strongly depends on the length of time under cultivation and the initial carbon content.

The current estimate of total carbon in the agricultural soils of Russia is 32.4 Gt. The most productive chernozem soils account for more than 60 percent of this total; chernozem soils are the black prairie soils of the grasslands of southern Russia and the Ukraine. On average, there has been a 24 percent reduction and a 10.2 Gt loss to the atmosphere since the onset of cultivation in these areas.

Over time, cultivated soils reach a new equilibrium at lower levels of organic carbon. The time required to reach equilibrium and the carbon content at this equilibrium point depend on the soil type, climatic conditions, and agricultural management practices. Long-term investigations at the Rothhamsted (England) Experiment Station indicate soil organic carbon has long-term stability. Jenkinson studied the soil carbon of unfertilized test plots that had been cultivated for more than 150 years and observed that soil carbon appears to reach equilibrium after approximately 45 years.

After eliminating agricultural soils in Russia considered too cold and wet for conservation tillage, the overall increase in soil carbon was estimated by Gaston and others to be slightly more than 10 percent of soil carbon equilibrium under conventional tillage practices (3.3 Gt). This point represents a new equilibrium condition for soil carbon. Establishment of this new equilibrium condition would be expected to take at least a decade. This is a single change, however—not an ongoing process. Once the new equilibrium is established, no additional sequestration of atmospheric carbon is expected.

Problems with Conservation Tillage

If conservation tillage has clear advantages in terms of soil erosion, long-term sustainability, and ability to prevent the addition of thousands of tons of carbon dioxide (greenhouse gas) to the atmosphere, why are only 7 percent of the world's agricultural lands converted to this management practice? The answers are complex. Conservation tillage is different. Farmers whose very survival depends on the success of their agricultural methods are often very reluctant to make changes in "what works." This is especially the case when the crop yields post-conversion could potentially decrease 5 to 10 percent in the first few years, according to Wagner. In many countries, technical agricultural education is not available or is very limited in its outreach. Conservation tillage requires significant technical and educational support. Lacking this support, very few farmers will be able to make the change on their own.

While on a large scale, tillage costs can be reduced, at smaller scales the costs of herbicide, fertilizer, and equipment far outweigh the benefits that accrue from conservation tillage. Herbicides must be applied in proper amounts and at critical times. There is very little room for error in selection and timely application of herbicides to control weeds. Conventional tillage buries weed seeds from the previous season. By comparison, not only are more weed seeds present in fields under conservation tillage, but debris on the surface also prevents mechanical removal of weeds. In a classic case of selection pressure, those few individual weeds that have a genetic resistance to specific herbicides are the ones that will survive and reproduce. Passing along this genetic characteristic produces weeds that are no longer subject to application of herbicides. The spread of resistant weeds has been noted in numerous scientific publications and even in *The New York Times* and *USA Today*. In addition to the cost of the chemicals, the need for very careful management regarding types and timing of application and the large-scale application of agricultural chemicals produce the risk of widespread contamination of watersheds and ecosystems. In a world where the sheer volume of food production could become slightly less important and the quality of food produced and the limitation of damage to the larger ecosystem are assuming increasing importance, complete control of weeds in agricultural systems by purely chemical means could become less attractive. In many parts of the world, herbicides are not readily available and information on their application is not widespread. In the absence of effective weed control by application of herbicides, conservation tillage is impossible.

Most agricultural equipment is built along traditional lines; conservation tillage requires a new approach to design and construction of tillage and planting equipment. Equipment that is able to push the residue out of the way, prepare an acceptable seedbed, place the seeds, and apply proper amounts of fertilizer and herbicide in a single pass is extremely challenging to engineer and manufacture. After several centuries, the moldboard plow and disc cultivator seem to be "mature technology" that do not pose a challenge to either manufacture or use. By comparison, most of the development of the technology required for conversion to no-till management is focused on very large-scale mechanized agriculture. Complexity in design and manufacture makes equipment for conservation tillage extremely expensive. As a consequence, adoption of conservation tillage often is inhibited by the cost and availability of equipment designed for conservation tillage. In Russia, where there are very few large private farms, the large collective farms are unlikely to be able to finance the required investment in equipment to convert their practices to conservation tillage. It is also highly unlikely that governments (the central Russian government or the governments of the individual republics or oblasts) will be willing to invest scarce resources in technology that, while increasing the long-term stability of agricultural production and preserving soil capital, produces very small increases in yield.

The culmination of herbicide reliance, the potential costs of conversion, and the small short-term benefits of no-till agriculture have led some farmers to believe that the practice poses an increased risk of failure. Many believe that the learning curve is too steep and fear that there is too much to lose if the practice is not successful, with little chance to gain directly from adoption of conservation tillage. In the United States, farmers who have converted to this practice have found themselves going back to the plow after fall harvesting and for weed control when herbicides do not seem to be working. Rattan Lal, the director of the Carbon Management and Sequestration Center at Ohio State University, and his colleagues agree that no-till agriculture is a positive force in conservation of soils, yet represents only a temporary solution to a permanent problem that will persist as long as soils are being heavily disturbed by agriculture.

The very mechanisms that help sequester carbon in agricultural soils through conservation tillage create limits on where this approach can be profitably applied. Soils that are considered marginal for cultivation because they are too cold or too wet are unsuitable for conservation tillage. In general terms, areas too cold, too wet, or with growing seasons that are too short are unsuitable for this practice.

The southernmost extremes of Russian agricultural lands are at equivalent latitudes to the northern Great Plains of the United States. Long cold winters and short warm summers characterize Russian agricultural lands. The extremely continental climate creates conditions in which cold soils are a significant limiting factor in agricultural production. In large areas, agricultural production is limited by short growing seasons and high soil moisture levels that restrict tillage operations in the early spring. In the northwestern parts of Russia, the intense winter cold is less important than the short span of the summers and the lateness of spring. It is reasonable to assume that areas where crop production is limited by low temperatures, high winter precipitation, and a short growing season will not be suitable for conservation tillage.

Suslov's *Physical Geography of Asiatic Russia* provides a very enlightening profile of marginal agricultural lands on the plains west of Irktusk. He describes fallow plowing of the partially frozen soil as a strategy to increase the temperature of the soil sufficiently to allow for crop production. While the poetic description of frost crystals sparkling like "newly minted kopeks" in the black soil is very colorful, it also suggests very strongly that conservation tillage with colder soils, higher surface moisture, and the insulating blanket of organic debris left on the surface would be impossible in this area.

Gaston and his colleagues estimated that 14 percent of agricultural lands in Russia are unsuitable for conservation tillage. If the remaining 447 million acres were to be converted to conservation tillage at the percentages anticipated for the United States and Canada (i.e., 45 percent), an additional 3.3 Gt of carbon would be sequestered in the soil and removed from atmospheric circulation. In terms of cost per ton of carbon removed from atmospheric circulation, it is impossible to justify this conversion. Instead, other reasons must be paramount for the conversion to take place. Long-term sustainability and preservation of the productive soil resources, coupled with the economies of production that accrue to individual operators in North and South America, must be the driving force behind conversion to conservation tillage.

—Gregory Gaston and Kacey Mayes

Further Reading

Astyk, Sharon. "Casaubon's Book." *Science Blogs*. June 21, 2010. Accessed September 15, 2010. http://scienceblogs.com/casaubonsbook/2010/06/gene_logsdon_on_no-till_agricu.php.

Boerboom, C. *Facts about Glyphosate Resistant Weeds. The Glyphosate, Weeds and Crops Series*. West Lafayette, IN: Purdue University Agricultural Extension Service, December 2006.

Cherdantsev, G. N. "Arable Land in the USSR in 1954" [Map]. In *A Geography of the USSR: Background to a Planned Economy*, by J. C. Cole and F. C. German. London: Butterworth, 1961.

Derpsch, R., and T. Friedrich. *Development and Current Status of No-Till Adoption in the World*. Rome: United Nations, Food and Agriculture Organization, 2009.

Fagan, B. *The Great Warming: Climate Change and the Rise and Fall of Civilizations*. New York: Bloomsbury Press, 2009.

Faulkner, E. H. *Plowman's Folly*. Norman, OK: University of Oklahoma Press, 1943, 161.

Gaston, G., T. Kolchugina, and T. S. Vinson. "Potential Effect of No-Till Management on Carbon in the Agricultural Soils of the Former Soviet Union." *Agriculture, Ecosystems and Environment* 45 (1993): 295–309.

Hobbs, J. A., and P. L. Brown. "Effects of Cropping and Management on Nitrogen and Organic Carbon Contents of a Western Kansas Soil." *Kansas Agricultural Experiment Station Tech. Bulletin* 144, 1965.

Horowitz, J., R. Ebel, and K. Ueda. *"No-Till" Farming Is a Growing Practice*. Washington, DC: USDA Economic Research Service, Economic Information Bulletin 70, November 2010.

Houghton, R. A., J. E. Hobbie, J. M. Melillo, B. Moore, B. J. Peterson, G. R. Shaver, and G.M. Woodwell. "Changes in Carbon Content of Terrestrial Biota and Soils between 1860 and 1980: A Net Release of CO_2 to the Atmosphere." *Ecological Monographs* 53, no. 3 (1983): 235–262.

Huggins, D., and J. P. Reganold. "No-Till: The Quiet Revolution." *Scientific American* (2008): 71–77.

Jenkinson, D. S. "The Rothamsted Long-Term Experiments: Are They Still of Use?" *Agronomy Journal* 83 (1991): 2–10.

Kern, J. S., and M. G. Johnson. *Impact of Conservation Tillage Use on Soil and Atmospheric Carbon in the Contiguous United States*. EPA/600/3-91/056. Corvallis, OR: Env. Research Lab., 1991.

Kobak, I. I. *Biotocial Compounds of the Carbon Cycle*. St. Petersburg, FL: Hydrometeoizdat, 1988.

Kobak, I. I., and N. Kondrashova. *Distribution of Organic Carbon in Soils of the World*. Trudy, GGI 320 (1986): 61–76.

Kononova, M. M. *Soil Organic Matter*. Moscow: Nauka, 1963.

Mann, L. K. "Changes in Soil Carbon Storage after Cultivation." *Soil Science* 142, no. 5 (1986): 279–288.

Medvedev, Z. A. *Soviet Agriculture*. New York: Norton, 1987.

Phillips, R. E., R. L. Blevins, G. W. Thomas, W. W. Frye, and S. H. Thomas. "No-Tillage Agriculture." *Science* 208 (1980): 1108–1113.

Priputina, I. V. "Lowering of the Humus Content of Chernozem Soils of the Russian Plain as a Result of Human Action." *Soviet Geography* 30, no. 10 (1989): 759–762.

Ryabchikov, A. M., ed. *Geographic and Zonal Types of Landscapes of the World* [Map]. Moscow: Moscow State University School of Geography, 1988.

Suslov, S. P. *Physical Geography of Asiatic Russia*, trans. by N. D. Gerchevsky. San Francisco: W. H. Freeman, 1961.

Symons, L. S. *Russian Agriculture: A Geographic Survey.* New York: Wiley, 1972.

United Nations, Food and Agriculture Organization. "Ag: Conservation Agriculture." Accessed December 2010. http:www.fao.org/ag/ca.

Wagner, Holly. "Ohio State Research." Ohio State University April 2004. Access September 15, 2010, http://researchnews.osu.edu/archive/notill.htm.

Carrying Capacity

The ability of the environment to indefinitely maintain a maximum number of organisms within a geographic area is called its *carrying capacity.* Living organisms depend on a variety of environmental components, such as air, water, life-supporting temperature ranges, nutrients, and soils, among others, plus services, such as waste decomposition mechanisms, pollination, and groundwater recharge. The renewal of these resources and services is time dependent; in other words, when used, these resources require time to regenerate to the original extent. If too many organisms exist within an environment, the resources and services may be used at a faster rate than they can be renewed. As a result, the supply becomes limited and the resources are less available to all individuals, leading to the demise of some. For this reason, carrying capacity can be viewed as the maximum number of living organisms that can be maintained indefinitely by the environmental components and services in a given geographic area without depleting the ability of those resources and services to be regenerated.

Measuring Carrying Capacity

The term "carrying capacity" was originally used in the 1800s in the context of payload carried by ships and other conveyances. It was then adopted to describe the limits of load-carrying animals, capacity of bipedal humans to carry things, the amount of pollen carried by bees, and even the amount of moisture carried by wind. It was essentially measured as the capacity of *Y* to carry *X*. The first time the term was used to describe environmental capability to support living organisms was in the context of increasing rabbit populations in New Zealand and their destruction of sheep pastures. It was then employed in terms of range productivity for the number of sheep that could be raised in Australian grazing lands if rabbit populations were reduced and at a given amount of rainfall. This term was later adopted by American ranchers and then was used for game management by the U.S. Forest Service.

Limits of Carrying Capacity

American naturalist Aldo Leopold's observations on animal population saturation and the limiting role of environmental factors in the decline of deer herds played a key role in wildlife managers taking up habitat manipulation, such as flooding, cropping and burning, predator control, and animal relocation, among other methods still practiced today. Initially, the limit of carrying capacity was impossible to enumerate because it could not be experimentally tested and replicated. In the 1950s, American ecologist Eugene Odum demonstrated that populations increased in a sigmoid (S-shaped) fashion and that there was a definite upper limit to growth. He observed this phenomenon when populations of fruit flies and flour beetles were tested under a constant laboratory environment. Odum called the maximum per-capita rate at which populations can grow the "intrinsic rate of natural increase" of organisms, which is attained when "environmental resistance" is absent. Environmental resistance is essentially the limits imposed by environmental factors on animal abundance. Odum's model, which could be modified to fit specific sites and species of interest, indicated that populations rose and fell around a fixed point. However, it was found difficult to use this model on many populations because the upper limit seemed to change periodically. Not being able to quantify the respective effects of the environmental and population factors was a problem. It is clear that while the intrinsic growth rate of populations may still be S-shaped, carrying capacity itself fluctuates for reasons that are not understood. Therefore, this upper limit could not be modeled at large scales and under variable factors.

Carrying capacity calculations are especially tricky when human populations are considered, because it is not known whether technology and adaptations can enable humans to raise the carrying capacity or whether human populations are susceptible to environmental resistance. The concept of carrying capacity must be addressed with caution because the upper limit that the concept claims to identify and model is dynamic, unpredictable, and uncontrollable. Notwithstanding the drawbacks of this concept, it has been used as a tool to explain the relationship between environmental allowances and living populations, especially in the contemporary politics of human–environment interaction. It is, therefore, worthy of consideration.

Although population growth is represented as an S-shaped curve that stops growing when it reaches carrying capacity, populations rarely remain stable. Instead, populations often overshoot this mark and then decline below the carrying capacity. This process is repeated through time, producing a fluctuating pattern. The fluctuation exists because of a negative feedback loop. A feedback loop includes the linkages between different parts of a system, with the effects of changes that occur in one part of a system being transmitted to another part through these linkages. For example, the number of deer in a given area affects the amount of grass that is eaten by the animals. The carrying capacity in that given area might be 50 deer. If 30 deer are present, there is plenty of food supply available for them, which allows their population to grow. As the deer numbers reach 50, growth does not abruptly stop, but rather overshoots the carrying capacity. When competition for grass increases, grass does not have enough time to regenerate. The decline in availability of grass induces increased competition and the eventual decline of the deer population. The negative response of grass to overgrazing can be seen as a negative feedback loop. The declining deer population does not stop when it reaches carrying capacity, but instead continues to decline until the grass has time to regenerate to support more deer. With the new availability of grass, deer populations start to grow, and the cycle is repeated. The magnitude and regularity of such fluctuations around carrying capacity depend on the environmental factors, the size of populations, the scale of the system, the time needed for regeneration, and unpredictable extraneous factors.

Enumerating the Carrying Capacity for Humans

The geographic limit of the human population encompasses virtually the entire planet. It is difficult to understand—let alone quantify—all of the large- and small-scale environmental factors on Earth; moreover, it is difficult to assess the multitude of complex interactions that humans have with the environment. Further, unlike deer, humans use environmental goods and services not just to feed themselves, but for activities that are beyond biological survival alone. For example, humans not only use water for drinking or washing (activities that are essential for survival), but also use water to make alcohol and paint or in amusement parks, the primary reason for which is enjoyment. Despite the complexities inherent in this process, several attempts at enumerating the carrying capacity for humans have been made.

Thomas Malthus, an 18th-century British scholar and economist, was perhaps the most influential in his argument that the human population grows exponentially, while expansion of food production follows a linear path. At some point, according to this theory, the human population will exceed the amount of available food, resulting in famine and a subsequent decline in numbers. Malthus applied his theory to the English population and predicted that it was growing beyond its means to produce enough grain to feed the people. However, the English population did not decline for various reasons, highlighting the fact that carrying capacity for human populations cannot be measured based in simple relationships.

Human beings also behave differently under various economic situations. Evidence shows that, rather than increase, the human population actually decreases when mortality rates decrease due to factors such as political stability, economic security, social equality, and better nutrition and health care. Some couples decide to have fewer children for various reasons, including the following: (1) There is lack of fear of losing their children before they reach adulthood; (2) with better economic opportunities for adults, income earned by children is less important for the family's well-being, thus making them a financial liability rather than an asset; and (3) the status of women in society and in the household gives them the freedom to choose when to have children. Additionally, as women seek economic opportunities outside the home, having fewer children has practical advantages. The effect of such negative feedback loops on declining populations is evident in many economically advanced countries in the world today, including Germany, France, and the United States.

Further, humans are not limited by their local environments and, therefore, are able to increase the carrying capacity of their population. Until approximately 10,000 years ago, the amount of nourishment available to humans was determined by the ability of the local environments to produce edible food. As a result, the carrying capacity of humans was perhaps not very different from that of deer. However, unlike deer, which do not actively encourage the growth of grass to keep their population alive, humans began modifying the environment to increase the goods and services available to them. As humans started replacing forest and grasslands with simple agroecosystems, they were able to increase the carrying capacity 10 to 50 times over. Humans also can make use of goods and services available in the neighboring environments by moving them over long distances. For example, modern urban areas produce immense amounts of sewage that are routed to distant aquatic environments, where the sewage is diluted and detoxified. The area required to perform this environmental service is many times larger than the environments in which the waste was generated.

Resource Optimists and Resource Pessimists

Although humans have been able to increase the carrying capacity, their propensity for excessive consumption of resources and services for nonessential purposes defies reason. To recall the definition of carrying capacity, it is the ability of the environment to indefinitely maintain a maximum number of individuals within a geographic area. The key word here is "indefinitely." Researchers warn that the world's current rate of consumption of resources is unsustainable. The concept of sustainability stresses the importance of resource use without compromising the ability of future generations to use them. At the present rate, it is expected that the human population will overshoot the world's carrying capacity within the next few generations.

Whether humans will change their behavior toward sustainable use of environmental resources and services seems to depend on one's point of view. One view argues that humans are different from other species in that they have the ability to modify their environment, whereas the counterposition suggests that humans are no different from other species and are destined for collapse sometime in the near future. These contrasting viewpoints are labeled as "resource optimists" and "resource pessimists," respectively.

While acknowledging that biological carrying capacity cannot be directly applied to humans, resource pessimists nevertheless argue that a Malthusian collapse at a global scale is impending. They support their argument by pointing out the following facts: (1) The global mining-based energy supplies are declining and cannot be replenished for tens of thousands of years; (2) against previous expectations that the oceans will provide an unending supply of food, the total number of fish caught is declining each year; (3) the human-induced increase in atmospheric carbon dioxide has affected our favorable climatic systems adversely, and even if all emissions were stopped abruptly, it will be hundreds of years before its effects wear off; and (4) environmental modification by humans has driven thousands of plants and animals to extinction, and the resultant loss of biodiversity is irreversible.

Resource optimists acknowledge that environmental problems exist, but they do not consider them severe enough to warrant becoming alarmed. They also accept that populations grow exponentially, but they deny that environmental resources and services grow linearly. They base their argument on the innovative nature of humans and the rapid growth of technology. When pushed to the brink, inaccessible environmental goods and services can be harnessed using improved technology, and pollution can be curtailed by increasing efficiency. These optimists base their argument on achievements such as the following: (1) reduction of the ozone hole by identifying and curbing the use of chlorofluorocarbons (CFCs); (2) when oil prices rose in the 1970s, consumer behavior modification shifted toward energy conservation, and profit motive drove oil companies to invent better technologies to drill for oil in previously inaccessible areas; (3) breakthroughs in genetics have allowed for the engineering and advantageous modification of food crops, and it is likely that loss of biodiversity can be partly reversed; and (4) innovation in battery technology is beginning to revolutionize the automobile industry for the production of electric vehicles.

Stretching Environmental Resources

Debates aside, the laws of conservation of energy and matter dictate that all materials that come from the environment have to go back into the environment. The fact remains that while humans can stretch environmental resources, they are completely dependent on the environment for their survival. The idea of modifying human behavior and innovation in response to emerging resource limitations is relevant. However, considering that populations fluctuate around a carrying capacity and that there is a lag time before actual collapse occurs, it is more critical to focus on anticipatory and proactive approaches.

—Bharath Ganesh-Babu

Further Reading

Hardin, G. "Cultural Carrying Capacity." Garrett Hardin Society. Accessed February 2011. www.garretthardinsociety.org.

Leopold, A. *Game Management*. New York: Scribner, 1933.

Odum, E. P. *Fundamentals of Ecology*, 3rd ed. Philadelphia: Saunders, 1971.

Sayers, N. F. "The Genesis, History, and Limits of Carrying Capacity." *Annals of the Association of American Geographers* 98, no. 1 (2008): 120–134.

Cereal Foods

Cereals are grasses cultivated for their starchy seeds, which are used for human food and feed for livestock. Wheat, rice, corn, rye, oats, barley, sorghum, and millet are the most important human food cereals. Cereals were the first food plant to be domesticated and replanted on a large scale. Grain storage structures have been found in the archaeological excavations of almost every major civilization. Wheat, rice, and rye are grown primarily for human food, whereas most of the corn, oats, sorghum, and millet sown is intended for cattle and poultry feed, plus human consumption. Cereals are high-quality, inexpensive foods that can be stored for long periods of time, handled as a bulk product, and shipped at low costs. Large quantities of cereals are exported from countries with cereal surpluses to areas where there is a demand for cereal.

Important Cereals

Rice has been an important wetland cereal crop for more than 5,000 years. In 2010, approximately 701 million tons was harvested in the world (of this figure, 440 million tons milled basis). Rice is the food staple for those who live in the humid tropics and subtropics. It requires warm temperatures and large quantities of water for growth. Paddy rice grows in 4 to 8 inches of water. Approximately 95 percent of the rice grown in the world is in South and East Asia. The major food crop for more than 50 percent of the world's inhabitants, rice is an intensively irrigated field crop, producing high yields per acre but requiring much hand labor in the planting and harvesting period. Its overall food value is very high.

Wheat has been a primary food crop for more than 6,000 years, and the varieties now grown are a mixture of at least three wild species. Hybrid wheat is sown in a wide range of climate conditions in more than 60 countries of the world. This dryland grass grows best under mild stress in areas of low rainfall (13 to 35 inches) characterized by cool, dry seasons. Spring wheat is grown in semi-arid regions, whereas winter wheat is grown in the drier margins of humid agroclimatic regions. In very dry agroclimatic regions, fields lay unplanted (fallow) for one year, and the moisture accumulated in that year helps support a spring wheat crop the next year. Drought is a persistent problem in the wheat-growing areas of the world, and a shortage of water limits yields. Excessive rainfall and high winds cause wheat plants to fall to the ground (lodge), delay wheat ripening, and interfere with harvesting. High humidity and excessive rain lead to the spread of wheat plant diseases that reduce yields. Wheat is an extensive dryland field crop with low yields per acre, and it is harvested by mechanized farm implements. More than 679 million tons of wheat was harvested in the world in 2010. The overall food value of wheat is very high.

Corn originated in the New World, requires a great amount of moisture and much heat in its life cycle, must be cultivated and tended, and is vulnerable to many corn diseases and insects. This high-value food crop was brought from the New World to Europe by Spanish explorers and then spread throughout the world by European merchants, settlers, and colonial administrators. Corn has been a food source for humans for at least 2,000 years. It is a food plant that can be grown in almost all warm and humid agroclimatic regions and at elevations ranging from sea-level fields to those sown above 10,000 feet. Corn offers the highest yield per acre of any grain crop now sown.

Rye is a hardy food grain related genetically to wheat and barley. It is believed to have come from a wild grass strain that grew as a weed in wheat and barley fields. Cultivated first as a food crop in Central Asia, rye was not a source of food in ancient Egypt or Greece. A very hardy plant, it is sown in areas with poor soils and in areas with cool summers and severe winters. It is the main food grain in many parts of northern Europe and cooler agroclimatic regions of Asia. Most rye is sown in the fall, remains dormant in the winter, grows to maturity in the spring, and is harvested in late spring and early summer. One of the less important human food crops, it is sown generally in areas where agroclimatic conditions are unfavorable for wheat, rice, or corn. Disease losses in rye are less than occur with other cereal food grains.

Oats is a cereal crop largely used as feed for livestock, although some is used for food by humans. Wild oats was a field grass in many different areas of western Europe. Their seeds were believed to have been distributed to other parts of the world as a weed mixed with barley seed. This cool, temperate agroclimatic crop is grown in a wide range of environmental conditions, particularly in northwestern Europe, northern United States, and southern Canada. In the United States, spring oats are sown in the Great Plains states and winter oats are sown in southern states. This cereal is planted in all months of the year in various regions of the world. Oats is less demanding in terms of soil nutrients than either wheat or barley and grows well in soils of minimum fertility and of high acidity. In the United States, approximately 90 percent of the oats harvested is used for animal feed.

Barley is one of the six most important cereal foods cultivated for human consumption. It is believed that barley was the first domesticated food grain; it has been found in archaeological sites dating earlier than 5000 BCE. Barley was an important food to Stone Age Swiss lake dwellers and a basic food to Egyptians, Hebrews, Greeks, and Romans. For centuries, it was the chief source of bread flour in western Europe. Barley, which is believed to have been domesticated in the Middle East, can be grown, in its many forms, in a great variety of agroclimates, including subtropical, temperate, and subarctic. Its growing season is short: for some varieties, 90 days, and in areas of extended summer sunlight, 60 to 70 days. In marginal agricultural regions, yields are small. Barley is typically harvested in the late spring or early summer. Insect pests, parasitic fungi, mildew, rusts, and blights damage barley plants and reduce yields. More than 150 varieties of barley are grown, and more than 60 percent of the world's barley crop is used for animal feed.

Sorghum is believed to have been domesticated in Africa. This cereal, which is grown for grain, is also known as milo, kafir corn, millet, Egyptian corn, jowar, jonna, Guinea corn, and koalang. It is a strong, tall grass that grows to a height of 8 to 16 feet in some parts of the world. Many varieties of sorghum are grown; some are used for cereal and others for forage or syrup. Sorghum is the principal cereal grain in Africa. It is a plant found in warm to hot agroclimatic regions with meager rainfall, and is very resistant to drought and heat. Sorghum's food content is similar to corn, but it is higher in protein and fat. In hot, dry agroclimatic regions, sorghum is substituted for corn as an animal feed grain.

Millet is a small seed grass first domesticated for human food in the dry and hot regions of Africa and Central Asia. Millet has many names: *proso* in Russia, Manchuria, and China; pearl or *bajri* in India; millet in Japan; and finger and little millet in North America. A millet plant normally grows from 4 feet to more than 10 feet tall. Pearl millet, a staple food crop in India and Africa, grows well in soils of minimum fertility and in areas of low moisture. *Proso* millet ripens in 60 to 80 days. It is used as poultry feed and birdseed in the United States and as a food grain in Asia and eastern Europe. Millet is sown in the late spring or early summer; young plants require very warm weather to flourish. Grain is harvested in the same ways as wheat and barley in the United States and as rice is in south Asia.

Nutritional Values

Wheat, rice, barley, rye, corn, oats, sorghum, and millet are similar in terms of their food value in the diet. They are deficient in calcium and, except for yellow corn, vitamin A. When these cereals are supplemented with calcium, vitamins, and minerals, they become more nutritionally complete. Cereals are the most important food crops in the world, while bread grains and potatoes are the least expensive basic foods. Consumer prices for cereal products are low compared to the cost of meat, milk, eggs, fruits, and vegetables. Prior to the modern Agricultural Revolution, which was marked by the invention and worldwide distribution of modern farm machinery and certified or hybrid seeds, there was a chronic scarcity of food in many parts of the world. Over time, however, wheat, rice, and corn production increased with the improvement of varieties, agrotechniques, herbicides, and fertilizers. More widely available grain and modern transportation networks have since reduced the potential for famines, though not the risk for hunger.

Wheat and rice are the two most important food cereal crops grown in the world. Wheat was the primary famine relief food in the 20th century. In 2007, climatic variations reduced the world's wheat harvest somewhat, but 612 million tons of wheat was reported to be harvested. This dryland, non-irrigated, extensively

American Farmers: Feeding the Hungry Quality Foods at a Modest Cost

Agriculture in the United States is an amazing success story. In 1940, one farmer fed 19 people; in 1960, 45; in 1980, 115; and in 2000, 130. In 2011, one farmer produces enough foodstuffs to feed 155 people. Today's farmers have achieved this gain with less land planted, less tillage, less herbicides, and less pesticides than in 2000.

Hard work, knowledge of the land, equipment, and seed or plant selection have enabled American agriculturalists to remain some of the most productive food producers in the world. They have responded to the changing dietary expectations in the world and to the new uses of grain in industry. Notably, the world demands for corn and spring wheat have led to increases in acres sown, yield, and production of these crops. Drought and inclement weather, however, did have some impact.In 2009, for example, 79.6 million acres of corn was harvested; in 2010, this acreage increased to 81.3 million. The average corn yield in 2009 was 165 bushels per acre, but dropped to 155.8 bushels in 2010. Total U.S. corn production in 2009 was 13 billion bushels, but decreased to 12.7 billion bushels in 2010. In 2009, U.S. spring wheat was harvested on 1.6 million acres; the same acreage was devoted to this crop in 2010. Average spring wheat yields in 2009 were 53 bushels per acre; by comparison, in 2010, they increased to 55 bushels per acre. Total U.S. spring wheat production in 2009 was 82.2 billion bushels; in 2010, it rose to 85.3 billion bushels. Farm incomes increased as production costs were reduced by purchasing fewer inputs made from fossil fuels and by maintaining high levels of production.

—*William A. Dando*

grown crop offers yields that vary annually, with drought being a constant threat to wheat's production. Wheat's significance can be observed by noting that, in most years, the five most significant agricultural imports worldwide are wheat, wine, soybeans, corn, and cheese. Although a great deal of rice is harvested in the world, a major portion of this crop is consumed in the countries where it is produced. The United States is the world's largest wheat exporter, whereas Thailand is the world's largest rice exporter. Wheat must be milled into flour for the creation of white bread. In this process, much of the most nutritious part of the wheat kernel is removed and processed into animal feed. Wheat flour is milled to satisfy the consumer's demand for white bread and to increase the shelf life of bread. It is enriched by adding nonfat milk solids, vitamin B, niacin, iron, and riboflavin.

Corn, corn meal, and corn grits are major components of the traditional diets of those persons who live in the American South, Mexico, Italy, Romania, and parts of Africa. Corn is low in niacin; thus, where corn products are the primary food, a niacin deficiency disease, pellagra, is common (see the "Deficiency Diseases" entry). Also, the milling and processing of corn produces a product for human consumption that is deplete of other essential nutrients. The rye grains used for bread are milled, using a different method and whole-meal flour is the basis for rye bread. Barley, when used in baking, is not milled; the end product is composed of whole barley. Oats, which is primarily an animal feed, must be dehulled, steamed, and rolled to make oat flour or oat meal for human consumption. Rolled oats and oatmeal are, in essence, the whole oat seeds.

Rice, which serves as the basic food for more than 3 billion inhabitants of the world, is grown in very humid tropical and subtropical agroclimates on swampy level land. Rice for human consumption is polished to improve its shelf-life and storage qualities, or cooked as whole rice because of its flavor. Dietary characteristics of polished rice resemble those of milled white wheat flour. Undermilled and steamed rice retain some of the vitamins and minerals lost through polishing, particularly thiamine. Thiamine deficiency affects the central nervous system and creates a debilitating disease called beriberi (see the "Deficiency Diseases" entry). Beriberi is widely seen in South and East Asia, where polished rice represents the principal food in residents' diets.

Global Cereal Supplies and Demand: A Sample Report for 2010

World cereal production varies greatly from year to year. Whereas industrialists can control their mill and plant environments because they are enclosed, agriculturalists cannot control their field or paddy environments and must depend on natural conditions for plant growth, development, and harvesting. National and international food planners and concerned not-for-profit agencies review the publications of the United Nations and private sources to determine the global cereal supply and estimated demand on an annual basis. The 2009 world cereal production was better than expected and slightly lower than the record 2008 production. The combination of a good world crop prospect and high carryover stocks of cereals from 2008 lessened concerns for spot shortages. World cereal inventories in 2010 reached an 8-year high—an important factor in global food security. After a period of market uncertainty and reduced consumer demands, the world cereal market has returned to a more normal situation. The one exception to this stable pattern is rice. International prices for rice have declined 30 percent below those for the corresponding period in the previous year. This decline in prices paid for rice, bigger harvests, and a sharp contraction in world cereal trade are lowering the global cost of imported cereals. The overall cereal import prices declined more than 10 percent in 2010.

Global production of wheat in 2010 was down slightly from previous forecasts. Wheat production in Asia was projected to significantly increase (6 percent or

Table 12. World Cereal Stocks (in million tons)

	2005	2006	2007	2008	2009 (estimate)	2010 (forecast)
Wheat	519	519	475	470	555	560
Rice (milled)	110	116	114	121	136	133

Source: FAO Prospects and Food Situation Preview. Rome: Food and Agriculture Organization of the United Nations, November 2009, No. 4.

more) in 2010, and North African harvests were forecast to be better than expected and double 2009's production. The U.S. wheat harvest was expected to drop about 10 to 11 percent; Russian and Ukrainian wheat harvests were better than expected; South America's wheat harvests were expected to produce slightly less wheat than 2009, largely a consequence of drought in Argentina; and prospects for the wheat harvest in Australia remained favorable.

In 2010, global rice production declined considerably, following weather anomalies and natural disasters in the rice-producing regions of Asia. Rice production forecasts for 2010 reported a slight reduction (possibly 2 percent) of the record 2008 rice harvest (see Table 12). The countries most affected by adverse weather conditions included India (excessive rain, then floods), Taiwan, Japan, Nepal, Pakistan, and the Philippines (earthquakes, cyclones, landslides, and flooding). In Africa, Egypt was anticipated to harvest a smaller rice crop due to the government's decision to reduce rice planting as a water-saving measure. Substantial harvest gains were projected for other rice-growing African countries. Rice production in South America and Europe was favorable, and production was expected to increase. Drought continued to reduce rice production in Australia.

Cereals as Comfort Foods

The basic ingredients for the world's comfort foods are derived from wheat, rice, corn, oats, barley, and rye. Wheat is the most important food grain of the inhabitants of the world's middle latitudes and dry, subtropical regions. It is the best bread-making grain, and for thousands of years, it has been the staple crop on which not only have the lives of individuals, but often the stability of nations depended.

The world production of rice exceeds corn production, and rice vies annually with wheat as the world's cereal production leader. Rice is the major food of those who live in South Asia and the Middle East. In these regions, rice is what white bread is to Europeans: It is the most desirable food. Although corn was not known to Old World agriculturalists until the discovery of the Americas, it is now one of the world's most widely produced cereals. Corn is grown in every country of the New World, from Canada to Argentina. It has become an important crop in Eurasia, from Portugal to China, and it is planted in Africa, from Egypt to the Cape of Good Hope. Numerous prepared foods are made from corn; it is an excellent animal food; and it is an important industrial crop.

Oats constitutes one of the largest world grain crops; it is used widely as a food for humans as well as a feed for animals. A hardy plant, it is a typical grain of the world's intermediate climates. The three chief centers of oats production are (1) northern United States and southern Canada, (2) northwestern Europe, and (3) central and eastern Russia.

Barley, which rates high in food value, is used for soups, porridge, and special breads. This cereal is used extensively for making malt, with the malt in turn being used in the manufacture of beer, ale, and whiskey. The principal use of barley is feed for domestic animals.

Rye as a bread grain is less satisfactory than wheat, and the bread produced from whole rye flour is dark brown with a crispy crust. Whole rye bread frequently is referred to as "black bread." It is a major food in Russia, Poland, Germany, and Slovakia. Rye's importance as a food crop has declined, whereas its role as an animal feed crop has expanded. Cereals from grasses that produce grain for food provided the basic food for most of humankind in the 20th century and will continue to do so in the 21st century.

—William A. Dando

Further Reading

Anderson, Eugene Newton. *Everyone Eats: Understanding Food and Culture*. New York: University Press, 2005.

FAO Prospects and Food Situation Preview, No. 4. Rome: Food and Agriculture Organization, November 2009.

Pearson, Debra. "The Effect of Agricultural Practices on Nutrient Profiles of Foods." In *Critical Food Issues: Environment, Agriculture, and Health Concerns*, edited by Laurel E. Phoenix. Santa Barbara, CA: ABC-CLIO, 2009, 177–193.

Vavilov, Nikolay Ivanovich. *Five Continents*, edited by L. E. Rodin, translated by Doris Love. Rome: IPGPRI; St. Petersburg: VIR, 1997.

Vavilov, Nikolay Ivanovich. "The Origin, Variation, Immunity, and Breeding of Cultivated Plants," translated and reprinted by K. Starr Chester. *Chronica Botanica* 13 (1949): 1–6.

Woolley, D. G. "Food Crops: Production, Limitations, and Potentials." In *Dimensions of World Food Problems*, edited by E. R. Duncan. Ames, IA: Iowa State University Press, 1977, 153–171.

Wortman, Sterling, and Ralph W. Cummings, Jr. *To Feed This World*. Baltimore, MD: Johns Hopkins University Press, 1978, 144–186.

Climate Change and World Food Production

Climate consists of all the different kinds of weather that occur at a place over time. It is made up of hot and cold weather, rainy and dry weather, and the extreme weather events that characterize the region. Climate can be defined for places of very different geographic size, ranging from a small town to the entire planet. Climate is much more stable than weather, but the weather systems that make up the climate of a region can change over time. For example, a region can grow warmer or colder over time, or wetter or dryer. In the United States, the climate of a place is described by the weather over a 30-year period. Thus climate change is considered to be a departure in weather that is different from the past decades. Climate change can occur over a period of years, decades, centuries, or even millions of years. It also can take place in a small area, such as part of a country, part of a continent, or over the entire planet.

Aspects of a Changing Climate

The fundamental basis of food production is agriculture. Because agriculture is based primarily on temperature and precipitation, changes in climate result in changes in food production. Many aspects of a changing climate can influence food production on the land masses:

- Atmospheric temperature
- Insect infestations
- CO_2 levels of the atmosphere
- Floods
- Glacial melting
- Droughts
- Rising sea level
- Plant diseases
- Times of first and last frosts
- Changes in stream flow

Somewhere between 10,000 BCE and 7,000 BCE, a new technology developed that marked one of the major transitions in the history of human life. Within this 3,000-year span, the deliberate planting and cultivation of food crops and the domestication of animals began. This phase was the beginning of what became known as the Agricultural Revolution. A form of gardening probably was practiced initially that eventually developed into true agriculture. Groups also must have settled on a seasonal basis to plant and harvest their gardens and then moved over their hunting territory at other seasons.

The Agricultural Revolution increased food availability and, in turn, the more stable food supply stimulated an increase in population. The human species was now free from many of the factors that had limited population growth and the human life span during the previous 3 million years. A self-propelling process began that has continued to the present day: The production of more food resulted in increased population, and increased population demanded more food.

Early agriculture was not just one agricultural system or approach. Rather, it took many different forms, generally involving some form of dryland crop cultivation, slash and burn, or herding. The crops grown varied from place to place, as did the animals that were raised.

In the early agricultural societies, surpluses were largely the result of improvements in plant and animal selection and breeding. Technology, such as fallowing, the use of natural fertilizers, and irrigation, increased food production. The Industrial Revolution resulted in more efficient crop production through mechanization and the creation of chemical fertilizers and pesticides. The Green Revolution, which greatly increased agricultural production, was based on manufactured chemicals and the application of large quantities of water.

Crop production in the United States has increased tremendously since World War II. Almost all of this gain in agricultural productivity has resulted from increased yields per acre, largely obtained through the use of chemical fertilizers and pesticides. In the United States, so many chemicals are now being used that the production of food has been transformed into an energy sink. In other words, more energy is being used to produce each ton of food than is obtained from that food. In essence, agriculture in the United States is a losing proposition in real terms of energy.

The planetary climate has changed continually through time. Sometimes it changed quite rapidly; at other times it has evolved very slowly. Since agriculture began, it has been subject to climate change. On the positive side, the very growth of the human population has paralleled a warming of the planet. Global warming opened new agricultural areas and improved the climate for agriculture over much of the mid-latitudes.

Climate Change

Climate change has been an element in the development of the biosphere as a whole and certainly a factor in agricultural success throughout history. By 11,700 years ago, only scattered areas of ice sheets remained in western North America. These ice sheets left behind large areas of glacial material that provided the basis for the development of very good soils, including some of the most productive soils on the planet today. Changes in climate that adversely affected agriculture then resulted in major setbacks for agriculture, at least regionally, if not for the majority of the planet. At about this time, the climate became colder. The remaining glaciers began to expand and some smaller glaciers developed again. This cold period, which lasted nearly 800 years and affected the region around the North Atlantic Ocean, was not a global event, however. Ultimately, it ended as fast as it began. In the space of some 50 years, temperatures on southern Greenland warmed as much as 12°F.

The island of Greenland provides an example of how climate change can affect crop production. The Vikings settled Greenland in the 9th century. The climate was certainly cold, but it was warm enough to support vegetation. The settlers were able to raise cattle, sheep, some grain, and a few vegetables. After some 300 years, the climate turned cold again and the settlements vanished. Since then, the climate has changed direction once again. Greenland is now warming nearly twice as fast as the global average. While it is by no means a tropical oasis, in recent decades sheep have been reintroduced into southern Greenland and hay crops and some vegetables can be grown.

A cold period known as the Little Ice Age affected the colonies in the Americas up until the 19th century. The year 1816 became known as the year without a summer. New England suffered from frosts in every month, even during the summer months. In Indiana, snow fell in May and June. Parts of the state had snow or sleet on 17 days in May. As a consequence of the enduring cold, very few crops were harvested north of the Ohio and Potomac rivers.

Without a doubt, climate is now changing and will continue to change in the future. The planet's climate is warming at a significant rate, and nothing is likely to slow this warming trend during the current century. If this is the case, then it should be apparent that it will be impossible to return the planet to the temperature it was in 1900 anytime in the near future. It is imperative that international leaders and decision makers recognize that current changes in climate are affecting agricultural production and will have a considerable impact on food production in the coming decades. It will be necessary to take into account climate change, both in the long-term and in the short-term, in the form of slowly changing temperatures and

Health and Climate Change

Global climate change is one of the most life-impacting of the many changes occurring in the world today. It reflects the increasing human domination of the Earth's physical environment, and it is a factor in the unprecedented expansion of population and human economic activities. Climate change has major consequences for the sustainability of ecological systems, for food production, and for human population health. Climate change in the past triggered disasters, affecting peoples and populations throughout the world. It caused crop losses and starvation, infectious famine diseases, social collapse, and even the disappearance of whole populations. Examples of the last effect include the mysterious demise of the Mayans in the Yucatan, the inhabitants of Mesa Verde in the American Southwest, and the Viking settlements in Greenland. Progressive deterioration of the climate led to a decline in food production, hunger, famine, and the loss of those who lived there.

Throughout pre-industrial Europe, food supplies were marginal, as the mass of people survived on monotonous diets of vegetables, grain gruel, and bread. When climate fluctuated in Europe during medieval times, widespread crop failures, food price increases, hunger, disease, and death followed on the heels of these changes. Animal diseases also proliferated; in 1315–1317, more than half of the sheep and cattle in Europe died.

At times, in the modern period, climate variations have caused regional food shortages, malnutrition, weakened individuals, susceptibility to infectious diseases, and low life expectancy. Various modeling studies estimate a slight downturn in food production in the latter half of the 21st century.

—*William A. Dando*

their ramifications and to examine the impact of the shorter-term changes that most frequently affect some sections of the planet but not the entire planet. The effects on food production will vary greatly, depending on the degree of the climate change, the geographical extent of the change, and the duration of the change.

Factors in Climate Change

The biggest factor in climate change is the rapid rise in global temperatures. Conservative estimates by the International Panel on Climate Change (IPCC) in its 2007 report projected an increase in global temperatures of 4°F to 6°F by the year 2100. Nearly every forecast of global temperature increases issued in recent decades has been an underestimate of what has actually occurred. Since the 2007 IPCC report was issued, new estimates have moved the range of temperature increase upward to as much as 14°F by 2100. Any increase in temperature will affect life on the planet, including agricultural production.

Because most plants evolved during cooler conditions associated with the ice ages, many plants are now growing near the upper limit of their optimal range. Crop yields start to decline when temperatures reach or exceed the optimal temperature range. This is the case in the tropics as well as mid-latitudes. The biological regions of today, such as the Great Plains and forests, originated in an environment that was cooler than today. If the planet warms very much, it will result in major changes in the regional systems. For example, many species of plants that are found in the tropical rainforests are already growing in a climate that is near their upper limit for growth. The warmer conditions are now reducing marginal grasslands to desert conditions due to greater evaporation and transpiration.

The transformation of ecosystems would, in all likelihood, result in massive human migrations with all of their resultant political and economic problems. The possibility of such changes occurring with further warming is very real. Depending on how rapidly warming occurs, the problems may set in sooner than currently anticipated. There will be a loss of much of today's plants and animals as well. It is possible that in the 21st century, warming will result in the extinction of many plants and animals. Changes in species structure may become great enough to make some ecosystems nonfunctional.

Carbon dioxide is a growth stimulant to green plants. Plants grow bigger and faster, contain more vitamin C and sugar, and are more disease resistant when grown in atmospheres with high concentrations of CO_2. However, recent studies indicate that this is a limited process in the real world. The net effect of warming temperatures will be to decrease agricultural production.

Rising temperatures also will result in more water-related stress for agriculture. Agriculture is now the largest user of water on a global basis. Crop production consumes some 70 percent of freshwater supplies on a global basis and approximately 80 percent in developing countries. Evaporation rates will increase as the climate warms, such that inevitably less of the world's fresh water will be available for irrigation. More than one-third of the global human population now lives in water-stressed regions; the percentage of the population living with water-related stress may increase to one-half by 2100.

Glacial ice represents the largest store of fresh water on the planet. This ice comprises a store of fresh water that has accumulated over the ages. Mountain glaciers alternately store and release water with the change of seasons. These actions tend to even out the flow of water throughout the year in the rivers carrying the melt water. At present, mountain glaciers are losing ice at a rapid rate. Over the Asian plateau, an estimated 6 percent of the ice has melted in recent decades. In some parts of the plateau, melting has reduced the ice pack by one-third. This melting often produces floods in the early summer months, followed by drought later in the summer. The loss of glacial ice in mountain regions around the world is already having a negative effect on the inhabitants of people living downstream at lower elevations, and these conditions will only increase in severity in the future.

In Asia, the ice melt at either end of the monsoon season lengthens the seasonal flow of water for irrigation. Seven of the major rivers in Asia have their source in the Himalayas, and most of the glaciers providing the water are melting rapidly. Nearly one-third of the global population relies on these rivers for their water supply. Melt water is often utilized downstream for irrigation of crops. Declining summer runoff is already affecting villages in China, India, Pakistan, Vietnam, Bhutan, and Nepal, which in turn is changing these countries' agricultural economies. One of the Himalaya glaciers, Gangotri, supplies 70 percent of the water in the Ganges River during the dry season. The Ganges, along with other glacier-fed streams, could dry up at times if the glaciers melt substantially. Certainly, with a shorter runoff season, the system of agriculture utilizing double cropping will break down, greatly reducing agricultural production. Once the ice is gone, the rivers will contain only the runoff from seasonal rain and snow; thus much less water will be available than is now in the streams.

In Africa, there are currently feuds over declining irrigation water supplies around the base of Mt. Kenya and Mt. Kilimanjaro. Some oases in China's western desert regions may dry up, as glacial melt water ceases to flow to feed the groundwater. The long-term result will be a huge drop in crop production associated with these glacier-fed rivers.

Major shifts in the general circulation of the atmosphere will occur as the Earth warms, along with changes in the length of the growing season in the mid-latitudes. In parts of North America and Europe, for example, the time of spring "green-up" came 5 to 6 days earlier in the last half of the 20th century. Changes in precipitation patterns will take place as well. The net effect of these changes on food production in North America cannot be predicted, though it appears certain that some changes will be offsetting, some will increase food production, and some will decrease it.

The combined effects of climate change and increased CO_2 levels may increase yields in some areas. For example, northern areas or areas where rainfall is abundant will experience warmer temperatures and a longer growing season—and hence increased production. Arable acreage in the northern Great Lakes states, the northern Great Plains, and the Pacific Northwest may increase. Improvements in crop yields might potentially offset some negative effects of climatic change in other areas. In many regions, the demand for irrigation is likely to increase as a result of higher prices for agricultural commodities. Farmers also may switch to more heat- and drought-resistant crop varieties, plant two crops during a growing season, and plant and harvest earlier. Whether these adjustments would balance the negative effects of climate change depends on the severity of the climate change.

An increased length of growing season may be offset by lower rainfall. In most regions of the United States, climate change alone could reduce dryland yields of corn, wheat, and soybeans. Losses from part of the country may range as high as 80 percent. These decreases would be primarily the result of higher temperatures, which will increase heat and water stress in crops. In southern areas where heat stress is already a problem, yields will drop even further. In areas where rainfall decreases, crop yields may decline. In response to the shift in relative yields, grain crop acreage in Appalachia, the Southeast, and the southern Great Plains may decrease.

Rising Sea Levels and Volcanic Eruptions

Rising sea level will greatly affect food production in some currently very densely populated areas. For example, in the coastal lowlands of Bangladesh, flooding of the land will result in the loss of agricultural land utilized by some 15 million people. This displaced population will need to relocate to other already crowded areas. The ability for people to adapt to climate change largely reflects economic factors. Adapting to the flooding by migrating is not a very practical option for the population of Bangladesh because these people are among the poorest in the world. In contrast, the wealthiest segments of the population will have the best chance to adapt.

Most volcanic eruptions do not alter global climate for more than 2 to 5 years. However, large volcanic eruptions in the past and potentially large ones in the future could significantly alter food production not just in local areas, but globally. In 2005, the Geological Society of London initiated a study of the potential impact of large volcanic eruptions.

Heat and Water Demands

Extreme heat waves can devastate crops. In 2003, summer temperatures in Europe averaged more than 10°F above normal. In response, corn yields in Italy dropped 36 percent below average; in France, yields of fruit fell 25 percent and wine production, 10 percent. Heat also affects the rate of plant pollination. A 3°F increase in temperature in rice-producing areas would cut rice pollination in half. In addition, rising temperatures will increase the frequency and extent of damage caused by plant diseases and pests.

Prolonged drier conditions in the western United States, coupled with increased demand for water from the Colorado River, may force a change in agricultural water use in this region. Researchers have determined that there is a 10 percent chance that Lake Mead and Lake Powell will run out of usable water by 2013. In 2007, the water level in Lake Mead dropped to its lowest level since the reservoir was first filled. The same study suggested that there is a 50 percent chance these reservoirs will be empty by 2021.

Abundant evidence from the past testifies to the negative impact of climate change on food production. That climate change will take place in the coming years is certain. The impact of climate change on the global food supply likely will increase, though how much is difficult to forecast.

—John J. Hidore

Further Reading

Dando, William A., and Vijay Lulla. "World Agriculture and Food Provisioning" In *Climate Change and Variation*, edited by William A. Dando. Washington, DC: National Council for Geographic Education, 2007.

GRAIN. "The International Food System and the Climate Crisis." *Seedling*. Accessed October 2009. www.grain.org/seedling/?id=642.

Hidore, John J. "Climate Change and the Living World." In *Climatology: An Atmospheric Science*, edited by John J. Hidore, John Oliver, Mary Snow, and Richard Snow. Upper Saddle River, NJ: Prentice-Hall, 2010, 219–230.

Hidore, John J. *Global Environmental Change*. Upper Saddle River, NJ: Prentice-Hall, 1996.

Speer, James H. "Vegetation Response to Global Warming." In *Climate Change and Variation*, edited by William A. Dando. Washington, DC: National Council for Geographic Education, 2007.

U.S. Department of Agriculture, Economic Research Service. *Household Food Security in the United States*. Washington, DC: U.S. Government Printing Office, 2009.

"Water: Our Thirsty World." *National Geographic Magazine*. Washington, DC: National Geographic Society, 2010.

Community-Supported Agriculture

Making its appearance in the United States in 1985, community-supported agriculture (CSA) is based on a cooperative relationship between a community and a local farmer, whereby individuals and families within a community buy

Food for Thought

Most U.S. community-supported agriculture (CSA) operations are located near urban areas in New England, the Mid-Atlantic states, and the Great Lakes region, with growing numbers in other areas, including the West Coast. CSAs have emerged in these locations because their urban centers serve as outlets for fresh produce, which can be marketed directly to consumers looking for locally grown foods. Interestingly, there is another trend in agriculture paralleling the emergence of urban CSAs. Between 2002 and 2007, the U.S. Census of Agriculture reported a 30 percent increase in the number of women serving as principal farm operators. The highest percentages of farms operated by females are found in urban counties in New England and on the West Coast. Increasing proportions of women are entering agriculture because alternatives such as CSAs afford them the opportunity to join what had long been a male-dominated profession. At the same time, the increased visibility of women as farm operators may encourage and facilitate the formation of new female-operated CSAs in urban areas, suggesting that the two trends may be self-reinforcing.

—Deborah Greenwood and Robin Leichenko

memberships or shares in a farm, in exchange for food produced on the farm. Originally intended as an alternative to chemically intensive, large-scale industrial agriculture, CSAs typically produce food using organic and sustainable production methods. Most CSA farms offer members organic fruits, vegetables, herbs, and flowers that are in season. Some farms will offer additional items such as milk, eggs, meat, and other goods. A CSA farm can typically support as many as 200 members on only a few acres of land, with the cost of an average membership ranging from $225 for a half-share to $500 for a full share, depending on the amount of labor the member is willing to volunteer on the farm and the quantity of produce received. One exception to the typical CSA farm is the Honey Brook Farm in Pennington, New Jersey. The largest in the United States, it farms 60 organic acres and offers 2,200 shares or memberships, representing nearly 4,000 people.

CSAs are designed to be beneficial for the farmer, the community, and the environment. They are intended to give consumers a voice and a choice in the way their food is produced and delivered to them. The foundation of the CSA relationship is a mutual commitment between the farmer and the community. Membership dues provide funds for the start-up costs of the farm before the season begins. In turn, the community members' investment is returned to them when they receive their weekly portion of the food the farm produces. CSAs also are intended to reestablish a sense of community and connection with this country's rural past, which some CSA members would like to recreate in today's information-intensive and rapidly moving society. In addition to instilling a sense of community, CSAs have beneficial impacts on human and environmental health, food security, and sustainability. For example, decreased usage of chemical pesticides and fertilizers contributes to improved water quality, while organic farming methods lead to less soil degradation as compared to conventional approaches.

CSAs and Urban Agriculture

Community-supported agriculture, as we recognize it today, began in Japan in 1971 and paved the way for the organic food movement. Born out of an opposition to the harmful chemicals used in farming, CSAs were initially seen as an alternative way to farm organically and safely. The Japanese version of the CSA, called Teikei, was the first known cooperative agreement between a farmer and families to provide them with produce in return for money and labor. Similar CSAs later emerged in Switzerland, influenced by agricultural practices and movements in Chile and France. In 1985, CSAs were introduced to the northeastern United States in Massachusetts and New Hampshire by way of Switzerland and Germany. The practice has taken hold more broadly, and CSA farms have since spread across the United States, numbering more than 1,000 today.

Within the United States and throughout the world, the rising popularity of CSAs reflects a broader shift from traditional farming in rural areas of some states, to alternative farming in more urbanized locations. Urban agriculture typically is defined as agricultural activity that occurs within the fringe areas of cities and entails either growing crops or raising animals for consumption or sale. Within the United States, the amount of farmland in metropolitan areas has increased substantially since the mid-1970s. Most American CSAs are located near urban centers in New England, the Mid-Atlantic states, and the Great Lakes region, with growing numbers being established in other areas, including on the West Coast. While large losses of farmland to urban development also have occurred, the importance of agriculture in urbanized areas has not diminished. To remain competitive and economically viable, urban farmers have intensified production, shifted their plantings to include higher-value crops, and identified new markets for their produce. All of these responses are reflected in the formation of CSAs.

Along with CSAs, related forms of urban agriculture include farm markets, school and community gardens, rooftop gardens, and specialty producers. Many of these efforts are based on the principles of CSAs. For example, community garden agriculture, such as the Greensgrow Philadelphia Project in Pennsylvania, brings culturally appropriate food to ethnic neighborhoods that are underserved by traditional grocery stores. Garden agriculture often contributes to the culinary community by providing high-quality fruits and vegetables to chefs in local restaurants. Specialty producers, who may grow vegetables and fruits for Asian, Latino, and other populations and restaurants, also are becoming increasingly visible in peri-urban areas located near cities with large ethnic populations. Another fairly recently introduced form of urban agriculture that is gaining popularity is the rooftop garden, which utilizes container beds for growing food. Rooftops represent a large portion of a city's unused surface area, making them ideal for gardening, due to their full exposure to sun and rain. Green roof farming also gives the residents of a building the opportunity and enjoyment of growing their own vegetables, herbs, and flowering plants, while providing an additional food source.

School gardens are becoming increasingly prominent in many urban areas. One example comes from New York City in the form of the 610 Henry Street garden, which is shared by Brooklyn New School (BNS) P.S. 146 and Brooklyn School of Collaborative Studies (BCS) P.S. 448. Starting agricultural education from the preschool level and continuing through high school, the Brooklyn school garden provides children with a hands-on learning experience in how to grow healthy foods, such as fruits and vegetables. In addition to growing food, the school focuses on community building among the students, who come from diverse ethnic backgrounds. Its unique worm-composting program provides the gardens with beneficial mulch used for fertilizer, the by-product of 40 pounds of worms that are fed with cafeteria food waste. In addition, the school's seed-to-salad program for first graders grows enough lettuce to supply the salad bar in the school cafeteria. By incorporating agricultural education into the school's curriculum via school gardens, the broad aim is to teach the students, from an early age, the importance of eating healthy food, and to make these habits an integral part of their lives, both within and outside the classroom.

Urban farms that reflect the principles of CSAs are also prominent in Latin America, Asia, Africa, and Europe. For example, a successful urban agricultural program exists in Havana, Cuba, where the national Ministry of Agriculture works in cooperation with the city's government to support and encourage local communities in the management of agricultural activities. The result is a successful urban farming program, with production consisting mainly of organiponicos—highly productive raised orchard beds that utilize biological pest control, organic pesticides, and manure to produce a diversity of crops. Paralleling the philosophy of the CSAs in the United States, this cooperative community effort in Havana brings together producers and consumers in the production of healthy, nutritious, locally grown food. In Shanghai, China, one of the world's fastest-growing cities, the Shanghai municipal government purposefully maintains and preserves urban farmland to provide city residents with locally grown wheat, rice, and vegetables. In South Africa, during apartheid, it was impossible for the black majority to farm in and around cities; since apartheid's demise, however, urban agriculture has experienced substantial growth and likely will become a permanent feature of the landscape. Another location that also lends support to the theory positing a positive influence of urbanization on agricultural practices is in the densely populated city of Brussels, Belgium. Within the Brussels metropolitan area, pressure on traditional farming has acted as an incentive for the development of a farming system that delivers food and other marketable products, while striving to enhance food security, environmental quality, and community health. Within Quebec, Canada, urban fringe areas located on the perimeters of the city contain many innovative, alternative forms of agriculture, such as organic, part-time farming, and local farming with on-farm entertainment called agritourism. In fact, more organic farms can be found in the metropolitan regions of Quebec than anywhere else in the province.

Environmental and Social Benefits of CSAs

Community-supported agriculture provides an alternative for farmers and consumers to organically produce healthier, non-chemically treated food. CSAs also allow producers to minimize "food miles"—that is, the distance food travels from farm to plate—thereby reducing greenhouse gas emissions from the burning of fossil fuels for food transportation. In addition, by bringing farming and nutritious foods into city neighborhoods, CSA farms are making local food systems more socially just. When CSAs are found in close proximity to urban farm markets, they can reduce the "food desert" phenomenon in some urban locations. Food deserts are residential areas with limited access to healthy, nutritious foods, such as fruits and vegetables. These areas, which are typically found in both low-income urban neighborhoods and remote rural areas, provide ready access to unhealthy foods from retailers such as fast-food restaurants, convenience stores, and gas stations, but tend to lack large grocery stores and fruit and vegetable markets. Food deserts contribute to serious nutritional problems such as obesity, diabetes, and heart disease among lower-income, and often minority, populations. In the United States, First Lady Michele Obama has made it her primary objective to help fight obesity, particularly in children, by encouraging and supporting local efforts to bring healthy, nutritious produce into urban food deserts. To demonstrate her affirmed support for community gardens and her campaign, she planted an organic vegetable garden on the White House grounds from which fresh produce is harvested.

A growing number of urban, CSA-type farms and farming organizations in the United States are also addressing the scarcity of healthy foods. Although slightly different in their methods of delivery, these groups all share a common goal—to bring social justice to low-income neighborhoods by giving them the opportunity to purchase fresh, healthy food from local farmers and provide better nutrition through food education and public health information. One such farm is the previously mentioned Greensgrow Philadelphia Project. This farm utilizes vacant land from an abandoned steel plant, covering an entire city block. It is now a permanent garden farm in the neighborhood and consists of a greenhouse, nursery, farm market, and retail nursery area. The farm grows and brings fresh, nutritious foods to poorer residents as well as supplies the Philadelphia restaurant community. The project represents a successful model for how small agricultural enterprises may become established in urbanized areas, providing nutritious, locally produced foods to low-income city dwellers. The project also provides agricultural education outreach to the community residents.

Another example is the Growing Power Farm organization based in Milwaukee, Wisconsin. The mission of this nonprofit organization is to provide access to healthy, safe, affordable food through the development of sustainable community food systems. Composed of multiple farms in urban and rural areas of Wisconsin and Illinois, members carry out their mission through hands-on training, education, and outreach to residents of diverse communities and, in particular, engage in youth development. Their two-acre urban farm and community food center in the city of Milwaukee produces fruits, vegetables, and herbs and raises goats, ducks, bees, and turkeys. Unique to the farm is an aquaponics system that farms tilapia and perch. Aquaponics is a method of farming that grows crops and fish in a recirculating-water system. The resulting goods are sold at local farm markets and restaurants and are popular with the ethnic community. The center's broader goal is to enhance social justice and to educate residents about nutrition and public health issues at the community level.

A third example is the New Brunswick Community Farm Market in New Jersey, which was conceived in 2009 to provide this low-income city neighborhood with the chance to buy healthy food from local farmers and learn about nutrition-related matters. The farm market sits on a mere one acre of vacant city land owned by Rutgers University and is run by the university and the New Jersey Agricultural Experiment Station. Local farmers have delivered fresh fruits and vegetables successfully to this underserved area through their participation in this urban project.

Achieving Sustainability: Promoting and Supporting CSAs

Community-based agriculture meets consumer demand for safe, nutritious, healthy foods that are locally produced while also establishing a connection between the community and the farmer. The success of CSAs demonstrates that agricultural programs can flourish in urban and peri-urban areas through cooperation and participation among the farmers, local government/policymakers, and community members, as they work together to move toward a more sustainable, socially just, and environmentally healthy food system. Importantly, these efforts often require local governments to formulate policies and programs that support farmers and various forms of urban agriculture including CSAs, community gardens, and organic agriculture. The success of such policies and programs can be seen in the cases mentioned earlier in Brussels, Belgium; Havana, Cuba; and New Brunswick, New Jersey.

Government support also can be beneficial in setting land-use and zoning policies that restrict development and preserve farmland, thereby ensuring that land is available for CSAs and other types of urban agriculture. The state of New Jersey's $1 billion initiative to preserve open space and farmland from development provides one example. Some municipalities in New Jersey employ "Transfer Development Rights" (TDRs) as a way to preserve farmland. Such a plan requires new housing developers to buy development rights or credits from farmers willing to preserve their farms, and then use these credits to build new housing in other areas. One such example is the 130-acre Suydam Farm in Franklin Township, New Jersey. This 300-year-old Dutch family farm—the oldest in Somerset County—survived development pressures from encroaching urbanization and was preserved as open space. The preservation strategy relied on a cooperative effort between the township, county, and state government's Agricultural Development Committee to contribute funds and purchase the development rights. The preserved farm will remain a part of New Jersey's rich agricultural heritage, and it also is an asset to the local community. Another approach, known as "New Urbanism," follows the same principle as TDRs and preserves farmland while creating a traditional town setting around it, where housing, schools, and businesses are nestled in close proximity to one another. Yet another tool is tax relief to farmers in the form of lower property taxes paid on land that is in production. In New Jersey, the Farmland Assessment Act was passed to protect farmland from being sold for development through lower assessment value.

An effective strategy to help achieve agricultural sustainability is to work through food policy councils. The Growing Power Farm, for example, is actively engaged in developing food policy initiatives in cooperation with the Chicago Food Policy Advisory Council (CFPAC). These partners' goal is to facilitate policy formulation that encourages and improves access for Chicago residents to healthy, safe, and affordable food through sustainable farming practices. With local governmental support and encouragement and the development of responsible food policy initiatives, CSAs ultimately can provide communities with greater control over their local food systems. Such efforts may not only enhance socioeconomic well-being, environmental quality, and health of residents, but also contribute to long-term efforts to achieve food security and sustainability.

—Deborah Greenwood and Robin Leichenko

Further Reading

Allen, Will. "Growing Power, Inc." 2010. www.growingpower.org.

Anderson-Wilk, Mark. "Does Community Supported Agriculture Support Conservation?" *Journal of Soil and Water Conservation* 62, no. 6 (2007): 126A.

Beauchesne, Audric, and Christopher Bryant. "Agriculture and Innovation in the Urban Fringe: The Case of Organic Farming in Quebec, Canada." *Tijdschrift Voor Economische En Sociale Geografie* 90, no. 3 (1999): 320.

Blanchard, Troy C., and Todd L. Matthews. "Retail Concentration, Food Deserts, and Food Disadvantaged Communities in Rural America." In *Remaking the North American Food System: Strategies for Sustainability*, edited by Clare C. Hinrichs and Thomas A. Lyson. Lincoln, NE/London: University of Nebraska Press, 2007, 210.

Brown, Cheryl. "The Impacts of Local Markets: A Review of Research on Farmer's Markets and Community Supported Agriculture (CSA)." *American Journal of Agricultural Economics* 90, no. 5 (2008): 1296.

Clancy, Kate, Janet Hammer, and Debra Lippoldt. "Food Policy Councils." In *Remaking the North American Food System: Strategies for Sustainability*, edited by Clare C. Hinrichs and Thomas A. Lyson. Lincoln, NE/London: University of Nebraska Press, 2007, 121.

Cone, Cynthia Abbott, and Andrea Myhre. "Community-Supported Agriculture: A Sustainable Alternative to Industrial Agriculture?" *Human Organization* 59, no. 2 (2000): 187.

Corboy, Mary. "Greensgrow Philadelphia Project." www.cityfarmer.org/greensgrow.html.

DeLind, Laura B., and Anne E. Ferguson. "Is This a Women's Movement? The Relationship of Gender to Community Supported Agriculture in Michigan." *Human Organization* 58, no. 2 (1999): 190.

Girardet, Herbert. "Urban Agriculture and Sustainable Urban Development." In *CPULS Continuous Productive Urban Landscape: Designing Urban Agriculture for Sustainable Cities*, edited by Andre Viljoen. Oxford, UK: Elsevier, Architectural Press, 2005, 32.

Harrison, Charles H. *Tending the Garden State: Preserving Agriculture in New Jersey*. New Brunswick, NJ/London: Rivergate Books, 2007.

Heimlich, Ralph. "Agriculture Adapts to Urbanization." *Food Review* 14, no. 1 (Jan-March 1991): 1.

Henderson, Elizabeth, and Robyn Van En. *Sharing the Harvest: A Citizen's Guide to Community Supported Agriculture*, rev. ed. White River Junction, VT: Chelsea Green Publishing, 2007.

Lyson, Thomas A. *Civic Agriculture: Reconnecting Farm, Food and Community.* Lebanon, NH: Tufts University Press, University Press of New England, 2004.

Pfeiffer, Dale Allen. *Eating Fossil Fuels: Oil, Food, and the Coming Crisis in Agriculture.* Gabriola Island, BC: New Society Publishers, 2006.

Schnell, Steven M. "Food with a Farmer's Face: Community-Supported Agriculture in the United States." *Geographical Review* 97, no. 4 (2007): 550.

Trauger, Amy. " 'Because They Can Do the Work': Women Farmers in Sustainable Agriculture in Pennsylvania, USA." *Gender, Place and Culture* 11, no. 2 (2004): 289.

Vandermeulen, Valerie, Ann Verspecht, Guido Van Huylenbroeck, Henk Meert, Ankatrien Boulanger, and Etienne Van Hecke. "The Importance of the Institutional Environment on Multifunctional Farming Systems in the Peri-urban Area of Brussels." *Land Use Policy* 23 (2006): 486.

Viljoen, Andre, and Joe Howe. "Cuba: Laboratory for Urban Agriculture." In *CPULs Continuous Productive Urban Landscapes: Designing Urban Agriculture for Sustainable Cities*, edited by Andre Viljoen. Oxford, UK: Elsevier, Architectural Press, 2005, 146.

Conservation and Sustainable Agriculture

Conservation is the careful preservation of an entity so that it can be maintained for a long period of time. Sustainability signifies the continued existence of an entity without losing its value in the future. Although humans depend on environmental goods and services for their survival and growth, their interaction with the environment points to an unsustainable trajectory owing to their rapid population growth, over-consumption, and myopic view of ecology and economics. This trend has been made apparent by recent shifts in climatic patterns, widespread loss of soil fertility, expansion of deserts, reduction of groundwater levels, decline of fisheries, large-scale extinction of plant and animal species, loss of agricultural diversity and food security, to name just a few developments. In the wake of these events, increasing advocacy for conservation and sustainability has emerged, especially related to food production and access.

When most humans lived in hunter-gatherer societies, about 12,000 years ago, they depended on their immediate environment for food. Because they lived in unmodified ecosystems, the survival and number of hunter-gatherer groups depended on their ability to obtain a sufficient supply of the edible energy flowing through the natural systems. Humans have a limited digestive capacity for breaking down plant matter, especially cellulose. Consumption of meat, which is more easily digested, is not efficient, as there is an erosion of energy at higher trophic levels in the food chain. Therefore, only a fraction of the total energy flow in the ecosystem can be utilized. The major effect of this low rate of energy availability is evident in the low population density found in these societies. When they are efficient in obtaining food from their environment, individuals are well nourished. Furthermore, they spend only a small part of the day obtaining food, and spend much of the time in leisure, ensuring a high return on investment of energy. Researchers have estimated that for every 10 to 20 kilocalories of food obtained, only 1 kilocalorie is spent performing the activity. Given this advantage, the reason that hunter-gatherers adopted agriculture is puzzling.

The Origin and Expansion of Agriculture

Three major hypotheses have been proposed to explain why hunter-gatherer societies might have adopted agriculture:

- According to the technical change hypothesis, as human technical capabilities in tool-making improved, technology-intensive activity such as agriculture became possible. Of course, some agricultural societies used extremely simple tools, whereas some hunter-gatherers used sophisticated tools.
- The coevolutionary hypothesis holds that agriculture evolved simultaneously with human evolution. As hunter-gatherers cut trees for building materials or fire wood, clearings opened up. Seeds of far-away plants found their way to the clearings through human waste. Taking advantage of the fertile human waste and the sunlight, edible plants started growing near the settlements. Proximate food availability boosted the population numbers, which led to the cutting of more trees and the opening up of more clearings. Over time, this positive feedback loop involving humans and plants led to the expansion of agriculture.
- The resource depletion hypothesis states that agriculture was a response to human population growth. As hunter-gatherer populations grew, local sources of edible energy became depleted. Because they had to go farther to find food, the return on investment of energy declined. Agriculture requires a high initial input and is, therefore, not efficient for small populations. However, as populations grow, less

effort is required to expand agriculture. In other words, the return on investment of energy for agriculture is high if the populations are large. When populations shrink, agricultural societies have been seen returning to hunting and gathering.

Whatever the reason for the emergence of agriculture, it is clear that agricultural systems are products of environmental modification. Often, tree species with low edible energy are replaced with cereal grasses that provide more edible energy per unit area for human consumption. While the most productive hunter-gatherer society may support 2 persons per square mile, even the most traditional agricultural society can support 40 to 400 persons per square mile. Nevertheless, relative to the hunter-gatherer approach, agriculturalism is associated with poorer nutrition, smaller body sizes, and increased diseases. Agriculturalists have changed or displaced hunter-gatherer societies in most of the world by engaging in transfer of knowledge and by expanding into those societies, either by conflict or by introducing new diseases. Moreover, agriculturalists have successfully expanded their range to temperate latitudes where the number of warm and sunny days—a critical requirement for crops—is limited. Modern hunter-gatherers are found only in frigid or arid environments. Few of these isolated societies exist elsewhere, because they have not come in contact with agriculturalists. The net effect of this unprecedented expansion is that more and more natural ecosystems have been replaced by modified agroecosystems. To understand the impact of agricultural systems on the natural environment, it is important to understand the environmental conditions required for agriculture and the economics of food production.

The Practice of Agriculture

The initial cost of replacing the natural systems with agricultural systems is very high. Removing trees and preparing the soil for growth takes tremendous effort, and often it requires a group of people, draft animals, or heavy machinery to accomplish. In addition, mature vegetation must be replaced with a selective set of desired seeds. Sowing seeds can be a labor-intensive task, involving the exposure and removal of soil, followed by planting of seeds and soil replacement. A wide range of tools, including simple sticks, animal-drawn plows, and machinery, are used to accomplish this task. In the case of traditional wet rice cultivation, seeds are broadcast on prepared clay fields, and later the seedlings are transplanted by humans in rows. In industrial societies, large machines use energy from fossil fuels to prepare the soil.

Exposure of soil to the sun, wind, and running water is one of the immediate consequences of this phase in agriculture. Previously intact soils continue to erode in agricultural areas. In the 1930s, for example, the Great Plains region in the

United States and Canada experienced a phenomenon called the "Dust Bowl," which was caused by the combined effect of a prolonged drought and use of farming methods that eroded the soils and dislodged the material, causing dust storms. It takes between 200 and 1,000 years for 1 inch of soil to form, but it can be eroded away in just a few seasons. Soil erosion causes loss of the fertile upper soil horizon, nutrient loss, and changes in the soil structure and its water-holding capacity. Over time, the arability of these soils inevitably declines. Estimates show that badly eroded areas in Illinois and Indiana have lost about 24 percent of their inherent initial productivity for corn. Further, eroded soil ends up in rivers, creating a large sediment load and rendering them unfit for living organisms (see the "Soil" entry).

One of the differences between a natural ecosystem and agricultural systems is that, in the absence of sufficient nutrients, only those plant species that can adapt to low-nutrient environments are able to occupy the natural ecosystem. With agriculture, however, humans usually force the growth of a select species, which may not be able to thrive without certain nutrients. This practice lowers the amount of edible energy generated and results in a low return on investment of energy. To maximize the energy derived from such environments, humans add nutrients to the soil. Moreover, when the edible portions of the crops are harvested, the soil nutrients locked in the food are removed and dispersed elsewhere. Some agriculturalists practice shifting cultivation, such as in the tropical forests, where farmers move from one patch of cleared forest to another in response to nutrient decline. In most other agricultural practices, artificial addition of nutrients is the solution to depletion of soil fertility, the cumulative effect of which is the modification of the natural biogeochemical cycling of the ecosystem. For example, it is estimated that 50 percent of the fertilizer applied to U.S. farmland is intended just to replace nutrients that are lost with soil erosion. In China, an estimated 30 percent of the nitrogen fertilizers and 22 percent of the potassium supplements are applied simply to replace nutrients lost through erosion.

Nutrient runoff from agricultural areas is considered to create heavy nutrient loading in water bodies, creating favorable environments for algae and phytoplankton to thrive near the surface. As the organic matter produced by the plankton sinks to the bottom, bacterial activity breaks down the material, utilizing the dissolved oxygen in the water. Through this process, increased phytoplankton results in decreased oxygen levels in the water bodies, creating an anoxic (zero-oxygen) condition, which in turn extinguishes other life forms. In the Gulf of Mexico, near the Mississippi River delta, a 6,000 to 8,400 square mile area of dead zone exists due to this reason.

Plants also require a supply of fresh water to survive. The distribution of naturally available water is uneven across the globe. The occurrence of specific types of plants in natural ecosystems reflects their response to precipitation and the availability of soil water. Plants draw water from the soil and circulate it within their systems before transpiring the water into the atmosphere through the openings on their leaves. In the moist tropics, which have cloudy skies through most of the year, tall trees with large leaves are successful because they compete for sunlight. However, the large surface area of leaves implies greater transpiration, which, in this case, is affordable due to the abundant supply of water available in this environment. In arid deserts, the plants that are able to conserve water by various strategies, including those with a smaller leaf area, are the most successful. Between these two extremes, plants with a wide variety of shapes, structures, and strategies naturally occur in response to precipitation.

Since agriculturalists have expanded their geographic range into areas with limited precipitation, and because most food crops are not resilient against water shortages, it has become necessary in many farming areas to supplement precipitation with artificial water supplies—that is, irrigation. Groundwater is pumped from aquifers, and surface waters from rivers and streams are dammed, diverted, and channelized to convey water to croplands. Natural water-cycling systems are modified and in many cases disrupted, resulting in collateral damage to biodiversity. The magnitude and impact of such irrigation methods vary depending on the scale of the agricultural system and the local geographic and climatic conditions. In the 1960s, the policies of the Soviet Union led to intense irrigation along the floodplains of the Amu Darya and Syr Darya, two rivers draining into the Aral Sea, which is situated between Kazakhstan and Uzbekistan. The combined effect of water diversion and arid climatic condition caused the Aral Sea to shrink to 10 percent of its original size within a span of 40 to 50 years. The resulting loss of aquatic biodiversity, along with a change in local climate, salinization of arable soils, and associated unemployment and poverty, has been called one of the worst environmental disasters in the world. Elsewhere, the large-scale damming of rivers for irrigation has been linked to the drowning of ecosystems; displacement of people; thermal anomalies in the water; trapping of sediments, nutrients, and pollution behind the dams; and increased evaporative loss of fresh water.

The clearing of vegetation by humans for agriculture is another form of disturbance of the natural ecosystem. Succession is a natural process in which new vegetation becomes established in clearings to take advantage of the sunlight and nutrients. As a result, agricultural crops are in constant competition with natural vegetation. Humans act as guardians to edible crops and remove competing "weeds" by hand. In larger agricultural areas, where manual weeding is impractical, chemicals or herbicides are used to kill naturally occurring plants. Moreover, because they represent an easily available food source, agricultural areas attract animals that can be formidable competitors to humans. To keep these "pests" at bay, traditional agriculturalists physically guard the crops or sometimes employ simple tools such as scarecrows, though these tools are impractical for managing large agricultural areas. Large-scale agriculturalists use chemical pesticides to kill pests and diseases. The use of chemicals on crops has been found to be detrimental to human health, as the residues from these chemicals inevitably find their way into the food supply. These chemicals also are found to flow through the food chain and adversely affect populations of living organisms.

Recent advancements in biotechnology have also led to the manipulation of food crops at the genetic level, to be resistant to diseases, herbicides, and pesticides (see the "Genetically Modified Foods" entry). Some practitioners have even inserted genes from poisonous animals such as snakes and scorpions into plants to make them resistant to insects and diseases. Genetically modified crops are highly controversial because the effects of these crops on agricultural diversity are expected by many to be disastrous. One of the fears is that the pollen from these crops could contaminate other species and reduce the already declining diversity of crops.

Green Revolution

The relatively high expense of labor and land, along with the costs of the machinery and energy needed to produce crops, favored the development of mechanized industrial agriculture in the United States in the mid- to late 20th century. In a phenomenon called the "Green Revolution," high-yielding crop varieties were bred to be grown in large-scale, mechanized-farmed lands. Such varieties have been manipulated to allocate more energy toward their edible parts. The net primary production is increased by artificially increasing the amount of nutrients and water. Even farm animals may be raised in such industrial production systems, where livestock are produced in confined areas called feedlots and are fed manufactured feed. Green Revolution agriculture and industrial farming require large amounts of inputs, transportation, and capital infrastructure. However, while the output per unit land or labor is high, these systems are unproductive when the output per unit energy is measured.

In the short term, the Green Revolution has certainly helped in feeding the growing world population. For example, during the 23-year span between 1950 and 1973, growth in the grain harvest equaled that achieved in the preceding 11,000 years. Between 1950 and 2008, grain yields went from 0.45 ton to 1.33 tons per acre. In a way, higher output ensured that a lesser amount of land was converted to agriculture. In the United States alone, nearly 200 million acres of land has reverted to natural ecosystems.

Nevertheless, the Green Revolution has proved to be largely detrimental to the natural environment. The energy used to produce edible plants and livestock has increased, which in turn has decreased the return on investment of energy. If the energy used to grow food is higher than the edible energy produced, then the system is considered inefficient. In the United States, it takes 10 kilocalories of energy from coal, oil, natural gas, and electricity to produce 1 kilocalorie of edible food—an inefficiency that was not perceived as cause for concern when energy supplies were cheap and seemingly abundant. The existing sources of energy are finite, however, and this makes the Green Revolution unsustainable. Further, the immense amount of methane and carbon dioxide emissions from industrial agriculture has had unintended consequences on the global climate. Nutrient loading in water bodies, as discussed earlier, is another consequence of the Green Revolution (see the "Green Revolution" entry).

One of the most serious and direct effects of the Green Revolution has been on agricultural diversity. Depending on market demand and yield expectations, one variety of a plant is often grown extensively. Such monoculture practices reduce costs and increase efficiency, but also render crops vulnerable to constantly evolving pests and diseases. Moreover, monoculture encourages the fading away of the diverse variety of seeds that were previously cultivated. Seeds of different varieties of edible plants act like genetic banks. If a disease out-competes one variety, then another variety, which may still be resistant to that disease, can help to redeem the loss. In a monoculture environment, in contrast, such a situation could be disastrous. While this risk is mitigated by the use of pesticides, the long-term prospects are unsustainable. Put simply, it is imperative to adopt sustainable solutions to feed the growing world population.

Conservation

The foremost challenge for future agriculturalists is to produce enough to sufficiently feed a growing population of 7 billion people today and eventually more than 9 billion people, while simultaneously conserving soil, water, energy, and biodiversity. To stabilize and reverse the unsustainable trajectory of modern agriculture, a multifaceted global strategy needs to be adopted by individuals, enterprises, and policymakers. It is necessary to increase land productivity by practicing conservation agriculture and precision farming methods with minimal soil disturbance, water-use efficiency, efficient production of protein, localization of agriculture inputs, and reduction in demand by controlling human population growth.

The basic adjustment toward sustainable agriculture should involve farming practices that consider the local topography, soil characteristics, climatic conditions, pests and diseases, available inputs, and individual farmer requirements. Adopting such a site-specific approach, rather than an enforced approach to agriculture, is the first step. Specifically, the following strategies can help sustainable agricultural practices.

First, crop varieties need to be chosen based on the suitability and limitations of the site. This strategy is preventive rather than reactive. It reduces the need for excessive nutrient inputs, pesticides, and dependence on artificial water sources, among other advantages. Although this sounds like common sense, industrial agriculture has been divorced from this pattern.

Second, crop and livestock diversification needs to be encouraged to enhance ecological and economic resilience. Monoculture systems are efficient and easy to manage, but they run the risk of sudden, large-scale collapse due to diseases and pest attacks. Diverse cropping not only spreads the economic risk and mitigates the effects of price fluctuation, but also allows for the evolution of a variety of plants, thereby strengthening biodiversity. After all, seeds that are locked away do not get a chance to grow and follow their evolutionary trajectories. Also, the rotation of diverse plants can be used to suppress weeds, pathogens, and pests by eliminating their competitive advantage. Cover crops help stabilize the agroecosystems, preventing soil and nutrient erosion, and mulch helps maintain soil water by improving infiltration. Chemical inputs can be avoided by encouraging diverse cover crops that create habitats for beneficial insects and arthropods that keep pests under control.

Specialized farms are a relatively recent phenomenon, introduced by pressures of industrial technology, government policy, and changing economic practices. In contrast, mixed farms that have raised crops and livestock in the same land are able to support optimal diversity. By maintaining pastures and foraging areas on slopes, but growing crops on flat terrain, soil erosion is prevented. Rotating forage areas and crops also helps to enhance soil quality because of the accumulating livestock manure. In addition, livestock can serve as buffers against crop failures, as they can sustain on crop residue and still provide income to farmers.

Third, maintaining soil health is crucial for sustainable agriculture. Proper soil, water, and nutrient management enhance the health and vigor of plants and help them develop pest and disease resistance. The need for material inputs, such as energy, water, nutrients, pesticides, and herbicides, is greatly reduced in such systems. The long-term health of the soil is maintained, thereby ensuring sustained output of food.

Fourth, development and application of efficient biological systems, requiring low levels of locally available material input, are the best ways to practice conservation-oriented agriculture. No single panacea exists for such systems, because one of the key words for sustainability is "local"—and what works locally for some farmers under certain conditions may not work for others. It is important to acknowledge the importance of localization and encourage research for specific systems. Practitioners of sustainable agriculture also must understand that the costs and benefits of different applications need to be weighed carefully. For example, in one case a few broad applications of synthetic herbicides on grapevines proved to be a better option than tillage, because they were less energy intensive and prevented soil compaction associated with a mower. Sustainable farming does not necessarily mean the absence of chemicals; rather, it should be understood as a knowledge-intensive agriculture.

Properly managed sustainable farms show increased crop yields when compared to industrial agriculture. A 2006 study conducted on 286 farms across 57 underdeveloped countries showed that, on the 91 million acres studied (3 percent of cultivated areas in developing countries), the average yield in those farms increased by 79 percent. All farms showed improved efficiency in water use. Furthermore, 77 percent of the farms showed a considerable decline (71 percent) in pesticide use and an increase in yield (42 percent). The study results should inspire a cautious optimism about the prospects for adopting sustainable practices, especially in underdeveloped nations.

—Bharath Ganesh-Babu

Further Reading

Brown, L. *Plan B 4.0: Mobilizing to Save Civilization.* New York: W. W. Norton, 2009.

Diamond, J. "Evolution, Consequences, and the Future of Plant and Animal Domestication." *Nature* 418 (2002): 700–707.

Pollan, M. *The Botany of Desire: A Plant's-Eye View of the World.* New York: Random House, 2001.

Pretty, J. N. "Agricultural Sustainability: Concepts, Principles, and Evidence." *Philosophical Transactions of the Royal Society* 363 (2008): 447–465.

Pretty, J. N., A. D. Noble, D. Bossio, J. Dixon, R. E. Hine, F. W. T. Penning de Vries, and J. I. L. Morison. "Resource-Conserving Agriculture Increases Yields in Developing Countries." *Environmental Science and Technology* 40, no. 4 (2006): 1114–1119.

SARE. "What is Sustainable Agriculture?" University of California at Davis: Sustainable Agriculture Research and Education. Accessed February 2011. www.sare.org/publications/whatis/whatis.pdf.

Shiva, V. *Stolen Harvest: The Hijacking of the Global Food Supply.* Cambridge, MA: South End Press, 2000.

United Nations, Food and Agriculture Organization. "Conservation Agriculture." Accessed February 2011. www.fao.org/ag/ca/.

Deficiency Diseases

The primary components of any diet are carbohydrates, proteins, and fats. Without calcium, potassium, salt, and various micronutrients, however, the human body is not able to grow, repair itself, or function properly. Deficiency diseases result from inadequate quantities of both macronutrients and micronutrients (see the "Protein and Protein Deficiency" entry). There is much concern about public health threats that result from dietary deficiencies of vitamins and minerals.

Vitamins are organic molecular compounds essential for converting carbohydrates into energy, turning proteins into tissue, synthesizing hormones and enzymes, or maintaining levels of trace elements in the blood. Thirteen vitamins are universally agreed upon. The five that have had the broadest public health consequences are vitamins A, B_1, B_3, C, and D.

Minerals are also essential to bodily health. Some, like calcium and phosphorus, are needed in relatively large quantities. Others, such as iron and iodine, are needed only as trace elements.

Some vitamins and minerals can be stored by the body for later use—for example, iron and vitamin D. Others are passed out of the body immediately if they are not used, such as vitamins B_1 and B_3. Table 13 lists nutrients that have been associated with major deficiency diseases, but it also calls attention to the problems of vitamin and mineral overload. A careful balance of the various nutrients is essential for health. Too little precipitates deficiency diseases, whereas too much may precipitate hypervitaminosis (too much of a vitamin) or toxicity. When a vitamin is missing in the diet, the condition is called avitaminosis.

Vitamin A Deficiency

Inadequate vitamin A, or retinol, results in night blindness, and then full loss of eyesight; it also inhibits bone growth. Although vitamin A deficiency (VAD) has been almost eliminated in the developed world, it has emerged as one of the leading causes of blindness in developing countries. It afflicts young children, perhaps half a million worldwide each year, especially after they are weaned off breastmilk. VAD may continue into the adolescent years, which partly explains why it is considered a major public health problem. Because vitamin A (or pro-vitamin A) is contained in

Table 13. Vitamin-Deficiency Diseases

Micronutrient	Deficiency	Normalcy	Overload
Vitamin A (retinol)	Vitamin A deficiency (VAD) Night blindness Blindness	Needed by the retina of the eye and by bones	Vitamin A toxicity (hypervitaminosis A)
Vitamin B_1 (thiamin)	Beriberi	Needed for cell metabolism	Excess amounts of water-soluble vitamins are quickly passed out of the body; hence, a daily supply is necessary
Vitamin B_3 (niacin)	Pellagra	Needed for cell metabolism	
Vitamin C (ascorbic acid)	Scurvy	Needed for growth and repair of tissue	
Vitamin D	Vitamin D deficiency Rickets (children) Osteomalacia (adults)	Needed to maintain levels of calcium and phosphorus in the blood	Vitamin D toxicity (hypervitaminosis D)
Calcium	Rickets (children) Osteomalacia (adults)	Needed by bones, muscles, and blood vessels	Hypercalciuria Hypercalcemia
Iron	Iron deficiency Iron-deficiency anemia	Needed to form hemoglobin	Iron toxicity
Iodine	Iodine-deficiency disorders Cretinism (children) Goiter (adults)	Needed to regulate production of thyroid hormones	Iodine toxicity

meat, dairy products, fruits, and vegetables, people eating a balanced diet should not suffer from VAD. As poverty increases the ingestion of starches, limits meat consumption, and minimizes market produce, VAD emerges. Sub-Saharan Africa and South Asia are the regions most widely affected by this deficiency. Fortified flour, milk, margarine, and even sugar may be used to make vitamin A more widely available. Produce with high levels of beta-carotene, identified easily by their bright orange and yellow colors (e.g., mangoes, papayas, carrots), are good sources of pro-vitamin A, which is easily converted to retinol in the body.

Beriberi and Pellagra

Of all the B-complex vitamins, only two are associated with widespread deficiency diseases. Inadequate consumption of vitamin B_1, or thiamin, results in outbreaks of beriberi. Inadequate consumption of vitamin B_3, or niacin, results in outbreaks of pellagra. Both of these diseases were widespread in the past. Today, they are associated mostly with poor nutrition, such as among alcoholics, in the developed countries of the world, and among the poorest populations (e.g., refugees and internally displaced persons) of the developing world. Each of these vitamins is necessary to metabolize carbohydrates. Fortunately, they are available in abundance as part of a balanced diet, although both may be lost from food during its processing and cooking. The enrichment of flour, bread, and pasta today assures that almost all men and most women get the necessary amount of thiamin and niacin. In the past, however, avitaminoses B_1 and B_3 were responsible for thousands of deaths. Pellagra has long been associated with corn (maize) diets, and beriberi, with rice diets, but only under circumstances specific to how corn and rice are prepared for human consumption.

Beriberi carries an Asian name, which is roughly translated as "I can't, I can't"—a sobriquet reflecting the fatigue typical of dry beriberi, which inflames the nerves and attacks the central nervous system. The other variety of the disease is wet beriberi, which affects the heart and circulatory system. Both forms are associated with difficulties in getting around because of swelling, pain, or paralysis in the legs and feet. In addition, both are potentially fatal. Because this disease is associated with rice monoculture, the largest areas affected by beriberi have been in East and Southeast Asia. It was the Dutch, in Indonesia, who first directed modern medical attention to beriberi's etiology and remediation. In fact, a Dutch doctor's pursuit of a cure for beriberi led to the discovery of vitamins in the 1910s, with the isolation of vitamin B_1 leading the way.

It is not rice that causes beriberi, but rather rice that has been milled to the point of losing its entire brown seed coat—that is, its reservoir of micronutrients, including thiamin. After steel rollers for milling rice were invented around 1870, this disease, which had been endemic but sporadic in rice-eating areas, became epidemic. In Japan, a leader in the adoption of industrial technology, beriberi (*kakké*) became "the national disease." Industrially milled white rice, because it could be processed cheaply in large quantities, became the preferred diet for military personnel and prisoners. Hence, beriberi played a role in weakening the Japanese army during the Russo-Japanese War of 1904–1905. In World War II, a diet of white rice, and the resulting beriberi, was the main cause of death among American and British prisoners in Japanese POW camps. With urbanization, beriberi also became a disease of Asian cities, its impact accentuated because white rice was preferred by the more affluent members of these populations.

Pellagra's name, an Italian derivative, describes its most visible symptom: *pelle* + *agra* means "sour skin." The first description of this disease was recorded by a Spanish physician in 1762. Physicians now use "the three D's" to diagnose pellagra: dermatitis, dementia, diarrhea. Untreated, they result in the fourth D: death within five years. Although this disease is easily treated today, it took until 1937 to confirm that pellagra could be cured by administering vitamin B_3 and prevented by enriching commercially available flour. Voluntary bread enrichment began in the United States in 1938, at which point pellagra virtually disappeared from the U.S. medical map. For a long time, pellagra was suspected to be caused by something *in* the diet (germs or toxins). Only in the 20th century was it shown to be the result of a micronutrient *absent* from the diet. That micronutrient was eventually shown to be nicotinic acid, which was rechristened niacin to void associations with nicotine.

Corn is a grain indigenous to the Americas, where it has been consumed for thousands of years without "sour skin." When it diffused to Europe and entrenched itself in the food culture of the western Mediterranean, this product arrived without the accompanying knowledge of how the indigenous Americans prepared it for consumption. Indian cultures, beginning in Mesoamerica, soaked corn in an alkaline solution (limewater), making it easier to grind and tastier. That process, known as nixtamalization, also made niacin and certain minerals available for absorption by the human body. When corn arrived in Spain and diffused from there, it came to be ground in mechanical mills simply because that is the way Europeans processed their cereal crops. As a result, niacin was never freed to do its work in the human body. Wherever diets became overly dependent on corn (i.e., in areas of monoculture), pellagra became a problem. Usually these areas were poor and rural. By comparison, residents of richer and urban areas had a variety of other niacin-rich foods at their disposal. The Iberian and Italian peninsulas especially were plagued by pellagra, a disease associated with corn diets and poverty.

It took an outbreak of pellagra in the southeastern United States to reveal its etiology. Pellagra always had been endemic in the American South, but the first epidemic hit in the early 1900s. Hundreds of thousands of people were affected. In 1914, the U.S. Public Health Service set up the Spartanburg (South Carolina) Pellagra Hospital to find a cure for the disease. The clinical setting allowed physicians to determine that a balanced diet or brewer's yeast led to a cure. The search, which began by looking for a toxin, ended with the discovery of a dietary deficiency. The relationship between pellagra and gender also was documented statistically during the first half of the 20th century: Women were twice as likely to be afflicted as were men. Part of the gender differential is accounted for by estrogen, which retards synthesis of niacin; another part may have been social—men and children were likely to have a more balanced diet because they ate first, in preference to women.

Scurvy

Avitaminosis C leads to scurvy, a disease of the flesh and connective tissue. Its chief symptoms are spongy gums, spots on the skin, and bleeding mucous membranes.

If left untreated, death ensues. When treated, however, scurvy is easily cured and entirely preventable. It was only in 1932 that the etiology of this disease was finally linked to the absence of ascorbic acid, or vitamin C, in the diet. Soon thereafter, vitamin C became commercially available, and scurvy was conquered in all except rare cases. Today, various foods are fortified with vitamin C, assuring adequate amounts in most peoples' diets, although debate continues about what constitutes "adequate amounts."

As long as uncooked or minimally processed fruits and vegetables are a part of the diet, adequate vitamin C is available to the body. Citrus fruits are the most famous source of ascorbic acid, but tomatoes, peppers, strawberries, and cantaloupes are also rich sources of this nutrient. In cold Arctic lands, uncooked meat has long served the same purpose among indigenous peoples. In the temperate realm, raw sauerkraut (pickled green cabbage) and kimchi (fermented cabbage and other vegetables) have held off scurvy during winter months. Exposure to the air, heat, and contact with copper, however, all degrade the vitamin C content of foods. Among modern-day populations, ascorbic acid is widely taken as a vitamin supplement.

Military populations, especially sailors, were most vulnerable to scurvy because they often had to subsist for long periods of time without fresh fruits and vegetables. Scurvy, in fact, inhibited the expansion of Europeans onto the world ocean. To sustain sailors on sea voyages of more than two to three months (maximum storage time for vitamin C in the body), food supplies had to include foods that contained vitamin C, and shipboard foods generally did not. It took four centuries, from Columbus through World War I, to first identify and then end the scurvy epidemic that plagued long sea voyages. The British found that oranges, lemons, and especially limes (which were more widely available in British dominions) needed to be a part of sailors' rations; hence, British seamen were called limeys. The Germans (and Dutch) sent crocks of sauerkraut on board their ships; hence, German seamen were dubbed krauts. Whether the foods provided consisted of citrus fruits or pickled cabbage, however, results were inconsistent because of variation in storage and preparation, and the root cause of scurvy remained elusive until the 1930s.

Rickets and Vitamin D Deficiency

Rickets is a bone disease caused by a deficiency of calcium, phosphorus, and/or vitamin D. Because calcium and phosphorus are widely available in a balanced diet, it is usually avitaminosis D that causes rickets, because vitamin D is required for absorbing calcium. While rickets—the result of extreme vitamin D deprivation among children—is rare today, vitamin D deficiency is a growing problem associated with osteoporosis, heart disease, and high blood pressure in the adult population. Somewhere between 70 percent and 75 percent of the world's population has been estimated to be vitamin D deficient, heralding a new pandemic both in the United States and around the world.

Only a few dozen cases of rickets are diagnosed in the United States each year, with many of these instances associated with long-term breastfeeding especially by malnourished mothers. Rickets is a children's disease in which the bones are not able to develop properly. Its chief symptom is bowed legs; the legs bend because they are not strong enough to carry the increasing weight of the body. Other skeletal deformities and fractures are also typical of rickets. Later in life, the same conditions may cause osteomalacia, or bone thinning. Among young children, when the skeleton is growing most rapidly, a lack of vitamin D in the blood means that calcium will not be delivered to the bones. Among adults, a lack of serum vitamin D means that calcium will be withdrawn from the bones. Milk and green, leafy vegetables are excellent sources of calcium, as are calcium supplements, but the calcium cannot be used by the body without the assistance of vitamin D.

Vitamin D is called the "sunshine vitamin" because it is synthesized in skin exposed to ultraviolet-B radiation. Dietary sources are limited to egg yolks, liver, and certain fish. Neither breastmilk nor cow's milk supplies vitamin D, even though both are good sources of calcium. For the skin to produce vitamin D, exposure to sunlight is required, but exposure is closely connected to environmental conditions and cultural practices. Elements of nature that affect how much vitamin D a person is able to synthesize include (1) season, (2) time of day, (3) cloud cover (4), latitude, and (5) skin color. Ideally, skin should be exposed when the sun is highest in the sky—namely, during summers and mid-days. In fact, the body is set up to manufacture vitamin D in the high-sun season and store it for the low-sun season. As latitude increases, it becomes ever more difficult to manufacture enough vitamin D; most sources theorize that light skins evolved to increase the body's vitamin D production. Melanin inhibits the dermal synthesis of vitamin D.

A region's social and economic geography may occasionally militate against sun exposure. In fact, the Industrial Revolution brought on the first epidemic of rickets in England. Coal-burning industries drew people to the cities where buildings also were heated with coal. Air pollution increased, exposure to sunshine decreased, and rickets prevailed from the 1700s through the early 1900s. Cod liver oil became popular as an antidote. All across Europe, rickets, or the "English Disease," became associated with coal-fired industrialization and urbanization.

More recently, urbanization in the Islamic realm has led to an increase in avitaminosis D among women who practice *purdah*. Courtyard houses, with interior outdoor spaces where women could unveil, have given way to apartment living, leaving few opportunities for women to be uncovered in the sunshine. In North

America, Europe, and Australia, women and men have taken to using sunblock when outdoors. From a micronutrient perspective, this practice is the non-Islamic world's equivalent of *purdah*. Nursing homes (and orphanages in some countries) house another population of sun-deprived individuals.

For anyone getting too little direct sun exposure (perhaps 15 minutes per day, or more for people with darker skins), dietary supplements are required. They take two forms: (1) fortified milk, orange juice, cereals, or baby formula and (2) tablets and capsules. Even so, a consensus is emerging that supplements cannot provide vitamin D in the quantities demanded by the body throughout the year. At the same time, recommendations to get more sun exposure are tempered by the known role of ultraviolet-B radiation in causing skin cancer. Those who insist upon more sun exposure also insist that the body cannot get adequate amounts of vitamin D through dietary sources alone.

Goiter and Iodine-Deficiency Diseases

Goiter, like rickets, has all but disappeared from developed countries. At the same time, like vitamin D deficiency, iodine deficiency has been increasingly acknowledged as a global public health problem. Extreme iodine deprivation results in goiter, which is an enlarged thyroid gland. Thyroid hormones must attach themselves to iodine atoms. Absent the iodine, thyroid hormone levels fall and the pituitary gland produces its own hormone to tell the thyroid gland to work even harder. The harder it works, the larger it becomes. The enlarged thyroid, which eventually becomes visible at neck level, is the characteristic symptom of goiter. It may take the form of a single swollen nodule or a multinodular goiter.

Goiter has been endemic in areas where iodine is not found in the soil—for example, in mountainous regions and areas far from the coast. Because the disease was so frequent in the English Midlands historically, it gained the appellation "Derbyshire Neck." The United States' "goiter belt" coursed through the landlocked Midwest. Even today, goiter is common in central Africa, central Asia, and Andean South America. In the past, the Alps were famous for goiter prevalence; today, Upper Egypt is one of the most well-known areas of endemic thyroid problems. As seafood consumption reaches areas far from coastal realms, however, goiter incidence declines. Nevertheless, public health initiatives have made iodizing salt (which does not spoil and is used by everyone) the most common method of combating goiter. Although iodine was experimentally added to salt in 1830, only in 1924 did Morton begin iodizing all salt.

In addition to goiter, iodine-deficiency disorders (IDD) continue to present a serious threat to brain development among the unborn, neonates, and children, and extreme iodine deprivation results in cretinism, a form of mental and physical retardation. Iodine deficiency is the leading preventable cause of mental retardation in the world. The Network for Sustained Elimination of Iodine Deficiency has as its motto: "Giving Children a Smart Start." As its goal, this organization is striving to provide 100 percent of the world's population with access to iodized salt, a goal that seems increasingly achievable. While only 20 percent of the world's people had access to enriched salt in the early 1990s, that figure had increased to 70 percent by 2010. Nevertheless, 1.6 million people remain at risk of IDD. Those who remain on the iodine frontier will be the hardest to reach because they are primarily poor and rural.

—Donald J. Zeigler

Further Reading

Bollet, Alfred Jay. *Plagues and Poxes*, 2nd ed. Demos, NY: Medical Publishing, 2004.

Carpenter, Kenneth J. *Beriberi, White Rice, and Vitamin B*. Berkeley/Los Angeles, CA: University of California Press, 2000.

Frankenburg, Frances Rachel. *Vitamin Discoveries and Disasters*. Santa Barbara, CA: Praeger/ABC-CLIO, 2009.

Holick, Michael F. *The Vitamin D Solution*. New York: Penguin Group, 2010.

Howson, Christopher P., Eileen T. Kennedy, and Abraham Horwitz, eds. *Prevention of Micronutrient Deficiencies*. Washington, DC: National Academy Press, 1998.

Jacobsen, Kathryn H. *Introduction to Global Health*. Sudbury, MA: Jones and Bartlett, 2008.

Desertification

Desertification traditionally was defined as the unwanted expansion of existing deserts; however, this definition is too limited. More broadly, desertification is the degradation of drylands, which include arid, semi-arid, and dry sub-humid regions. Arid and semi-arid regions are the largest biomes on Earth, covering more than one-third of the world's land area. The ecosystems of these biomes are extremely vulnerable to overexploitation because low rainfall slows the regeneration of vegetation. The global average of dryland degradation is approximately

33 percent of the total geographic area (TGA). More than 70 percent of the world's drylands, or approximately 9,000 million acres, is degraded.

Causes

The causes of desertification include socioeconomic factors such as international trade patterns, national irrigation infrastructures, and regional land-tenure laws, as well as physical problems associated with deforestation, erosion, soil salinization, poor grazing practices, and climate change. The ongoing problem of climate change is linked to changing temperatures as well as to changing precipitation patterns. In agriculture, both the amount and the timing of precipitation are extremely important. Poor agricultural practices include overgrazing, overcultivating, inappropriate irrigation practices, and numerous activities that degrade the quality or quantity of soil. As the global population grows, the land devoted to growing food must keep pace and farming must be conducted sustainably.

Drylands have adapted to be able to respond rapidly to normal climatic variations. In most of these arid and semi-arid regions, precipitation has pronounced seasonal variability as well as wide fluctuations over years and even decades. Over time, dryland ecology has developed strategies to cope with this variability in precipitation, and the flora and fauna respond quickly. As an example, the vegetation boundary south of the Sahara Desert migrates up to 200 kilometers (124 miles) with annual fluctuations in moisture regimes. However, drought occurs when rainfall is significantly below normal recorded levels for an extended period of time and can initiate or aggravate desertification. Drought kills vegetation, which in turn exposes soil to erosion. The Intergovernmental Panel on Climate Change (IPCC) predicts that drought frequency and intensity will increase with climate change.

People, too, have developed numerous coping strategies to adapt to changing weather patterns, such as shifting agriculture and nomadic herding. Yet, during the past few decades, these strategies have been challenged by changing economic and political situations, such as an increase in population density and growth in the number of settled communities. If nomads can no longer move the herds to follow the grasslands because a different ethnic group has settled the land they once used for grazing, conflict and degradation can ensue.

Overgrazing and overcultivation of land also result from stressed economic conditions. Often, unregulated access to land causes some people to overexploit resources to increase individual gains at the expense of the larger community. Poor people, especially poor women, are usually denied access to the most fertile land, but instead are consigned to the most fragile areas with the most tenuous resources. When their economic constraints offer them few alternatives, people tend to salvage what they can from the scant resources available, even though the land is degraded as a result.

Other forces are far beyond the control of the individual. International economic systems and structures can lead to the exploitation of land for short-term gains. For example, an economy based on cash crops for export typically does not return sizable profits to the local community; as a consequence, little cash remains for land restoration. Similarly, local markets do not receive as much benefit when certain taxes are imposed, thereby exacerbating overexploitation of the land.

Education and environmental awareness play enormous roles in desertification mitigation. For example, during the Dust Bowl of the 1930s in the United States, farmers in the Midwest used plows that were designed for the more temperate latitudes of Western Europe; this technology was used during a severe drought. Overtilling exposes soil to further erosion and loss of organic nutrients. Similarly, some of the ranchers leasing land from the Bureau of Land Management (BLM) in the western United States employ antiquated practices that have resulted in overgrazing and desertification. If rangelands are not partitioned by fencing or other means, and if the cattle are not rotated between various grazing areas to allow untrammeled vegetation to regenerate, lands can become seriously degraded. The compaction from hooves and eating grasses down below the buds precludes rejuvenation of grasslands. Further, the compaction of the soil results in impoundment of surface water and increased evaporation, such that precipitation then tends to run off rather than percolate down through the soil, and groundwater is not recharged. Less infiltration of rainwater results in less vegetation and, over time, former grasslands become desert.

Population growth is often proposed to be the primary cause of desertification, but a clear statistical relationship between these two states does not exist. Obviously, a higher population density exerts greater pressure on local resources. This pressure also can be indirect. For example, growing urban populations place increased demands on food grown in less densely populated rural areas. Even so, the causes of desertification are far more numerous and complex than any single variable, such as population growth, can explain. For example, a decrease in population can result in desertification because fewer people remain to properly manage the land. This process is evident in many of the once fertile hillside terraces in Yemen, which fell into disrepair when a substantial portion of the farmers migrated to oil-producing countries nearby. Other areas that are densely populated have managed to avoid such degradation, such as around the city of Kano, Nigeria.

Some causes of desertification are also consequences. Armed conflict causes degradation of the land and can result from degradation. Thc fighting in the Darfur region of the Sudan has led to an influx of heavy concentrations of refugees who cannot help but overwhelm the resources of their tent settlements. When a greater number of people, especially those from a different ethnic background, move into an area and compete for already scarce resources, conflicts tend to arise or worsen. Increased rates of degradation can lead to conflict and heightened conflicts further degrade the land.

Effects

During the process of desertification, less productive soils result when eroded and exposed topsoil is blown away by the wind during deflation or washed away by precipitation as runoff. Both the soil's physical properties and its biochemical composition are altered for the worse by these factors. Rills, gullies, and ravines often appear in soil; in turn, the soil's nutrients are removed by the erosional forces of wind and water. When land becomes compacted and encrusted at the surface, water tables drop. The capillary action of soils in arid regions pulls minerals, including salts, up from below. Evaporation leads to salt pans or salt flats or simply the salinization of soils, and vegetation is damaged or destroyed. The loss of vegetative cover is both a consequence and a cause of desertification. Loose sand and high winds can sandblast plants during a process referred to as abrasion, or bury them, or expose their roots causing necrosis. When pastures are overgrazed, edible plant species are replaced with inedible woody shrubs in a process called bush encroachment. Once vegetation disappears, transpiration also vanishes locally, leaving less atmospheric moisture available for cloud formation and precipitation.

The ecological services and complexity of croplands, pastures, and woodlands are diminished once they give way to deserts. Of the 25 global "biodiversity hotspots" assessed by Conservation International, 8 are located in drylands. It is estimated that dryland soils contain more than one-fourth of all the organic carbon stores in the world. Between 1900 and 1950, approximately 15 percent of dryland rangelands were converted to cropland. In these regions, soil used for farming is tilled and left bare for extended periods, leaving them more susceptible to erosion and nutrient loss. Desertification, therefore, results in the loss of biological diversity, ecological services, and economic productivity.

Some of the indirect consequences of desertification occur outside the degraded area. Degraded land often causes flooding downstream, a decrease in water quality, sedimentation of rivers and lakes, and siltation of reservoirs and navigation routes. Sedimentation of waterways wreaks havoc on certain aquatic species. Desertification also increases the risk of dust storms, whose effects can be devastating. The seemingly incessant dust storms of the Dust Bowl, for example, caused people to pack windowsills with oil-soaked rags in an attempt to keep the dust out of their homes. Cloths were worn over the nose and mouth in an attempt to breathe.

Dust storms are common in Iraq, where (as occupying military forces have found, to their great chagrin) expensive machinery is damaged, visibility is reduced, and metals corrode in the wake of these onslaughts. Dust storms also contribute to mental stress. Dust storms worsen certain health problems, including eye infections, respiratory disease, and allergies. Dramatic increases in the frequency of dust storms were recorded, not only during the Dust Bowl years in the United States, but also in the Virgin Lands region in the former Soviet Union in the 1950s and in the African Sahel during the 1970s and 1980s.

Desertification and Agriculture

The relationship between degraded soils and food productivity is complex. Crop yield is affected by numerous factors, such as the temperatures and precipitation, disease and pests, genetics, farming practices, and market forces. However, because food production is undermined by soil loss and degradation, desertification is considered an enormous concern for the global community. Achieving a nutritionally adequate diet for the world's increasing number of people will require tripling food production over the next 50 years. Even under good circumstance, this increase would be a challenge. If desertification is not slowed and reversed, crop production in desertifying areas will continue to decrease. Malnutrition, starvation, and ultimately famine could follow. Areas that have high levels of poverty are those usually struck by famine. Drought and desertification often initiate a food crisis, which is then made worse by the lack of efficient food distribution policies and methods, or the inability to buy what is available leads to starvation.

Although few standardized data are available on the economic losses resulting from desertification, a study conducted by the World Bank suggested that the depletion of natural resources in one Sahelian country was approximately equal to 20 percent of its annual gross domestic product. Globally, annual income lost in the areas immediately affected by desertification totals approximately $42 billion each year. The indirect economic and social costs incurred outside the affected areas, including the suffering and influx of "environmental refugees," are likely to be much higher.

Mitigation

Numerous attempts to combat desertification have been implemented; however, most methods address the symptoms of sand migration and not the root causes of land degradation such as overgrazing, unsustainable farming, and deforestation. In developing countries that are experiencing the highest rates of desertification, many local people use trees as firewood for heating and cooking—a practice that can lead to deforestation and create a downward spiral that actually worsens their poverty. Solar ovens and efficient wood-burning cook stoves are endorsed as a way to relieve deforestation for fuel wood; however, these techniques are generally prohibitively expensive in the same regions where they are needed.

Techniques to counter desertification center on two approaches: (1) providing water, either through drilling wells or extending water pipes over long distances, and (2) hyper-fertilizing or stabilizing soils. Stabilizing, or fixating, the soil has been successfully achieved through the use of shelter belts, woodlots, and windbreaks. Windbreaks are constructed by planting rows of trees or bushes perpendicular to the prevailing winds and have been effective in reducing soil erosion. They were widely used during the mid-1980s in the African Sahel.

An additional mitigation technique is to spray nano clay, a petroleum product, over semi-arid cropland. Where nano clay can be attained at a reasonable cost, it is applied to seedlings and its coating impedes the loss of moisture as well as hinders the chance of the seedling being blown away. In addition to applying synthetic and organic fertilizers, adding nutrients to the soil and restoring its fertility is often accomplished by plants. Legumes, for example, take nitrogen from the air and fix it in the soil, thereby making it available for root uptake. Crops such as barley, beans, and dates are the most widely used with this method of fertilization.

Organic waste material from housing construction is also beneficial for soil enrichment. Materials such as hazelnut shells and bamboo have been made into biochar, or *Terra preta nov*, through pyrolysis. Biochar is used to enrich planting spaces for nutrient-demanding crops. Other mitigation methods are used as well, such as piling stones around the base of trees to help collect morning dew and retain soil moisture. Additionally, artificial grooves dug into the soil may capture rainfall and trap wind-blown seeds.

Even though desertification has received some attention by the news media, most people remain unaware of the expansion of deserts and the enormous extent of environmental degradation of productive lands caused by this trend. At the local level, some individuals and governments continue to thwart desertification. Sand fences are used throughout the Middle East and the United States, similar to the way snow fences are used in the north. Straw grids of approximately a square meter in area serve to decrease the surface wind speeds. Trees and shrubs planted next to the grids are sheltered by the straw until their roots can take hold. However, certain studies show that tree planting enhances the depletion of local water supplies. To stabilize dunes where water is available for irrigation, shrubs planted on the lower one-third of a dune's windward, or stoss, side tend to stabilize the dune. The vegetation both decreases the wind speed at the base of the dune and provides a root network that serves as an anchor to inhibit dune migration. Higher wind speeds at the crest of the dune tend to level it, so trees are then planted on top of these flattened surfaces.

Desertification and Poverty

The areas most prone to desertification are typically plagued by poverty, hunger, and political instability. The poor suffer the most because their livelihoods depend largely on agriculture, livestock, and forests. In India, for example, 25 percent of the TGA has degraded into desert, another 69 percent is classified as dryland, and the main source of income is agriculture. More than 250 million of the world's inhabitants are directly affected by desertification, and about 1 billion people in more than 100 countries are at risk of becoming affected in the near future. These populations include many of the world's poorest, most marginalized, and politically disenfranchised people. The Executive Secretary of the United Nations Convention to Combat Desertification (UNCCD) predicts that without proper action and monitoring mechanisms, in both developing and developed countries, approximately 50 million people could be displaced from their homes by desertification and land degradation within the next decade.

The International Year on Deserts and Desertification in 2006 provided a chance for the global community to address the pressing problem of desertification along with the resulting loss of soil and fertile land. In April of that year, the focus at the International Geneva Symposium was the disastrous effects of desertification, poverty, and human rights. Those in attendance emphasized the need to protect land and soil to achieve the Millennium Development Goals (MDG), which included a 10-year strategic plan (the Strategy) to combat desertification.

Governance

Global leaders long have been aware that desertification is a major environmental, economic, and social problem for many countries in all regions of the world. In 1977, the United Nations Conference on Desertification developed a Plan of Action to Combat Desertification. Despite this and other efforts, in 1991 the United Nations Environment Programme concluded that even though there were examples of success in certain locations, the problem of land degradation in arid, semi-arid, and dry sub-humid areas had intensified.

Desertification remained a focal point for the attendees of the United Nations Conference on Environment and Development (UNCED), or the Earth Summit, which took place in 1992 in Rio de Janeiro. At the summit, a new approach to desertification was adopted that promoted sustainable development at the local level and supported pertinent international partnerships. In addition, the United Nations General Assembly was asked to establish an Intergovernmental Negotiating Committee to prepare a Convention to Combat Desertification through the United Nations by June 1994. The committee efficiently negotiated an agreement; a working resolution was adopted on June 17, 1994, in Paris; and the United Nations Convention to Combat Desertification (UNCCD) was formed. This convention's absolute governing body is the Conference of the Parties (COP), which held its first session in Rome, Italy, in October 1997.

Small strides were made, and nearly a decade later in May 2008, a top-level international policy dialogue on the task of "Coping with today's global challenges in the context of the Strategy of the United Nations Convention to Combat Desertification" took place in Bonn, Germany. The dialogue was designed to coordinate ideas from a number of stakeholders on the Strategy (the 10-year strategic plan) and to increase the commitment of relevant parties and decision makers. The dialogue was attended by delegates from the 193 countries that are parties to the UNCCD, including ambassadors, ministers, intergovernmental organizations, UN agencies, nongovernmental organizations (NGOs), and people from the private sector. The dialogue occurred in three phases, each of which included presentations by and discussion among participants.

Those attending the meeting in Bonn decided that it is necessary to draw in the scientific and technological community more intensively and to devise indicators that can be used at national levels and beyond. Further, the monitoring and reporting processes from the parties is to be mainstreamed so that both the countries affected by desertification, drought, and land degradation and their development partners can measure the effectiveness of the programs developed. The UNCCD established, therefore, a Committee on Science and Technology (CST) that advises the COP. The delegates made great strides toward guiding the ninth Conference of the Parties (COP_9) which was held in Buenos Aires, Argentina, in 2009.

Action Programs

The UNCCD has advocated the development of national, subregional, and regional action programs in the countries most affected by desertification. Criteria for developing these programs are specified for the treaty's five regional implementation annexes: Africa (where desertification is most severe), Asia, Latin America and the Caribbean, the Northern Mediterranean, and Central and Eastern Europe. The UNCCD emphasized the need for these programs to adhere to a democratic, bottom-up approach that endorses popular participation by creating an environment that enables local people to stop and reverse land degradation through self-help. The role of government is to create and foster this enabling environment by decentralizing authority; empowering women, farmers, and pastoralists; and improving land ownership and occupancy systems. Additionally, it is considered important to recognize the role of NGOs and private enterprise, and to facilitate their coordination with governments to better prepare and apply the action programs. While the action programs will be fully incorporated into national policies, they need to remain flexible enough to evolve with changing circumstances.

The UNCCD represents an enormous step forward in the battle against desertification, but it is only a beginning. While governments continue to regularly assess the action programs, they are also focusing on raising awareness and providing training and education in both developing and developed countries. Desertification can be reversed only if broad-based changes are made in local and international behavior. Slowly and incrementally, these changes will eventually lead to sustainable land use and food security for the growing global population. In this sense, reversing desertification is merely a part of a much broader objective: the sustainable development of countries affected by drought and desertification.

Paradigm for Progress

While solutions and remedies for desertification at the local level are occurring, the paradigm for the future of sustainable development needs to be fundamentally changed. The poorest and hungriest people suffer most from the effects of climate change and desertification. The poorest countries, particularly those in Africa, suffer disproportionately from freshwater shortages and increased food costs. Recent demands for bio-fuels, many of which are proposed to be grown on arable lands, are encouraged by public policies focusing on energy production, even though this approach could further drive up food prices. As was realized long ago, unsustainable production and consumption in wealthier nations is the primary cause of continued environmental degradation.

A human rights–based approach to desertification would seek equity, striving to ensure that no peoples or countries shoulder a disproportionate share of the negative effects of environmental degradation. The IPCC's (2007) latest report on climate change notes that developed countries bear responsibility for global climate change. The developed nations, then, as the main consumers of energy and primary producers of greenhouse gases, are largely responsible for resolving and mitigating their impacts on the poorest countries and peoples. While environmental laws fall short of protecting human rights, linking environmental protection and human rights provides the broad-sweeping and forward-thinking approach necessary to combat desertification, eliminate malnutrition, ensure fair water resource allocation, and protect the overall well-being of people and the planet.

—Mary Snow and Richard Snow

Further Reading

Annan, K. *The Millennium Ecosystem Assessment.* [Initiated by the United Nations Secretary General Kofi Annan in 2000, and supported by governments that are parties to four multilateral environmental conventions.]

Berry, L., J. Olson, and D. Campbell. "Assessing the Extent, Cost and Impact of Land Degradation at the National Level. Overview: Findings and Lessons Learned." *UNCCD Global Mechanism* (2003).

Bridges, E. M., I. D. Hannam, L. R. Oldeman, F. W. T. Penning de Vries, S. J. Scherr, and S. Sombatpanit, eds. *Response to Land Degradation.* Enfield, NH: Science Publishers, 2001.

ICARDA/ICRISAT. *Oasis Challenge Programme Proposal.*

"Millennium Ecosystem Assessment." *Ecosystems and Human Well-being: Desertification Synthesis. World Resources from Land Degradation to Land Health.* Washington, DC: United Nations Convention to Combat Desertification (UNCCD) Brief for Policymakers Institute, 2005. www.millenniumassessment .org/documents/ document.355.aspx.pdf

"Millennium Ecosystem Assessment." In *Current State and Trends Assessment.* Washington, DC: World Resources Institute, 2005, Chapter 22: Dryland Systems. www.millenniumassessment.org/documents/document.291.aspx.pdf

Reynolds, J. F., et al. "Global Desertification: Building a Science for Dryland Development." *Science* 316, no. 11 (2007): 847–851.

Shah, J. *DNA: Daily News and Analysis.* Mumbai, India, December 25, 2009.

Thomas, R. J. "Addressing Land Degradation and Climate Change in Dryland Agroecosystems Through Sustainable Land Management." *Journal of Environmental Monitoring* (in press).

Winslow, M., B. I. Shapiro, R. Thomas, and S. V. R. Shetty. "Desertification, Drought, Poverty and Agriculture: Research Lessons and Opportunities." ICRISAT/ ICARDA, 2004. www.oasisglobal.net/SoAOct04.pdf.

www.ipcc.ch/publications_and_data/ar4/wg1/en/ch7s7-5-1-1.html

www.unccd.int/library/menu.php?newch=l5

Drought

Drought is an extended episode of unusually dry weather due to insufficient precipitation, reduced stream flow, or inadequate glacier or snow melt. It is capable of causing severe consequences, such as water shortages and crop damage, which are among the root causes of food scarcity and famine. In fact, the majority of devastating crop failures that lead to catastrophic famines are the result of drought. The basis of humans' food supply is agriculture, which is directly dependent on the weather. Major famines have occurred throughout the world from the time agriculture began to spread around the globe.

Drought affects the quality and quantity of crop yields and the food supply for domestic animals. The resultant loss of milk and meat products amplifies the negative impact of drought. The duration of the drought, the magnitude of the affected geographic area, and the extent of the moisture deficit contribute to the significance of drought. Drought is a normal, recurring feature of climate in many parts of the world. In the future, as the climate becomes warmer and more variable, an increase in droughts and floods is highly probable. Rising global temperatures cause higher rates of evaporation, which can result in extreme precipitation events in those areas where there is sufficient surface moisture. Conversely, increased evaporation in dry regions with limited surface moisture could lead to an escalation of widespread droughts that are potentially more intense and of longer duration than normal.

Drought Types

At least four different types of drought can be distinguished—meteorological, hydrological, agricultural, and socioeconomic. If the water supply falls well below the expected average when compared to an area's typical amount of precipitation, it is considered a *meteorological drought*, which can last for weeks or even years. Because the climatic conditions that cause precipitation deficits vary substantially from one place to another, meteorological drought should be considered in light of the particular affected region. When an area endures a shortfall in precipitation, snow melt, or glacier melt, such that a pronounced decrease in local lake, reservoir, and river levels occurs or groundwater or stream flow is significantly reduced, the event is considered a *hydrological drought*.

Agricultural drought combines characteristics of both meteorological and hydrological drought, but also incorporates an assessment of the negative impacts on crops and livestock by considering parameters such as precipitation deficits, soil moisture shortages, evapotranspiration rates, and water table levels. When considering agricultural drought, one must take into account the amount of water necessary to sustain a particular crop or animal, as agricultural systems are highly variable.

Socioeconomic drought involves various aspects of meteorological, hydrological, and agricultural droughts and their respective effects on supply and demand. When the supply of a particular economic good is exceeded by the demand for that good due to a lack of adequate water supplies, it is considered a socioeconomic drought.

Many definitions of drought have been proposed. Drought is initiated when there is inadequate precipitation over an extended period of time, leading to a water deficit for certain aspects of society and the environment. Although drought is a natural occurrence, anthropogenic activities often aggravate its negative effects. Recent droughts around the world and the resulting economic and environmental impacts emphasize how all societies are susceptible to drought. The interplay between human actions and the natural world often contribute to the impacts of socioeconomic drought, which include famine and food shortages resulting in malnutrition, disease, and death.

Drought Monitoring

Devising a universally accepted quantitative index for drought is challenging because every geographic area has its own criteria for drought. Nevertheless, several indices are highly regarded and useful for measuring and predicting drought.

The Keetch-Byram Drought Index (KBDI) was developed for the U.S. Department of Agriculture's Forest Service using mathematical models to predict wildfire potential based on soil moisture and other drought conditions. The KBDI ranges from 0, representing normal conditions, to 800, indicating extreme drought. It is based on a soil moisture capacity of 8 inches of water. The higher the KBDI, the greater the potential for wildfire due to the lower moisture content of the vegetation.

The Palmer Drought Index (PDI) assesses soil moisture supply and demand through the use of an algorithm that approximates evapotranspiration and soil recharge rates, using current precipitation and temperature data. With the PDI, which has been shown to be very effective in determining long-term drought, a value of 0 represents normal conditions, –2 signifies moderate drought, –3 indicates severe drought, and –4 represents extreme drought. The National Oceanic and Atmospheric Administration (NOAA) publishes drought maps using the PDI on a weekly basis, and a modified PDI is used to develop a generalized map of abnormally dry crop conditions, which is beneficial for predicting potential food shortages.

In addition to quantitative indices, some institutions have designed their own applications to map droughts:

- The Global Drought Monitor is an online geographic information system (GIS) that monitors the severity of droughts worldwide with daily updates. The GIS aids humanitarian relief organizations by warning of potential food, water, and health problems. The Global Drought Monitor also benefits the general public, government, and industry by improving awareness of drought and its impacts.
- The North America Drought Monitor (NA-DM), a cooperative effort involving drought experts in Canada, Mexico, and the United States, is designed to monitor drought across the North American continent on an ongoing basis. The program was initiated in 2002 and is part of a larger effort to improve the monitoring of climate extremes on the continent.
- The National Drought Mitigation Center (NDMC) helps people and institutions develop and implement measures to reduce societal vulnerability to drought, stressing preparedness and risk management rather than crisis management. Most of the NDMC's services are directed at agencies that are involved in drought and water supply planning. The NDMC supports the U.S. Drought Monitor, which is an online mapping application synthesizing multiple indices and impact scenarios.

Recent Droughts

While many parts of the world have experienced an increase in extreme precipitation events during recent years largely as a result of warmer temperatures and increased evaporation, the areal extent of drought has increased as well. Currently, nearly half of the planet is affected by prolonged drought. Additionally, both the intensity and the duration of droughts are on the rise. Warming temperatures are connected to changes in snowfall trends, storm tracks, and atmospheric circulation patterns, all of which have an influence on moisture availability and can contribute to drought. Since the 1970s, drought has been more common in the tropic and subtropical regions of the world, and it is suspected that climate change is an underlying cause of this trend.

Since the mid-1950s, a widespread drying trend has been observed over large sections of the Northern Hemisphere landmass, including much of Canada, Alaska, North Africa, and Southern Eurasia. The primary cause of this drying trend is a decrease in precipitation over land since the 1950s. In addition, increased rates of surface warming in recent years are likely to have played a role in the warming. Areas considered to be very dry, based on the PDI, have more than doubled since the 1970s, due to a decrease in precipitation related to the El Niño–Southern Oscillation (ENSO), followed by increases in surface temperatures.

After several years of adequate precipitation, Australia experienced an extreme drought during 2002. While Australia has witnessed droughts in the past, daytime temperatures during the 2002 drought were much higher than in previous events. The average annual high temperature for 2002 was 1°F warmer than during the 1994 drought and 2°F warmer than during the drought in 1982. Three years of reduced rainfall caused the long-term drought to continue through 2005, accompanied by record high maximum temperatures over Australia during 2005.

Elsewhere, methods of measurement such as stream flow, soil moisture, lake levels, and precipitation totals reveal that drought conditions for the western United States, southern Canada, and northwest Mexico began in 1999 and lasted into 2004. For the first time on record, the Colorado River experienced 5 successive years of below-average stream flow from 2000 to 2004. The area undergoing moderate to extreme drought in the western United States rose to more than 20 percent in 1999 and remained there until the end of 2004.

Drought Impacts

Numerous environmental impacts may occur as a result of drought. The drying up of lakes, wetlands, and other wildlife habitat can lead to the loss of plant and animal species. Scorched soils can affect biological productivity, which often takes years to recover. Parched vegetation, from grasslands to forests, is more susceptible to wildfires, which jeopardize air and water quality while simultaneously causing damage to the landscape, leading to the loss of biodiversity, and placing both human and wildlife populations at risk.

The use of water has risen dramatically during the past few decades as the global population and economies have continued to grow. Agricultural production and the water necessary for domestic and industrial activities are obviously affected by drought because of the reliance of these sectors on surface water and groundwater supplies. However, natural ecosystems and the people they support are also suffering from a lack of water in many of the world's river basins. An estimated 1 to 2 billion people live in these water-stressed regions. Of course, vulnerability to drought largely depends on where an individual lives. A person's level of education, income, and age also are contributing factors. Environmental degradation, the level of disaster response, and health-sector preparedness are other elements that determine susceptibility to drought.

When farmers are faced with a decrease in crop and livestock production due to drought, those who provide goods and services to farmers are affected as well. These repercussions can take the form of reduced income, decreased tax revenue, higher unemployment rates, and higher costs for food, services, and energy. These impacts often include health issues, water-use conflicts, safety concerns, and economic inequalities. Migration of these effects poses a serious challenge for many developing countries suffering from drought. Poor rural farmers often are forced to move to large urban areas in an effort to find some form of subsistence. Unfortunately, they often end up living in overcrowded shanty towns on the outskirts of metropolitan areas where they are likely to endure food, water, and shelter shortages due to the increased demand on the social infrastructure, which in turn leads to increased poverty and economic instability.

Many individuals living in drought-prone areas have diets that are mineral and vitamin deficient. Thus micronutrient deficiencies are common as a result of inadequate food consumption that can accompany drought. When malnutrition is present, the odds of contracting an infectious disease rise significantly. The lack of rainfall and the resultant stagnation and contamination of lakes, canals, and streams can exacerbate the risk of contracting an infectious disease. Droughts can not only cause physical and economic harm, but also are capable of producing mental problems, including post-traumatic stress, depression, anxiety, and long-term behavioral disorders.

Drought Projections

The European heat wave of 2003 killed more than 35,000 people and typifies the type of prolonged extreme heat event that is anticipated to become more prevalent as the global climate becomes warmer. A warmer climate is forecast to generate wetter winters and drier summers across most of the middle and higher latitudes of the Northern Hemisphere, which will produce a higher potential for summertime drought. Because of the higher rates of evaporation and the enhanced capacity of a warmer atmosphere to hold water vapor, there is likely to be an increase in extreme precipitation events intermingled with dry spells of longer duration. Additionally, in those places where precipitation totals are predicted to decline, the decrease in moisture is expected to be more severe.

A recent study published in the *Proceedings of the National Academy of Sciences* projects that if carbon dioxide concentrations reach 600 parts per million (ppm), several regions of the world, including the Mediterranean, South Africa, and the American Southwest, will face major droughts similar to that experienced in the U.S. Dust Bowl during the 1930s. According to current projections, carbon dioxide levels will reach 550 ppm by the year 2035.

A report by Australia's Bureau of Meteorology predicts that the country's agricultural sector will endure much higher temperatures and increased drought due to climate change, with serious implications for the country's food supply. The report suggests that extremely high temperatures that previously occurred every 25 years or so could become common by the year 2030. Those regions of Australia enduring exceptionally low rainfall are projected to double, with the probability of drought increasing within the next few decades.

In the United States, researchers at the Lamont-Doherty Earth Observatory examined the output of numerous climate models based on predicted temperature, precipitation, and evaporation data in the Southwest. They concluded that the region will endure a dramatic increase in drought conditions by the year 2021, with all of Texas and parts of New Mexico, Arizona, western Mexico, the Yucatan Peninsula, and Central America experiencing extreme drought.

The United Kingdom's Met Office Hadley Centre modeled the occurrence of drought events, based on a global average temperature that is 7° F higher than preindustrial temperatures (1860–1891). The results indicate that land surfaces located between 60 degrees south latitude and 60 degrees north latitude will experience 62 percent more drought in the future and undergo drought conditions 14.5 percent of the time, compared to 10 percent during the preindustrial period. These models also suggest that the Mediterranean will see an increase in drought risk from 10 percent to 20 percent; the Amazon Basin will undergo an increase in drought risk from 10 percent to 25 percent; and Southeast Asia will suffer an increase from 10 percent to 18 percent over the previous preindustrial period.

Projections for the 2090s show a net overall drying trend for the planet. The proportion of land surface in extreme drought is projected to rise from the present 3 percent to 30 percent by the end of the 21st century. Additionally, the number of extreme drought events per 100 years is expected to increase by a factor of 2, and the average length of drought is likely to increase by a factor of 6 by the 2090s. Likewise, higher temperatures and a decline in summer precipitation across southern Europe are expected to cause a reduction in summer soil moisture and lead to increased drought frequency and intensity.

Drought Mitigation

Faced with the threat of more frequent and intense droughts in the future, governments and other institutions are taking measures to mitigate these insidious disasters. In an attempt to aid administrators, businesses, and farmers, the Drought Watch Service was organized by Australia in 1965. To support its efforts, instruments to record diurnal rainfall rates were strategically placed across the country.

Once the data are compiled and analyzed, the results are made available to the public, and those areas with rainfall deficits are published in the *Monthly Drought Review*, allowing officials to make contingency plans to provide relief from drought to their citizens. The Drought Watch Service issues a drought watch when precipitation totals for 3 consecutive months fall within the lowest 10 percent of recorded rainfall. If precipitation rates are within the lowest 5 percent, the area is classified as undergoing a severe rainfall deficiency. For regions already considered arid, the time frame is 6 successive months. The drought watch lasts until above-average rainfall returns for at least a 3-month period.

In the United States, the National Integrated Drought Information System (NIDIS) was established by an act of Congress in 2006. It is supported by diverse organizations such as the U.S. Geological Survey, the U.S. Army Corps of Engineers, the Environmental Protection Agency, the U.S. Department of Energy, the National Science Foundation, and the National Aeronautics and Space Administration. The primary purpose of the NIDIS is to monitor, forecast, and mitigate drought through the use of an early warning system, designed to educate and increase public awareness regarding drought.

Global attempts to advance environmental education and specifically drought are being addressed by the United Nations. Foremost among these efforts are soil conservation and reforestation programs to reduce land degradation in drought-prone areas. The United Nations also is seeking to reduce poverty through the implementation of programs aimed at offering alternative ways of earning a living that can reduce damage to delicate arid areas.

—Richard Snow and Mary Snow

Further Reading

Burke, E. J., S. J. Brown, and N. Christidis. "Modelling the Recent Evolution of Global Drought and Projections for the 21st Century with the Hadley Centre Climate Model." *Journal of Hydrometeorology* 7 (2006): 1113–1125.

Cook, E. R., C. Woodhouse, M. Eakin, D. M. Meko, and D. W. Stahle. "Long-Term Aridity Changes in the Western United States." *Science* 306 (2004): 1015–1018.

Douville H., F. Chauvin, S. Planton, J. F. Royer, D. Salas-Melia, and D. Tyteca. "Sensitivity of the Hydrological Cycle to Increasing Amounts of Greenhouse Gases and Aerosols." *Climate Dynamics* 20 (2002): 45–68.

Parry, M. L., O. F. Canziani, J. P. Palutikof, P. J. van der Linden, and C. E. Hanson, eds. *Contribution of Working Group II to the Fourth Assessment Report of the*

Intergovernmental Panel on Climate Change. New York: Cambridge University Press, 2007.

Piechota, T. C., H. Hidalgo, J. Timilsena, and G. A. Tootle. "The Western Drought: How Bad Is It?" *Eos* 85 (2004): 301–1308.

Solomon, S., D. Qin, M. Manning, Z. Chen, M. Marquis, K. B. Averyt, M. Tignor, and H. L. Miller, eds. *Contribution of Working Group I to the Fourth Assessment Report of the Intergovernmental Panel on Climate Change*. New York: Cambridge University Press, 2007.

Evidence for and Predictions of Future Climate Change

Numerous paleoclimate reconstructions from land and sea provide a detailed record of natural climate variability prior to the Industrial Revolution of 1850 CE. Understanding the full spectrum of past natural climate variability can help humans anticipate how future changes may affect their lives. To gain this understanding, all concerned must first focus on how climate change was documented during the Earth's most recent interglacial period (the Holocene), which was also the time when humans began to cultivate crops and manage food supplies. A second focus must be on accurate predictions of future climate change, based on powerful and complex climate models. Such work is critically important because the unprecedented increase in greenhouse gases and temperature over the past 1.5 centuries is certain to continue, causing stress to agricultural supplies both now and in the future.

The abrupt abandonment of cliff dwellings through the American Southwest at the end of the 13th century is one of the great mysteries on Earth. Why did the ancestral Pueblo people suddenly leave these intricately built homes, like the ones so perfectly preserved in Mesa Verde National Park? Cleverly carved into hidden cliffs, these homes were safe, comfortable for their time, and beautifully decorated with cliff art. Theories abound. One of the explanations that makes the most sense is that climate change drove the cliff people away. Droughts dried up rains that once fed crops above the cliffs. Without the food that had sustained generations of residents, the cliff people vanished. Some sources contend that the world's inhabitants today may be poised on the precipice of a similarly radical change.

Let the reader travel to the future and imagine what some of the ecosystems around the world will look like. The stakes are high. If humans continue on their current path of failing to reduce carbon emissions, they can expect a drastically different Earth in the not so distant future.

First Stop: Tucson, Arizona, in the Year 2100

In the recent past, Tucson was already an arid desert that fostered little life. In fact, each spring, soaking thunderstorms and churning rivers of melting snow triggered an awakening in this unique, dry landscape. Intricate yellow blossoms leapt out from a sea of giant saguaros. The irregular bent arms from these strange, tall cacti created almost comical shapes against a backdrop of orange and purple sunsets. A stunning variety of birds nested in the saguaros, while slithering rattlesnakes and wild hogs called *javelinas* found food and shade on the desert floor. In nearby communities around the American Southwest, crops from chilies to pecans to wine grapes and olives thrived, thanks to irrigation fed by snow melt and relatively mild temperatures.

In 2100, if humans continue on their current path of radically altering the environment, many of the plants and animals that once thrived in this unique desert will have withered away. The snow melt that fed creeks and rivers will be gone. The average summertime high temperature will exceed 100ºF for six months of the year. Without water, hundreds of large farms in southern Arizona will no longer be able to grow the crops that used to thrive here.

Second Stop: Perth, Western Australia, in the Year 2100

The capital city of what is today the fastest-growing state in Australia continued to attract young, affluent immigrants until the latter part of the 21st century. People came for the high-tech jobs that were abundant in Perth and the mining jobs found in the mineral-rich areas to the north. They stayed for the water-based recreation, stunning scenery, near-perfect Mediterranean climate, and access to plentiful fresh produce and world-class wine. Things changed in the latter part of the century, when the average high temperature exceeded 90°F for four consecutive summer months. Increased aridity dried the already severely limited surface water supply, and the high cost to desalinate water and cool homes drove humans to consider alternative places to live.

By 2100, Perth's port city, Fremantle, had finally given up on trying to save the historic central business district from the half-mile inundation of shallow water from the Indian Ocean. Businesses and the historic limestone homes near the shore were condemned. The once-thriving wine industry in the Swan River Valley had long since stopped producing grapes of any quality, and most of the farmers left after years of unpredictable rain and irrigation water.

Geologic Time Morphing into Hyper-fast Global Changes

Consider this startling fact: Atmospheric carbon dioxide (CO_2) levels are currently 25 percent higher than at any time in the past 800,000 years. Yet, within a microsecond in geologic time, just 90 years, these levels are expected to double again. Change that should take millions of years is now occurring at breakneck speed.

What will these skyrocketing carbon levels mean for humans? Will they cause more aridity, higher average temperatures, and more extreme weather patterns? In their new climate, how will humans grow enough food to sustain their growing populations? And without the snowmelt and rains that sustain them, how will humans quench their thirst? To answer these questions, it is critical to understand where humans came from and how swings in climate change have always governed human habitation.

Historically, humans have maintained a fragile balance between food and climate. The ability to grow a wide array of crops in traditionally cold regions of northern Europe and Greenland during Medieval times (i.e., the 10th to 14th centuries) was directly related to the unusually warmer summer temperatures and expanded growing season. When temperatures cooled leading into the Little Ice Age, crops failed and humans were forced to move or alter their diet. The recent history of unprecedented global warming suggests that humans are once again facing a balancing act with regard to global agricultural productivity. Future increases in temperature, coupled with reduced moisture availability, will directly affect how future societies are able to nourish an ever-expanding population.

Whereas some have suggested that moderate increases in global temperatures may provide overall benefits to global agricultural productivity, especially if excess CO_2 acts as a carbon fertilizer, climate scientists know that any warming will negatively affect low-latitude, developing countries. What is more, significant warming will cause serious damage to the global system.

A look back at the record of natural climate variability over the past 12,000 years helps to better elucidate humans' future dilemmas. These relatively recent years, in geologic terms, are absolutely relevant to the current debate, because it was during these times that humans began to cultivate plants and establish a more agrarian society. Such a perspective is essential to understand how current climatic conditions and those predicted for the future may influence humans' ability to grow and distribute food on a global scale. Humans also must examine and consider future climate predictions up to the year 2100, as increased concentrations of greenhouse gases drive warmer and warmer temperatures farther into the Earth's atmosphere and add energy to the global atmospheric and hydrologic circulation.

Some Perspective from the Distant Past

Reconstructing climate variability before humans dominated the biosphere gives climate scientists insights into the world's current situation. Over the past 2.6 million years (the Quaternary period), the Earth's climate system has oscillated systematically between glacial modes, generally lasting 100,000 years, and interglacial modes, lasting on average 12,000 years. Thanks to numerous, high-resolution ice core records from Greenland, Antarctica, and elsewhere, climate scientists know a great deal about climate variability during these most recent glacial and interglacial periods. During a glaciation, the Earth's climate can only be categorized as cold, dry, and highly variable. Some of the largest known shifts in global temperature occurred at the end of the last glacial period, as global temperatures spiked by 13°F in only 20 years. To put this change into perspective, since the Industrial Revolution some 160 years ago, Earth's temperature has increased by only 1.1°F, yet the rise has caused dramatic changes to numerous ecosystems.

In comparison to the large swings in global temperature that punctuated the end of the last glaciation, the Holocene period has been a time of relative climatic stability. The large continental ice sheets, which greatly influenced the unstable glacial climate system, mostly melted away 10,000 years ago. Their departure left behind a calmer world that was full of changes, but on a far smaller temperature scale compared to the past glacial fluctuations.

Climate scientists have a good idea of how the Earth's climate varied over the past 12,000 years, based on numerous high-resolution temperature and precipitation records from both land and sea. Proxy records such as tree rings, lake sediment, and ice core data document that the Holocene period was constantly beset by subtle global changes in temperature, precipitation, monsoon dynamics, and the El Niño–Southern Oscillation that persisted for several centuries. Most notably, the Northern Hemisphere experienced a sustained cold and dry event approximately 8,200 years ago that lasted for 400 to 600 years. A flood of fresh water and glacial ice into the North Atlantic Ocean caused this climatic cooling, which now is known as the "8.2 ka" event. Should a cold climate anomaly of this magnitude occur today, the consequences to agricultural productivity would be substantial, especially in higher-latitude regions that now have a short summer growing season.

Most of the climate variability during the Holocene can be attributed to small changes in the Earth's orbital forcings (precession, obliquity, and eccentricity), solar variability, land ice coverage, and changes in oceanic circulation. For example, the Early Holocene climate transformation, from 12,000 to 9,000 years ago, was a time of rapidly increasing temperatures that caused glaciers to melt and release massive quantities of fresh water into the oceans. Humans would have faced a variety of challenging climate shifts during this time, including sustained drought, as the Earth transitioned to a more stable mode. The relatively warm Hypsithermal interval, from 9,000 to 5,000 years ago, encompassed a time of maximum solar radiation in the Northern Hemisphere. During this time, summer temperatures exceeded today's record-high values by more than 3ºF to 6ºF in certain high-latitude regions of the Northern Hemisphere. Had humans been farming in these regions, the growing season certainly would have been expanded by weeks, allowing for cultivation of crops that currently require a more temperate climate. The most recent "Neoglacial cooling," from 5,000 years ago until the Industrial Revolution, had been a time of steadily decreasing temperatures as the Earth's orbital cycles gradually shift toward glacial conditions.

Human Migration Triggered by Chaotic Weather

For more than 10 millennia, humans have cultivated grains, raised livestock, and created large, complex societies, in part based on easy access to potable water and the ability to grow and manage food supplies. In Asia first, and later in the eastern Mediterranean, humans discovered the skills necessary to plant and maintain crops such as rice, millet, wheat, and barley. Much later in what is present-day Central America, humans grew corn, tomatoes, and a variety of tubers, just to name a few crops. During all of this time, the Earth experienced significant variations in temperature and precipitation, but it remained generally favorable to widespread agriculture.

Climate scientists know few details of how early to mid-Holocene climate oscillations directly influenced early human societies. From 9,000 to 6,000 years ago, summer temperatures in the high-latitude Northern Hemisphere were 3°F to 6°F warmer than today because of the 6 to 8 percent increase in solar radiation. This warming spawned regular, intense, and long-lasting monsoonal rains that persisted in Africa, India, Asia, and the southwestern United States. One can imagine that regular rainfall from 9,000 to 6,000 years ago allowed some early farmers to generally succeed with summertime crops. However, it is presumed that the ever-present multi-decadal droughts caused some agrarian societies to collapse or at least migrate to more fertile regions.

Climate scientists now know much more about how late Holocene variability affected humans. Three of the largest and most complex non-European societies—the Maya, Khmer, and Anasazi—appear to have essentially vanished after three different multi-decadal droughts.

Today, humans face an unexpected situation. For the past 6,000 years, global surface air temperatures have been gradually and consistently cooling, as the Earth inevitably shifts toward the beginning of the next glaciation. Things began to change in the middle of the 19th century when temperatures actually began to rise. This warming trend continues today. Paleoclimatic evidence suggests that global temperatures currently are warmer than at any time since humans began to grow food and that the rate of temperature increase is unprecedented during any interglacial period over the past 2.6 million years.

Climate Change and Mass Migration: The Mayans

Elaborate ruins in Tulum cling to cliffs overlooking the turquoise Caribbean Sea along Mexico's Yucatan Peninsula. Just a couple of hours away, deep in the jungle, Chichen Itza is home to an even more remarkable set of ruins. Very suddenly, a little more than a thousand years ago, the Mayan people, who once lived throughout the Yucatan, abruptly abandoned their low-elevation homes. For almost 2,000 years, the Mayan civilization had thrived there, developing a highly advanced and organized culture. No one knows exactly what happened to drive them away, but one of the most convincing theories is that climate change forced the Mayans to flee.

A series of devastating multidecade droughts may have forced a mass migration of the Mayans to wetter and higher elevations. Subtle changes in nearby ocean currents may have caused dramatic climate changes on land. The people living in the affected areas may have faced a stark choice: leave or die. While these droughts were regional events that likely were interspersed with wet years, they were substantial enough to cause widespread crop loss, famine, disease, and cultural imbalances. Climate change may have put an end to one of the most advanced societies in the New World.

If humans could listen to the ghosts of Tulum and Chichen Itza now, they might warn of highly relevant climate catastrophes. The Mayans were not alone in their fate. Climate change may have triggered other mysterious mass migrations over the last 15,000 years. The voices of these people, echoing among the homes they once occupied, may provide cautionary tales for humans today.

—Michael W. Kerwin

To all but a few climate change skeptics, humans are to blame for this most recent change. Since the mid-19th century, the burning of fossil fuels has increased the Earth's concentration of greenhouse gases. In 2007, the Intergovernmental Panel on Climate Change listed 18 different greenhouse gases, but noted that 3 are the most important in terms of future climate change: carbon dioxide, methane, and nitrous oxide. Today, the concentration of atmospheric carbon dioxide is rising by more than 1 part per million per year. Coincident with this rise in greenhouse gases, the Earth's temperature at the surface has increased by 1.1°F, at a faster rate than any time in the past 12,000 years. Evidence suggests that this warming trend is gaining speed. The year 2010 tied 2005 for the warmest year on record, and all of the past 34 years have exceeded the 20th-century average for temperature. Winter minimum temperatures also have increased over the past 160 years. In the central Rocky Mountains of the United States, this increase in winter temperatures is one of the reasons that pine bark beetles have killed as many as 92 percent of certain tree species.

An Unrecognizable World

Predicting the magnitude and spatial extent of future trace gas–induced climate changes can be accomplished only with state-of-the-art, global earth system models that can simulate the complex interactions among the oceans, atmosphere, and biosphere. Accurate predictions of future climates at local and regional scales are essential, as citizens and policymakers prepare for drastic changes in the Earth's climate system.

In some ways, predicting future changes in Earth's climate is a little like predicting the weather. Both types of models attempt to connect changes that may take place in the upper atmosphere with the rest of Earth's climate system. In reality, millions of data points interact with one another in an effort to simulate conditions over land, in the ocean, and in the atmosphere. The big difference is the time scale involved. Whereas most weather models predict changes for the coming week, predictive climate models tend to focus on the next 50 to 100 years.

But how do climate scientists know that these models work? With weather models, one can simply wait a week and see how accurate the forecast was. With climate models, climate scientists do not have that luxury. To gain confidence in the results of climate models, many different models are compared to one another to evaluate their performance. These models also are asked to simulate current and past conditions on Earth that are known, with the results being used for model validation.

The consensus results of dozens of the most powerful climate models are sobering. By the year 2100, for example, coastal regions will experience an increase in sea level, ranging from 20 to 40 inches and resulting in an inundation of salt water up to more than one-half mile from the shore. Away from the coast, cities—particularly those in arid to semi-arid interior regions—can expect summer temperatures as much as 8°F warmer on average than today's record-high levels. Temperate mountain ranges below 15,000 feet will no longer contain permanent snow, the primary source of water for numerous large cities adjacent to these mountains.

The direct effects of these changes are obvious. With an estimated 600 million people who currently live within 6 miles or so of the ocean, sea level rise is certain to cause unprecedented negative impacts. Shorelines around the world can expect submergence, flooding, coastal erosion, ecosystem changes, and increased salination, all of which will cause mass migrations on a scale never seen before. In the United States alone, the most likely rise in sea level by 2100 is expected to flood at least 9 percent of the land in 180 of the largest coastal cities.

Unfortunately, these direct impacts may be only the beginning of the challenges presented to future leaders. Indirect effects such as the need for additional air conditioning in large, arid cities will stress the world's already limited supplies of petroleum, coal, and natural gas. The expansion of tropical ecosystems and conversion of dry coastal areas to wetlands will exacerbate the occurrence and spread of infectious diseases. Agriculture may no longer be feasible in some regions, as freshwater supplies dwindle in response to higher evaporative demand. In fact, among all of the potential consequences of global warming, agriculture may be the most important, as a geographically and climatically different world attempts to feed a global population that may exceed 10 billion people.

Can Predictive Models Be Trusted?

Predicting future climate variability is among the most complicated tasks for climate scientists. Unfortunately, the relationship between greenhouse gases and surface temperature is not linear. Instead, any change in climate forcing, including an increase in atmospheric carbon dioxide, will be enhanced or dampened by numerous other forcings and feedbacks that together combine to create the climate of a certain region. For example, an increase in atmospheric carbon dioxide will cause the upper troposphere to absorb more outgoing long-wave radiation, resulting in a fraction of a degree warming in one level of the Earth's atmosphere. To predict how this atmospheric perturbation might affect the surface, however, one must consider a huge array of interconnected systems, including the temperature and circulation of the oceans and atmosphere, the type and amount of precipitation that reaches the Earth's surface, the type and coverage of vegetation and the way that this vegetation cycles nutrients such as nitrogen and carbon, the amount and thickness of snow and ice on the Earth's surface, and, of course, humans' alterations of the surface of the Earth. For example, the construction of large cities in arid regions has caused significant local heating that is unrelated to the trace gas–induced global warming.

Even with rigorous testing and validation, relying upon climate models to precisely predict the climate 90 years from now is a venture fraught with uncertainty. Models must make assumptions about Earth systems that are difficult to predict. The concentration of CO_2 in the atmosphere is one of the most important variables in any predictive climate model, yet it remains unknown how future societies will limit carbon emissions or whether such regulations will have any impact. Developed countries, for example, are likely to soon adopt restrictions on CO_2 emissions, which may alter the overall concentration of atmospheric CO_2. At the same time, how developing countries moderate their emissions of carbon and how the Earth's land and oceans absorb or release carbon into the atmosphere are equally important factors.

Likewise, the effects of certain negative or unknown climate feedbacks can further complicate predicting the future. Clouds, for example, have always been difficult to deal with in climate models. Do excess clouds block more incoming solar radiation or absorb more outgoing long-wave radiation? Another complicating factor is tropospheric ozone, which is a by-product of fossil fuel burning. Its composition and duration pose a challenge for climate modelers because the gas is short-lived, yet has a large impact in the atmosphere. In addition, emissions of methane, carbon monoxide, and nitrogen oxide strongly affect the formation of ozone.

What Can Be Expected?

Even with global cooperation and serious carbon mitigation, the outlook for Earth's climate is sobering. The concentration of atmospheric CO_2, which today hovers around 380 parts per million, is expected to at least double by the year 2100. This change alone will stress the Earth's climate system in unpredictable ways because such high levels have not been seen for at least 2.6 million years and probably much longer than that.

Surface air temperatures are expected to increase globally by 2ºF to 4°F by 2100, with average summer temperature increasing by 7ºF or 8°F in semi-arid continental interior locations. These changes will ensure that all land areas experience fewer cold days and nights by 2100. Instead, all land areas will experience warmer days and nights and can expect more frequent and sustained heat waves in the future.

Sea level in 2100 will be higher by 20 to 40 inches. This change will directly affect 10 percent of the human population—namely, those who now live within 6 miles of the coast. Snow cover in the Northern Hemisphere will continue to decrease, and mountainous areas below 15,000 feet can expect to have no permanent snow by the year 2100.

The impact of changes in precipitation is harder to gauge. Climate models show only insignificant changes in precipitation over land areas, with some areas getting moderately wetter while others become drier. Nearly all the models predict that heavy precipitation events (both snow and rain) will increase in the future. Likewise, the frequency and severity of drought will increase, as will the intensity of tropical cyclones.

The future of the world is in the hands of those who live on Earth today. Without action, humans will invite chaos. If humans begin to intervene, they may be able to control some climate change, thereby influencing their ability to feed themselves and withstand radical weather events. Even with small changes, however, many of Earth's systems will never be the same. For example, an atmospheric CO_2 concentration of 750 parts per million by 2100 will commit the Earth to 8 to 10 feet of sea level rise and temperature increases over the next several centuries, even if all future emissions were halted.

—Michael W. Kerwin

Further Reading

Cline, W. R. *Global Warming and Agriculture: Impact Estimates by Country.* Washington, DC: Center for Global Development and Peterson Institute for International Economics, 2007.

Easterling, W., et al. "Food, Fiber, and Forest Products." In *Climate Change 2007: Impacts, Adaptation, and Vulnerability*, edited by M. L. Parry et al. Cambridge, UK: Cambridge University Press, 2007, 273–313.

Fischlin, A., et al. "Ecosystems, Their Properties, Goods and Services." In *Climate Change 2007: Impacts, Adaptation, and Vulnerability*, edited by M. L. Parry et al. Cambridge, UK: Cambridge University Press, 2007, 211–272.

Matthews, H. D., and K. Caldeira. "Stabilizing Climate Requires Near-Zero Emissions." *Geophysical Research Letters* 35(2008): L04705, 10.1029/2007 GL032388.

Meehl, G. A., et al. "Global Climate Projections." In *Climate Change 2007: The Physical Science Basis. Contribution of Working Group I to the Fourth Assessment Report of the Intergovernmental Panel on Climate Change*, edited by S. Solomon et al. Cambridge, UK: Cambridge University Press, 2007, 747–846.

Overpeck, J. T., B. L. Otto-Bliesner, G. H. Miller, D. R. Muhs, R. B. Alley, and J. T. Kiehl. "Paleoclimatic Evidence for Future Ice-Sheet Instability and Rapid Sea-Level Rise." *Science* 311 (2006): 1747–1750.

Plattner, G. K., et al. "Long-Term Climate Commitments Projected with Climate Carbon Cycle Models." *Journal of Climate* 21 (2008): 2721–2751.

Schneider, S. H., et al. "Assessing Key Vulnerabilities and the Risk from Climate Change." In *Climate Change 2007: Impacts, Adaptation and Vulnerability. Contribution of Working Group II to the Fourth Assessment Report of the Intergovernmental Panel on Climate Change*, edited by M. L. Parry et al. Cambridge, UK: Cambridge University Press, 2007, 779–810.

Solomon, S., G. K. Plattner, R. Knutti, and P. Friedlingstein. "Irreversible Climate Change Due to Carbon Dioxide Emissions." *Proceedings of the National Academy of Science USA* 106 (2009): 1704–1709.

Weiss, J. L., J. T. O. Overpeck, and B. Strauss. "Implications of Recent Sea Level Rise Science for Low-Elevation Areas in Coastal Cities of the Conterminous U.S.A." *Climatic Change*. doi: 10.1007/s10584-011-0024-x.

Famine

Demands for food by humans are daily, while the biological production of food is seasonal. The seasonality of food production and the world's unequal distribution of arable land place a premium on efficient food storage facilities, food preservation techniques, and food distribution systems. In traditional cultures, the most serious food shortages occur in the pre-harvest season, when substantial labor input is required to till the soil and plant crops and when stored food from the last harvest runs low. At that time there is little food available or the food that is available is sold at an inflated price. Pre-harvest hunger occurs most frequently among the poor. According to ancient chroniclers and modern records, every society has people who suffer from hunger. Hunger in any form damages the moral and economic base of society. Acute or chronic hunger produces a segment of a nation's population that is physically and psychologically damaged and creates a core of resentful people. Prolonged hunger can set an environment for social unrest and revolution; prolonged mass starvation and famine lead to submission, resignation, and death.

Defining Famine: 1798–1980

There is a great difference between hunger, undernutrition, malnutrition, and famine. Even so, "famine" is a term applied to many occurrences and conveys thoughts and impressions not related to massive deaths by starvation. Famine, as a descriptive term for a devastating social hazard, carries emotional overtones. It is improperly applied many times to describe a brief food shortage or lack of a specific food item. Predictions of famine somewhere in the world are voiced so frequently that they have a negative impact for relief of a real famine (see Figure 11). To some degree, individuals cease to question "what is a famine" and accept famine as a natural hazard, equivalent to a violent storm, a massive flood, or a devastating earthquake.

Thomas Malthus, a British political economist, concluded in 1798 that famine was a natural event by which population was held in balance with food supply. To him, famine was a strong and constantly operating check on population—a great restrictive law. Malthus's philosophy and definition of famine became the basis for government action in times of famine for nearly a century and a half.

In the mid-1960s and 1970s, Malthus's definition of famine and his philosophy were seriously challenged by many. Agronomist William Paddock and Foreign Service Officer Paul Paddock contended that when a person collapses from lack of food, the dividing line between hunger and starvation is passed; when entire families or communities die because of intense starvation, then a famine is occurring. Many researchers and writers concluded that famines were synonymous with crop failures produced by abnormal climatic conditions or severe weather variations, including Miloslav Rechcigl, Chief of Research and Institutional Grants Division, AID. Some suggested that famines were periods of extreme want which led, at times, to the reemergence of untamed instincts. Those who believed this theory included decision makers and church officials, such as Father Noel Drogat, a spokesperson for the Society of Jesuits. K. Malin, a Soviet Marxist, defined famine as an "inherent by-product of capitalism." Cecil Woodham-Smith, an award-winning biographer, contended that famine was nothing more than a great hunger. These authors wrote in a way to imply that famine was a term understood by all and needed no standard definition.

Figure 11. Dando scale of famine severity

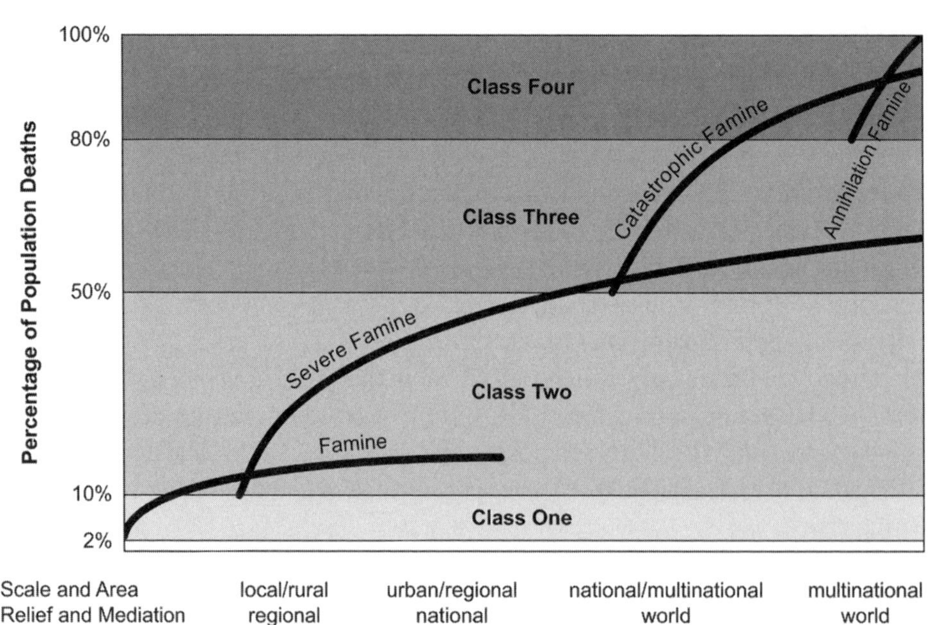

Source: William A. Dando.

Josue de Castro, Chairman, Executive Council, Food and Agriculture Organization of the UN, defines famine as epidemic hunger affecting great masses, while Donald Hughes, a specialist in economic development, describes famine as a vicious cycle of hunger and disease created by poverty. Biologists Maarten J. Chrispeels and David Sadava write that famines are periodic hungers caused by natural disasters. K. Chang, a specialist on Chinese culture, suggests that famine is a combination of undernourishment and drought. B. H. Farmer, of Cambridge University in England, writes that famine now means less than one or perhaps two square meals per day. Others have defined famine as a crop failure or a period of want.

Don and Patricia Brothwell, archaeologists and linguists, explain famine as "severe malnutrition." They conclude that populations close to or exceeding the carrying capacity of a territory or place are most susceptible to famine. At Ohio State University, F. Deatherage equates famine with chronic food shortages. He observes that insufficient food and malnutrition are a way of life for many people. The Deputy Director of FAO in Rome's Nutrition Division, B. M. Nicol, professes that famine and starvation are synonymous. At a symposium dealing with nutrition and relief operations in times of disaster, William H. Foege, from the U.S. Department of Health, Education, and Welfare, declared that famine was an epidemic of malnutrition. A member of the Department of Agriculture at the University of Oxford in England, G. Masefield, implies in his writings that famine is a prolonged shortage of food resulting in increased human death rates. Standard dictionaries define famine as a scarcity of food, producing hunger and starvation, a lack of food in a place, or a time of starving. These published statements or definitions are interesting and do reflect concern for famines, but are nevertheless incomplete.

G. Masefield, although using the FAO famine definition in his work, has developed a very concise famine definition. He states that famine is a wide and prolonged shortage of food resting in an increased human death rate. He also notes that a famine does not become a famine if it is relieved in time. Mirian E. Lowenberg and her colleagues contend that famine is acute, extreme shortage of food within a region. Death becomes part of plant geneticist Jack R. Harland's famine definition, consisting of a crop failure that results in starvation and death. Senator Joseph D. Tydings of Maryland links famine with population and believes that nature uses famine as a culling tool.

Biologist Paul E. Ehrlich and physicist Harrison Brown conclude that overpopulation and famine are components of a place's population–food balance and suggest that both terms have nearly the same meaning. Jean Mayer, a nutritionist, expands his definition of famine to include place and time; he reports that a famine occurs in a definable area and has a finite duration. In a study of world problems, Brian

Ferris and Peter Toyne define famine as "a temporary, but severe, local shortage of food." They believe that famines have a definite spatial and temporal aspect. Richard G. Robbins, Jr., a historian, makes a strong distinction between crop failure and famine. He notes that a nation or region can experience a series of bad harvests and yet not experience a famine. Conversely, famine may occur in times of good harvests when human decisions deprive areas and people of needed food. In a classic article published in the *International Encyclopedia of the Social Science* in 1968, M. K. Bennett defines famine as a "shortage of total food so extreme and protracted as to result in widespread persisting hunger, notable emaciation in many of the affected population, and a considerable elevation of community death rate attributable, at least in part, to deaths from starvation."

In 1976, the editor of this book, a geographer at the University of North Dakota at that time, wrote a definition of famine based on an analysis of thousands of famines in a time span of nearly 6,000 years. He defined famine as "a protracted total shortage of food in a restricted geographical area, causing widespread disease and death from starvation." This famine definition or variations of this definition has been used by world relief agencies, governmental agencies, private foundations, and religious hunger/famine relief agencies for decades.

Defining Famines: 1980–2010

In the 1980s, public interest in famine issues waned. U.S. governmental agencies were pressed with a myriad of social and Cold War issues, a conservative approach to funding research on hunger and famine diminished the volume of literature on this topic, and a different philosophy underlying hunger and relief activities led to a reexamination of famine definitions in the 1980s. A group of economists began writing and publishing on the topic of the impact of poverty in creation of social ills. Amartya Sen focused upon starvation, market mechanisms, and entitlements. John Osgood wrote on the challenge of famine and described famine as "the regional failure of food production or distribution systems, leading to sharply increased mortality due to starvation and associated diseases." He contended that famine occurrence had an economic basis, and at times was created by economic imbalances that affected the entire economic system of the famine event.

Geographer Brian Murton, in his essay on famine, under the heading of "Definitions and Dimensions of Famine," states that hunger is not a single uniform experience. Rather, it is a range of vulnerability related to the human food need in the life cycle, household food insecurity, food supply shortfalls, and massive hunger crises affecting large numbers of people in a definable region, resulting

Famine Intensity and Magnitude Scales

The word "famine" has many meanings. It is an emotional term with cultural implications and political connotations. The ambiguity about whether a famine is occurring has prompted renewed interest in qualitatively and quantitatively defining this term. Disputes have taken place between those sources who would define a famine as an event with many people dying of starvation versus a process that can be traced on a famine continuum, from a food shortage to widespread death.

The Indian Famine Codes, devised by the colonial British in the 1880s, defined three levels of food deprivation: (1) near scarcity, (2) scarcity, and (3) famine. Famine was defined by a 140 percent rise in food prices, movement of hungry people in search of food, widespread death from starvation, and death from diseases. The Punjab Food Code stated that death is the sole criterion for declaring a famine.

In Africa, a Kenyan famine early warning system identified three levels of food crisis: (1) alarm, (2) alert, and (3) emergency. Famine indicators were rainfall levels, price of foodstuffs, physical condition of livestock, rangeland conditions, and numbers enrolled on work-for-food projects. The World Food Program (WFP) and the Food Security Assessment Unit (FSAU) devised a system for northeast Africa with four levels: (1) non-alert, (2) alert, (3) livelihood crisis, and (4) humanitarian emergency.

To many, famine intensity is a social problem. Those suffering from food stress normally try to cope through market structures, such as selling possessions for food and relying on family and community support structures. It is only when social structures collapse and large numbers of people are suffering from undernutrition, malnutrition, and starvation that there is a famine.

—William A. Dando

in a substantial increase in mortality. Murton notes that the large number of deaths is the manifestation of a famine in most people's minds.

S. Millman, in his article titled "Hunger in the 1980s: Backdrop for Policy in the 1990s," contends that hunger and famine are three phases in a continuum: food shortage, food poverty, and food deprivation (see Figure 12). A famine occurs when total food supplies within a region are insufficient to meet the needs of a population. Food poverty occurs when there are no food shortages; food deprivation, when there is no food poverty; and food shortages, when there is no famine.

In the 1990s, an evolving perspective cast famine not as a discrete event but rather as merely the tip of the iceberg of underlying social, economic, and political decisions or processes. Specifically, B. Curry questioned whether famine is a truly discrete event or an extension of chronic hunger. Activist and moralist Massod Hyder, writing in a medical handbook, defines famine as "a serious shortage of food affecting a region or large groups of people over an extended period of time

Figure 12. The modern starvation cycle

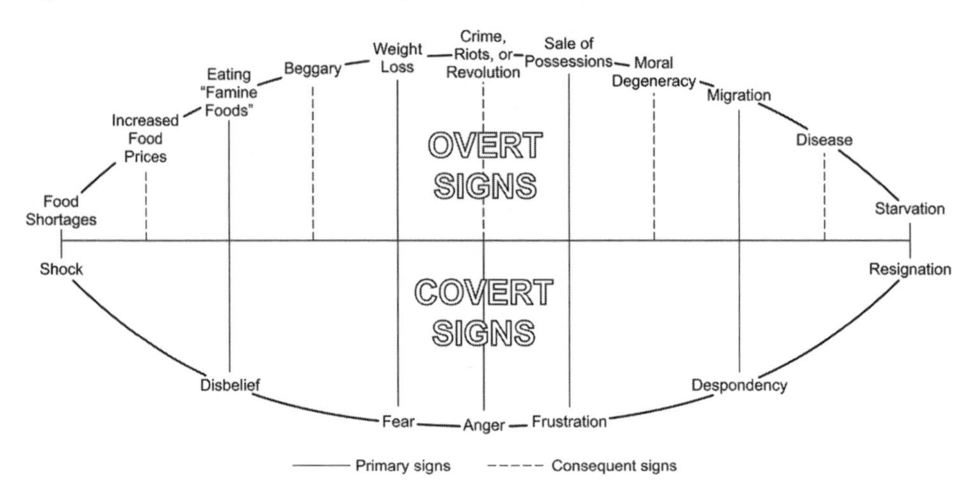

Source: William A. Dando.

resulting in hunger, suffering, and marked increase in death rates." Also, Hyder stresses that famine events are viewed by most individuals as an acute period of non-availability of food, rather than as the absence of timely action to remedy this lack. Citing the Bible, and specifically Joseph's story of the seven-year famine, Hyder remarks that famines—at least most of them—have been preventable. As in Joseph's time, famines are not caused simply by lack of food, and the paradox of famine in the modern world is that it occurs in the midst of plenty.

In their article on famine, starvation, and fasting, E. M. Widdowson and A. Ashworth lament that despite human science and technological advancements at the beginning of the 21st century, famine and starvation abound. They define famine as "a prolonged scarcity of food causing many in a district or country to die from starvation and/or disease." These authors distinguish famine from hunger by citing the coping strategies used to deal with each scenario. Also, they associate famine with ineffective political structures, armed conflicts, environmental degradation, and ethnic or religious unrest. Widdowson and Ashworth strongly stress that a major thrust in famine prevention and mitigation should be eliminating poverty; they contend that food security has become the money "cry" word and that humanitarian assistance has evolved into an industry.

K. P. West states that famine has affected humankind, shaping history, influencing world demography, and influencing human lives from antiquity. He defines famine as widespread inaccessibility of food leading to mass numbers of starved individuals. West stresses inaccessibility to food is not equivalent to non-availability of food within a famine region. He observes that all definitions of famine include a statement of widespread inaccessibility to food leading to mass deaths by starvation. West explains that many tend to view famine as a continuum, on which famine is the end product of a process that begins with persistent hunger; it is catastrophic, distinct, and a human tragedy of unparalleled proportion. West endorses Amartya Sen's contention that starvation is a matter of some people not having enough to eat, and not a matter of there being not enough to eat.

In a snapshot of a page that appeared on the Global Policy Forum's Alert Net on April 29, 2010, Alex Whiting questions, "Why is famine so hard to define?" He notes that the lack of an agreement on an internationally acceptable famine definition makes it difficult for governments, donors, and not-for-profit organizations to know whether a real famine actually is occurring.

Famine Characteristics, Types, and Historic Regions

Famines differ and have identifiable characteristics. All famines are unique, varying in their duration, size, intensity, and death count. Famines are not restricted to a particular climatic region, a specific cultural or racial area, or an identifiable economic group in a place. They can be classified by degree: (1) famine; (2) severe famine; and (3) catastrophic famine. They can also be mapped and scaled: (1) local; (2) urban; (3) regional or zonal; (4) national; and (5) international. Finally, they can be classified by cause: (1) physical; (2) transportation; (3) cultural; (4) political; (5) overpopulation in marginal agricultural land; and (6) contrived, planned, or unplanned urban.

1. *Physical famines* occur in regions or zones where the physical environment is not conducive for intensive or extensive forms of sedentary agriculture. Humans, recognizing the limitations placed upon agriculture by nature and utilizing the technology available at the time, have developed agrotechniques that enable them to temper or adjust to natural hazards in all but their extreme forms. This type of famine occurred in Egypt in the early years of its civilization (about 4000 to 500 BCE) and in ancient Israel (about 2000 to 1000 BCE).
2. *Transportation famines* take place in highly urbanized, commercial, or industrial food-deficit regions dependent upon distant food sources and supplies delivered by a well-developed transportation system. Famines occur when the transportation of food to the place, area, or zone is severely disrupted. Such famines took place in the Roman Empire from 501 BCE to CE 500.
3. *Cultural famines* occur in both food-surplus and food-deficit regions, induced by archaic social systems, cultural prejudices, or cultural discrimination.

Famines classified as cultural famines have taken place throughout the world but were very common in Western Europe, from CE 500 to 1500.

4. *Political famines* take place in areas that are normally self-sufficient in basic foodstuffs but where regional politics or political systems determine food production, food distribution, and food availability. Political famines occurred in places and countries primarily located in Eastern Europe from about CE 1501 to 1700; in the Soviet Union between 1918 and 1947; and in China from 1959 to 1961.
5. *Overpopulation famines* occur in drought-prone and flood-prone, overpopulated, marginal agricultural regions or zones where farmers or herders have few farm implements and minimal food-producing herd animals, and they plant primarily the same crop every year. This type of famine took place most often in Asia during the CE 1701 to 1970 period.
6. *Urban famines* occur in large cities or metropolitan areas when a severe food shortage reduces the amount of food supplies reaching the urban markets and the available food for human consumption is severely reduced or nonexistent. Uncontrolled, mismanaged, or unmanaged urbanization can lead to mass starvation and. at times, famine that claims a tremendous number of lives. Urban famines are caused by transportation disruptions or blockades, war and military policies, revolutions, strikes, or international, national, or regional food market-based decisions. During World War II, for example, the German blockade of Leningrad in the Soviet Union and the German military decision to punish the Dutch led to urban famines that claimed the lives of millions.

Using a database of more than 6,000 famines occurring in approximately 6,000 years (4000 BCE to 2000 CE) and utilizing computer mapping software to regionalize famine over time (temporally) and space (the world) where a high frequency of a single famine type was cited, six historical famine regions may be identified, along with one region expected to experience numerous famines in the future (see Figure 13):

1. Famine Region I—Northeast Africa and the Middle East, between 4000 and 500 BCE.
2. Famine Region II—Throughout the Roman Empire, including portions of Europe, the Middle East, and North Africa, between 501 BCE and 500 CE.
3. Famine Region III—Western Europe, including France, Germany, England, Ireland, Scotland, Wales, Denmark, and Sweden, between 501 and 1500 CE.
4. Famine Region IV—Eastern Europe (specifically Poland, Hungary, Livonia, and Western Russia), from 1501 to 1700 CE.

Figure 13. World famine regions, 4000 BCE–2001 CE

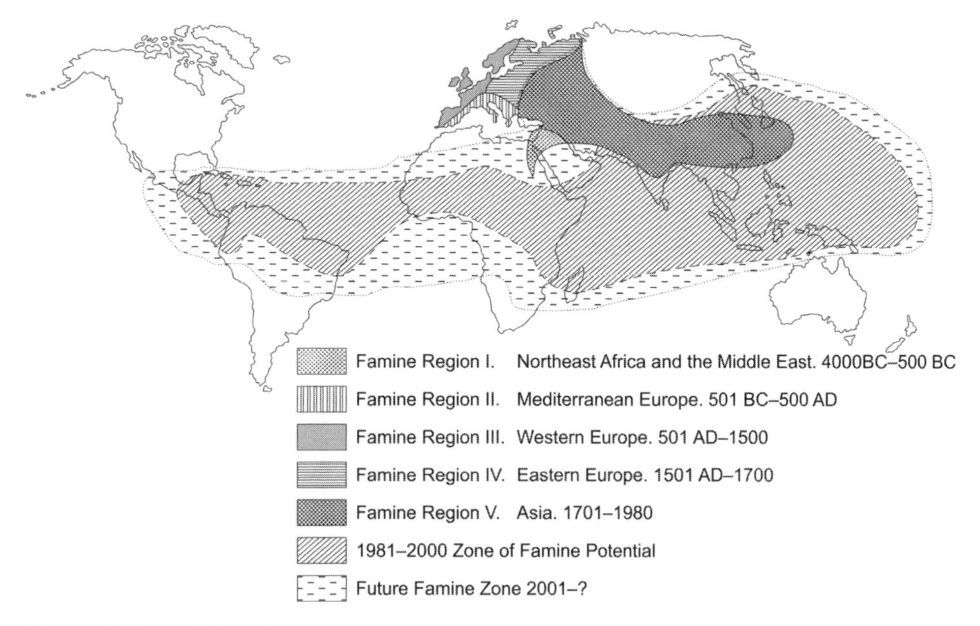

Source: William A. Dando.

5. Famine Region V—Asia, focused upon India, China, and Russia/Soviet Union, from 1701 to 1980 CE.
6. Famine Region VI (zone of high potential famine occurrence)—Not conclusively defined but focused on a wide globe-girdling belt that spans the world, primarily along the equator, including many developing nations of South America, Africa, and Asia, from 1981 to 2000 CE.
7. Future Famine Zone—An expanded Famine Region VI that includes portions of Eastern and South Asia, Africa, South America, and numerous cities of megamillion population within a globe-girdling belt around the equator, where political turmoil and authoritarian regimes may prevent humanitarian relief action and create famines.

Famines have claimed the lives of humans for more than 4,000 years. They now can be averted whenever national governments are sufficiently concerned about the welfare of their most vulnerable citizens and when international action can be mobilized to avert a horrendous food-deficit disaster. From a life-taking calamity that destroyed the hopes and dreams of millions of the world's citizens, famine in the 21st century has become a preventable human cultural hazard.

—William A. Dando

Further Reading

Curry, B. "Is Famine a Discrete Event?" *Disasters* 16 (1992): 138–144.

Dando, William A. "Biblical Famines, 1850 B.C.–A.D. 46: Insights for Modern Mankind." *Ecology of Food and Nutrition* 13 (1983): 231–249.

Dando, William A. *The Geography of Famine*. New York: John Wiley & Sons, 1980, 57–67.

Dando, William A. "Famine: Problems in Definition." *Bulletin of the Association of North Dakota Geographers* XXIX (1979): 41–48.

Dando, William A., and Caroline Z. Dando. *A Reference Guide to World Hunger.* Hillside, NJ: Enslow Publishers, 1991, 45–53.

Field, John Osgood. *The Challenge of Famine*. West Hartford, CT: Kumarian Press, 1993.

Hyder, Massod. "Famine." In *The Oxford Illustrated Companion to Medicine*, edited by Stephen Lock, John M. Last, and George Dunea. Oxford, UK: Oxford University Press, 2001, 1301–1303.

Millman, S. "Hunger in the 1980s: Backdrop for Policy in the 1990s." *Food Policy* 15 (1990): 277–285.

Murton, Brian. "Famine." In *The Cambridge World History of Food*, edited by Kenneth K. Kiple and Kriemhild Conu Ornelas. Cambridge, UK: Cambridge University Press, 2000, 1411–1427.

Sen, Amartya. "Starvation and Exchange Entitlements: A General Approach and Its Application to the Great Bengal Famine." *Cambridge Journal of Economics* 1 (1977): 33–39.

West, K. P. "Famine." In *Encyclopedia of Human Nutrition*, 2nd ed., edited by Benjamin Caballero, Lindsay Allen, and Andrew Prentice. Oxford, UK: Elsevier Academic Press, 2005, 169–177.

Widdowson, E. M., and A. Ashworth. "Famine, Starvation, and Fasting." In *Encyclopedia of Food Science and Nutrition*, edited by Benjamin Caballero, Luiz C. Trugo, and Paul M. Finglas. Oxford, UK: Elsevier Science, 2003, 2243–2247.

Famine Early Warning Systems Network (FEWS NET)

The Famine Early Warning Systems Network (FEWS NET) comprises a group of agencies with the primary mission of identifying natural, cultural, economic, and political conditions that have the potential to lead to famines and other forms of deficiencies in food demands throughout the world. Once these life-threatening

FEWS-NET

Famine Early Warning Systems Network (FEWS-NET) is an assemblage of agencies and private-sector companies that monitors and addresses conditions that may lead to hunger and famine throughout the world. As nations continue to enter the global economy, reciprocity among its citizens is becoming the manner in which the world does business. Trade agreements among countries promote a more even distribution of goods and services that enable societies to advance and prosper. The applications of open competition, cutting-edge research in science and technology, the rules of law and individual property rights, modern medicine, a consumer society, and an insatiable work ethic have been the applications propelling Western civilization. Cooperative arrangements among nations, developed through national and international agencies, will enable a more open and fair market system that will enhance worldwide enterprise. The Earth has a finite reserve of resources, and they are not evenly distributed. The rapid growth of emerging nations will require those who have an abundance of a particular resource to share or trade to even out the distribution for all concerned.

—Danny M. Vaughn

conditions have been identified, each contributing agency promises to provide data, give advice to those in leadership roles, and document the necessary steps to practitioners to optimally mitigate problems associated with famine before they can become a serious catastrophe.

The original organization was created and funded by the U.S. Agency for International Development in 1985. It was renamed the *Famine Early Warning System* following the 1984–1985 famines in Ethiopia and Sudan, in which more than 1 million people died. In 2000, the name was changed to the *Famine Early Warning Systems Network* as part of a move to improve food security information systems and develop early response mitigation procedures so as to reduce the potential for famine and food shortages throughout Africa and other regions of the world. FEWS NET has regional offices throughout Africa, including Ouagadougou, Nairobi, and Pretoria, and national offices in Guatemala, Haiti, and Afghanistan.

A partnership, together with a cooperative agreement to share information, was created between the U.S. Geological Survey, the National Oceanic and Atmospheric Administration's Climate Prediction Center, the U.S. Department of Agriculture's Foreign Agricultural Service, and a private-sector company to manage field operations, with additional linkages between regional and international organizations. A FEWS NET software tool, known as Population Explorer, is employed to gather useful population data, and it seeks to answer fundamental questions such as the following: What is the population density within a 1 square kilometer area? How many men and women live there? What is the age distribution? What is the estimated limit of population growth given an understanding of the available resources? The software also is being used as an early warning instrument for disaster preparedness and response and to aid in planning health, educational, and other public services. Each of the partnership agencies, private companies, and other regional and international organizations contributes to these issues by assessing a number of variables that potentially could contribute to famine and increase the severity of food security levels.

The U.S. Geological Survey's *Early Warning and Environmental Monitoring Program* provides support activities to FEWS NET. In association with the food security monitoring efforts, scientists investigate and report on climate change; management of natural resources; assessment of water resources; identification of potential hazardous conditions, including procedures for mitigation; and change detection throughout the natural environment.

A number of data acquisition and dissemination portals provide valuable information throughout several world regional locations. The South Asia portal is a source of rainfall data for the mainland areas of Southeast Asia. The South Central Asia portal focuses on temperature, rainfall, snow cover, and vegetation data for Afghanistan, Pakistan, and Tajikistan. The Central America, Mexico, and Caribbean data portal provides rainfall, vegetation, and crop monitoring information for Haiti, Hispaniola, and Mexico. The Middle East data portal makes available remotely sensed imagery and regional spatial data to Yemen. The Africa data portal, formally called the Africa Data Dissemination Service (ADDS), provides remotely sensed imagery and spatially modeled data through a geographic information system (GIS). Daily and 10-day meteorological data, focusing on information important for agricultural predictions and crop yields, are broadcast to regional and continental scale locales throughout Africa.

The FEWS NET has two software portfolios that provide detailed data for decision support when dealing with issues of environmental concerns, and specifically the potential for famine. The first, *Early Warning Explorer*, is interactive and Web-based; it provides worldwide, visual map data delineating continental-scale estimates in rainfall, land-surface temperature, and total perceptible water, with a critical focus on various climatic anomalies. The *Decision Support Interface* software is another Web-based tool that provides data on weather and crop conditions throughout drought-prone regions. Using spatial modeling rules, maps are developed to identify and illustrate specific environmental criteria that can contribute to drought conditions and ultimately affect agriculture and regional food security.

The *International Weather and Climate Monitoring Project* of the Climate Prediction Center of the National Oceanographic and Atmospheric Administration is also a partner of the FEWS NET; it is funded in part through the U.S. Agency for International Development. Although the program was originally developed to focus on sub-Sahelian Africa, it now covers the entire African continent, plus Afghanistan, Central America, the Caribbean, the Mekong River Basin, and most of southern Asia. Daily precipitation estimates are reported using high-resolution, remotely sensed imagery coupled with ground-based data recordings of local and regional precipitation events. Additional spatial and tabular data are incorporated into a geographic information system to model potential scenarios for developing conditions that may contribute to famine and, by extension, affect food security throughout previously identified regions of the world.

As a worldwide partnership, the FEWS NET also combines its resources and data with local and regional information from the Mesoamerica Famine Early Warning System (MFEWS) and the Asia Flood Network (AFN) to disseminate up-to-date maps, tables, and predictive spatial models. Hydro-meteorological data, modeled in a spatial context, offer a powerful means of assembling large volumes of information that, when analyzed in geospatial modeling software programs (GIS), yield the most comprehensive decision-support system for forecasting conditions leading to regional droughts and other naturally occurring disasters. Climate extremes pose a real threat to agriculture and ultimately food production, which, by extension, can be a catalyst for large-scale regional famine.

Geomatics, Spatial Literacy, and FEWS NET

Geomatics encompasses several related mapping disciplines and spatial analytical tools that, when combined, function to collect, store, manipulate, revise, analyze, and deliver information (output in printed or digital format). This information is referenced spatially, based on location through the use of a geographic coordinate system. Remote sensing, GIS, global positioning systems (GPS), computer-assisted cartography, land surveying, and photogrammetry have been integrated to form the applied mapping sciences. These disciplines collectively provide a suite of research tools used by local, regional, and national agencies throughout the world as a means of identifying and resolving geospatial (*geo*: "earth"; *spatial*: "any feature/object that can be measured that takes up space") issues, and their impact on human societies and the natural environment.

The planet Earth functions as a dynamic set of integrated systems, driven by an internal energy source, created through the radioactive decay of terrestrial materials (elements, minerals, magma, and rocks), with its principal external energy source being the sun. A *system* is any number of components or parts that, when assembled, function as a unit. Atmospheric circulation, weather conditions, global climate, ocean circulation, weathering, and erosion of the planet's surface are processes driven by both internal and external energy sources. A *process* is any action (an energy exchange) that results in some change in the physical or chemical properties of material phenomenon (any event or observable facts). Energy exchanges occur between the principal systems of the atmosphere, hydrosphere, lithosphere, and biosphere; they form the basic foundation for the Earth system's science. The full nature and scope of understanding of the interactive processes that drive this dynamic planet have been the life work of scientists seeking to answer a number of questions centering on the *what*, *where*, *why*, *how*, and *when* of phenomena. Understanding of the Earth system's science and knowledge of spatial literacy are imperative in the context of the natural processes driving the extreme weather and longer-term climate conditions that can ultimately contribute to catastrophic events leading to drought and famine.

Spatial literacy is an understanding and appreciation for the location, distance, direction, pattern, shape, size, and position of objects, along with the associations they share within and among them. Spatial literacy is fundamental to geographic inquiry and applicable to most other academic and applied disciplines. Spatial literacy has special significance to the principal mission of FEWS NET and its partners, as they seek to identify natural, cultural, economic, and political conditions that have the potential to lead to famine and other problems in food requirements throughout the world.

Earth surface feature relationships are distributed throughout space and time, and any formal investigation into these relationships is called *geospatial analysis*. In geospatial analysis, the distribution and association of Earth and environmental phenomena are evaluated by isolating significant variables. Models are used to identify active processes, concepts, and the manner in which humans interact with or are affected by an event. Given the complexity of natural and human dimensions, the GIS can provide a mechanism that will isolate and simplify multivariate phenomena.

Any multivariate analysis must begin with a structured method of inquiry; it is initiated by identifying a number of questions pertinent to an issue. In the case of famine and food security, questions to be addressed might include the following possibilities:

1. Where are the critical poverty-stricken regions throughout the world?
2. Where do the people of these regions acquire their present food sources?
3. Which types of food are readily available, and which types of food can be developed, given the resources that are present throughout a region?
4. How much of a society's food supply depends on regional agricultural practices?

5. How do the seasonal weather and intermediate- and longer-term climate affect the production of food supplies?
6. Which best practices in agricultural development need to be identified and implemented in a particular region, based on soil fertility, local weather conditions, available technology, cultural and religious concerns, and the overall ability of a society to adapt to change?

Initial field observations and measurements are often necessary to identify questions of concern and to understand the need for additional data collection methods. Once they have been completed, more sophisticated data collection operations can be undertaken through detailed field measurements and the processing of remotely sensed imagery. These efforts are followed by the creation of complex tabular databases to be used in spatial modeling procedures developed within a GIS. In the final analysis, specific data sets, enhanced imagery, and complex spatial models are built and presented to a collective group of decision makers for final approval and implementation.

Skills Necessary for Accurate Spatial Analysis

Competent spatial analysts in all support agencies and private research groups are necessary for addressing the challenges of hunger and famine throughout the world. Particular skills required include the following:

- Understanding the nature of interactive natural processes operating throughout the subsystems of the atmosphere, hydrosphere, lithosphere, and biosphere
- Comprehending the impact of human behavior on the natural environment and from cultural, economic, and political perspectives
- Geographic reasoning applications to the basic tenets of spatial issue inquiry (i.e., content, size, shape, pattern, density, cluster, location, distance, and direction of Earth objects throughout space and time)
- Establishing evidence-based summaries, conclusions, and decision-making criteria that will provide optimal information to resolve Earth and societal issues that can impact the future health and vitality of populations

Through the association of FEWS NET, geospatial analysts form a critical partnership in identifying, understanding, explaining, and mitigating the problems that directly can affect famine and the security of world food supplies.

—Danny M. Vaughn

Further Reading

Humanitarian Early Warning Service. Accessed January 2011. www.hewsweb.org/home_page/default.asp.

"National Weather Service, Climate Prediction Service and Other Inter-agency Partnerships Focusing on Potential Environmental Disasters Throughout the World." Accessed January 2011. www.cpc.ncep.noaa.gov/products/fews/.

Reliefweb. Accessed January 2011. www.reliefweb.int/rw/dbc.nsf/doc100?OpenForm.

U.S. Agency for International Development (USAID) Famine Early Warning Systems Network (FEWS NET). Accessed January 2011. www.fews.net/Pages/default.aspx.

Farm Adjustments to Climate Change

Climate change is a process that all farmers now must consider and plan for, every agricultural year. How they address those changes will directly affect the outcome of world food production because farming adjustments to change are farm specific and personal. Farm adjustments involve many variables that affect the ability of producers to grow food and animal feed for farm use and for local, regional, and global consumer needs. Variables that farmers must consider when making adjustments include climate, farming costs, technology, transportation, and governmental policies. Focusing solely on climate change does not provide enough information for farmers to make informed decisions that will ensure adequate global food production in the future.

Practical Responses

Weather, a daily event, is of extreme concern to a farmer. It can change the outcome of a year's work in an instant. When hail destroys a crop or when a severe dry period occurs and not enough pasture is available to support a large herd of livestock, farmers must adjust their decision making. Climate—that is, the average weather of a place over a long time period—is important to farmers because of the long-term adjustments that will be necessary, depending on whether the Earth is cooling or heating. Uncertainty is common among farmers regarding this issue because unusual, crop-impacting extremes have affected yields in recent years. It has been wetter than normal in some areas and drier in others. An indicator of climate warming is the migration of animals and plant diseases poleward and the spread of desert-type plants.

Climate change and the rate at which it changes may necessitate planting different crops or crop varieties developed for conditions present in a different climate. A drier climate will be more harmful to farming operations than a cooler climate. Planting time for fall-sown crops may need to be delayed, while early frosts and poor soil conditions could delay some spring planting by a few weeks.

Farmers believe that climate change is such a slow process that they will have ample time to make the necessary adjustments to farming operations. Adjustments involve how farmers plan and carry out work activities, each year. For example, in the spring of 2010, soil testing in the western Great Plains showed a full moisture profile, so astute farmers planted their fields accordingly. In July, high temperatures and increased plant evapotranspiration resulted in the complete depletion of soil moisture. Farmers then had to immediately alter their farming plans.

Another factor in food production may be in the quantity and quality of pasture grasses and nutritional change as a result of climate variation. For instance, Buffalo Grass is currently the staple prairie grass in western Kansas. This variety is quite drought tolerant and once supported large herds of bison. There is no hybrid research on Buffalo Grass, and the grass is not being planted. If not overgrazed by cattle, Buffalo Grass can provide forage for livestock for many decades, thereby making a major contribution to food production. If the climate continues to warm, prairie grasses may become more important as a source of animal food. Many dryland farmers and ranchers believe that more research on natural grasses should be conducted to improve grass quality.

One strategy that dryland farmers are already using is no-till farming, in which the stubble or plant residue is left on the soil's surface rather than plowing or disking it into the soil. New crops are planted directly into the stubble. Seeding genetically modified (GM) herbicide-tolerant plants makes it possible and practical for growers to control weeds by applying herbicides rather than plowing. Nevertheless, no-till farming remains a controversial practice. It is embraced by many farmers, but other farmers disagree with its value. No-till farming saves moisture and helps prevent erosion. The long-term effect of chemicals is yet to be determined, however, and must be considered. Chemical fertilizers have become very effective but are now very expensive. These inputs are at least three times more expensive than they were just five or six years ago.

Another issue related to no-till farming is carbon sequestering. By planting new grasses or by practicing no-till farming, farmers ensure that carbon is sequestered in the soil, offsetting emissions from industries. The purchase of carbon credits, through the Chicago Climate Exchange, is more likely to benefit farmers where moisture is more plentiful than their counterparts in dry agricultural regions.

The concern is whether no-till farming and carbon sequestering are simply political agendas or whether they have long-term value in increasing food production.

To cope with increased temperatures and slower pasture grass growth, some farmers practice intensive grazing. With this strategy, farmers keep cattle on a certain portion of the land for a period of time and then move them to another spot. Intensive grazing has been very effective in decreasing weeds because the native grass chokes them out. As climate changes, this practice is seen as a very viable way to preserve the nutrients in the soil needed for pasture maintenance and grass growth.

One important adjustment farmers may have to make is in the breed of livestock that will be raised for food. Agricultural animals used for food include cattle, swine, and poultry. Animals convert inedible plants to human food. The quality of meat produced, resistance to disease, nutritional value, and the amount of land needed for production will all be considerations of importance as climate changes. Each breed is noted for certain characteristics that make the breed either more or less desirable for production.

If climate continues to change, farms and ranches currently supporting herds of Angus and Hereford cattle may find it necessary to raise Texas Longhorns or Brahman breeds or perhaps a hybrid breed such as Brangus. Brangus combines the hardiness of the Brahman and the quality of the meat of the Angus. Numerous breeds are available, and farmers will be able to slowly sell older livestock, replacing them with more suitable breeds, depending on climatic conditions and availability of pasture. Some ranchers have been raising bison for many decades and find that the native grasses are excellent feed for both bison and cattle. Most farmers and ranchers say that, on a pound-for-pound basis, cattle provide more food to consumers than bison. If livestock type adjustments become necessary as climate changes on the western Great Plains, raising bison instead of cattle may be a logical step.

The Agriculture Institute of Salina (Kansas) does research on alternative crops, especially grasses. Perhaps replacing old existing grasses with new drought-resistant types may sustain agriculture and food production in drier regions. If water is no longer available or too expensive on some currently irrigated farms and ranches, there may be renewed interest in prairie grasses. In general, grasses requires less moisture than does cropping.

Some farmers are experimenting with an Ethiopian grain, called teff or *eragrostis*. This nutritious grain is high in protein, carbohydrates, and fiber. Teff, a dryland, grass-type plant, may be a viable alternative to other crops as the climate changes. A gluten-free grain, teff is known to be both flood and drought resistant. It is a new crop in many portions of the world. Teff has been grown outside of Ethiopia for decades on a small scale and on farms on the western Great Plains since 2005. Teff field days are held in dry agricultural areas of the United States, with seed growers and county agents being made available to answer questions about this specialty crop.

Technology

Technology makes farming easier than ever before, in both complex and simpler ways. For example, a simple strategy is eliminating much of the cold-weather problems associated with feeding cattle. Forty years ago, feeding hay to cattle required the farmer to get out of the truck or farm cart, cut the baling twine, and then throw the 60- to 80-pound bales onto the ground. In contrast, today a farmer hauls the large round bales to the cattle. Without getting out of the truck, a farmer activates a hoist that lifts the bale and deposits it onto the ground. The farmer never has to leave the inside of the truck. In addition, weather forecasting technology is now able to predict bad storms early enough that farmers can get hay out to cattle a day or two in advance, thereby preventing loss of livestock.

Agriculture policy has changed cattle rearing in other ways. Because of mad cow disease, the paperwork involved in documentation of each animal's vaccinations is time-consuming and cattle must be implanted with microchips carrying vaccination information. Moreover, vaccinations are very expensive, adding yet another cost to food production. Large-animal veterinarians are scarce. The increased costs incurred because of veterinarian transportation to treat livestock herds increases the cost of food production, with these costs ultimately being passed along to the consumer.

Seed companies are continually developing new seed varieties for staple crops. Some varieties are drought resistant, whereas others are flood resistant. An example, in flooded areas of Bangladesh in 2007, a variety of rice was grown that was capable of growing as high as 10 or 12 feet as the flood waters rose. Farmers using that variety of rice produced a crop even though they were experiencing the greatest-magnitude flood in a decade. Grain variety research is essential to meet farming needs and food needs as climate evolves.

Today, farmers are often college graduates who have majored in agricultural economics. The toughest decision wheat farmers make is deciding which crop variety to plant so as to produce the highest yield. Crop tours help farmers make prudent planting decisions and climate change will make attendance at such events even more important.

It takes approximately 12 years to develop a new wheat variety because it must meet stiff criteria to be widely accepted: It must be high yielding, have excellent quality for bread making, and have good resistance to disease and insects. Hundreds and thousands of wheat lines are tested, but only a few achieve the breeder's

Henry A. Wallace

U.S. Vice President Henry A. Wallace, born in 1888, was a third-generation farmer. His first contribution to agriculture came when he was only 15. At a science fair, Wallace proved that corn yield was unrelated to the appearance of the corn. His interest in agriculture took him to Iowa State College, where he studied agriculture and graduated in 1910. Wallace's chart showing the ratio of corn feed to hog weight became instrumental in increasing farm profits. His discovery led to the establishment in 1926 of the first commercial hybrid seed company, the Pioneer Hi-Bred Corn Company.

Wallace was appointed Secretary of Agriculture during Franklin D. Roosevelt's administration, and he served as FDR's vice president from 1941 to 1944. After an unsuccessful run for president as a candidate of the Progressive Party, Wallace returned to agriculture, where he continued to develop hybrid seeds and a new chicken that had less body weight and laid more eggs. Wallace was stricken with Lou Gehrig's disease (amyotrophic lateral sclerosis); he died from this disease in 1965.

—*Kay Weller*

set goal. Many state universities and seed growers, such as Kansas State University and Monsanto, respectively, now are collaborating to improve wheat breeding programs. They also will be able to genetically fingerprint new varieties.

Monsanto is an example of an agricultural company that is applying innovations and technology to help farmers worldwide produce more food while conserving the land. The company aims to help farmers grow crops sustainably so these farmers can continue to farm and produce healthier foods, better animal feeds, and more fiber, while also reducing agriculture's impact on the environment. Estimates are that by 2050, our planet must double its food production to feed an anticipated population of 9.3 billion people. This increase will put more pressure on both water and energy supplies. Monsanto and other such companies plan to develop improved seed, thereby doubling the year-2000 yields of corn, soybeans, cotton, and spring-planted canola. Improved certified seeds are being developed to use one-third fewer key resources per unit of output while lessening habitat loss and improving water quality.

Henry Wallace, founder of Pioneer, a well-known seed company, used a science-based approach to increase productivity and profitability through technological innovations. His company now serves food producers in 70 countries. It produces seed for alfalfa, canola, corn, mustard, pearl millet, rice, sorghum, soybeans, sunflower, and wheat. Pioneer strives to develop products with traits and performance standards ideally suited for the growing conditions and local cropping practices in which the seed will be grown. In a period of climate uncertainty, this technology will help farmers adjust their decision making to match current trends.

Climate change and the introduction of new grain varieties may result in the need for new and different farm equipment. Equipment is very expensive, with some tractors costing $250,000 or more. Farmers often lease large equipment rather than purchase it. The rising cost of equipment limits who can assist in farming operations because of the risk of damage, especially to computerized equipment. New tractors come equipped with global positioning system (GPS) units. These GPS units can be useful to the farmer as they sense, calculate, and determine how much pesticide or herbicide to apply to any given land parcel. Given the high costs of machinery, some farmers now specialize in providing a specific service to other farmers, such as cutting and baling their hay. Relieved of doing time-consuming work with poor equipment, a farmer can devote more time to other important farm tasks that he or she is equipped to do. Each hour that farm machinery sits idle costs the farmer $50 in depreciation; thus keeping farm implements running is essential to successful farming management. Owning one piece of equipment and hiring jobs out is a farming adjustment that likely will become increasingly important in the future.

Adjustments in farming practices associated with climate change will also take many other forms. For example, petroleum is the basis for many pesticides and herbicides. Petroleum security is necessary for the sustainability of food production, with the resulting foods being sold at affordable prices. Pesticides are more toxic than herbicides but are well regulated. Many new chemicals are supposed to be less toxic than in the past, so perhaps their more benign profiles will help prevent the leaching of toxic chemicals into groundwater.

What national and international decision makers do will drastically affect the adjustments that farmers must make in a period of climate change. Political agendas are not always in the best interests of either the producer or the consumer. Indeed, some farmers believe that many decision makers are more concerned about their prospects for reelection to office than about issues associated with food production. Bankers are also decision makers affecting farming practices. They now require farmers to purchase crop insurance, which increases production costs and, in turn, is passed on to the consumer. Bankers also control the opportunity a farmer has when considering the purchase of new equipment. Most politicians and bankers have little understanding of farming.

Farmers' Concerns and Challenges

Of serious concern to farmers, investors are buying huge acreages at inflated prices, making it more difficult for the family farmer to expand operations. In most cases, the absentee owners have little agricultural knowledge, and some do not care whether the land is cropped. They may instead use much of the land for financial shelters or recreational hunting purposes. Rarely do they buy farm equipment from local dealers or seed and fertilizer from local suppliers. They contribute little to local communities and rural life. Absentee ownership of land has become a serious issue, but it has received little attention outside farming communities. Agricultural land should be retained for food production, not recreation.

Another issue that rural dwellers face is the fact that many older farmers own the land and are unable to farm it, yet many younger farmers cannot afford to buy or rent the land. Land is a resource. Ownership of the land includes a responsibility and an obligation to be wise stewards of the land.

Another concern that farmers and consumers will likely face as climate changes relates to the rising costs of food production and distribution. As the cost of technology, such as farm machinery, rises and the costs of pesticides and herbicides needed for higher yields increase, eventually the consumer will have to pay more for food. These rising food costs will make it more challenging for people living in many regions of the world to maintain a diet that supports their physical and emotional needs. Global health, as a result of increasing food prices, should be a major concern of decision makers as they consider the effects of climate change.

Climate change is such a slow process that the farmer has adequate time to adjust to most challenges. In contrast, food terrorism is a more rapid-onset concern of many rural dwellers. If crops are wiped out by a new type of biological weapon that can eliminate multiple crops, a farmer presently has no way to respond. For example, if a car with a trailer—a common sight on many U.S. highways—proceeds along Interstate-80 from Chicago to Cheyenne, dispersing a biological or chemical agent over a wide area of this country, this act could eradicate or reduce corn and wheat crops over several states. Corn and wheat and their by-products are components in many foods. Eradication or yield reduction of these crops could result in huge losses of food and animal feed. On a global scale, rice is the staple of the diets in many countries. Eradication or a reduction of that crop in even one country might potentially bring on a horrible regional famine. Damage from food terrorism can be inflicted quickly, leaving the farmer with no way to respond to consumers' needs and recover the loss in farm income.

Many farmers believe that water availability will be the greatest challenge that they face in the foreseeable future, and climate change may exacerbate that problem. Many groundwater aquifers are under stress, due to overuse and contamination from chemicals. Irrigation technology, such as drip hoses or center pivot irrigation, reduces the amount of water used and enables farming in places that would otherwise be unsuitable for food production. Some aquifers are now salty. Only the application of advanced technology will make it feasible to distill this water and make it useful for agriculture. Currently, water for food production remains relatively inexpensive in many food-producing countries of the world, but any scarcity of irrigation water will result in higher costs for food production. These costs will ultimately have to be borne by the consumer. If many people worldwide are seriously affected by their inability to secure basic food items, the number of hungry people in the world will increase.

Farmers are very resourceful and will adjust their decision making to climate change. Events and factors that are out of the control of farmers will make it increasingly difficult for them to produce food to sustain a growing world population. In the developed nations of the world, the number of farmers is declining; in the United States, farming is being relegated to large agribusinesses. Industrial farms, state farms, and farm factories may become so large and be managed by so few parties with agricultural experience that food security will be at risk. Climate change is just one of the many adjustments and challenges the farmer will face. Fortunately, it is a challenge that may be met based on changing technology and prudent decision making on the part of food producers.

—Kay Weller

Further Reading

"Breeds of Livestock." Accessed August 2010. www.ansi.okstate.edu/breeds.

Glover, Jerry, et al. "Increased Food and Ecosystem Security via Perennial Grains." *Science* 328 (June 25, 2010): 1638–1639.

International Service for the Acquisition of Agri-Biotech Applications. Accessed July 2010. www.isaaa.org/kc/.

"Iowa Crop Performance Tests." Accessed July 2010. www.croptesting.iastate.edu/downloads/06ICIAflyer.pdf.

Johnson, Angie. "K-State Develops Quality Wheat Varieties for Kansas." 2003. www.iso.k-state.edu/media/webzine/0204/wheat.html.

Monsanto. 2010. www.monsanto.com/responsibility/sustainable-ag/default.asp.

Mutel, Cornelia Fleischer, and John C. Emerick. *From Grassland to Glacier.* Boulder, CO: Johnson Printing, 1992.

Opie, John. *Ogallala: Water for a Dry Land*, 2nd ed. Lincoln, NE: University of Nebraska Press, 2000.

Peterson, Rorik. "Water Woes: Structure and Agency Influences on Adaptive Capacity in Southwest Kansas." *Geographical Bulletin* 47 (2006): 73–84.

Pimentel, David. "Water Resources: Agricultural and Environmental Issues." *Bioscience* 54 (2004): 909–918.

Pioneer. Accessed July 2010. www.pioneer.com/web/site/portal/menuitem.cc20eec90551c318bc0c0a03d10093a0.

"Planting Decisions Make Difference for Kansas Wheat Producer." *The Hill City Times* 133 (June 23, 2010): 18.

Postel, Sandra L. "Entering an Era of Water Scarcity: The Challenges Ahead." *Ecological Applications* 10 (2000): 941–948.

Railey, Karen. "Whole Grains: Teff (Eragrostis)." Accessed June 2010. http://cheetday.com/teff.html.

Rosegrant, Mark W., Cai Ximing, and Sarah A. Cline. "Will the World Run Dry? Global Water and Food Security." *Environment* 45 (2003): 25–36.

Swaffar, Steve. "Carbon Credits, What's It All About?" 2008. http://kfb.org/views/swaffarcarbon.htm.

"Teff Field Day on August 5." *Hill City Times* 134 (July 28, 2010): 4.

Tilman, David, et al. "Forecasting Agriculturally Driven Global Environmental Change." *Science* 292 (2001): 281–284.

Food Aid Policies in the United States: Contrasting Views

Today, most Americans believe that more fortunate people, relief organizations, aid foundations, and the U.S. government should assist the hungry in securing sufficient, safe, and nutritious food to meet their dietary needs and preferences. Dietitians assess the nutritional adequacy of food-insecure people's diets. Federal agriculture officials count the number of food-insecure Americans. In November 2009, DeParle and Gebeloff reported that more than 36 million people participated in the federal government's Supplemental Nutrition Assistance Program (SNAP). Policymakers champion laws that they believe will reduce the harmful bodily effects of low-cost, high-calorie foods.

The concerned citizens performing these actions attempt to solve a life-impacting problem. In the United States, many people attribute this problem to an individual's lack of income. They often presume that those individuals living on low or fixed incomes are unable to secure access to enough food to live active and healthy lives. In 1989, the U.S. government policymakers formally adopted these understandings. They adopted the concept of "food insecurity" to describe the condition that exists whenever "the availability of nutritionally adequate and safe foods or the ability to acquire acceptable foods in socially acceptable ways is limited or uncertain" (Anderson 1990, 1560). This definition includes a person's inability to purchase appropriate food from stores, growers, restaurants, or other similarly organized groups.

Concerned people gain knowledge of an individual's hunger through their daily experiences. These experiences are influenced by their distinct social relations. Through these interactions, they access a variety of goods and services, including money, employment opportunities, food supplies, social networks, transportation providers, government, firms, housing, and other resources. Concerned people accessing goods and services form distinct understandings about the inherent goodness or badness of specific populations, resources, and service providers.

Concerned individuals tailor their actions to a distinct hunger knowledge. While this knowledge is spatially and socially diverse, it is based on three common understandings, which form the basis for individuals' beliefs regarding the way other people ought to behave. The "liberal tradition" contends that government policymakers should help to ensure equal opportunity of success for everyone. The "Marxist-oriented radical policy economy tradition" argues that low-income laborers should act together to minimize the differential effects of capitalism. Finally, the "conservative tradition" contends that all people should help others to help themselves.

The following section identifies, contrasts, and compares these primary traditions. In so doing, it highlights unexpected parallels across knowledges generally identified as very different. These diverse understandings provide a basis for thinking about the long-term causes and effects of individuals' hunger and food insecurities.

Liberal View

Those persons professing and adopting a liberal perspective are guided by a liberal tradition. This tradition is grounded in the philosophical beliefs of liberalism. These beliefs identify people as contributing citizens and human beings. According to the liberal perspective, people, as citizens and living products of our society, have certain universally recognized, inalienable rights. They exercise these rights in distinctly defined territories. Within their personal space, all individuals maintain equal political standing and moral worth. Persons, by exercising their rights as citizens of the United States and acting in their personal space, effect change in the social, political, and economic systems influencing their daily lives.

Researchers have traced the origins of modern liberal understandings to policymakers' practices and rhetoric. A. O'Connor (2001) and B. O'Connor (2004) note that liberal food-aid interests have been traced to 19th-century concerns regarding the economic rights of average citizens. B. O'Connor and others also have traced its language to U.S. President Franklin D. Roosevelt's attempts to garner public support for government relief and food-aid expansion in the 1930s. Finally, liberal philosophy and liberal food policy for the poor and those in need were solidified in

President Lyndon B. Johnson's policies designed to advance opportunities for low-income people and racial minorities.

In the United States, traditional liberals believe that the poor lack opportunities to improve their social standing and economic status. At times, they even suggest that these problems are transmitted culturally from parent to child. Nevertheless, these liberals generally insist that the poor are interested and willing to pursue improved employment opportunities, food supplies, and job training programs; they simply need to be provided with an equal opportunity to access these goods and services.

Liberals base much of their social policies on their understandings of exploitive and benevolent capitalism. They contend that a person's hunger sometimes results from the indirect costs of a capitalist, free-market economy. Within the labor market, liberals believe that economic transformations will diminish the importance of some people's skills, causing these individuals to experience problems of unemployment or underemployment. With their loss in wages, some workers will not receive enough income to purchase adequate provisions for themselves and their families.

As part of this tradition, liberals believe that government ought to enact policies to ensure all citizens of the United States have access to food supplies, regardless of their race, gender, sexual orientation, and geographic location. Those given the responsibility for managing these policies should help all people secure equal access to food supplies. Finally, low-income workers should take advantage of these opportunities to improve their education and work skills. In doing so, every citizen of the United States can effect change within the systems contributing to persistent problems of hunger.

Contemporary liberals advocate for equal opportunities. In policymaking, they support reforms designed to enhance everyone's socioeconomic advancements, including policies designed to reduce discrimination on the basis of a person's personal characteristics or geographic location. They also support program improvements which they believe will provide the poor with long-term economic benefits. As part of this effort, liberals advocate the establishment of programs such as job training and education programs to provide the poor with marketable skills.

Radical View

Those who profess a radical view with respect to food security in the United States are guided by the Marxian tradition. Like members of the liberal school, individuals acting in this tradition believe that all people should enjoy inalienable rights. Similar to the stance taken by conservatives, radicals believe that those suffering from food deprivation and social injustice learn of their rights through social groups—for example, via households, families, communities, government, and private property owners. These organized groups are a manifestation of locally specific activities.

Radicals also believe that individuals should exercise their rights in a capitalist society. They argue that the practices of land ownership, income property, and the means of production prompt society to divide itself into two distinct social classes. Members acting in these classes perform unequal duties of owners and laborers. Laborers promote themselves as a commodity available for owners' purchase, whereas owners purchase laborers' time, efforts, thoughts, and anxieties at a particular wage or price. Laborers perform activities of labor to produce owners' goods and services; owners derive income from their sales of these items.

According to radicals, in such a society laborers occupy a subordinate economic position. To these radicals, members of the working or laboring class are challenged to achieve social equality. As Peet describes, economic insecurity or poverty inevitably results from the sale of labor. Such an economic relationship is perpetuated by the owning class. These owners instigate labor policies and human rights policies to ensure that this insecurity persists. In turn, the persistence of these inequalities enables owners to accumulate wealth (i.e., labor surpluses). The owners' historical accumulation of these surpluses enables them to gain an increasing share of society's total wealth.

Moreover, radicals insist that impoverished laborers retain an ability to improve their positions, which they can realize by engaging in individual and organized collective political actions. Through these actions, laborers make powerful people aware of capitalism's contradictions. With this newfound awareness, some policymakers will become interested and encouraged to champion widespread social and economic reforms. An outcome of this process should transform the political and economic systems perpetuating problems of poverty and hunger.

Today, radicals advocate for the legal rights of low-income laborers. They make distinct policy contributions in accordance with their unique particular understandings. Radicals engaging in a socialist perspective advocate for policies to transfer money and/or ownership from the wealthy to the working people. Advocates engaging in a strict Marxian interpretation call for social and economic revolution. Supporters of particularly vulnerable groups (e.g., women, racial minorities, and the elderly) emphasize economic and social policy reforms to improve the socioeconomic conditions in which its members live today and whose children will live in the future.

Conservative View

Those persons adopting a traditional conservative perspective are guided by the conservative tradition. Like liberals and radicals, conservatives presume every citizen of the United States has unalienable rights. Americans protect and advance these rights through their participation in distinctly organized social groups. Conservatives emphasize these views and programs and organize groups bonding like-believers in unchangeable social traditions. Those advocating distinctive conservative points of view mold their distinct character and direct the functions of conservative social policies.

Contemporary traditional conservatives believe that those acting in organized social groups are responsible for the group members' economic fates. These fates may be enhanced by elected officials acting in government. Pinker notes that traditional conservatives understand well and value the effects wrought by elected conservatives acting in government. They respect the legislators who focus on enacting systematic provisioning solutions. They believe that these systems represent a practical and effective expression of an organic model of society. Such conservatives accept a need to reform the systems usually enacted by liberal congressional representatives, and they adopt such reforms when a majority of societal members believe these improvements are necessary for enhancing civic virtues, public order, and morality.

In this tradition, conservatives suggest that those struggling with poverty have the ability to improve their level of living: They just need to work together to help their fellow poor. Through these actions, individuals ought to positively transform existing provisioning systems. Hence, it is believed that those suffering from hunger, hunger-related diseases, and loss of jobs will be more capable of improving their own economic status.

Those professing a traditional conservative understanding champion gradual, systematic reforms that are designed to enhance an individual's freedom of choice. Such choice is enacted in a gradual and evolutionary process—one designed to promote local and state decision making in service provisioning. This process, they believe, will enable the federal government officials to act in a more traditional governing role.

In the 21st century, those who live and vote in the United States increasingly are less likely to adopt strict, traditional conservative understandings. Most Americans have great reservations about radical and conservative understandings of poverty. The widespread adoption of these ideas has advanced a hybridized conservative understanding. In 2000, Larner noted that writers sometimes used the terms "neo-conservatism" and "neo-liberalism" interchangeably to describe these understandings.

Neo-conservative View

Those adopting a neo-conservative perspective accept aspects of both the radical and conservative traditions. These people believe that everyone maintains certain rights. Neo-conservatives realize these rights are basic in a distinctly organized capitalist society. Within this society, members participate in organized social groups, which themselves maintain distinct traditions. A tradition of liberal government defined historically and geographically distinct social contracts; in other words, people and groups acting in particular places performed these contracts in accordance with locally specific social traditions and norms.

Authors have traced the origins of contemporary neo-conservative poverty understandings to individuals acting in scholarship and policy. This perspective's interests have been traced to radical scholars' theories and critiques of the U.S. government's domestic poverty policies. The language used in these critiques persuaded other radical and liberal scholars and policymakers to adopt these views. In turn, the widespread adoption and persistence of neo-conservative understandings influenced individuals acting in domestic and international policy spheres.

Neo-conservatives believe that liberals successfully championed a series of enabling government policies. Congress's enactment of these policies enabled all people to access a centralized system of government subsidy. For example, liberal lawmakers ensured that qualified low-income laborers would receive generous quantities of assistance and the means to improve their own economic condition. Many low-income laborers took advantage of provisioning programs to better their economic conditions. Some who were unable or were less successful have strived to retain their income and benefit packages.

Neo-conservative individuals agree that the poor who cannot or will not break free of the bondage of poverty tend to rationalize their economic conditions and enjoy their benefits—namely, their conditions of employment, personal finances, and dependency. In perpetuating social-dependent behaviors, neo-conservatives argue, the poor act in ways that encourage the development of a subculture of dole living, pathological behaviors, cultures of dependence, and destructive social environments. Over time, children's imitation of their parents' behaviors encourages the intergenerational persistence of women-headed families, poor work ethics, low aspirations, living for instant gratifications, and violence. In the public sector, government employees' imitations of colleagues' behaviors provide for a persistent culture of worker entitlement and strong public employee associations.

Finally, neo-conservatives understand that a head-of-the-household's inability to rise above the demeaning life in poverty interferes in the local, state, and national production and consumption of goods and services. Government employees' proclivity to artificially direct resources to underperforming populations results in many of the poor remaining unemployed. The poor's voluntary unemployment then shrinks the number of laborers available for producing goods and services. As labor supplies dwindle, owners find it increasingly expensive to hire well-qualified workers. The government shifts its provisioning tax burden among

Understandings of Food Assistance

Since the inception of food assistance programs, many people have called for reforms of these efforts. These debates emerge from an individual's knowledge or experience with hunger. While this knowledge is spatially and socially diverse, it also describes many common understandings of food assistance. These common understandings and perceptions are recognized in liberal, radical, and conservative traditions. For example, people who recognize a liberal tradition contend that the government should ensure that all people are capable of feeding themselves. In contrast, people who adhere to a Marxist-oriented radical political economy tradition assert that low-income workers should act together to minimize the effects of capitalism. Finally, some people believe that all individuals ought to help others feed themselves; such people advance the ideas of the conservative tradition. To many people, perception is reality.

> When perception is stronger than mindfulness, we recognize various appearances and create concepts . . . to describe these apparent realities. We then take these concepts to be actually existing things and begin to live in the world of concepts, losing sight of the underlying, insubstantial nature of phenomena. . . .
>
> Of course, we need [these] concepts, and in many situations they serve us very well. We use various concepts as convenient designations, but if we understand that the words do not refer to solid "things" that have independent existence in themselves, then we stay free in the use of them. Problems only arise when we forget that they are constructs of our own mind, and impute a reality to them that they do not inherently have.

—*Joseph Goldstein in* Insight Meditation: *The Practice of Freedom.* Boston: Shambhala Publications, 1993

fewer and fewer participants. With these added expenses, for-profit firms and nonprofit groups are less likely to advance creative and innovative provisioning solutions.

With a set point of view in regard to social issues, neo-conservatives advocate for reforms to reduce the size, scope, and attractiveness of existing "safety net" provisioning programs. By implementing these neo-conservative reforms, they believe, lawmakers will positively transform the poor's self-discipline and morality. The poor will be better poised to take advantage of all possible opportunities to improve their standard of living, including employment, education, job training, and financial investment tools.

Those adopting a neo-conservative perspective advocate for policies to reform the overall global performance of domestic provisioning programs. Neo-conservatives adopting more traditionally conservative views support progressive tax reforms and financial incentives designed to fund traditional social institutions, such as heterosexual marriages, male-headed families, and private property ownership. Neoconservatives accepting more radical understandings advocate for dramatic, systematic reforms that would support the domestic and international roll-out of workfare programs, market-based alternatives, and draconian provisioning practices.

Overlapping Themes

In the United States, many people believe that hunger remains a serious national problem. Rather than being the result of food shortages or widespread crop failures, some people suggest, this problem arises from a person's condition of poverty. According to this view, hunger arises when people lack the ability to purchase their daily food needs in economic transactions. In 2009, U.S. Department of Agriculture researchers estimated that approximately 14.7 percent of U.S. households were limited in their ability to secure unlimited and certain access to adequate quantities of food.

To solve this problem, concerned citizens, decision makers, and national lawmakers draw upon their personal knowledge of hunger. This knowledge is spatially and socially diverse. The diversity is distinguished in the three major traditions: The liberal view maintains that lawmakers have a duty to ensure all people enjoy an equitable access to food; the radical view contends that low-income people ought to rise above their status and act together to transform systems contributing to their persisting food insecurity; and the conservative view maintains that all people should act to help each other help themselves.

Using this knowledge and remembering the first line of the Declaration of Independence, lawmakers act to transform policymaking and constantly improve provisioning practices. Liberal anti-hunger researchers draw upon national aggregates to emphasize the individual and household effects of government assistance programs. Radical political economists examine the effects of declining incomes, manufacturing, and government subsidies on peoples' bodies. Conservative and neo-conservative community food-security researchers focus upon establishing mechanisms to enable individuals, households, and communities to provide for their own food needs.

The outcome of these actions provides for distinctly different food and hunger policy solutions. Contemporary liberals advocate for reforms to improve a person's ability to access food supplies. They support policies and programs designed to promote equitable access in the production and consumption of food supplies. Radical political economists advocate for the legal rights of low-income laborers. Among food security supporters, many radicals support an expansion of entitlement and non-means-tested provisioning programs. Conservatives and neo-conservatives champion reforms to reduce the scope and function of government assistance. They lobby and act to shift food and hunger relief aid decision making to state and local governments and nonprofit organizations.

—Ann Myatt James

Further Reading

Allen, P. "The Disappearance of Hunger in America." *Gastronomica: The Journal of Food and Culture* 7, no. 2 (2007): 19–23.

Allen, P. "Reweaving the Food Security Safety Net: Mediating Entitlement and Entrepreneurship." *Agriculture and Human Values* 16 (1999): 117–129.

Anderson, S. A. "Core Indicators of Nutritional State For Difficult-To-Sample Populations." *Journal of Nutrition* Supplement 120, no. 11 (1990): 1555–1600.

Brown, M. "Liberalism." In *The Dictionary of Human Geography*, 4th ed., edited by R. J. Johnston, M. Watts, D. Gregory, and G. Pratt. Malden, MA: Blackwell, 2000, 446–448.

Clark, B. *Political Economy: A Comparative Approach*, 2nd ed. Westport, CT: Praeger, 1998.

Cochran, C. E., L. C. Mayer, T. R. Carr, and N. J. Cayer. *American Public Policy: An Introduction*, 5th ed. New York: St. Martin's Press, 1996.

Collins, S. D. *Let Them Eat Ketchup! The Politics of Poverty and Inequality.* New York: Monthly Review Press, 1996.

DeParle, J., and R. Gebeloff. "Food Stamp Use Soars, and Stigma Fades." *The New York Times*, November 28, 2009. www.nytimes.com/2009/11/29/us/29foodstamps.html?ref=us.

Jansson, B. S. *The Reluctant Welfare State: A History of American Social Welfare Policies*. Belmont, CA: Wadsworth, 1988.

Kodras, J. E. "Breadlines." In *Geographical Snapshots of North America*, edited by D. G. Janelle. New York: Guilford Press, 1992, 103–107.

Larner, W. "Neo-liberalism: Policy, Ideology, Governmentality." *Studies in Political Economy* 63 (2000): 5–25.

Lobao, L., and G. Hooks. "Public Employment, Welfare Transfers, and Economic Well-Being Across Local Populations: Does a Lean and Mean Government Benefit the Masses?" *Social Forces* 82, no. 2 (2003): 519–556.

Mancino, L., and C. Newman. *Who Has Time To Cook? How Family Resources Influence Food Preparation*. Washington, DC: Economic Research Service, U.S. Department of Agriculture, 2007. www.ers.usda.gov/Publications/ERR40/.

Mink, G., and R. Solinger. *Welfare: A Documentary History of U.S. Policy and Politics*. New York: New York University Press, 2003.

Nord, M., and A. Coleman-Jensen. *Food Security in the United States*. Washington, DC: Economic Research Service, U.S. Department of Agriculture, 2010. www.ers.usda.gov/Briefing/FoodSecurity/.

O'Connor, A. *Poverty Knowledge: Social Science, Social Policy, and the Poor in Twentieth-Century US History*. Princeton, NJ: Princeton University Press, 2001.

O'Connor, B. *A Political History of the American Welfare System: When Ideas Have Consequences*. Lanham, MD: Rowman & Littlefield, 2004.

Peck, J. *Workfare States*. New York: Guilford Press, 2001.

Peet, R. "Inequality and Poverty: A Marxist-Geographic Theory." *Annals of the Association of American Geographers* 65, no. 4 (1975): 564–571.

Pinker, R. "The Conservative Tradition of Social Welfare." In *The Student's Companion to Social Policy*, 2nd ed., edited by P. Alcock, A. Erskine, and M. May. Malden, MA: Blackwell, 2003, 78–84.

Piven, F. F., and R. A. Cloward. *Regulating the Poor: The Functions of Public Welfare*. Updated edition. New York: Vintage Books, 1993.

Schram, S. F. *Words of Welfare: The Poverty of Social Science and the Social Science of Poverty*. Minneapolis, MN: University of Minnesota Press, 1995.

Starkey, L. J., K. Gray-Donald, and H. V. Kuhnlein. "Nutrient Intake of Food Bank Users Is Related to Frequency of Food Bank Use, Household Size, Smoking, Education and Country of Birth." *Journal of Nutrition* 129, no. 4 (1999): 883–889.

Vaisse, J. *Neoconservatism: The Biography of a Movement*. Cambridge, MA: Belknap Press of Harvard University Press, 2010.

Food Assistance Landscapes in the United States

Hunger is a problem of food consumption. People experience food insecurity when social impediments inhibit their reliable access to food, and hunger follows if their caloric and nutrient intakes are insufficient and inadequate. This is an obvious point, but as the economist and philosopher Amartya Sen pointed out in 1982, in scholarly explanations, government policies, and popular imaginations, the issue of hunger often succumbs to a "fetishism of the food commodity." In reality, food per se (i.e., the existence of supplies, the productivity of agriculture, and the pounds collected by charities) becomes the primary focus of hunger analysis and solutions, rather than the relationship between people and food. It is the latter concern that is actually the key to explaining and addressing food insecurity and hunger. In the

How Do Poor People in the United States Avoid Hunger?

In the United States today, an unprecedented number of people receive food assistance. While many of them work, their incomes are too low in relation to their needs for food and other basic necessities to be able to afford these items.

The largest source of food assistance is the federal government, which spends more than $80 billion annually and operates 15 programs. In 2010, more than 40 million people participated in the largest of these programs, known as "food stamps." In 2001, only 18 million people participated.

In addition, tens of millions receive food assistance from a private system made up of nonprofit charitable providers of food, such as food pantries and soup kitchens. While the considerable growth of this private system in recent decades has enabled millions of people to avoid hunger, it also has given rise to concerns that it has pulled attention away from the inadequacy of existing public programs. Moreover, concerns have arisen that private food assistance is also insufficient, uneven, and not accountable to criteria of equitability.

—Andrew Walter

United States, the vast majority of people cannot directly access an adequate quantity (not to mention quality) of food by producing their own, due to lack of land, limited knowledge of food growing, living conditions, and increasing time spent at work. Most Americans are able to acquire enough food by exchanging their incomes for it. Many people, however, have incomes that are too low in relation to the cost of food as well as other costs of living, including housing, health care, transportation, and education. Unable to meet their food needs to a sufficient extent through production and exchange, these people often rely on various modes of transfer, such as family and friends, churches and charities, and a range of government programs. In a context of high unemployment and low stagnant wages, the food assistance provided in these ways is vitally important.

The landscape of food assistance in the United States is a diverse and dynamic one. This variable nature is illustrated by two articles published in *The New York Times*. In 2001, an article focused on the growing significance of private modes of food assistance. It reported that, in the previous year, more people had received assistance from private charitable food providers than had participated in federal food assistance programs. Eight years later, the *Times* ran a second article that revealed a disruption of that trend, as participation in the food stamp program climbed to unprecedented levels and surpassed, again, the assistance sought from private charities. Even as the need for public food assistance grows, this landscape is shaped by a constellation of longer-term and deeper trends that have shaped the terms under which poor and hungry Americans receive food assistance.

The Lay of the Land: The Public–Private System of Food Assistance

According to researchers, federal programs provide approximately 90 percent of the combined public and private food assistance in the United States. The U.S. Department of Agriculture (USDA) administers 15 programs through which one in four people in the United States received food assistance in 2009. Altogether, the federal government spent $80 billion on these programs, a considerable increase over the 1999 figure of $31.5 billion, reflecting growth in need, based on information from the Economic Research Service in 2010.

The largest of federal food assistance programs, in terms of cost, is the Supplemental Nutrition Assistance Program (SNAP), formerly known as the Food Stamp Program. In 2010, 40.3 million people per month received benefits through this program, a 136 percent increase over the number served in 2000. Any individual who meets the nationally uniform income-based eligibility criteria is entitled to SNAP benefits, making SNAP the only program offering "universal" food assistance. In 2010, the average monthly benefit was $133 per person.

All other federal food assistance programs either are targeted to specific "vulnerable" groups or provide benefits that are not guaranteed to eligible individuals. These targeted food assistance programs provide benefits for low-income infants and children, their mothers, the elderly, and the homeless. Low-income children receive food through the School Lunch and School Breakfast programs, which provide cash and commodities to schools on the condition that they, in turn, offer free and reduced-price meals to low-income students. Of the two, School Lunch is the larger program; it served almost 32 million students per day in 2010. During that same year, School Breakfast provided food to 12 million students per school day. A third high-profile program through which the federal government supplements the food intakes of targeted populations is the Special Supplemental Nutrition Program for Women, Infants, and Children (WIC), which provides assistance to low-income children and their mothers. WIC benefits are available to low-income pregnant or postpartum women, infants, and children younger than age 5 who can prove residency in whatever state is administering the program. In 2010, more than 9 million individuals received monthly assistance through WIC. Two additional federal programs target children, elderly, and other vulnerable people. The Child/Adult Care Food Program provides cash and commodities to state and local agencies that offer meals at adult and child day care programs as well as emergency homeless and other rehabilitative shelters. More than 3.4 million people per day accessed food through this program in 2010. The Commodity Supplemental Food Program purchases food and provides it to approximately 519,000 children and elderly people per year through local public and private agencies.

A third type of food assistance program, which is neither universal nor targeted, is the Emergency Food Assistance Program (EFAP), which provides food commodities to states for distribution to regional and local private nongovernmental food banks and soup kitchens making up the emergency food assistance system (EFAS). While the amount of food allocated to each state is determined by the projected size of its low-income and unemployed populations, it is difficult to ascertain how many operating agencies and programs received this food and, moreover, how many different individuals received food from the local nongovernmental sources.

The EFAS represents the "private charities" mentioned in the *New York Times* article cited earlier. It consists of an extensive, decentralized, and growing network of mostly nonprofit organizations that are initiated, funded, administered, and operated at the local level. Four different types of agencies are most significant to the operation of the EFAS. The two most common, food pantries and emergency kitchens, serve a "retail" function by providing food directly to needy individuals and families. Food pantries offer grocery products for consumption away from the distribution site, while emergency or "soup" kitchens serve meals that are consumed at the distribution site. The other two types of agencies, food banks and food rescue organizations, serve a wholesale function by supplying food to food pantries and emergency kitchens, which in turn distribute the food to individuals and families. Food banks and food rescue organizations differ in their size, scale of operation, and the nature of the food they distribute. Food banks typically (but not only) distribute nonperishable foods obtained in bulk from food manufacturers, wholesalers, retailers, other food banks, and USDA commodity programs. Food rescue agencies supply perishable and already-prepared foods obtained as leftovers from restaurants, periodic banquets, and farmers and growers. As a result of these differences, food banks tend to distribute a larger volume of food and operate at a larger scale than food rescue organizations.

The materialization and growth of the EFAS constitutes a significant change in the larger public assistance landscape. The extent and growth of this network of private food assistance providers was documented by a comprehensive government study in the early 2000s, according to Ohls, Cohen, Saleem-Ismail, and Cox (2002). The study revealed that there were more food pantries than any other type of EFAS organization, followed by emergency kitchens, food banks, and food rescue organizations. In general, the EFAS is a relatively young and rapidly growing system. Four out of every 5 food pantries and emergency kitchens, at least 9 out of every 10 food banks, and every food rescue organization had been established after 1981. Much of this growth, moreover, occurred after 1991. Altogether, the cost of transferring food to individuals and families through the organizations making up the EFAS is estimated to be approximately $2.3 billion per year—a cost that includes the value of the food, volunteer labor, and operating expenses of food pantries and banks, excluding emergency kitchens and food rescue organizations. In addition to the federal government, which provides much of the food distributed, and private organizations, the EFAS is assisted considerably by an organization called Feeding America, formerly known as America's Second Harvest. Feeding America is the largest private nonprofit food assistance organization in the United States. The federal government's survey of the EFAS, the most recent systematic national sample, found that approximately 80 percent of all food banks, which supplied 80 percent of all food pantries and emergency kitchens, were affiliated with Feeding America. Furthermore, another study found that Feeding America provided 26 percent of all food distributed to households in the United States through food pantries.

While systematic participation data are not as readily available for the EFAS as for federal food assistance programs, it is clear that food assistance through private organizations plays an important role in enabling individuals and families in the United States to access food. A body of evidence now exists that shows a large number of low-income households receiving private food assistance and doing so with increasing frequency. According to a 2010 report published by Feeding America, 37 million different people received food distributed through its network during the previous year. In comparison, the number served in 2001 was 23 million. Additionally, based on a national survey called the Food Security Survey, federal government analysts reported that the number of households using food pantries increased from approximately 3 million to 5.6 million between 2001 and 2009.

Millions of people receive billions of dollars' worth of food assistance through this public–private system of food assistance, enhancing the quantity and quality of food they and their families are able to obtain and consume. Nevertheless, the persistent lack of assistance received by many poor and food-insecure people remains a major concern, especially in relation to SNAP benefits. An estimated 33 percent of eligible individuals did not participate in this program in 2008. Participation rates vary significantly across the states, which administer the program—ranging from 94 percent in Maine to 46 percent in Wyoming, a spatial variation only partly explained by economic and demographic patterns. Other factors affecting participation rates include social stigma, physical inability (e.g., among the elderly, lack of public transportation), and administrative rules, requirements, and "red tape" that impose barriers to participation and cause delays in receipt of aid, such as requiring that SNAP recipients be fingerprinted. While eligibility criteria are used to specify vulnerable populations, it is possible that someone meeting these criteria is not food insecure or hungry. Nevertheless, many people, though identified as food insecure and at risk for hunger by government and other researchers, do not receive federal food assistance. According to recent estimates by the Economic Research Service in 2009, slightly more than 40 percent of all food-insecure households did not participate in at least one of the three largest federal food assistance programs (SNAP, WIC, School Lunch), while one out of five food-insecure households with children did not receive federal food assistance.

While the EFAS has grown remarkably over the past 25 years, it does not appear to be functionally adequate to serve as a reliable and sufficient source of food on its own. To begin with, the system does not have the capacity to provide assistance to a substantial portion of the needy population. According to a Feeding America report, 27 percent of the food pantries and 10 percent of the emergency kitchens in its network have at some time turned away people due to a lack of resources, such as food supplies, operational funding, or volunteers/staff, with slightly less than half of both types of organization attributing this denial of aid to low food supplies. Moreover, the system is often haphazardly organized, temporally unstable, and highly uneven geographically. In general, these problems relate to the difficulty of matching resources needed in a supply-driven system. Dependent on donated food and money as well as volunteered labor power and time, food banks, pantries, and soup kitchens literally take what they can get, regardless of their clients' needs. Few have the cash resources to purchase additional supplies to cover those needs that are unmet by donations. The upshot is that food assistance offered through the EFAS is characterized by a high degree of instability over time and variation from place to place in terms of capability and client accessibility.

Moreover, while it is not possible to determine with precision the value of the support obtained by those who use the EFAS, due to a variety of reasons, including principally local variability in record keeping, it is evident that the individual benefit is not that substantial. Based on Daponte and Bade's estimate in 2000 of the value of food provided through the EFAS over the year and the USDA's EFAS client survey, the average yearly EFAS benefit per person is less than $10. This figure can be explained by the following calculation: $1,421,500,000 of food per year divided by 12 months = $118,458,333 food per month; that figure divided by 12,500,000 clients = $9.50 food per client per month.

The Changing Landscape: The Depoliticization, Commodification, and Privatization of Food Assistance

Any discussion of food assistance in the United States that does not address the social forces configured into and continuing to shape the public–private system is incomplete. An effective understanding of the landscape of food assistance must, therefore, give attention to the ways in which food assistance is both a reflection of and a social, economic, and political force within the larger society. In this regard, numerous close observers of the food assistance realm have identified the changing political status of hunger and food assistance as one of the fundamental dynamics characterizing the relationship between food assistance and American society, with considerable implications for the nature of citizenship and democracy in the United States. Since the 1970s, hunger has been "depoliticized"—a term describing the process whereby popular understandings of hunger and responses to it lost their moorings in notions of rights, justice, and public accountability.

This phenomenon began when civil and economic rights movements, while focusing on hunger and poverty, forged a political arrangement in which the legitimacy of the federal government was linked to the occurrence and alleviation of hunger. In this way, these movements achieved for Americans a "right to food" that had both legal and democratic (political) bases. For example, the Poor Peoples' Campaign in 1968 involved visits to the USDA's head offices in Washington, D.C., where economic rights activists protested the inadequacy of federal food assistance programs. When President Richard Nixon took office in 1969, he and his Secretary of Agriculture were reportedly warned that they would pay a significant political cost unless they responded effectively to address the question of hunger (Kotz 1969). Indeed, Nixon appeared to accept a public responsibility for hunger, declaring that its occurrence "in a land such as ours [was] embarrassing and intolerable." Moreover, he linked adequate nutrition to the effectiveness and strength of the democratic polity, observing that "More is at stake than the health and well-being of [hungry] American citizens. Something like the very honor of American democracy is involved" (quoted in Kotz 1969, 193–194). Around this time, the food stamp program (now known as SNAP) became a national program with universal benefit levels and eligibility criteria, thereby creating a guaranteed opportunity for food assistance for American households.

By the late 1990s, however, things had changed in terms of hunger's political status. As various close observers have pointed out, the occurrence of hunger and food insecurity in the United States in the first decade of the 21st century no longer constituted a scandal and received little attention from political leaders.

Scholars on this political trajectory have identified two key underlying processes. The first is described as the "commodification" of the right to food. The right to food, like any other social right, is said to be "commodified" when people lose access to genuine alternatives to market dependence. In the United States, the commodification of social rights, including the right to food, is linked to a reform movement that swept through virtually all social welfare programs during the 1980s and 1990s. The signature moment in this process was the passage in

1996 of the key piece of welfare reform legislation known as the Personal Responsibility and Work Opportunity Reconciliation Act (PRWORA). PRWORA wrought a general restriction of cash assistance by establishing time limits and new rules on program participants, primarily those requiring a show of work effort by recipients. While the focus was on cash assistance, welfare reform also affected federal food assistance. The food stamp program (SNAP) was partially marketized through the tightening of links between food assistance and employment. At the same time, welfare reform left SNAP as the only universal assistance program in the country; that is, it is the only program in which individuals meeting eligibility requirements have a legal claim on benefits. Thus, as *The New York Times* recently reported, SNAP has become "the safety net's safety net," as increasing numbers of Americans without work and cash assistance of any form (e.g., welfare, unemployment insurance) rely on SNAP benefits alone, which are designed to be a stop-gap and partial form of assistance. The likelihood that many of these people are seeking assistance through the EFAS draws our attention to the second force underlying the depoliticization of hunger in the United States—the privatization of food assistance.

The privatization of food assistance refers to the shift of responsibility for hunger relief from the state to private, often nonprofit, organizations. This process is reflected in the growth of the EFAS and in the active role played by political leaders to encourage that growth through policy, funding, and rhetoric. In 1998, Poppendieck provided the most elaborate analysis of the connections between this trend and the depoliticization of hunger. In short, she argued that the growth and large-scale institutionalization of charitable food undermines collective efforts to protect, restore, and enhance the right to food by dissipating political and moral pressures that might otherwise lead to more fundamental solutions. The EFAS, she argued, operates as both a cultural and political force, and in both ways it undermines public action to prevent hunger. Culturally, emergency food provides a moral safety valve by reassuring concerned citizens that effective action is being taken. Indeed, many people become attached to this form of food assistance because, through it, they are able to participate in solving a problem that concerns them. Similarly, private food assistance serves as a political safety valve in two ways. First, it allows political leaders to dismantle rights-based programs across different levels of government, but primarily at the federal level, by mitigating the consequences of the program elimination when it occurs. Moreover, the availability of EFAS aid weakens the advocacy process by distracting advocates from the destructive implications of privatization, as it draws them into the time- and labor-intensive work of establishing and sustaining charitable food organizations. Second, because of the visibility of the emergency food system, as well as its apparent efficiency and the breadth of social attachment to it, politicians are able to claim that they are addressing the problems of food insecurity and hunger when they fail to support public policies or government action to address hunger. In other words, precisely by relinquishing their responsibility for those conditions, they are able to present themselves as providing a solution.

The overarching implication of these cultural and political effects is that the growth of the EFAS does not simply widen the overall channel of food assistance. Poppendieck argued that there is a cost to "kindness"—namely, such charity ultimately takes the form of an erosion of the public's discursive and material commitment to the equal protection of all citizens from hunger, a cornerstone of any right to food. At the same time, the presence of the EFAS generates tacit acceptance of an unequal distribution of wealth and a decline in justice because standards of equity and equality do not apply to charity. There is no mandate or requirement of charitable givers to provide food assistance in an equitable manner. The net result is that as the EFAS has grown, food assistance has become less of a public concern for the citizenry and government and more of a private concern for individuals, such as clients, volunteers, and donors.

The political and social reverberations of commodification and privatization across the food assistance landscape ultimately play out in the ways in which they transform citizenship in the United States. In relation to food assistance, privatization and commodification erode notions of social obligation and common humanity embodied in public approaches. Instead, food-insecure citizens are considered customers or fortunate recipients and not members of a mutually dependent community with justifiable claims. According to Sen (1999), this perception in turn shapes public expectations of elected officials and government programs, as well as the private sector, and ultimately the relationship between democracy and hunger in the United States.

—Andrew Walter

Further Reading

Becker, E. "Shift from Food Stamps to Private Aid Widens." *The New York Times*, November 14, 2001. www.nytimes.com/2001/11/14/nationa/14HUNG.html.

Brown, L. "Fat and Hungry: Will Political Leaders Ever Get It Right?" Keynote Address to the Pennsylvania Hunger Action Center. Philadelphia, PA: Pennsylvania Nutrition Education Network, May 14, 2003.

Daponte, B. O., and S. L. Bade. *The Evolution, Cost, and Operation of the Private Food Assistance Network*. Discussion Paper No. 1211-00. Madison, WI: Institute for Research on Poverty, 2000.

DeParle, J., and M. Gebeloff. "Food Stamp Use Soars, and Stigma Fades." *The New York Times*, November 28, 2009. www.nytimes.com/2009/11/29/us/29 foodstamps.html?scp=6&sq=deparle+safety+net&st=nyt.

DeParle, J., and M. Gebeloff. "Living on Nothing But Food Stamps." *The New York Times*, January 3, 2010. www.nytimes.com/2010/01/03/us/03foodstamps .html?scp=7&sq=deparle+safety+net&st=nyt.

Economic Research Service. *The Food Assistance Landscape: FY 2009 Annual Report*. Washington, DC: United States Department of Agriculture, 2010.

Economic Research Service. *Food Assistance and Nutrition Program Final Report: Fiscal 2009 Activities*. Washington, DC: United States Department of Agriculture, 2009.

Esping-Anderson, G. *The Three Worlds of Welfare Capitalism*. Cambridge, UK: Polity Press, 1990.

Folbre, N. "No Free Lunches: Getting Food Aid to Eligible Families" *The New York Times*, December 7, 2009. http://economix.blogs.nytimes.com/2009/12/07/ no-free-lunches-getting-food-aid-to-eligiblefamilies/?scp=10&sq=supplemental %20nutrition%20assistance%20program&st=cse.

Hershkoff, H., and S. Loffredo. *The Rights of the Poor: The Authoritative ACLU Guide to Poor People's Rights*. Carbondale, IL: Southern Illinois University Press, 1997.

Kim, M., J. Ohls, and R. Cohen. "Hunger in America 2001." Mathematica Policy Research, 2001. www.mathematica-mpr.com/pdfs/hunger2001.pdf.

Kodras, J. E. "Economic Restructuring, Shifting Public Attitudes and Program Revision: The Politics Underlying Geographic Disparities in the Food Stamp Program." In *Geographic Dimensions of United States Social Policy*, edited by J. E. Kodras and J. P. Jones. London: Edward Arnold, 1990, 218–236.

Kotz, N. *Let Them Eat Promises: The Politics of Hunger in America*. Englewood Cliffs, NJ: Prentice-Hall, 1969.

Mabli, J., R. Cohen, F. Potter, and Z. Zhao. "Hunger in America 2010." Mathematica Policy Research, 2010. http://feedingamerica.org/faces-of-hunger/hunger -in-america-2010/hunger-report-2010.aspx.

Nord, M., A. Coleman-Jensen, M. Andrews, and S. Carlson. *Household Food Security in the United States, 2009*. Economic Research Report Number 108. Washington, DC: U.S. Department of Agriculture, 2010.

O'Brien, D., K. Prendergast, E. Thompson, M. Fruchter, and H. Torres Aldeen. *The Red Tape Divide: State-by-State Review of Food Stamp Applications*. Chicago, IL: America's Second Harvest, 1997.

Ohls, J., R. Cohen, F. Saleem-Ismail, and B. Cox. *The Emergency Food Assistance System—Findings from the Provider Survey. Volume II: Final Report.* Food and Nutrition Research Report No. 16-2. Washington, DC: Economic Research Service, U.S. Department of Agriculture, 2002.

Poppendieck, J. *Sweet Charity? Emergency Food and the End of Entitlement.* New York: Penguin Books, 1998.

Riches, G. "Advancing the Human Right to Food in Canada: Social Policy and the Politics of Hunger, Welfare, and Food Security." *Agriculture and Human Values* 16 (1999): 203–211.

Sen, A. "The Right Not to Be Hungry." *Contemporary Philosophy* 2 (1982): 343–360.

Theodore, N. "Welfare Reform, Work Requirements and the Geography of Unemployment," *Urban Geography* 22 (2001): 490–492.

Tiehen, L. "Use of Food Pantries by Households with Children Rose During the Late 1990s." In *Food Review.* Washington, DC: Economic Research Service, U.S. Department of Agriculture, 2002.

Food Diffusion

Diffusion of food is the spreading or scattering widely, from a single source or place, of a food or a method of preparing, presenting, and eating the food. This process is carried out by letting other humans know that the food or method exists, is available, tastes good, and has nutritional value. Initially, food diffusion was a survival mechanism, then it became a means of cultural dietary transfer, and now it is a quality of life and health-enhancing tool. The four main elements in the diffusion of foods are (1) innovation, (2) communication channels, (3) time, and (4) the dominant social system.

Food Innovation

A food innovation is one or more additions made in a traditional diet or way of preparing and presenting a food. The need, ease of preparation, taste, nutritional value, and costs, combined with the way the food innovation is communicated among members of a social system, determine whether the innovation will be accepted. Some food innovations diffuse slowly; others spread very rapidly. The factors that determine a specific food innovation's rate of adoption are related to the food's image, its place in society's food balance, and its dietary compatibility. Social prestige, convenience, and consumer satisfaction also are important factors.

The greater the perceived value of the food and the greater a consumer's desire for the food, the more rapid the adoption rate.

A food innovation must complement existing values, past experiences, and acquired taste to become widely accepted. Most quickly adopted new foods are inexpensive to grow or buy. They blend well with traditional foods or newly acquired tastes and are uncomplicated to prepare. If individuals or groups see or taste the food innovation, they are more likely to adopt it. Visibility and taste stimulate inquiries about a new food and experimenting with the new food. Food innovations that are perceived by members of society as contributing something positive to the society's basic diet and are simple to grow, prepare, and eat will be adopted more rapidly than other food adoptions.

Communication Channels

Food diffusion is promoted by providing information or news about a new food or ways to prepare it, which may be learned by others speaking, writing, actions, or example. At first, food was diffused along major transportation routes, rivers, lakes, and sea ports, and from nodes of political and military power or decision making. Information about new foods was shared by merchants, travelers, and military groups with political leaders, decision makers, and wealthy trend-setters. Throughout history, interpersonal channels were very effective in forming and changing attitudes toward the new food and influencing the decision to adopt or reject the new food. In the 20th century, mass-media channels became effective modes of informing consumers and creating knowledge of new foods.

Most members of a society are skeptical about new foods, fearful of their long-term impact upon them or their families, and reluctant to give up land to an unproven food that could be used for producing foods they know. In most instances, individuals evaluate a new food through subjective reports of peers who have tried the new food and adopted the new food innovation into their diet. Food diffusion is essentially social in nature, driven by word-of-mouth, and encouraged by recognition of the new food's value.

Time

New food adoptions involve individuals or groups in a source region, a means of diffusing a new food or food preparation technique to a new society, and time for individuals in the receptor society to become accustomed to the new dietary item. The last component—time—is a critical element in the diffusion of new foods that encompasses three aspects: (1) the time or stage in history when the diffusion process began; (2) the time for the new food to get to the new consumers; and (3) the time it takes for the new food to filter from the initial point of new food introduction to most or all members of a social group. Time is a factor in all stages of the innovation diffusion and adoption process. It is required to develop or create the new food; communicate information about the new food; transport samples, seeds, or plants; and then overcome any remaining uncertainty about a food innovation or diet inclusion consequences.

Within a social system, five adapter categories or classifications of society members are distinguished based on how they accept new foods: (1) innovators or trend-setters; (2) early adopters or opinion leaders; (3) early majority or liberals; (4) late majority or conservatives; and (5) laggards or those isolated from social networks.

- *Innovators*, who account for 2.5 percent of the people in the adopting culture, are the most venturesome, more cosmopolitan, and more financially well-off members of the society. They have the social and financial resources to absorb the loss if the new food or idea is rejected by most of their society.
- *Early adopters*, who represent 13.5 percent of the food-receiving society, are more integrated into the social system. They have the greatest degree of opinion leadership in most societies. This segment of society generally makes the most careful and judicious new food adoption decisions.
- The *early majority*, 34 percent of society, adopts new food items before the average person. They interact well with their peers but seldom hold positions of opinion leadership. They are an important link in the new food adoption process.
- The *late majority*, another 34 percent of society, adopts the innovation only after the new food is tested by others and more than half of their clan or society. Members of this group approach innovations with skepticism and caution. Pressure from peers normally is necessary to motivate acceptance of a new addition to their diet.
- *Laggards*, the last 16 percent of a cultural group to adopt the diffused food, have no influence on decisions made regarding anything new. Many are social loners or isolated from their clan, cultural group, or society.

Time is a dimension in acceptance of the new food in all five adapter or adoption categories. The rate of adoption is measured by the number of people who adopt the new food in a given time period.

The Social System

A social system consists of individuals, informal groups, and organizations engaged in making decisions to accomplish common goals. They have boundaries or fuzzy borders that limit the area into which a food innovation or new food item diffuses. Each social system residing in a specific and place-determined agroclimatic region has food and food preparation norms. In this environment, some established food patterns can be modified by group members who are able to influence attitudes within the social system. Acceptance of the diffused new food is a self-sustaining process if the innovators and early adopters become the critical mass. These two adapter categories are instrumental in getting the food innovation adopted and sharing the new food or food innovation with another social system or group. Through a "domino effect," one social system accepts a new food, then influences other social systems positioned near them to also accept the new food. Like a row of falling dominos, the new food spreads from a single source to other places, regions, countries, continents, and possibly the world.

Food Plant Diffusion: 10,000 BCE–1500 CE

The source regions from which food plant and food animal domestication originated were few; they occupied approximately 10 percent of the Earth's surface. Ninety percent of the Earth's surface did not contribute significantly to human food choices. Most of the food crops and food animals now grown or reared were domesticated more than 4,000 years ago or before. Until the discovery of the New World (North and South America), the diffusion of foodstuffs was slow. After the discovery of the New World and the period of great discoveries, food diffusion rapidly increased. A great exchange of plants, in particular, and animals took place between Europe and North and South America, Africa, Australia and New Zealand, and South and East Asia. Emigration of European peoples and the prosperity of European populations were major factors in the dispersal of food crops. Today, North America, Europe, and Australasia—the most technologically and socially advanced modern agriculture continents—base their agricultural economies on food plants and animals domesticated elsewhere

Prior to and during the time of the Roman Empire, 10,000 BCE to 400 BCE, considerable exchange of plants and animals took place between Europe, North Africa, Central Asia, Southwest Asia, and Northwest India; between India, China, and Malaysia; and between China, Central Asia, and Southwest Asia. After the fall of the Roman Empire, the Arabs became the most important agents of plant diffusion. The Islamic Empire, by the 1400s CE, extended from Spain and Morocco in the west, to the Indus River, northwestern India, and Indonesia in the east. Members of this culture spread oranges, limes, lemons, and citrons into the western Mediterranean lands. Later, the Crusaders linked the eastern Mediterranean lands to western Europe and made known citrus fruits north of the Alps. Rice, cotton, and sugarcane were grown in the Middle East prior to the establishment of the Roman Empire. When the Arabs conquered Mesopotamia, all of these crops were brought to North Africa, Spain, and portions of southern Italy. In the 700s CE, Arab traders settled along the east coast of Africa and introduced rice, citrus fruits, coconut palms, cucumbers, and mangos. In a remarkable period of Chinese history, during the Han Dynasty (202 BCE to 220 CE), northern Chinese included wheat, barley, and peas in their diets. These foods were diffused from Southwest Asia, along with millet and grapes from Turkestan. Those living in south China received bananas, rice, yams, sugarcane, and tea, diffused from Southeast Asia.

Food Plant Diffusion After 1500 CE

In the early 1400s CE, Portuguese mariners explored and settled offshore islands along the African west coast. They introduced sugarcane, bananas, and grapes to those living on these islands and along the coast. After the discovery of the Americas by Columbus in 1492, the Spanish conquest of the New World began. Ships containing plants and animals from Spain regularly sailed to the country's new colonies; on their return voyages, they transported American plants and animals to Spain. In 1564, the Spanish crossed the Pacific Ocean, from Mexico to the Philippines, bringing crops that originated in the Americas to Asia but also European crops and animals directly from Spain. European colonists carried the food plants and transported the food animals they knew to the Americas, to South Africa, and to Australia and New Zealand. The exchange of plants and animals was accelerated during the early colonial period.

Food crops grown in Europe prior to 1500 CE originated in Southwest and Southeast Asia and were confined largely to countries on the northern coast of the Mediterranean Sea. Potatoes and corn from America were adopted throughout Europe after 1500. At the same time, indigenous grasses and root crops became more diffused throughout Europe. Potatoes and corn yielded more food per acre than the grain crops grown in Europe at that time, and helped sustain the burgeoning European population. Potatoes were first grown in Spain in 1570, in Italy in 1587, in England and Germany in 1588, and in the Scandinavian countries in the 1700s. They emerged as the staple food crop in Ireland in the early 19th century. Corn was introduced to Spain by Columbus and then was spread throughout the Mediterranean countries by Spanish merchants. The Turks, who learned about corn through their Spanish contacts, introduced corn to Egypt in 1517, and to the Middle East and then southeast Europe (i.e., in the Balkan nations they occupied). Tomatoes were brought from Mexico to Spain in 1535 and initially were a garden ornamental. When tomatoes were improved to their current form in the 1800s, they became a vital component of the diets in Mediterranean countries. Beets were grown in Mediterranean countries in the late 400s and early 500s CE. In northwest Europe, sugar beets' sugar content was enhanced in the 1700s by breeding, from 6 or 7 percent to 20 percent.

Prior to 1500 CE, food crops in Africa south of the Sahara Desert basically consisted of sorghums, various millets, African yams and rice, bananas, and Asiatic rice. American food plants were adopted rapidly and widely after 1500. Corn was introduced to North Africa by the Spanish and Turks. This crop was brought to West Africa by the Portuguese in 1525; by the mid-1800s, it was grown throughout Africa. Sweet potatoes, yams, and manioc were introduced to West Africa by the Portuguese in the late 1500s, along with peanuts. Cacao was taken to Sao Tomé in the 1600s and to the Gold Coast of Africa in 1879. Coffee was indigenous to Africa; it was grown in Ethiopia, in the Congo Basin, and at selected sites on the African west coast. Coffee bean seeds were taken to various European colonies in the West and East Indies in the 1500s and 1600s; they were improved through selective breeding and reintroduced to West Africa in the 1870s.

There have been few changes in major food crops grown and consumed in Asia since 1500 CE. Food crops had been diffused between India and China through Burma and Vietnam. India's food crops also were acquired historically from Africa, Southwest Asia, and Southeast Asia. The basic Indian diet was stabilized before 1500 CE. Corn, chili peppers, tomatoes, and sweet potatoes were introduced by the Portuguese in the early 1500s, peanuts in the 1600s, and potatoes and manioc in the late 1700s. The Dutch brought coffee plants to Ceylon in the late 1600s. Tea was taken to India and planted in Assam after the abolishment of the East India Company's monopoly over the Chinese tea trade in 1833. The basic Chinese diet was stabilized many centuries before the discovery of the New World; food crops from America had their greatest impact on diets in southern China. Peanuts were seeded and grown in the 1530s, corn in the 1550s, and sweet potatoes in the 1560s. These crops diffused to China from India through the efforts of the Portuguese and from the Philippines and East Indies by the Spanish and Portuguese.

The colonization of North and South America after the discoveries of Columbus in 1492 led to the diffusion of Old World food plants to the New World and to the diffusion of New World food plants to the Old World. Wheat was taken to Hispaniola in the 1490s, and rye, oats, barley, and rice prior to 1512. In the early 1500s, yams, sorghum, pearl millet, and cowpeas were brought from West Africa to South America and to the West Indies by Spanish and Portuguese settlers. The Spanish and Portuguese also took basic items in their diets to the New World—particularly grapes, figs, olives, and citrus fruits. Sugarcane, bananas, and rice were grown by the Portuguese on the plantations of their West African colonies.

These crops then were taken to the New World, with the first American sugar being shipped to Spain in 1512. The French and Dutch introduced coffee to their New World possessions in the early 1700s.

North America contributed few significant food crops to the diets of early European settlers. Corn, squash, and beans were grown from south of the Great Lakes to the Gulf of Mexico. During the 1700s, Spanish settlers introduced wheat, grapes, and citrus fruits to the Southwest and to Florida. English settlers in the American South brought peanuts and manioc during this same time period. The first potatoes planted in South Carolina in 1674 came from Ireland. Asian rice was brought to the Carolinas in 1694 from Madagascar. All of the common food crops grown in England had made their way to New England by the early 1600s. Russian immigrants brought durum wheat to North Dakota and hard red winter wheat to central Kansas in the 1870s, 1880s, and 1890s. Soybeans were diffused by various means from China in 1804. Additions to the standard American diet in the late 1800s and in the 1900s came from the development and improvement of existing food crops.

Australia and New Zealand had no basic indigenous food crop plants. All food crop plants grown in Australia and New Zealand were carried to these countries by settlers who arrived after 1788. Tropical food plants, such as bananas, sugarcane, rice, and citrus fruits, were all grown by the early 1800s. Grape vines were planted in 1789, but the wine industry did not develop until the 1840s. English wheat did not do well in either Australia or New Zealand, and wheat was not successfully grown until the introduction of Mediterranean varieties. Although New Zealand had taro and sweet potatoes that were carried there in the period of Maori settlement, colonists from England brought their food crops with them; by the 1840s, most common English food crops were grown there.

Polynesia, Micronesia, and Melanesia, located in the southwestern quarter of the Pacific Ocean, remained isolated until the 1500s CE. Southeast Asian root plant foods and sweet potatoes supported a diet that was composed primarily of foods from the sea. Rice was not grown there. The Spanish did introduce American food crops to the populations who dwelled in Guam and the Marquesas and other islands located on the trade route from South Asia to the Philippines and Mexico. In the 1700s, after the exploration voyages of James Cook and Louis Antoine de Bougainville, corn and manioc were introduced to the newly discovered islands. Coconuts, brought from Malaysia by early settlers, were and remain the most important cash crop grown on many islands in the southwest Pacific.

Livestock Diffusion

The first cattle diffused from the Middle East to Egypt in approximately 5000 BCE. Cattle rearing of these longhorn breeds eventually expanded along the coastal

Features of World Food Consumption

The first feature that characterizes world food consumption is the great changes in diets and dietary expectations that have occurred since the 1960s and the considerable differences between developed and the developing world in calories consumed per day. This gap is more than 1,000 calories between, for example, Mozambique at the lower end and Belgium at the higher end of the caloric scale.

The second feature is the predominance of plant foods in diets. Approximately 80 percent of the world's total supply of calories is derived from plants, along with 70 percent of the protein and 50 percent of fats.

The third feature is that livestock products continue to provide a greater proportion of calories in developed countries than in developing countries. The fourth feature is the greater dietary diversity found in developed countries, as well as the larger amounts of sugar, vegetables, fruit, and vegetable oils consumed in these countries relative to developing countries.

The fifth feature is the proportion of calories derived from starchy staples, particularly rice, wheat, and potatoes—a factor that is inversely related to income. Some nutritionists claim that wheat flour produces 8 times as many calories per acre as beef or lamb, and potatoes more than 20 times as much.

The sixth feature is the hierarchy of food prices in most parts of the world. Calories derived from cereals and roots are the least expensive per calorie, whereas the most expensive calories are derived from fruits and vegetables. The seventh feature is that the poorer households of the world spend a higher proportion of their income on food.

Finally, the eighth feature of world food consumption is the growing dependence on imported foods in developing countries, especially in the large urban centers in developing countries. In some major urban centers of the developing world, the choice of food does not depend on the foods produced locally, but primarily on the economic capability of the city or country to import food.

—William A. Dando

lands of North Africa and into Spain and western Europe. Spanish and Portuguese settlers brought longhorn cattle to the Americas in the 1500s; today's Texas Longhorns are descendants of these animals. Shorthorn cattle entered Europe through the Balkans from Turkey. In western Europe, the Jersey and Guernsey cattle were descendants of these shorthorn cattle. Shorthorn cattle eventually displaced the longhorns in all of North Africa and the Sahel (i.e., the savanna area south of the Sahara Desert). Zebu cattle spread from India in 4000 BCE into Southwest Asia, Southeast Asia, East Asia, and Africa. Cattle, sheep, and goats were diffused to northern China at an unknown date.

North and South America had no important domesticated food animals before the arrival of the Europeans. Sheep, cattle, and goats were brought to the Indies by Columbus on his second voyage. Cattle were introduced into Mexico in the early 1500s, and ranching quickly became an important agricultural activity. Mexican longhorns were introduced into Texas in the 1600s and California in the 1700s. Settlers in New England and the southeastern Canada brought cattle from northwest Europe in the 1600s. Herefords were brought from Europe in 1817, Brown Swiss in 1869, Aberdeen Angus in 1873, and Frisians in 1875. Herefords, Shorthorns, and Aberdeen Angus were introduced to Argentina and Uruguay in the 1800s. The Indian zebu was imported into Brazil beginning in the 1870s.

Water buffalo were first used for plant growth agricultural tasks in India in approximately 3000 BCE. By 2500 BCE, they had spread to Mesopotamia. Diffusion westward was slow. Records show that water buffalo were used as draft animals in the Jordan River Valley of Palestine in the 700s CE and in Egypt and in the Balkans by 1200 CE. Water buffalo were used for field work and, in some cases, as a source of milk in China, Thailand, the Philippines, Indonesia, and Pakistan.

Sheep and goats were domesticated before cattle and pigs. These animals were taken by early agriculturalists to China, Europe, Africa (especially northern Africa), and India. In 1500, the Portuguese took sheep to Brazil and the Spanish brought them to the West Indies.

The wild boar was found in many areas of western Europe, the Middle East, Russia, and China. Originally a forest animal, the domesticated pig remained a major food source in the areas where its domestication originally took place in China, western Europe, and eventually the Americas.

Poultry Diffusion

Domestic poultry are descendants of the jungle fowl of India. The date of their first domestication has not been determined with certainty. Although the chicken is the most important form of poultry for human food consumption, turkeys (from North America) and geese are raised primarily for meat, and ducks for meat and eggs. Diffusion of chickens, ducks, and geese was rapid; chickens and geese were used as food sources in South and East Asia, the Middle East, Palestine, and Egypt before 1000 BCE.

Ancient Greeks and Romans were very successful poultry farmers and created distinctive breeds of their own. Over the centuries, five major chicken breeds were developed: (1) English, (2) Asiatic, (3) American, (4) Mediterranean, and (5) Continental European. In many developing countries of the world, the chickens raised belonged to no definite breed. Today, chicken breeds vary greatly, because some chickens might be selected by farmers specifically for egg production, whereas others may be chosen for meat quality and amount, feed consumption, plumage and color, or disease resistance.

Ducks also diffused widely and quickly and are raised on a limited scale in almost all countries of the world. Use of turkeys for meat increased in international acceptance after World War II. Turkeys generally are eaten at festivals, but diet-conscious consumers have demanded that more turkey white meat be made available in advanced countries of the world. Geese also diffused widely; these birds are raised in small numbers on farms throughout the world and are an important food component of European diets.

Future Diffusion of Foods

Humankind is searching continually for new foods. As the world becomes linked more closely and as the international media spread knowledge of food, diets, and health, citizens of the world are more food conscious. People everywhere want improved and more varied diets. Concomitantly, agricultural researchers are continuously striving to improve the quality of traditional foods and create new foods. Food innovations, food recipes, food perceptions, and food needs will stimulate the spread of food types in the 21st century.

—William A. Dando

Further Reading

Barstow, C. *The Eco-Foods Guide*. Gabriola Island, BC, Canada: New Society Publishers, 2002.

Broek, J. O. M., and J. W. Webb. *The Geography of Mankind*. New York: McGraw-Hill, 1968.

deLaubenfels, D. J. *A Geography of Plants and Animals*. Dubuque, IA: Wm. C. Brown Company, 1970.

FAO Production Yearbook, 23rd ed. Rome: United Nations, 1969.

Food: The Yearbook of Agriculture 1959. Washington, DC: U.S. Government Printing Office, 1959.

Grigg, D. B. *The Agricultural Systems of the World*. London: Cambridge University Press, 1974.

Nabhan, G. P. *Where Our Food Comes from*. London: Island Press/Shearwater Books, 2009.

Purnell, L. D., and B. J. Paulanka. *Transcultural Health Care*. Philadelphia: F. A. Davis, 2003.

Rogers, E. M. *Diffusion of Innovations*, 4th ed. New York: Free Press, 1995.

Food, Famine, and Popular Culture

Food is one of humanity's great pleasures. While essential to life, it can be much more than mere energy; food can be a source of comfort, a sensual pleasure, and a hallmark of culture. Generally, culture is thought of as the shared beliefs and practices characteristic to a particular human population, which are passed and taught from one generation to the next. These shared beliefs and practices include all aspects of life as members of that population live it, from gender roles to holiday rituals. Food is an important aspect of culture, reflecting the geography of a people, their history, their religion, and their social practices.

Culture within an area can be divided into the culture of the common people and the culture of the higher classes or castes ("high culture"). Popular culture refers to the everyday culture of the common citizen. Popular culture is often associated with mass media and consumption, especially since the early 20th century, and there are certainly strong ties between them. In reality, popular culture has existed for thousands of years.

When exploring aspects of food and culture, most researchers consider the foods of the common people. These common people are also the first and hardest hit when food shortages occur, suffering from malnutrition and starvation, and sometimes dying of famine. As a result, their culture—popular culture—includes these experiences of famine foods, of survival stories, of migration, and of loss.

Food and Culture

The foods of the common people have changed considerably since our human ancestors left the forests and began walking upright on the savannas. In their ecological niches, humans developed food traditions: what to eat and what not to eat, what to prepare, who does the preparation, what to serve when, and how to maximize available foods. Migration took humans into new regions with new food sources and new traditions developed. The development of agriculture focused production on select species of plants and animals, reducing food diversity but improving food security. All cultures developed their essential everyday cuisine, largely based on carbohydrates (grains such as wheat, rice, and corn, in particular).

For instance, the common diet of Egypt was based on bread and beer (wheat and barley), enriched by a wide variety of vegetables and legumes, and supplemented by protein sources such as fish, meat, and poultry.

Food came to be codified in each culture. The French have long been known for their cuisine and for their great appreciation and pleasure of it (gastronomy). In Mandarin Chinese, the common greeting, "Chi fan le ma," translates not as "Hello," but rather as "Have you eaten rice [food] yet?" In some cultures, members have a social obligation to their communities to take turns hosting feasts, such as with the potlatch of the Native Americans in the Pacific Northwest or the pig feasts of Papua New Guinea, whose purpose is the redistribution and reciprocity of wealth.

Food and Globalization: A Look at Chocolate

Globalization brought diverse peoples into contact with one another and introduced them to new food sources, more than doubling the known foods (see the "Food Diffusion" entry). While new foods may enrich and diversify diets, they can also please the senses and serve as a status symbol and a novelty until they are fully incorporated into a culture.

Chocolate is an excellent example of this intertwining of food, popular culture, and globalization. Cacao, the basis of chocolate, was domesticated in Central America, and its use can be traced back to between 600 and 400 BCE. Cacao became a significant component of life in the region, offered to the gods, drunk by royalty, used in marriage rituals, and given to pregnant women. An important item of trade, its "beans" were used as currency. Much of what is known of cacao's use in Central America is derived from its depiction on extant material culture, particular on pottery and murals. When cacao was brought back to Spain by explorers in the early 16th century, it was exclusively for royalty.

Over the next century, culinary and medical uses (particularly for stomach ailments) of cacao spread to France and England; again, these uses were initially reserved for royalty, then eventually trickled down through the upper classes to the lower classes. The great demand for cacao led European countries to establish cacao plantations in their tropical colonies outside of CentralAmerica (i.e., Caribbean, Africa, Southeast Asia). As production increased, cacao became affordable to ordinary European citizens, who consumed it largely as a beverage (hot chocolate). Demand for chocolate was so great that shady business people would dilute it with other substances (sometimes poisonous) to extend it, leading other business people to advertise the purity of their chocolate in newspapers. Chocolate houses in London or cafes in Paris provided stylish locations to consume hot chocolate and discuss the latest politics and intrigue. Fine porcelain sets and silver sets were produced and sold for drinking chocolate at home. Recipe books were printed and distributed for hot chocolate, candies, and desserts.

Today, chocolate products can be found throughout the world, with each culture having its own chocolate traditions. Chocolate has become a globalized popular culture fixture, with greater links made between popular culture and food through the media of television, film, and music, all appealing to the masses. Product placement began with silent films, with one of the earliest instances being the mention of Hershey bars in *Wings* (1927). Drinks, foods, and candies are now part of the planning for films, both in its production and its advertising. Manufacturers pay to have their products placed in films, and films are promoted in candy and cereal aisles of grocery stores, such as *Shrek* candies and cereal.

Food and Popular Culture

The great significance of food in human life has resulted in numerous examples of food in popular culture, touching on nearly all aspects of life. In the United States and Europe, most children are familiar with old tales of *Goldilocks and the Three Bears* (Goldilocks eats the baby bear's porridge) and *Hansel and Gretel* (children are lured to a gingerbread house by a witch). Young children in the United States learn to read with Dr. Seuss's *Green Eggs and Ham* (1960; a character refuses to try the dish but eventually likes it) and play the simple board game *Candyland* (1949). Thanks to advertising, most American children can recognize the signs and logos of restaurants such as McDonald's, Burger King, and Taco Bell before they can read. And in China and Southeast Asia, children know the logo for White Rabbit candies, a milk taffy.

In popular music, food is a constant presence. Children in the West learn elementary songs such as "Pease Porridge Hot," "The Muffin Man," or "Shortnin' Bread." In classical music, the great Italian composer Gioachino Rossini was also a gourmand, and composed such pieces as *Quatre Hors d'Oeuvres* (four "appetizers"; the fourth movement is "Le Beurre"—butter) and *Quatre Mendiants* (assortment of fruits and nuts). In early 20th-century America, blues legend Bessie Smith sang "Gimme a Pigfoot and a Bottle of Beer" (1933). Pop love songs tend to focus on sweetness, such as the Archies' "Sugar, Sugar" (1969) and Sarah McLachlan's "Ice Cream" (1999).

Since the development of popular motion pictures in the early 20th century, food has maintained a significant presence in films, sometimes playing a major role in the plot. Spectacular eating scenes can be found in *The Adventures of Robin Hood* (1938; Errol Flynn eating a massive leg of lamb), *Lady and the Tramp*

(1955; romantic spaghetti), *Tom Jones* (1963; a flirtatious shared meal at a tavern), and *9½ Weeks* (1978; sensual grazing out of the refrigerator). Cafeteria grazing and food fights were elevated to new levels after John Belushi's Bluto showed the way in *Animal House* (1978).

Food was a major element in *Babette's Feast* (1987), where in late 19th-century Denmark a cook/housekeeper uses her lottery winnings to prepare a feast of appreciation for simple villagers. In the slacker comedy *Harold and Kumar Go to White Castle* (2004), two stoners spend the movie trying to find a White Castle hamburger restaurant to satisfy their marijuana-induced cravings. *Julie & Julia* (2009) tells the stories of chef Julia Child and her introduction to French cooking and of Julie Powell, who develops a successful blog and eventually a book, based on cooking her way through *Mastering the Art of French Cooking* (Child was one of the book's authors).

On a serious note, *Alive* (1993), based on a true story, captured the struggle of a Uruguayan soccer team to survive after a plane crash in the Andes, with some of the men resorting to cannibalism. Likewise, the great popular interest in vampires and zombies in books and movies represents a fascination with breaking one of the great human taboos: the consumption of other humans.

Currently, there is tremendous interest in food and cooking in the Western world. Cooking magazines such as *Bon Appétit*, *Food and Wine*, *Saveur*, and *Cooking Light* do tremendous business with their magazines and associated cookbooks. Television shows such as *Iron Chef* (Japanese originally; the highly successful American version is a spin-off) and *Top Chef* have made celebrities out of chefs. The great success of these programs has resulted in spin-offs: *The Next Iron Chef*, *Top Chef Masters*, and *Top Chef: Just Desserts*. The Food Network, a network focused on cooking shows and food programs, was launched in 1993; it recently expanded to a second channel, The Cooking Channel (2010). Perhaps it is no wonder that Western society has developed obesity problems given people's obsession with food.

Famine and Culture

While the somber topic of hunger and famine seems to be the antithesis of popular culture, these topics can also be found in popular culture. Historically, all peoples at risk of hunger and famine include these experiences in their culture. Traditions of "famine foods" are passed down from generation to generation, and stories are widely told about past hunger times.

Besides oral traditions, references to ancient famines can be found in major cultural texts such as the *Epic of Gilgamesh*, the Jewish Tanakh, and the Christian

Bible. Some of the earliest accounts of famine are found on ancient Egyptian steles and bas-reliefs. Development of mass printing increased the available information about the world, as books and other printed materials, such as magazines and newspapers, became accessible to the common people. A frequently cited statistic on famine—that between 108 BCE and 1911 CE, more than 1,828 famines occurred in China—was undoubtedly based on China's great historical literary resources (woodblock printing beginning in the first century CE). In China in 1406, Zhu Su compiled *Herbal for Relief from Famines*, a compendium of plants that could be eaten in famine times, complete with illustrations. In the Western world, Gutenberg's Bible, printed in 1455, represents an important cultural document as well as an account of ancient famines.

In 1729, Jonathan Swift famously wrote a satire on Irish hunger: *A Modest Proposal, for Preventing the Children of Poor People in Ireland from Being a Burden to Their Parents or Country, and for Making Them Beneficial to the Public.* This document was first published as a pamphlet, inexpensively produced and widely available at local shops. In it, Swift states (tongue firmly in cheek): "[A] young healthy child well nursed is at a year old a most delicious, nourishing and wholesome food, whether stewed, roasted, or boiled, and I make no doubt that it will equally serve in a fricassee, or a ragout . . ." Swift wrote this satire for a broad Irish audience chafing under English rule who would immediately recognize the humor. His work was, through and through, popular culture. After 280 years, Swift's *A Modest Proposal* is still in print; it is widely recognized as a masterpiece of satirical writing, yet at the same time documents the historical food insecurity of Ireland's lower classes.

The famine with probably the greatest impact on popular culture was another Irish hunger time: The Irish Potato Famine (see the "Great Irish Famine: 1845–1850" entry) killed an estimated 1 million people between 1845 and 1850 and resulted in the immigration of another million people to countries such as the United States, Canada, and Australia. For those who lived and remained in Ireland, the catastrophe resulted in songs; poetry; "famine landscapes" of abandoned cottages, famine roads, and mass graves; and stories passed down through generations. As years passed, literature was written and paintings created about the famine.

For those who emigrated and their descendents ("the Irish diaspora"), the Irish Potato Famine became the linchpin of their cultural identity. Today, a tremendous global market exists for all sorts of cultural materials related to the famine: Books, both fiction and nonfiction, and documentaries on this topic are produced regularly. Many Irish Americans take tours of Ireland to visit their family's ancestral landscapes. Nevertheless, relatively few dramatic films have been made based on the Irish famine (or on any famine, for that matter) compared to films addressing other human catastrophes, such as war or natural disasters. This omission may

Famine Foods

Famine foods are resources to which people turn during food shortages to prevent death by starvation. They are edible to humans and available when other sources are not. During the Irish Potato Famine, for example, some kelp species were consumed by coastal Irish populations. In New England 200 years ago, lobster and other shellfish were considered poverty foods and their consumption concealed because of the stigma attached to eating these items.

Famine foods may be undomesticated species gleaned from the wild (e.g., nettles or young ferns), may taste bad or cause distress after consumption (e.g., acorns), or traditionally may not be consumed by humans (e.g., rice hulls). They may be off-limits for cultural reasons such as religious prohibitions (e.g., pork for Jews and Muslims), seed stock saved for future plantings, or because they are culturally viewed as companion animals, such as dogs, cats, or horses.

These food sources have long been assumed to be less desirable and the foods of last choice, the foods of desperate people. However, recent work on nondomesticated plant species has found some to have unusually high nutritional values. Scientists now are exploring these plants for possible development into new crops. Some of these plants also are getting new attention from gourmands interested in eating locally and seasonally, with locally foraged regional plants such as dandelion greens, nettles, or mushrooms having a certain cachet.

—Christina E. Dando

be due to the great human tragedy that occurs with hunger and famine—it may be too somber a topic for entertainment. Alternatively, it may reflect the slow unfolding of hunger and famine over long periods of time, such that the tale lacks the dramatic urgency of, say, a tsunami or a rebellion, and its long duration is more difficult to capture on film.

In 2005, the United Nations World Food Programme took an unprecedented step: It attempted to use popular culture and the latest technology to teach children about global hunger. Program organizers developed a computer game, entitled *Food Force* (available free online as a download), in which players join the World Food Programme and attempt to airdrop food to hungry residents on an imaginary Indian Ocean island. In the first six weeks after its release, the game registered more than 1 million players.

Famine and Global Culture

As globalization has linked our world, particularly over the past 200 years, humans have gained both more and more-immediate knowledge about what is going on in our world. News of hunger and famine now reaches almost all corners of the globe. Over the past 100 years, more developed countries have begun to reach out and attempt to assist other areas of the world facing food shortages. This awareness of hunger and the desire to assist in some way has become part of popular culture as concerned individuals seek to motivate others to assist with these global crises.

The first major benefit concert was held in 1971—The Concert for Bangladesh, organized by John Lennon and Ravi Shankar to provide relief for Bangladeshi refugees, whose country had been devastated by war and weather (cyclone). The Ethiopian famine of 1984–1985 (see the "Ethiopian Famine: 1984–1985" entry) affected global popular culture when musicians in Great Britain used music to bring greater public attention to the disaster. A BBC news report on the Ethiopian situation inspired musicians Bob Geldof and Midge Ure to organize a U.K.-based relief effort they termed "Band Aid" in 1984. Band Aid initially focused on the production of a record single that could be used to raise relief monies. Entitled "Do They Know It's Christmas?," the song involved many of the most popular British and Irish musicians of the day in its production and featured the chorus, "Feed the world; let them know it's Christmas time." In July 1985, the organizers followed up the single with a major concert event, Live Aid, held simultaneously at Wembley Stadium in London and John F. Kennedy Stadium in Philadelphia. Band Aid and Live Aid together raised an estimated $60 million, all of which was earmarked for food assistance. Band Aid/Live Aid inspired musicians in the United States and Canada to follow suit by releasing two singles: USA for Africa's "We Are the World" and Northern Light's "Tears Are Not Enough." Besides raising funds, these ventures began a tradition of celebrities wielding their status in an effort to bring greater attention to issues they view as important.

Thanks to Band Aid, USA for Africa, and other organized efforts, greater public attention was brought to the crisis in Ethiopia, which became part of the global shared consciousness. In what has become a classic moment of stand-up comedy, American comedian Sam Kinison ranted in 1985 about the situation:

> . . . But I'm not trying to make fun of world hunger. Matter of fact, I think I have the answer 'cause I've spent a lot of time working it out. You want to stop world hunger?
>
> Stop sending 'em food. Don't send these people another bite, folks.
>
> You want to send them something, you want to help?
>
> Send them U-Hauls. Send them U-Hauls, some luggage, send them a guy out there who says, "Hey, we been driving out here every day with your food, for, like, the last thirty or forty years, and we were driving out here today

across the desert, and it occurred to us that there wouldn't be world hunger, if you people would LIVE WHERE THE FOOD IS!

YOU LIVE IN A DESERT! YOU LIVE IN A F—ING DESERT!

NOTHING GROWS OUT HERE! NOTHING'S GONNA GROW OUT HERE!

YOU SEE THIS? HUH? THIS IS SAND.

KNOW WHAT IT'S GONNA BE A HUNDRED YEARS FROM NOW? IT'S GONNA BE SAND!

YOU LIVE IN A F—ING DESERT! GET YOUR KIDS, GET YOUR STUFF, GET YOUR S—T, WE'LL MAKE ONE TRIP, WE'LL TAKE YOU TO WHERE THE FOOD IS!"

While Band Aid sought to bring attention to the crisis, Kinison's monologue reacted to the great public awareness of the crisis and sarcastically offers a practical solution. Kinison's screaming profanities, played for laughs, reveal a lack of understanding of the complexities of hunger and famine issues: If it were only so simple as moving to a new place (although it could be argued that this is, indeed, what 1 million Irish did—never mind the million people who perished). But Band Aid/Live Aid can be similarly viewed, like Kinison's rant, as a superficial response. Like a Band Aid bandage, the aid was a simple patch to a complex problem that did not address the underlying problems and unintentionally may have interfered with healing. Such is often the case with popular culture and serious issues such as famine, presenting the proverbial double-edged sword: Is the greater public awareness of famine that can come from popular culture worth the superficial conceptions that are often created?

—Christina E. Dando

Further Reading

Bremner, Robert H. *Giving: Charity and Philanthropy in History.* New Brunswick, NJ/London: Transaction Publishers, 1994.

Civitello, Linda. *Cuisine and Culture: A History of Food and People.* Hoboken, NJ: John Wiley & Sons, 2004.

Grivetti, Louis E., and Howard-Yana Shapiro, eds. *Chocolate: History, Culture and Heritage.* Hoboken, NJ: John Wiley & Sons, 2009.

King, Carla, ed. *Famine, Land and Culture in Ireland.* Dublin: University College Dublin Press, 2000.

Ó Gráda, Cormac. *Famine: A Short History*. Princeton NJ: Princeton University Press, 2009.

Rosenberg, Tina. "What Lara Croft Would Look Like If She Carried Rice Bags." *The New York Times*, December 30, 2005. www.nytimes.com/2005/12/30/opinion/30fri4.html?_r=1&scp=8&sq=lara%20croft&st=cse

Schroeder, Fred E. H. *5000 Years of Popular Culture: Popular Culture Before Printing*. Bowling Green OH: Bowling Green University Popular Press, 1980.

Food Poisoning

Food poisoning, also known as foodborne illness, is a result of consuming organisms or toxins in contaminated foods. More than 250 different foodborne illnesses are known. The World Health Organization (WHO) estimates that annually more than 3 million premature deaths and 1.5 billion cases of diarrhea in children can be attributed to consuming contaminated foods. The Centers for Disease Control and Prevention (CDC) estimates that among the 290 million residents in the United States, there are approximately 76 million incidents of illness, 325,000 hospitalizations, and 5,000 deaths annually due to foodborne illnesses.

Food poisoning symptoms vary in severity and duration, depending on the type and amount of toxin ingested and the health status of the individual. Poisons in foods enter the body through the gastrointestinal (GI) tract. The most common symptoms of food poisoning are nausea, vomiting, abdominal cramps, and diarrhea. Body water and electrolytes are lost during episodes of vomiting and diarrhea, which in turn leads to dehydration. Signs of dehydration include dry mouth, lightheadedness, and decreased amount of urination. The urine becomes concentrated and looks dark in color. Those individuals who are most susceptible to dehydration include children, the elderly, and people with a suppressed immune system. Other symptoms of chronic food poisoning include stunted growth in children, kidney and liver failure, brain and neural disorders such as Guillain-Barré syndrome, reactive arthritis, and death.

Food Poisoning and Famine

Numerous studies indicate that people die of food poisoning during times of famine. Extreme hunger leads to desperate actions to end the pain of chronic hunger. Many people eat available foods that may be potentially contaminated. Food poisoning coupled with nutritionally compromised immune system can lead to chronic illness and ultimately death.

Tracking of Food Poisoning

Food poisoning illness is widespread and an increasing public health problem across the globe. Data for such illnesses are scarce, however, as many countries do not have established surveillance programs to report, identify, and track outbreaks. Those countries tracking foodborne illness rates show that it is increasing around the world. To prevent needless deaths, there is a growing movement to adapt food preparation and storages techniques to reduce the risk of food poisoning.

Causes of Food Poisoning

Many types of agents cause food poisoning, including bacteria, parasites, molds, viruses, and harmful and heavy metals. The incubation period for each agent varies. Some common types of food poisoning start within 2 to 6 hours, whereas others may not appear until days after the exposure.

Bacteria

Bacteria are single-cell microorganisms that exist on their own or as parasites in a host organism. They are present in virtually all types of environments, including soil, water, air, organic matter, and bodies of animals and humans. Most bacteria are harmless and beneficial, but some are harmful and toxic. Under the right conditions, bacteria grow quickly. One bacterium can reproduce every 20 minutes. Thus, in 16 hours, one bacterium can multiply to 70 trillion bacteria.

Food poisoning results from eating food that has been contaminated by harmful bacteria and or their toxins. Bacteria can infect the intestines or produce toxins. Gastrointestinal inflammation with infectious bacteria leads to diarrhea and the inability to absorb nutrients and water. By comparison, bacterial toxins poison the GI tract causing nausea and vomiting. In both situations, a chronic loss of body water and electrolytes may lead to kidney failure and death. It may also cause malnutrition, as nutrients cannot be absorbed.

Bacillus cereus

Bacillus cereus causes two types of foodborne illnesses: emetic and diarrheal. The emetic form has a rapid onset of 1 to 6 hours and is characterized by nausea, vomiting, and abdominal cramps. This form is linked to the consumption of contaminated rice, potato, pasta, and other starchy foods. The second type occurs after an incubation period of 8 to 16 hours and is characterized by either

The Five-Second Rule

A common belief applied when food drops on the floor is the "five-second rule." This rule is applied with conviction by its believers, who state that it is safe to eat any food that drops on the floor if it is picked up within five seconds. This rule became part of some cultures to the extent that some scientists decided to test it. They found that this "rule" is a myth and that bacteria get picked up with the food, no matter how fast it is removed from the floor.

Here are five facts to remember for those who are tempted to apply the five-second rule:

1. Bacteria can attach themselves to food instantly.
2. Foods with high moisture content are more likely to pick up larger numbers of bacteria.
3. Non-carpeted floors may be cleaner than the carpeted ones, but they still harbor bacteria.
4. Clean and dry floors also are covered with bacteria.
5. Newly washed floors are only as clean as the mop that was used to clean it.

—Kausar F. Siddiqi and Sara A. Blackburn

small-volume or profuse and watery diarrhea. Foods commonly associated with diarrheal *B. cereus* poisoning include meats, fish, milk, and vegetables.

Campylobacter jejuni

Campylobacter jejuni causes an infection of the small intestine known as *Campylobacter* enteritis. *C. jejuni* is responsible for one of the several types of travelers' diarrhea. Even a small number of bacteria—as few as 500 to 600—can produce detrimental effects in the GI system. Other symptoms include nausea, headache, muscle ache, abdominal pain, and fever, and, on rare occasions, Guillain-Barré syndrome.

C. jejuni are carried by healthy cattle and flies on the farm. These bacteria can also be present in raw chicken, raw milk, and nonchlorinated water. Cooking chicken properly, pasteurizing milk, and chlorinating water can destroy these pathogens.

Clostridium botulinum

Clostridium botulinum grows in anaerobic (without oxygen) conditions and causes botulism. Just a few nanograms of the toxin can be fatal. After the body's exposure to the toxin, flaccid paralysis progresses symmetrically downward, usually starting with the eyes and face, then moving to the throat, upper body, and extremities. When the chest becomes completely affected, asphyxiation occurs, resulting in death.

C. botulinum produces both spores and toxins. The spores are heat resistant, whereas the toxin is destroyed when heated to 176°F (80°C). Any type of nonacid food (pH greater than 4.6) is conducive to growth of *C. botulinum* and its botulism toxin. Consequently, this toxin can be present in a wide range of foods, such as canned foods including asparagus, beets, corn, green beans, peppers, ripe olives, spinach, tuna, chicken, chicken liver, and liver pate. It can also be present in ham, luncheon meats, sausage, stuffed eggplant, and smoked and salted fish

Clostridium perfringens

Clostridium perfringens grows in anaerobic conditions and produces spores. It is commonly a benign component of the normal flora in the human and animal intestines. The spores remain in soil that contains human or animal fecal matter.

The most common cause of food poisoning by *C. perfringens* is leaving prepared foods out at an improper temperature, which allows these bacteria to flourish. Symptoms of *C. perfringens* poisoning include diarrhea and severe abdominal cramping. A serious, often fatal form of this disease is necrotic enteritis or pig-bel disease. In this case, death is caused by infection and necrosis of the intestines, which results in septicemia (blood poisoning). Foods generally implicated in *C. perfringens* poisoning are meats, meat products, and gravy.

Escherichia coli O157:H7

Escherichia coli bacteria normally live in the intestines of human beings and animals without causing any problems. Certain strains of these bacteria can cause severe food poisoning, however. One strain that is of particular importance is *E. coli* O157:H7; ingestion of just small amounts of this pathogen causes illness. Symptoms include nausea, vomiting, severe abdominal cramping, watery diarrhea that can progress to bloody diarrhea, hemolytic uremic syndrome (breakdown of red blood cells and kidney failure), and death.

Foods often implicated in *E. coli* O157:H7 outbreaks include ground beef, alfalfa sprouts, lettuce, unpasteurized fruit juices, dry-cured salami, cheese curds, and game meat. Cross-contamination of food by fecal matter is the primary cause of this poisoning.

Listeria Monocytogenes

Listeria monocytogenes is found in the soil, silage, and other environmental sources. These bacteria cause a serious illness called listeriosis. Most healthy adults do not show any symptoms of this disease; instead, those most susceptible to this illness are pregnant women, newborn children, the elderly, and persons with a compromised immune system. Symptoms of *L. monocytogenes* poisoning include nausea, vomiting, fever, chills, and backache. The most serious form of this disease is accompanied by meningitis (inflammation of the membranes that cover the brain and the spinal cord), encephalitis (inflammation of the brain), and septicemia (blood poisoning). In pregnant women, infection of the fetus is common and can lead to spontaneous abortion, a stillborn baby, or a newborn with birth defects.

L. monocytogenes can grow in temperatures as low as 37°F (3°C); therefore, these pathogens are able to grow in refrigerated foods. Foods most likely to carry *L. monocytogenes* bacteria include raw milk, cheeses (especially soft-ripened), ice cream, raw or cooked poultry, raw and smoked fish, luncheon meats, hot dogs, raw vegetables, and raw meats.

Salmonella spp.

Salmonella spp. (spp. means "species of") bacteria are commonly present in animals, especially swine and poultry. They are also present in water, soil, insects, food processing and preparation surfaces, animal fecal matter, and raw meats, fish, and poultry. These bacteria are facultative anaerobic organisms, meaning that they can grow with or without oxygen. In acute cases, *Salmonella* spp. cause nausea, vomiting, abdominal cramps, diarrhea, and fever. If this infection becomes chronic, then arthritic symptoms may follow three to four weeks after the onset of acute symptoms.

Foods commonly associated with *Salmonella* poisoning are raw meats, poultry, eggs, milk and dairy products, fish, shrimp, frog legs, yeast, coconut, cake mixes, cocoa, chocolate, sauces, salad dressings, cream-filled desserts, and toppings.

Shigella spp.

Shigella spp. are anaerobic facultative bacteria that are commonly found in the intestines and fecal matter of humans and warm-blooded animals. *Shigella* produces a toxin that prevents water from being reabsorbed into the gastrointestinal tract, which results in watery diarrhea. As few as 10 *Shigella* bacteria can cause an infection called shigellosis. Symptoms of shigellosis include abdominal cramps, diarrhea, chills, fever, fatigue, dehydration, and blood, pus, or mucus in stools.

Foods commonly associated with shigellosis include salads (chicken, macaroni, potato, shrimp, and tuna), milk and dairy products, poultry and raw vegetables, and any food contaminated with fecal matter containing these bacteria. Unsanitary handling of food and use of water with fecal bacteria are the most common causes of contamination.

Staphylococcus aureus

Staphylococcus aureus is a facultative anaerobic bacterium that grows on food and produces a heat-stable toxin. A dose of less than 1.0 microgram of the toxin can produce symptoms within one to six hours of ingestion; the onset of symptoms of staphylococcal poisoning is generally acute and rapid. Symptoms typically include severe nausea, acute abdominal cramping, vomiting, and diarrhea. In more severe cases, headache, muscle cramping, and transient changes in pulse rate and blood pressure may occur. Foods that generally cause staphylococcal food poisoning include meat and meat products, poultry and egg products, custards, salads (such as chicken, egg, macaroni, potato, and macaroni), bakery products (such as cream filled pastries, cream pies, and chocolate éclairs), sandwich fillings, and milk and dairy products.

Vibrio spp.

Three species of *Vibrio* are connected with foodborne illness: *V. cholera, V. paranhaemolyticus*, and *V. vulnificus*. Symptoms of *Vibrio* spp. food poisoning typically include nausea, vomiting, abdominal cramps, diarrhea, headache, fever, and chills. *Vibrio* spp. are resistant to salt, which explains why they are commonly found in seafood. Foods associated with this type of foodborne poisoning are raw, underprocessed, improperly handled, and contaminated fish and shellfish. Notably, these bacteria are found in clams, crabs, lobster, oysters, and shrimp (Table 14).

Table 14. Food Poisoning Bacteria

Bacteria	Onset of Symptoms	Signs and Symptoms	Common Foods
Bacillus cereus (diarrheal)	8–16 hours	Abdominal cramps	Meats, fish, milk, and vegetables
Bacillus cereus (emetic)	30 minutes –6 hours	Nausea, vomiting, and abdominal cramps	Rice, potato, pasta, cereal, grains, and other starchy foods
Campylobacter jejuni	2–5 days	Nausea, abdominal cramps, watery diarrhea, muscle ache, and headache	Raw chicken, raw milk, and nonchlorinated water
Clostridium botulinum	12–72 hours	Nausea, vomiting, and abdominal cramps	Sausage, meat and seafood products, and canned vegetables
Clostridium perfringens	8–22 hours	Diarrhea and intense abdominal cramps; severe case: death due to septicemia	Meat, meat products, and gravy
Escherichia coli O157:H7	12–72 hours	Nausea, vomiting, severe abdominal cramps, bloody diarrhea, kidney failure, and death	Ground beef, alfalfa sprouts, lettuce, unpasteurized fruit juices, dry-cured salami, cheese curds, and game meat

Table 14. (Continued)

Bacteria	Onset of Symptoms	Signs and Symptoms	Common Foods
Listeria monocytogenes	9–48 hours for GI symptoms; 2–6 weeks for throughout the body	Nausea, vomiting, fever, chills and backache; severe: meningitis, encephalitis, septicemia, and birth defects	Raw milk, cheeses (especially soft-ripened), ice cream, raw or cooked poultry, raw and smoked fish, luncheon meats, hot dogs, raw vegetables, and raw meats
Salmonella spp.	6–48 hours	Acute: vomiting, abdominal cramps, diarrhea, and fever Chronic: arthritic symptoms	Raw meats, poultry, eggs, milk and dairy products, fish, shrimp, frog legs, yeast, coconut, cake mixes, cocoa, chocolate, sauces, salad dressings, cream-filled desserts, and toppings
Shigella spp.	12–50 hours	Abdominal cramps, diarrhea, chills, fever, fatigue, dehydration, and blood, pus or mucus in stools	Salads (chicken, macaroni, potato, shrimp and tuna), milk and dairy products, poultry, and raw vegetables
Staphylococcus aureus	2–7 hours	Vomiting, watery diarrhea, possible low-grade fever	
Vibrio vulnificus	2–48 hours	Nausea, vomiting, abdominal cramps, diarrhea, headache, fever, and chills	Fish, clams, crabs, lobster, oysters, and shrimp

Molds

Molds are microscopic fungi that live on plants and animals. An estimated 300,000 species of fungi exist. A few of the molds produce toxic chemicals known as mycotoxins, which can cause illness to humans exposed to these poisonous substances.

Molds can easily establish themselves on crops, such as fruits, grains, and nuts, both while these crops are in the field and after their harvesting. Types of mycotoxins include aflatoxin, fumonisin, ochratoxin A, patulin, trichothecenes, and zearalenone. Long-term, low-level exposure to mycotoxins is linked to cancer.

- Aflatoxin is a cancer-causing poison. Foods associated with aflatoxin poisoning are peanuts, tree nuts, and corn.
- Fumonisin causes kidney and liver disease. It is generally found in corn.
- Ochratoxin A causes kidney disease. Foods linked to this type of poisoning include cereals, coffee, and wine.
- Patulin causes genetic mutations. Foods that can contain patulin include fruit juices, especially apple juice.

- Trichothecene, when ingested in high doses, causes severe vomiting and diarrhea. This mycotoxin can be found in cereals.
- Zearalenone disrupts the endocrine system—the body system that includes the endocrine glands, which produce different hormones that are essential for the normal functioning of the body. This mycotoxin can be found in cereals.

Parasites

Parasites range in size from single-cell organisms to worms that are visible to the naked eye. They can be present in food or water and can cause illness and sometimes death.

Parasites require a living host to survive. They are often excreted in the feces. As a consequence, they can be transmitted from one host to another, either animal or human, through consumption of food or water contaminated by the feces.

Washing hands with warm water and soap before handling food, before eating, and after using the bathroom, changing a baby's diaper, and handling animals is the most effective way of preventing parasites from spreading.

Viruses

Viruses are much smaller in size than bacteria and require a living host—either human or another type of animal—to survive. Three viruses are of primary concern when dealing with food: hepatitis A, Norwalk, and fotavirus. Viruses do not multiply in foods, but rather are transferred from one food to another by the person handling the food or by food being exposed to a contaminated water supply.

Proper hand washing, especially after using the toilet, is the most important factor in preventing the spread of foodborne viruses.

Hepatitis A Virus

Hepatitis A virus (HAV) is a foodborne virus. The onset time for infectious hepatitis is 15 to 50 days after consuming the contaminated food. The disease is characterized by sudden onset of fever, nausea, abdominal discomfort, fatigue, and loss of appetite. Colonization with HAV causes a liver disease known as infectious hepatitis. Advanced stages of the disease are marked by enlargement of the liver and jaundice, a yellowing of the skin.

A person who is infected with HAV may not show any symptoms for as long as six weeks after initial acquisition of the virus. Individuals remain contagious from one week before the onset of symptoms until two weeks after the symptoms appear; during that period, they may transmit the virus by handling food with unwashed hands.

Foods commonly associated with HAV exposure include salads, cold cuts, sandwiches, fruits and fruit juices, milk and milk products, vegetables, raw or lightly cooked oysters and clams, iced drinks, and bakery products.

Norwalk Virus

Norwalk virus and Norwalk-like viruses cause nausea, vomiting, abdominal pain, and diarrhea. The illness occurs 24 to 48 hours after consuming the contaminated food or water.

The most common source of Norwalk outbreaks is contaminated water; other sources include shellfish, salad ingredients, and raw or inadequately steamed clams and oysters.

Rotavirus

Three groups of rotaviruses (groups A, B, and C) are known to infect humans. Rotavirus A, the most common, causes viral gastroenteritis. Symptoms include vomiting, watery diarrhea, and low-grade fever. Rotaviruses are transmitted via the fecal–oral route. A person with rotavirus diarrhea excretes a large number of viral organisms that can be readily transmitted through contaminated hands. For example, infected food handlers may transmit the virus by handling food preparation tools or food that does not require any further cooking, such as salads and fruits.

Food Poisoning Prevention

Many steps can to be taken to prevent food contamination and poisoning. Washing hands with warm water and soap before handling food, before eating, and after using the bathroom, changing a baby, and handling animals is the most effective way of preventing foodborne illness from spreading. These measures are explored in the "Food Safety" entry.

—Kausar F. Siddiqi and Sara A. Blackburn

Further Reading

DeWaal, Caroline Smith, and Nadine Robert. Global and Local: Food Safety Around the World. Washington, DC. Centers for Science in the Public Interest, 2005. Accessed August 2010. www.cspinet.org/new/pdf/global.pdf.

McSwane, David, Nancy Rue, and Richard Linton. *Essentials of Food Safety and Sanitation*, 2nd ed. Upper Saddle River, NJ: Prentice Hall, 2000.

Mokyr, Joel, and Cormac O'Grada. "What Do People Die of During Famines: The Great Irish Famine in Comparative Perspective." *European Review of Economic History* 6 (2002): 339–363.

Todar, Kenneth. "Todar's Online Textbook of Bacteriology." Accessed September 2010. www.textbookofbacteriology.net.

U.S. Food and Drug Administration. "Bad Bugs Book." Accessed September 2010. www.fda.gov/Food/FoodSafety/FoodborneIllness/FoodborneIllnessFoodbornePathogensNaturalToxins/BadBugBook/default.htm.

U.S. Food and Drug Administration. "Foodborne Illness-Causing Organisms in the U.S." Accessed August 2010. www.fda.gov/Food/ResourcesForYou/Consumers/ucm103263.htm.

Webb, Patrick, and Andrew Thorne-Lyman. "Entitlement Failure from a Food Quality Perspective: The Life and Death Role of Vitamins and Minerals in Humanitarian Crises." Research Paper No. 20006/140, United Nations University, 2006. www.fao.org/righttofood/KC/downloads/vl/docs/AH529.pdf.

Food Policy Debates: Global Issues of Access

In particular times and places, citizens seemingly unite about food accessibility issues—issues that arise from their concerns of uniformly applied policies. For example, Poppendieck (1986) described citizens' activities in conjunction with Depression-era issues of economic and agricultural policies. Bentley (1998) and Levenstein (2003) detailed citizen complaints arising from mandatory war rationing programs. Maney (1989) related individuals' actions against perceived racial inequalities among food assistance and agricultural policies. In addition, Trattner (1999) detailed widespread public support behind issues of recent U.S. welfare reforms.

Normally, individuals voicing their concerns develop a series of claims that are based on people's knowledge. A liberal tradition contends that elected government officials ought to help ensure that everyone maintains an equal opportunity of success. A Marxist-oriented radical policy economy tradition argues that laborers should act together to minimize the differential effects of capitalism. Finally, a conservative tradition contends that everyone ought to act to help each other.

Throughout the world, concerned people engage various methodological tools to measure issues of securing food access. These issues concentrate on individuals' inability to secure adequate access to food supplies. The outcomes of this process then afford proponents with evidence to support their policy claims. This evidence usually highlights the adverse physiological and social effects of individuals' limited accessibility to food.

Global Measures of Access

Concerned persons may choose to employ a variety of methods to measure food access. These methods concentrate on two primary measures. Researchers employing absolute measures attempt to define an absolute, basic minimum standard. Generally, these standards remain constant over time. In contrast, citizens citing relative measures define their evidence based on a condition of comparative disadvantage. Individuals employing this standard assess people against some condition that is continually changing in time.

In many places, researchers use absolute measures, which maintain minimum thresholds or lines. Experts generally develop these measures based on an assessment of individuals' basic physiological needs. In so doing, they assume that all people maintain some minimum subsistence level—a level that is measurable and quantifiable. When measured against this standard, people who are unable to securely access a minimum quantity of food are considered disadvantaged or deprived. Researchers adopting absolute measures may assess a person's or household's income, caloric intake, body measurement, or some other quantifiable standard. These values represent minimum quantities of money, calories, or body size, beyond which individuals are thought to risk living without adequate quantities of food.

For many years, Americans measured issues of food access based on household income. This absolute measure was developed by Mollie Orshansky of the Social Security Administration, who constructed poverty thresholds based on the nation's lowest-cost food plan. This food plan, which was prepared and priced by the U.S. Department of Agriculture, established the minimum quantity of food necessary for an adequate, nutritious diet. In her work, Orshansky assumed that people need to maintain a basic level of subsistence. To access these goods, they must maintain a minimum level or standard of income. If some people do not obtain this level of income, these people are categorized as economically disadvantaged in their ability to securely access adequate food.

While absolute values are valuable, many concerned parties prefer to adopt relative measures when assessing food security. These measures are based on comparisons of people within the broader society. Individuals employing these measures assume that people act in organized social groups. Within these groups, some people are seen to maintain comparatively fewer resources. Although these individuals may be physically able to survive, it is believed that their inability to secure adequate access prevents them from fully participating in society.

Food Policy Debates: Issues of Access

Around the world, many concerned people measure issues of food access, especially in relation to children. They employ these methods as a means of comparing the challenges faced by people in different times and places—that is, to assess the effects of hunger. They are interested in these effects because they see them as essential steps to help improve the lives of children. This concern was reflected in the Preamble of the Recommendations of the Board of Inquiry into Hunger and Malnutrition in the United States in 1968:

> There must be a commitment by the nation to the proposition that every child has the right to an adequate diet. What do we mean by a "commitment"? We mean more than a statement by the President, or the preamble of a law. We mean that there be an organized set of laws and executive policies framed to achieve this objective . . . With a realistic and sense of resolve, we must say that all our children shall eat well.
>
> [Further,] [t]here must be a similarly resolute commitment to the proposition that every adult shall have the means to obtain an adequate diet. . . . When we speak here of a resolve [we mean] every adult be enabled to provide food for himself [sic] and his [sic] dependents, we are asking the nation not merely to make, but to keep, its promises.

—Preamble, Recommendations of the Board of Inquiry into Hunger and Malnutrition in the United States (1968)

Source: Citizens' Board of Inquiry into Hunger and Malnutrition in the United States. 1968. *Hunger, U.S.A.* Washington, DC: New Community Press.

—Ann Myatt James

Relative measures take a variety of forms, with these forms usually being specific to people residing in particular times and places. For example, such measures may include median incomes, food expenditures, proximate distances, travel times, consumption, vehicle ownership, assistance program participation, public opinion, and other factors. Generally, these factors are observable, although they need not be countable. As the relative measure rises or falls within the population, the measure under which a person may be determined to be disadvantaged varies in time and space.

Patterns of Access: U.S. Rural and Urban Locales

Researchers employ these tools to measure an individual's ability to secure access to food across different populations, where each population is distinguished by its geographic location. In assessing these populations, one method is to compare published access accounts of rural and urban areas in the United States. In so doing, some general trends and patterns can be discerned that suggest individuals experience a variety of issues; these issues, in turn, are influenced by their distinct geographic and social locations.

In the United States, researchers compare residents located in rural and urban places. The Census Bureau distinctly characterizes these places. According to its definitions, "rural areas comprise open country and settlements with fewer than 2,500 residents," whereas urban areas comprise "larger places and densely settled areas around them." Using slightly different terminology, researchers may refer to these types of places as nonmetropolitan and metropolitan areas. They often use the terms "rural" and "nonmetro," and "urban" and "metro," interchangeably.

Observers assessing nationwide conditions also describe geographical patterns in people's ability to securely access food supplies. Miller (2005) indicated that impoverished people reside in the most urban and rural locations. Other researchers have indicated that residents of these locations often have more limited access to supermarkets and healthful foods. These issues are highlighted in geographic concentrations of food insecurity, and their concentrations tend to overlap places of persistently impoverished populations—in the United States, those groups are concentrated in rural Appalachia, the Mississippi Delta, the Texas–Mexico border, and native tribal lands.

Researchers studying the details of particular regions and/or states have often found that nationwide patterns give way to qualitative differences. Scholars describe a diverse range of possible factors affecting these differences. White (2007) reviewed the influence of socioeconomic and cultural effects upon national shifts in food retailers. Cohen (2003) examined the influence of post–World War II suburbanization and legacies of discrimination in influencing low-income people's residential locations. Bluestone and Harrison (1982) detailed the consequences resulting from global shifts in U.S. industrial activities. Wolch and Dear (1993) explained homelessness as a process involving global and local economic trends, social forces, and a politics of rejection and apathy. Schram (1995), Yapa (1996), and O'Connor (2001) theorized the ways in which researchers' language is implicated in policies affecting the poor's daily experiences.

Concerned residents assessing the details of their neighborhood tend to find that rural and urban residents experience variations in their daily efforts to secure food. In 1998, Eisinger noted that residents of less populated places were poorly served by nonprofit food providers, such as food pantries and soup kitchens. Labor market scholars indicate that rural residents have higher rates of low-wage employment and receive lower earnings for service work. While rural and urban people can improve their income with government subsidies, rural residents generally make use of these programs less frequently. Some scholars indicate that such infrequent use may result from people's inability to access assistance offices and information. Shaw (1996) supported this conclusion, finding that small communities and non-densely populated areas maintain fewer opportunities for individuals to access public services, such as transit systems.

Observers have also taken note of rural and urban people's distinctly personal issues. According to researchers, these issues cut across rural and urban populations, and they vary according to the person's location in distinctly organized social groups. Maxwell (1996) reviewed studies reporting that individuals' access to food in a household is linked to the control they have over household resources and income. Many feminists find women's disadvantaged labor market position hinders their ability to secure similar household incomes. Social researchers conclude that low-income individuals' negative feelings about and experiences with assistance programs are influenced by assistance workers' actions. They also observe that these actions vary according to numerous factors, including the workers' political and social views.

Diverse Physical and Social Outcomes

Through their assessments, researchers highlight the adverse physical and social outcomes of individuals attempting to secure adequate food access. In communicating these outcomes, they often presume that some level of similarity exists among distinctly organized populations. Hence, they extend and generalize their findings across geographic space.

Throughout the world, researchers find malnutrition to be the biggest risk factor for physical illnesses. Among both children and adults, malnourishment reduces the body's defenses against a range of diseases, including lower respiratory infections, malaria, measles, diarrhea, pneumonia, and other diseases. Undernourished people infected with HIV/AIDS have been found to develop full symptoms of the disease more quickly. Underweight pregnant women are believed to be increasingly vulnerable to obstructed labor, anemia, and post-delivery infections. The adverse effects of these physical conditions are heightened by a variety of problems, such as poorly functioning sanitation systems, chronic diseases, parasitic infections, and unsafe water supplies.

Individuals experiencing conditions of chronic hunger often experience a variety of mental and emotional health issues—for example, demoralization, anxiety, and acute stress. Persons experiencing these conditions may consider suicide. In 2003, public policy researchers reported that adult Oregonians living in households with hunger were more likely to report mental health problems. Over a two-year period, 22 percent of these adults reported that they seriously had considered suicide.

In developed countries, researchers find that low-income people are more likely to experience conditions of hunger and obesity. Among both children and adults, significantly overweight people are believed to be at higher risk of developing type 2 diabetes, pulmonary complications, sleep apnea, metabolic syndrome, cardiovascular diseases, cancer, and arthritis. Children stigmatized for their weight are increasingly likely to experience low self-esteem, negative body image, and depression.

Researchers highlighting accessibility issues have also pointed out outcomes related to individuals' social, economic, and political well-being. Soubbotina (2004) described undernourishment as a critical component in promoting an individual's persistent poverty. Malnourishment adversely influences children's school attendance and educational attainment; it also contributes to lower wages and reduced earning capacity. United Nations researchers report that poor nutrition among women adversely affects their educational and employment opportunities.

Global Implications

Around the world, many people continue to experience difficulties accessing adequate food supplies. In 2010, the United Nations reported that 925 million people experienced issues of chronic hunger. Agricultural researchers in the United States reported that 50.2 million Americans lived in food-insecure households. The researchers working in these organizations independently measured these conditions based on absolute numbers of calories and self-reported issues, respectively.

Both rural and urban residents are concerned about their ability to secure access to adequate quantities of food. These residents travel varying distances to food suppliers, which may include relief agencies, food retailers, agricultural fields, and cooperative storage sites. In addition, people experiencing food accessibility issues seek help from others, such as individuals acting in their social networks, tribal nations, organized groups, and families. Those challenged to secure food access often cultivate more accessible supplies, such as agricultural markets, gardens, bartering relations, and wild fruits, nuts, and vegetables.

Researchers suggest that individuals experience a diverse range of experiences related to accessing food supplies. These experiences result in a diverse range of outcomes, and these outcomes vary according to the person's distinct geographic and social location. Generally, it is believed that these issues adversely influence people's physical, mental, and social well-being.

Outcomes of this process afford citizens with the evidence they need to advance reforms of uniformly applied policies. These actions may both enable and constrain particular people's ability to access food supplies. In the United States, it generally is believed that Depression-era policy reforms temporarily enabled more unemployed people and farmers to access assistance programs. World War II policies directed people, food, and resources to warfronts. Cold War policies shifted domestic farm surpluses and antipoverty researchers toward developing nations. Civil rights policies gave many racial and ethnic minorities a more equal chance of accessing cash and food assistance. Following a recent earthquake, U.S. agriculture officials formulated policies to shift quantities of domestic agriculture surpluses toward Haitian residents.

Many critics question the appropriateness of these reforms of uniformly applied policies. They assert that concerned citizens ought to support reforms that recognize the spatially contingent nature of hunger. Through their recognition, they would support their findings of individuals' diverse issues, experiences, and outcomes. These issues arise from a variety of complex interconnections, afforded by individuals' unique geographic locations in their household, neighborhood, county, state, region, nation, and global society.

—Ann Myatt James

Further Reading

Bentley, A. *Eating for Victory: Food Rationing and the Politics of Domesticity.* Urbana, IL: University of Illinois Press, 1998.

Bluestone, B., and B. Harrison. *The Deindustrialization of America.* New York: Basic Books, 1982.

Cohen, L. *A Consumers' Republic: The Politics of Mass Consumption in Postwar America.* New York: Alfred A. Knopf, 2003.

Desjarlais, R., L. Eisenberg, B. Good, and A. Kleinman. *World Mental Health: Problems and Priorities in Low-Income Countries.* Oxford, UK: Oxford University Press, Inc., 1995.

Eisinger, P. K. *Toward an End to Hunger in America.* Washington, DC: Brookings Institution Press, 1998.

Food and Agriculture Organization, United Nations. "The State of Food Insecurity in the World, 2005: Eradicating World Hunger—Key to Achieving the Millennium Development Goals." January 16, 2006. ftp://ftp.fao.org/docrep/fao/008/a0200e/a0200e.pdf.

Food and Agriculture Organization, United Nations. "The State of Food Insecurity in the World: Addressing Food Insecurity in Protracted Crises." October 6, 2010. www.fao.org/docrep/013/i1683e/i1683e.pdf.

Food Research and Action Center. "Obesity, Food Insecurity and the Federal Child Nutrition Programs: Understanding the Linkages." October 2005. www.frac.org/pdf/obesity05_paper.pdf.

Glasmeier, A. K. *An Atlas of Poverty in America: One Nation, Pulling Apart, 1960–2003*. New York: Routledge, 2005.

Haynie, D. L., and B. K. Gorman. "Determinants of Poverty Across Urban and Rural Labor Markets." *Sociological Quarterly* 40, no. 2 (1999): 177–197.

Hobbs, K., W. MacEachern, A. McIvor, and S. Turner. "Waste of a Nation: Poor People Speak Out About Charity." *Canadian Review of Social Policy* 31 (1993): 94–104.

Iceland, J. *Poverty in America: A Handbook*, Vol. 2. Berkeley, CA: University of California Press, 2006.

Levenstein, H. *Paradox of Plenty: A Social History of Eating in Modern America*. Berkeley, CA: University of California Press, 2003.

Maney, A. L. *Still Hungry After All These Years: Food Assistance Policy from Kennedy to Reagan*. New York: Greenwood Press, 1989.

Maxwell, S. "Food Security: A Post-modern Perspective." *Food Policy* 21, no. 2 (1996): 155–170.

Miller, K., B. Weber, L. Jensen, J. Mosley, and M. Fisher. "A Critical Review of Rural Poverty Literature: Is There Truly a Rural Effect?" *International Regional Science Review* 28, no. 4 (2005): 381–414.

Nord, M., and A. Coleman-Jensen. "Food Security in the United States: Measuring Household Food Security." November 16, 2010. www.ers.usda.gov/Briefing/FoodSecurity/measurement.htm.

O'Connor, A. *Poverty Knowledge: Social Science, Social Policy, and the Poor in Twentieth-Century US History*. Princeton, NJ: Princeton University Press, 2001.

Oregon Center for Public Policy. "Oregon Hunger Issues: Finding #2 from OCPP's Analysis of a State Health Survey." December 11, 2003. www.ocpp.org/2003/issue031211-2.pdf.

Poppendieck, J. *Breadlines Knee-Deep in Wheat: Food Assistance in the Great Depression*. New Brunswick, NJ: Rutgers University Press, 1986.

Sanchez, P. A. "Hunger in Africa: The Link Between Unhealthy People and Unhealthy Soils." *Lancet* 365 (2005): 442–444.

Schram, S. F. *Words of Welfare: The Poverty of Social Science and the Social Science of Poverty*. Minneapolis, MN: University of Minnesota Press, 1995.

Soubbotina, T. P. "Beyond Economic Growth." World Bank, June 2004. www.worldbank.org/depweb/english/beyond/beyondco/beg_06.pdf.

Shaw, W. *The Geography of United States Poverty: Patterns of Deprivation, 1980–1990*. New York: Garland, 1996.

Trattner, W. I. *From Poor Law to Welfare State: A History of Social Welfare in America*, 6th ed. New York: Free Press, 1999.

U.S. Department of Agriculture, Economic Research Service. "Measuring Rurality: What Is Rural?" March 22, 2007. www.ers.usda.gov/briefing/Rurality/WhatIsRural.

U.S. Department of Health and Human Services. "The 2008 HHS Poverty Guidelines: One Version of the [U.S.] Federal Poverty Measure." January 23, 2008. http://aspe.hhs.gov/poverty/08Poverty.shtml.

Weinreb, L., C. Wehler, J. Perloff, J., R. Scott, D. Hosmer, L. Sagor, and C. Gundersen. "Hunger: Its Impact on Children's Health and Mental Health." *Pediatrics*, September 18, 2006. www.uri.edu/fhn/handouts/fall06/Weinreb,%20Hunger%20Its%20Impact%20on%20Children%27s%20Health%20and%20Mental%20Health.pdf.

White, M. "Food Access and Obesity." *Obesity Reviews* 8, Supplement 1 (2007): 99–107.

Wolch, J., and M. Dear. *Malign Neglect: Homelessness in an American City*. San Francisco: Jossey-Bass, 1993.

Yapa, L. "What Causes Poverty? A Postmodern View." *Annals of the Association of American Geographers* 86, no. 4 (1996): 707–728.

Food Safety

Food safety protects food from contamination. It is estimated that sanitation problems are 5 percent equipment and materials and 95 percent human error. Since 1945, more than 400 million people are believed to have died due to hunger and poor sanitation. Many acute and long-term diseases such as diarrhea and several forms of cancer are due to unsafe foods. More than 200 diseases are transmitted through food. The World Health Organization (WHO) reports that food and water safety issues cause 2.2 million deaths every year, 1.9 million of whom are children.

During times of famine and extreme hunger, many people eat foods that may be unsafe. The pathogens from contaminated food can grow in the intestines and cause illness. When coupled with a nutritionally compromised immune system, chronic illness and ultimately death may occur. It is important to keep foods safe to prevent this outcome.

Potential Food Safety Hazards

Potential food safety hazards can be biological, chemical, or physical in nature. Food becomes hazardous for consumption when it is acquired from unsafe sources or is stored improperly. Inadequate personal and food-handling hygiene, insufficient control of time and temperature, and cross-contamination are well-known causes of food degradation. Universally recognized and accepted sanitation procedures used at critical points and at appropriate times break the chain of food contamination.

Biological Food Hazards

Biological food hazards include pathogenic bacteria, molds, parasites, and viruses or toxins from biological sources. The human body may harbor organisms in the mucus, feces, vomit, and skin. In turn, germs spread through sneezing, coughing, and poor personal hygiene. The primary means of spreading pathogens are the hands, food-contact surfaces, utensils, and cleaning cloths.

Chemical Food Hazards

Chemicals that can cause illness when consumed are considered chemical food hazards. Cleaning supplies, pesticides, sanitizers, polishes, machine lubricants, and toxic metals may leach into food if the food is not prepared properly or stored properly.

Physical Food Hazards

Physical food hazards include foreign matter that may cause illness or injury to the consumer. Objects such as broken glass, hair, stones, shotgun pellets, metal fragments, bone chips, jewelry, insects, plastic, and wood can be introduced into food products during any stage of production or preparation.

Keeping foods safe is essential to human health. By managing food safety, it is possible to break the food contamination chain and prevent the spread of diseases. Several international organizations have developed guidelines for safe food practices that are tailored to the needs of the populations served. WHO recommends five steps as means of keeping food safe: (1) keep oneself and the work surface clean; (2) keep the raw foods separate from the cooked foods; (3) cook foods thoroughly; (4) keep food at safe temperatures; and (5) use safe water and raw materials.

HACCP

The Hazard Analysis Critical Control Point (HACCP) system was developed for the National Aeronautics and Space Administration (NASA) by the Pillsbury Company. NASA and the U.S. Army Natick (Massachusetts) Laboratories assigned the Pillsbury Company the task of developing a strategy to ensure food safety for the astronauts on their first manned space missions. As part of this effort, scientists and engineers studied problems that could contribute to foodborne illnesses. It was essential to produce safe foods for space travelers, of course, so critical control points in food production and service that could lead to foodborne illness were identified. Procedures to eliminate them were initiated—and HACCP was established. The food industry broadly adopted HACCP in the late 1970s.

The guidelines for HACCP application were defined by the Codex Alimentarius Commission in the Codex Alimentarius Code of Practice. This Commission implements the Joint Food and Agriculture and the World Health Organization Food Standards Program. As HACCP has evolved over the years, it repeatedly has been proven effective. HACCP now is internationally recognized as the best system for ensuring food safety.

—*Kausar F. Siddiqi and Sara A. Blackburn*

FAT TOM is an acronym to help individuals remember the importance of food, acidity, time, temperature, oxygen, and moisture in preventing food contamination. HACCP is an acronym for hazard analysis and critical control points. It is important to select one method for maintaining food safety and to follow it consistently. The key universal points for food safety are discussed next.

Key Universally Accepted Safe Food Handling Procedures

The infectious dose for some pathogens is very small. Infections result from the direct transfer of the pathogen from surfaces via hands or food to the mouth, nasal mucosa, or eyes. Personal and food-preparation hygiene by individuals breaks the chain of food contamination. Hence, cleanliness practiced during food handling, preparation, storage, serving, and eating is critical.

Several widely accepted techniques are used to prevent food contamination. Do not prepare food when sick. Coughing and sneezing spread germs, so cover the nose and mouth with a disposable tissue to prevent germs from spreading when coughing and sneezing. Avoid working with food if hands have open cuts or sores. Do not smoke or chew gum near food. Wear clean clothes and keep hair and beards covered when preparing food to reduce the risk of contamination.

Hand Washing

Hand washing with soap and water is an effective way of preventing transmission of infection. It reduces the chance of food contamination from the hands and can decrease the incidence and cycle of diarrhea, respiratory infections, skin disease, and eye infections. Cleaning hands by either hand washing or using an alcohol hand sanitizer are effective ways of reducing the risk of contaminating foods.

Proper hand washing is the thorough washing of hands with soap, under clean and running water. The accepted procedure for hand washing with soap is as follows:

1. After wetting the hands and applying soap, rub the hands together for 15 to 30 seconds, making sure that the nails and the areas between the fingers and thumb are clean.
2. Rinse the hands thoroughly under running water to remove all soap.
3. Dry the hands with a clean dry towel.
4. Turn off the faucet with a dry towel or elbow; avoid touching the faucet with clean hands.

The hands must be washed:

1. Before and after preparing or handling food
2. After using the toilet, handling any feces, or changing a diaper
3. After handling raw food, such as meat, poultry, or fish
4. Before and after eating, feeding, or giving care to another person
5. After contact with contaminated material, such as cleaning cloths, garbage cans, or any other unsanitary surface
6. After coughing, sneezing, or blowing the nose
7. After smoking
8. After handling animals or people
9. After contact with blood or other body fluids, such as vomit
10. Before and after taking care of wounds

A waterless hand sanitizer can be used when soap and water are not available. These alcohol-based sanitizers work by destroying harmful microorganisms. A sanitizer is not effective if food, dirt, or other material is visible on the hands, as it does not remove dirt. All visible dirt should be removed from the hands before using the sanitizer. Thoroughly wet the hands with the sanitizer and rub them briskly until they are dry.

Food Preparation

In food preparation, cleanliness and sanitation are the watchwords. Sanitize work surfaces before preparing and handling food. Disinfect food preparation surfaces using a chlorine bleach solution, ethanol, or ultraviolet light. Use separate cutting boards for raw meats and fruits and vegetables. Clean and sanitize dishes and utensils. Use separate utensils to prepare each food. Do not reuse utensils that have been licked, such as tasting spoons.

Store foods in clean areas, off the floor, and at a proper temperature to prevent contamination and growth of pathogens. Label a prepared food item to indicate when it was made and when it is best to use the product ("best before" date). Properly dispose of uneaten food and packaging. If refrigeration is unavailable, do not use foods that need to be refrigerated; this step reduces the chance that microorganisms will multiply.

Dishwashing and Storage

Proper washing of dishes, silverware, glassware, and pots halts the path to infection. Note that cooking equipment such as knives and can openers can also be sources of bacterial contamination. It is important to follow standard procedures when washing food preparation equipment by hand. If a dishwashing machine is not used, a three-compartment sink should be used. The first compartment is for washing, the second for rinsing, and the third for sanitizing.

The steps for proper foodware washing are as follows:

1. Scrape, rinse, or soak all items before washing.
2. Place items in the first sink with the detergent solution. The water temperature should be at least 110°F. Use a brush, cloth, or scrubber to loosen remaining soil. Replace the detergent solution when the suds are gone or the water is dirty.
3. Spray-rinse or immerse items in the second sink. The water temperature should be at least 110°F. Remove all traces of food and detergent. If using the immersion method, replace the water when it becomes cloudy or dirty.
4. Immerse items in the third sink, which should be filled with hot water heated to at least 180°F or containing a chemical-sanitizing solution. Proper personal protective equipment should be used to avoid injury. If chemical sanitizing is used, the sanitizer must be mixed at the proper concentration. The water must be at the correct temperature for the sanitizer used. Items must be immersed for at least 30 seconds, depending on the type of chemical used.

5. Air-dry all items on a drain board to prevent contamination. Towels should not be used to dry items.
6. Clean items should be stored covered or protected to prevent contamination. Glassware can be stored inverted. Silverware is stored so that handles can be grasped.

Proper Dishwashing Sink Setup

Compartment 1: Wash	Compartment 2: Rinse	Compartment 3: Sanitize
110°F Soapy water	110°F Clear water	180°F Clear water or Chemical sanitizer

Chemical Sanitizing Setup

Chemical Solution	Concentration Level	Minimum Temperature	Minimum Immersion Time
Chlorine solution	25 ppm (~½ tsp household bleach per 1 gallon water) 50 ppm (~1 tsp household bleach per 1 gallon water) 100 ppm (~1 ½ tsp household bleach per 1 gallon water)	120°F 100°F 55°F	10 seconds 10 seconds 10 seconds
Iodine solution	12.5 ppm (47.5–94.5 mg/gallon)	75°F	30 seconds
Quaternary ammonium solution	200 ppm maximum	75°F	30 seconds

If chemical sanitizing is used, the sanitizer must be mixed at the proper concentration. The water must be at the correct temperature for the sanitizer to be effective.

Garbage and Trash Receptacles

Dispose of garbage and trash promptly. Keep garbage cans and garbage holding areas clean. Wash hands after handling garbage.

Domestic Animals and Flying Insects

Animals and flying insects can be sources of contamination. It is best to keep them out of food preparation areas.

Safe Water and Raw Materials

Water, including ice, and raw materials may be contaminated with hazardous materials. Choose foods that are processed for safety (e.g., pasteurized milk). Use safe or clean water for hand washing, washing fruits and vegetables, cooking, drinking, making ice, and washing utensils and food contact surfaces. Water can be made safe by bringing it to a rolling boil, chlorination (3 to 5 drops of chlorine per liter of water), filtration, and ultraviolet treatment.

Microorganisms grow well in a nutrient-rich environment. Foods prone to such growth include protein-rich foods, such as meat, fish, poultry, eggs, and milk. Acidity or alkalinity is measured as pH on a scale of 0 to 14.0. A pH of 7.0 is considered neutral; a pH higher than 7.0 is alkaline, and one lower than 7.0 is acidic. The lower the pH, the higher the acidity; the higher the pH, the lower the acidity. Foodborne pathogens thrive in a near-neutral environment at a pH in the range of 6.6 to 7.5. By comparison, most do not grow at pH levels less than 4.6.

Primarily, the pathogens in food are aerobic, requiring oxygen for growth. Very few pathogens require anaerobic (without oxygen) conditions. *Clostridium botulinum* is one organism that grows and produces the botulism toxin in an oxygen-free environment. Improperly preserved canned foods, especially home-canned products, are typical sources of botulism. When meat, fish, poultry, beans, pasta, and other foods are canned, the oxygen is removed in the canning process. This process makes these foods shelf-stable, such that the foods do not require refrigeration in the canned state.

Water is an essential nutrient for pathogen growth. The amount of water in a food is a factor in determining the extent to which the food is perishable. The amount of water available for use is measured as water activity (a_w) on a scale from 0 to 1.0. Bacteria, molds, and yeasts grow best in foods with a_w between 0.86 and 1.0; meat, produce, and soft cheeses have a_w in this range.

Foods that are preserved with salt and sugar, such as jams, jellies, and beef jerky, have lower a_w due to the addition of salt and sugar. Decreasing the a_w level inhibits the growth of bacteria and makes the product more shelf-stable (i.e., it does not require any refrigeration until it is opened). Pathogens have difficulty growing in foods with a_w lower than 0.85; these foods include dry pasta, crackers, cookies, flour, powdered milk, and the like.

Separating Raw and Cooked Foods

Separating raw foods from cooked foods prevents cross-contamination, reducing the risk of transferring pathogens from one food to another. Improper handling of raw meats, fish, and poultry can easily result in contamination throughout the food preparation area and the food.

Prevent cross-contamination by using separate cutting boards and knives for (1) fresh produce; (2) raw meat, fish, and poultry; and (3) cooked products. Use a separate plate for cooked meat, fish, poultry, or eggs rather than the one used for the product in its raw state.

When shopping, keep the raw meats, fish, and poultry separate from other foods and carry them in separate shopping bags. Keep them separate in the refrigerator on the bottom shelf, in a tray, to prevent juices from dripping into other foods.

Marinades, batter, or breading for coating raw meats, fish, or poultry should not be used on cooked food without boiling or cooking them first. Discard the leftover marinades, batter, or breading used for meat, fish, or poultry.

Cooking Foods Thoroughly

Cook food thoroughly to destroy harmful microorganisms and to prevent them from multiplying. A food thermometer should be used to measure food temperature. Cook food to an internal temperature of at least 158°F to ensure food safety. Seafood should be heated to at least 145°F. Meats and leftover casseroles should be cooked to at least 165°F.

If a thermometer is not available, cook meat and poultry until juices run clear, not pink. Soups and sauces should be heated until they come to a rolling boil. Eggs should be cooked until the egg white and yolk are firm. Flesh of fish with fins should be opaque and separate easily with a fork. Shells of clams, oysters, and mussels should open upon cooking, while scallops must be milky white or opaque and firm. Shrimp, lobster, and crab should have pearly white and opaque flesh.

Microwave ovens cook food unevenly, leaving cold spots. Stir the food to make sure that it is heated throughout to a safe internal temperature.

Keeping Foods at Safe Temperatures

Most pathogens grow at temperatures between 41°F and 140°F—a range called the temperature danger zone (TDZ). Microorganisms flourish between 70°F and 120° F, especially in an environment with neutral pH and in high-protein foods. Bacteria grow very slowly at temperatures less than 40°F, and growth is stopped completely at 0°F or lower. These pathogens also grow very little at 140°F and higher, and are destroyed at temperatures of 180°F and higher. Foods should not be allowed to stay in the TDZ for more than 2 hours, and for only 1 hour on a very hot day. After cooking, food should be maintained at a temperature of 140°F or higher for serving, as this practice retards the growth of bacteria.

Chilling food by using a refrigerator or freezer is a method of controlling bacteria from multiplying. The refrigerator temperature should be set at a maximum of 40°F and the freezer at 0°F. The freezer temperature should be at or below 0°F; industrial freezers should be set at –6°F. Refrigerate or freeze perishable foods, such as meat, fish, poultry, eggs, milk, and other dairy products within 2 hours of purchase or use and within 1 hour when the temperature is 90°F or higher. Always marinate food in the refrigerator.

Defrost frozen foods in the refrigerator. In cases where food is needed promptly, use other quick methods of defrosting; for example, defrost frozen foods in the microwave or immerse them in a sealed package in cool water and change the water every 30 minutes until the food is defrosted. Do not defrost any food in hot water. Never defrost food at room temperature, as the surface temperature of the food will be at unsafe levels while the internal part or the food is still defrosting. Defrosted foods can be refrozen if thawed in the refrigerator; otherwise, they must be cooked before freezing.

Divide larger amounts of food into smaller packages or shallow containers to reduce the amount of chilling time. Avoid overstuffing the refrigerator or freezer to allow cold air to circulate around the food. Transport perishable foods in a cooler packed with ice or cold packs to maintain proper temperature and prevent spoilage. Maintain cold foods at 40°F or less until serving time.

Food Storage

It is important to store food properly. Avoid leaving foods in storage for extended periods of time. Rotate food using a FIFO (first in, first out) practice. Store newly received food behind the older stock to ensure the use of older food first.

Keep dry-storage areas clean and dry, between 50°F and 70°F. Store items in a well-ventilated area away from the wall and ceiling, and at least 6 inches off the floor. Store fresh and frozen foods in the refrigerator and freezer, respectively. Label food products, including leftovers, with the product name and date and the expiration dates clearly visible. To ensure food safety, follow a common-sense rule: When in doubt, throw it out.

—Kausar F. Siddiqi and Sara A. Blackburn

Further Reading

Arduser, Lora, and Douglas Robert Brown. *HACCP & Sanitation in Restaurants and Food Service Operations: A Practical Guide Based on the USDA Food Code*. Ocala, FL: Atlantic Publishing Group, 2005.

"Food Safety Lessons: FAT TOM." September 2010. Iowa State University: www.extension.iastate.edu/foodsafety/Lesson/L4/L4p1.html.

National Restaurant Association. *ServSafe Essentials*, 5th ed. Chicago, IL: National Restaurant Association Education Foundation, 2010.

"The USDA/FDA 2009 Food Code." August 2010. www.fda.gov/Food/FoodSafety/RetailFoodProtection/FoodCode/FoodCode1997/UCM054471.

World Health Organization. "Prevention of Foodborne Disease: Five Keys to Safer Food." December 2010. www.who.int/foodsafety/consumer/5keys/en/index.html.

Food Sources

In November 1996, the world's leaders gathered in Rome to discuss the right to food for all the world's inhabitants. These leaders pledged their nations' support and made their commitment to a program to provide food security for all, to eradicate hunger, and to reduce the number of undernourished people by 50 percent in 2015. The World Food Summit Plan of Action included a list of objectives and actions to achieve those goals. On April 11, 1998, the United Nations' Commission on Human Rights reaffirmed the fundamental right of everyone on Earth to be free of hunger and expressed the need for common understanding of what "food" is and what the "minimum nutrition standard" is. To achieve food security for all, the world's food supply must be adequate and foodstuffs commonly available should be culturally acceptable.

Agriculturalists and fishermen of the world—those concerned with producing crops, raising livestock, and catching fish—must continually strive to supply the food needs of a growing world population from a limited number of food crops; a limited number of food-producing livestock, poultry, and insects; and a limited number of food fish.

Significant Food Crops

1. Cacao—cocoa
2. Cereals—barley, buckwheat, corn, millet, oats, rye, sorghum, wheat
3. Fruits—apples, apricots, bananas, berries, cherries, dates, figs, grapes, grapefruit, lemons, limes, mulberries, olives, oranges, papaya, peaches, pears, pineapple, plantains, plums, prunes, raisins, tangerines
4. Nuts—almonds, black walnuts, Brazil nuts, cashews, chestnuts, English walnuts, hazelnuts, peanuts, pecans, pistachios
5. Oils—coconut oil, cottonseed oil, olive oil, palm oil, peanut oil, rapeseed oil, sesame seed oil, soybean oil, sunflower seed oil

6. Pulses—beans, chickpeas, cowpeas, lentils, lupines, peas, pigeon peas, vetch
7. Starchy roots—cassava, sweet potatoes, taro, yams
8. Sugar—date sugar, sugar beets, sugar cane, maple sugar
9. Vegetables—asparagus, cabbage, carrots, cauliflower, green beans, green peas, lima beans, melons, onions, potatoes, pumpkins, radishes, sweet corn, tomatoes
10. Wine—from grapes and other fruits

Significant Food-Producing Livestock, Poultry, and Insects

1. Meat—buffalo, camel, cattle, goats, llamas, pigs, poultry (chickens, ducks, geese, turkeys), rabbits, reindeer, sheep, yak
2. Milk—cows, camel, goats, sheep, water buffalo, yak
3. Eggs—chickens, ducks, geese
4. Honey—bees

Significant Food Fish

1. Anchovies
2. Bass
3. Carp
4. Catfish
5. Cod
6. Croaker
7. Flounder
8. Haddock
9. Hake
10. Halibut
11. Herring
12. Mackerel
13. Menhaden
14. Perch
15. Salmon
16. Sardine
17. Shad
18. Sturgeon
19. Tilapia
20. Trout
21. Tuna

Multiplying Food Source Diversity

Food sources must, in most cases, be converted into nourishing, attractive, and culturally acceptable edible items. Over the ages, humans have experimented with, devised, created, and consumed a rich range of foods and an almost unlimited wealth of dishes. The world's food preparers have created a treasury of simple and elaborate fare and have contributed, to the world society, items that nourish the body, stimulate the senses, and create moods. The world's physical geography is made up of countless micro-ecosystems and micro-climates with great contrasts and great opportunities as food site sources. From the subarctic regions of the north to the tropical rainforests in a globe-girdling belt around the equator; from the rugged mountains to the low hills, fertile plains, grasslands, and deserts; and from the oceans and seas to lakes and rivers, the world's physical environment nurtures varying eating habits and differing food staples. At one time, many regions of the world were isolated from others, and isolation engendered many different cultures. Each culture and each cultural group brought a unique legacy to the development of the world's food base. In more recent years, transportation and communication ties have steadily improved, facilitating the movement of foodstuffs from one part of the world to another.

Fortunately, purely local conversions of site-specific food sources into regional staples have survived. Sometimes local food sources and preparation of these local foods are enriched by ideas from outside. New foods and new food creations emerge from imported ideas and imported ingredients. The world's menu of foods has been enlarged constantly. For those peoples whose nations and cultures are linked to the changing world, a lifetime of discovery with something for everyone is in place. For example, consider the remarkable variations in topography, climate, culture, and cuisine found in Europe. Here, food sources and food preparation can still be classified into separate groups according to region. Located on the other side of the Eurasian continent and covering a vast segment of southeast Asia, China is a nation of two large basic food regions. The traditional divisions are the rice eaters of the central provinces and the south and the wheat and other grain eaters of the north. Dividing a nation that covers an area of 4 million square miles and a history going back more than 5,000 years into only two regions, however, does not acknowledge the countless food sources, food combinations, and food subtleties.

Farming in China has always been intensive. Millions of Chinese spend their lives working on an acre or two of cultivable land in their struggle to produce vital food. A region's primary crop largely determines what is eaten there. Nothing goes to waste in a Chinese farming community; every animal contributes to the production of food. Chickens and pigs dispose of kitchen scraps and leftovers and forage for themselves. The recurring miseries of drought and flooding have created memories of hunger and famine that are not easily erased. Frugality and survival have led to experimentation with unusual ingredients that are blended into delectable dishes. Virtually nothing is excluded from consideration as a potential food dish ingredient. Freshness is a primary requirement for all Chinese dishes. For the poorest Chinese family, a meal may consist of only cooked rice or cooked noodles and at times some meat, fish, or vegetables. Pork is the meat of choice and is the basic ingredient in 7 out of 10 meat dishes. The number of Chinese beef dish recipes is relatively small, as are the number of dishes incorporating lamb, mutton, or goat meat. To many Chinese, the chicken is the pig of the poultry world. Chinese cooks utilize almost every part of a chicken into a nutritious food source, including the feet, blood, bone marrow, and almost all the organs. Even the toughest bird is turned into a savory chicken stock used in sauces and soups. The magnitude of basic food sources, the ingenuity of frugal Chinese cooks, and the memories of hunger and famine years blend together to make a wealth of food dishes and a nation whose people treasure food diversity.

Converting Food Sources into Items Fit to Eat

Agricultural food sources frequently need to be processed to change their form or purity before they can be used as a source of nutrition for humans. Much of the final food processing is done in a household during meal preparations. Food must be ground, peeled, washed, pitted, diced, skinned, mixed, and then cooked. Cooking is universal throughout human cultures and has much to do with the nutritive value and safety of food. Basic cooking terminology includes the following items:

1. Bake—cook by dry heat, usually in an oven
2. Barbecue—roast meats slowly on a spit or rack over heat and seasoned with a sauce
3. Baste—moisten foods to prevent drying while roasting
4. Beat—work a mixture smooth with a rhythmic movement
5. Blend—mix two or more ingredients thoroughly
6. Boil—cook in a boiling liquid
7. Braise—brown meat or vegetables in a small amount of hot fat
8. Broil—cook over an open fire or under a flame (direct heat)
9. Brush—coat food, usually with butter or egg, using a small brush
10. Candy—cook fruit or vegetables in a heavy sugar syrup
11. Caramelize—melt sugar slowly over low heat until sugar is liquid, brown, and caramel flavored
12. Coat— roll foods in flour, nuts, sugar, crumbs, or some other material until all sides are evenly covered
13. Coddle—cook slowly and gently in water just below the boiling point
14. Combine—mix all ingredients together
15. Cook—prepare food by applying heat
16. Cream—beat shortening or butter with other ingredients until smooth
17. Crisp—make firm and brittle in very cold water or fry in fat until crisp
18. Devil—add to a food a hot seasoning, such as mustard
19. Dredge—coat food with a dry ingredient, such as flour

20. Flambé—cover a food with brandy or cognac, then ignite it to burn off the alcohol and serve flaming
21. Fold—combine two ingredients very gently with a whisk or spoon
22. Fricassee—braise with a liquid poultry or meat that has been cut into small pieces
23. Fry—cook in a small amount of fat on a stove
24. Garnish—decorate any food
25. Glaze—cover meat, fish, or fruit with a coat of syrup or jelly
26. Knead—press dough hard with the heels of hands until the dough becomes stretched
27. Marinate—soak meat in lemon or tomato juice or other acidic liquid to enhance flavor and tenderize it
28. Mix—stir ingredients until they are thoroughly combined
29. Pan-fry—cook on a stove or fire in a hot, uncovered skillet with little or no fat
30. Peel—remove or strip the outer coverings of vegetables or fruit before using
31. Pot-roast—cook meat in a small amount of fat or oil, then in a small amount of liquid in a deep, heavy, covered pot
32. Puree—force fruits or vegetables through a sieve or blender until smooth and pulpy
33. Roast—cook in an oven by dry heat
34. Sauté—fry foods until golden and tender in a small amount of fat
35. Scallop—arrange foods in layers, usually in a cream base, in a casserole dish, then bake
36. Scramble—mix foods gently while cooking
37. Sear—brown the surface of meat over high heat
38. Simmer—gently cook in a liquid, somewhat below the boiling point
39. Skewer—thread foods on a wooden or rod skewer and then grill or broil
40. Steam—cook over simmering water
41. Stew—cook a mixture of foods very slowly in a pan with just enough liquid to cover the foods
42. Stir—mix ingredients together
43. Toast—brown and dry the surface of foods with heat
44. Toss—tumble foods lightly with a lifting motion
45. Whip—rapidly beat a food or foods to incorporate air and increase volume

Electronic cooking is done by radiant energy in the form of high-frequency radio waves, stimulating molecular activity that generates heat within a food item.

Cooking depends on the fuels available for heat. In many parts of the world, securing cooking fuel is difficult and expensive, and searching for fuel requires much time. Wood is often very scarce and the cutting down of trees for firewood results in rapid soil erosion. In some areas of the world, animal dung is used for fuel in cooking. Basic foods are cooked because heating causes chemical and physical changes. The flavors and appearance of food items are enhanced and chemical compounds in the cooked item are transformed into new compounds. Enzymes are inactivated; proteins are denatured, making them more easily digested; and amino acids are made available to the consumers as nutrients.

Cooking also makes grains and beans softer and easier to digest, kills microorganisms, and decreases the chance of foodborne illness. Women in most cultures select and prepare food. They learn from their mothers and then pass down to their daughters simple and easy-to-follow cooking practices that transform one basic food into many diverse and appetizing dishes. A general theme in traditional food preparation is to keep the amount of cooking water small, cook foods until just tender, use low cooking temperatures, and hold food that has been cooked only briefly before serving.

Foods for the Future

At the beginning of the 21st century, advances in agricultural productivity and human ingenuity have not enabled the world to be free of hunger, famine, and malnutrition. Global demands for cereals are projected to increase by 41 percent between 1993 and 2020, for meat by 63 percent, and for root crops by 40 percent. Cereal needs can be met by increasing yields, reducing harvest and post-harvest losses, and expanding the sown area. Cereal production in developing countries will necessarily be augmented by imports from the developed countries of the world. Cereal export sources in 2020 are projected to include the United States (60 percent), the European Union (16 percent), and Australia (10 percent). The gap between production and demand in basic food sources will result in the following trends:

1. Increased grain price uncertainty and volatility
2. Increased pressure upon limited water resources
3. Increased demands for fertilizers, herbicides, and pesticides
4. Increased investments in agricultural research and modern biotechnology
5. Reduced waste of basic foods

Weed Out Hunger

Agricultural suppliers in the United States are working collectively to weed out hunger. In Terre Haute, Indiana, the Halex GT Corn Herbicide Food Drive, an annual event, collects food and cash for the Terre Haute Catholic Food Bank and Food Kitchen. Suppliers are conducting drives in other parts of the country at retail locations, supermarkets, trade shows, and agricultural events. Also, working with other groups and activists, they have lobbied Congress to boost the reimbursement rate for school lunches. They applauded the Agricultural Department's announced plan for the first nutritional overhaul of students' meals in 15 years. President Barack Obama signed a bill expanding access to free lunch programs and increasing the federal reimbursement for free school lunches by 6 cents per meal. In addition, the U.S. Department of Agriculture has established a program to fund suppers for food-at-risk children who are attending after-school programs in communities where at least 50 percent of the households fall below the poverty level.

The number of Americans who live in food-insecure households rose from 36 million in 2007 to 49 million in 2008, according to the USDA's Economic Research Service. This population includes 17 million children, up from 12 million in 2007. One in four children in the United States is food insecure and one in five lives in poverty.

—*William A. Dando*

Public and private investment in agricultural research is crucial to future world supplies of basic foods. Developing countries do not have funds to invest in agriculture, and what minimal research that is done is not directed toward the needs of poor small farmers. Efforts to improve staple food crop productivity on small farms must be accelerated in the early 21st century. Advances in modern science and technology, when applied to agriculture, offer a powerful instrument for world food security. There are great opportunities for modern agrotechnology and biotechnology to help meet the world's increasing food needs, make recommendations for more efficient use of arable land, devise means to reduce food losses, adapt methods for adjusting to climate change and weather pattern variations, and assist national leaders in better managing agricultural resources. Expanded research in agriculture would, in the long term, conserve biodiversity, protect fragile ecosystems, and strengthen international food safety.

At the grassroots level, changes in food sources and recommended new foods are being made constantly in all parts of the world. For example, in a survey conducted by the Canadian Restaurant and Foodservice Association (CRFA) among

400 chefs in 2010 to identify food trends, 10 were identified as major foods for the future:

1. Ancient grains, including kamut, spelt, amaranth, and teff
2. Gluten-free beer, made from flaxseed, millet, teff, and soybeans
3. Vegan food dishes (no meat, fish, poultry, dairy products, or eggs)
4. Organic alcohol and organic foods (i.e., no chemical fertilizers, pesticides, herbicides, or fungicides used in the growth cycle)
5. African foods and flatbreads
6. New cuts of meat
7. Gluten-free food options
8. Middle Eastern foods
9. Quinoa (pronounced "keen-nwa"), a grain crop with a source of complete protein
10. Nontraditional fish—including herring, roe, octopus, sea cucumbers, jellyfish, and Red Sea urchin

The results of this survey reflect the merging of traditional and nontraditional foods in a modern multicultural world. Scientific research and the bonding of peoples in the world into a "world economy" and a "world food system" have expanded knowledge of food sources and food preparation and have increased nutrient source diversity. Enough food is available on Earth to provide food security and food diversity for all.

—William A. Dando

Further Reading

Anderson, J. *A Geography of Agriculture*. Dubuque, IA: William C. Brown, 1970, 40–41.

Barstow, Cynthia. *The Eco-foods Guide*. Gavriola Island, Canada: New Society Publishers, 2002, 13–20.

Corey, Helen. *Food from Biblical Lands*, 2nd ed. Terre Haute, IN: Helen Corey Press, 1990, vii–viii.

Eide, A. "The Human Right to Adequate Food and Freedom from Hunger." In *The Right to Food in Theory and Practice*, edited by the Food and Agriculture Organization. Rome: Food and Agriculture Organization of the United Nations, 1998, 1–5.

Lai, T., J. Ram, D. Perkins, and P. Cook, eds. *Hong Kong and China Gas Chinese Cookbook*. Hong Kong: Pat Printer Associates, 1978, 17–18.

McCall's Cookbook. New York: Random House, 1963, 6–14.

Pinstrup-Andersen, Per. *The Future World Food Situation and the Role of Plant Diseases*. Paper presented at the joint meeting of the American Phytopathological Society and the Canadian Phytopathological Society, Montreal, Canada, August 8, 1999. www.apsnet.org/education/feature/foodsecurity.

Pimentel, D., and M. Giampietro. "Food, Land, Population, and the U.S. Economy." November 21, 1994. dieoff.org/page55.htm.

Stainsby, M. "10 Foods for the Future." *Vancouver Sun*, March 10, 2010. www.vancouversun.com/health/foods+future/2665725/story.html.

Genetically Modified Foods

Genetically modified foods or genetically modified organisms (GMOs) are mostly crop plants created for human and animal consumption. They have had their genetic material (DNA) modified in the laboratory to create new species with more desirable characteristics. For example, genetic engineering techniques have been applied to develop GM plants with greater resistance to herbicides, pests, and disease. Genetic engineering permits scientists to transfer genes among very different species that would not interbreed in nature and to incorporate nonplant genes into crops. Most recently, this technique has been extended to animal production. For example, in 2006, a pig was engineered to produce omega-3 fatty acids through the expression of a roundworm gene.

Genetically modified foods are the newest form of agricultural intensification. The need for more intense food production is obvious: The world is currently adding 80 million new mouths to feed per year (an additional 2.7 billion new people are anticipated to be born in the next 40 years) and living standards around the world are rising, which requires more grain to be grown to produce more animal products and other sources of protein. The world is already home to more than 1 billion people who are hungry or malnourished. In addition, increasing pressure is being placed on agricultural lands to produce more biofuels. Initially, intensification meant the greater use of fertilizers, pesticides, and irrigation to increase yields—the amount of food produced per acre (or hectare). But that in itself was not enough, and indeed most people recognize our need to reduce dramatically the environmental impacts of traditional farming practices, which have caused such widespread damage to soils, ecosystems, watersheds, and even the atmosphere.

The first gene revolution involved cross-breeding and the development of hybrids. Cross-breeding means selecting desired qualities in a species of plant or animal and combining those traits within that same or a genetically very similar species. This approach to increasing and improving food supplies has proved very successful, resulting in insect- and disease-resistant plants, larger ears of corn, bigger tomatoes, and greater yields. However, cross-breeding has three major drawbacks. First, it is slow, taking agricultural scientists many years to produce a commercially valuable new crop variety. Second, it can combine traits only from the same or genetically very similar species. Third, it is not long before pests and diseases mutate to reduce the effectiveness of these new hybrids.

To counteract these disadvantages of cross-breeding, scientists have recently developed a second gene revolution involving a kind of genetic engineering that does not restrict plant or animal breeding to similar species. Instead, it permits the transfer of genes among very different species or even from nonplant sources into a plant or animal. For example, pieces of the DNA of a common daffodil and from a soil bacterium were combined into conventional strains of rice to produce a more nutritious rice strain–one containing more iron and beta carotene, which the body can convert to vitamin A, a nutrient needed to prevent blindness and increase resistance to common childhood infections. As Oxford University economist Paul Collier put it, "Genetic modification offers both faster crop adaptation and a biological, rather than chemical, approach to yield increases."

Genetically engineered foods are being touted by some as a major part of the solution to world hunger problems. It is, therefore, rather surprising that a study published by the United Nations in 2009 (*The Environmental Food Crisis*, dealing with how to feed the world population) barely mentions GMOs. Thus far, the major application of genetic engineering has been to increase resistance to disease, viruses, and insect pests in crops. Nevertheless, genetic engineering has already been extended to animals. Fish, such as salmon, have been metabolically engineered to mature more quickly. In addition, pigs bearing spinach genes have been developed that produce lower-fat bacon while other pigs have been engineered to absorb plant phosphorus more efficiently (which then reduces the phosphorus content of their manure, which in turn does less damage to the environment). Other farm and laboratory animals have also been modified, including GM cows that produce casein-enriched milk, ideal for cheese-making; goats that churn out spider silk in their milk; and mice that produce healthy fish oils. The potential to do more is enormous.

Current Use of GMOs

The first commercially grown genetically modified crop was a tomato, though it was later taken off the market as a commercial failure. Today, the crops with the highest levels of genetic modification are soybeans, cotton, corn (maize), and canola (rapeseed). On a worldwide basis, an estimated 77 percent of the soybeans grown have been modified to incorporate a herbicide-resistant gene; in the United States, the percentage modified is even higher, 93 percent. Almost half (49 percent) of all pest-resistant cotton grown for cottonseed oil worldwide has been genetically modified, and 93 percent in the United States. Corn has been modified in several different ways: to provide resistance to herbicides and insects and to enrich the grain with beta carotene, vitamins, and folate. Throughout the world, 26 percent of all corn currently grown is genetically modified; in the United States, the proportion is 86 percent. In the United States, 95 percent of sugar beets and 93 percent of canola are genetically modified (although only 9 percent of sugar beets in the world are engineered in this manner) to produce resistance to herbicides and, in the case of canola, to increase its medium-length fatty acids and thus modify its oil content. Moreover, genetically engineered microorganisms are routinely used as sources of enzymes for the manufacture of a variety of processed foods, including alpha-amylase from bacteria, which converts starch to simple sugars; chymosin from bacteria or fungi, which clots milk protein for making cheese; and pectinestarase from fungi, which improves the clarity of fruit juice.

Crops that have been modified to be resistant to pests, disease, or herbicides include not only soybeans, corn, canola, cotton, and sugar beets, but also wheat, walnuts, potatoes, peanuts, squashes, tomatoes, tobacco, peas, sweet peppers, lettuce, and onions.

Although most GM crops are grown in North America, developing countries are increasingly adopting this approach to agriculture. Brazil, Argentina, India, China, Paraguay, Mexico, Uruguay, and South Africa have all embraced transgenic crops. So have other countries in the developed world, such as France, Germany, Spain, Bulgaria, Romania, and Australia. In addition, developing countries are using GM strains to a much greater extent; for example, 87 percent of cotton grown in India for cooking oil and animal feed is now genetically modified to reduce losses to insect predation.

Benefits/Advantages of GMOs

Genetic engineering remains in its infancy. So far scientists have concentrated most intensely on producing insect-resistant crops, because of the enormous crop losses from insect pests. *Bacillus thuringiensis* (Bt) is a naturally occurring bacterium that produces crystal proteins that are lethal to insect larvae; by transferring these Bt crystal protein genes into corn and cotton, scientists have created crops that are able to produce their own pesticides against insects. In the future, GM crops will likely be developed that can combat a whole range of pests, diseases, and viruses, along with tempering the constraints of poor climate and soil conditions. The main goal is to increase yields, but use of GMOs also can help the environment and reduce costs of food production, most significantly, in the following ways:

- *Pest Resistance.* Genetically modified crops that contain Bt reduce and may even eliminate the need for chemical pesticides, the use of which creates potential health hazards as well as polluting runoff and poisoning water supplies.

- *Herbicide Tolerance.* Crop plants genetically engineered to be resistant to powerful herbicides can reduce the amount of weed-killer now sprayed in large quantities on the fields. For example, Monsanto has developed a strain of soybeans that are genetically modified to remain unaffected by the company's herbicide product, Roundup. These soybeans require only one application of weed-killer instead of multiple applications, which both reduces production costs and limits dangerous agricultural waste runoff.
- *Disease Resistance.* Plant biologists are working on plants that internalize resistance to a wide variety of viruses, fungi, and bacteria that destroy or reduce crop yields.
- *Cold Tolerance.* Recognizing that unanticipated frost can destroy sensitive young plants, scientists have been working on introducing an antifreeze gene from cold-water fish into such plants as tobacco, potatoes, and strawberries. In the future, it is hoped that scientists can extend the growing areas of crops into higher latitudes.
- *Drought and Salt Tolerance.* The potential for this development is enormous, especially with the great growth of populations in water-short environments and the anticipated increase in drought-prone environments because of climate change. Creating plants that can withstand long periods of drought or high salt content in soil and groundwater will enable people to grow crops in formerly inhospitable locations. It is thought that efforts to create genetically engineered crops may be able to respond to increased drought more quickly than older, more traditional methods of developing new species.
- *Enhancing Nutritional Characteristics.* Malnutrition is common in many parts of the world, and it is hoped that genetically enriching plants with greater protein, vitamin, and mineral content could alleviate nutrient deficiencies.
- *Pharmaceuticals.* Medicines and vaccines are often costly to produce and require special storage conditions. Recognizing this fact, researchers are working to develop edible vaccines in such crops as potatoes and tomatoes on the grounds that these vaccines will be much easier to ship, store, and administer than traditional injectable vaccines. One idea, for example, is to develop bananas that produce human vaccines against infectious diseases such as hepatitis B.
- *Phyto-remediation.* Trees as well as traditional crops can be engineered to help clean up contaminated soils and groundwater. For instance, poplar trees have been genetically engineered to remove heavy metal pollution from contaminated soil.

To date, more than 40 plant varieties have been genetically modified to provide not only these qualities but several others, including modified ripening characteristics (e.g., in tomatoes and cantaloupes), improved flavor, increased shelf life and hardiness, and allergen-free products. Geneticists have even created no-tears onions and caffeine-free coffee plants. There are also plans to grow fruit and nut trees that provide yields years earlier than non-GM varieties and to develop plants that produce new plastics with unique properties. Although very few genetically modified whole fruits and vegetables are currently available at produce stands, most highly processed foods such as vegetable oils or breakfast cereals contain tiny percentages of genetically modified ingredients because so often the raw ingredients have been mixed into one processing stream from many different sources. This relationship is especially likely to hold true in the United States, where soybean derivatives are added to a great variety of processed foods.

Risks/Disadvantages of GMOs

Many people and organizations have raised serious concerns about GM foods, criticizing agribusinesses for pursuing profit over safety and governments for failing to exercise adequate regulatory oversight. These concerns center on three categories.

Environmental Hazards

There is considerable anxiety over the fact that Bt toxins kill not only crop-damaging pests but also many other species of insect larvae indiscriminately. It is also feared that insects will develop resistance to Bt or other crops that have been genetically modified to produce their own pesticides, just as other species of both plants and insects have developed resistance to pesticides such as DDT in the past.

Another concern is that crop plants engineered for herbicide resistance and weeds will cross-breed, resulting in the transfer of the herbicide resistance genes from the crops into weeds, thereby creating super-weeds. Other introduced genes also could cross over into nonmodified crops planted nearby. Indeed, field trials have shown that such gene transfer does occur. One 2002 study demonstrated that transgenes had spread from the United States to traditional corn varieties in Mexico, while a 2004 study revealed that conventional varieties of major U.S. food crops have been widely contaminated. As recently as August 2010, it was reported that escaped GM strains of canola had been found growing wild along roads in North Dakota. In addition, the U.S. National Academy of Sciences reported that glyphosate (herbicide)-resistant weeds, which developed as an outgrowth of the use of engineered crops, have had substantial negative impacts.

These concerns can be addressed, at least to some extent, by making plants with pollen that is sterile or does not contain the introduced gene and/or by requiring buffer zones around fields of GM crops. Some critics doubt whether these ideas will work, however. The problem is that once GM genes have escaped into the wild, they cannot be recalled.

Human Health Risks

Because genetically modified foods have existed for only a short time, many are concerned about the long-term unintended consequences of altering nature. One issue is allergenicity—that is, the risk that introducing a gene into a plant may create a new allergen or cause an allergic reaction in susceptible individuals. To date, such fears have not been realized. Indeed, a 2008 review published by the Royal Society of Medicine found no reports of adverse health effects from GM foods that had been consumed by millions of people worldwide for more than 15 years. No epidemiological studies have been conducted to determine whether engineered crops have caused any harm to the public, however. Even so, the overwhelming body of scientific evidence indicates that GM crops are safe.

Economic Issues

Serious questions have arisen about the costs of using GM seeds and about patent infringement. Clearly, agribusinesses that have invested heavily in agribiotechnology need to recoup their costs and ensure a profitable return on their investment. Yet the price for such GM seeds may well be beyond the reach of farmers, especially in the developing world. Also, because many new plant genetic engineering technologies and GM plants have been patented, patent enforcement becomes a real issue. Farmers who raise these plants are required to give up control over their seeds (and even the kind of pesticide that is effective) and not use their harvested seeds the following season as they have in the past, but instead depend totally on a single, large, for-profit corporation such as Monsanto.

In India, thousands of Indian cotton-growers were reported to have committed suicide after their GM crops failed. They fell into debt and were no longer able to afford new Bt seeds and the chemical pesticides required. As a result, they lost their familial holdings.

Some people claim that there are less expensive ways of delivering vitamins and minerals to people than by genetically modifying crops such as rice and corn. In addition, consuming locally grown food rather than food trucked in from places hundreds of miles away and ensuring better distribution of resources and wealth might go further toward solving the world's food needs than spending huge amounts of money on GM foods.

Regional Differences in the Use of Genetically Modified Foods

Worldwide, there is considerable disagreement over the merits and the risks attached to the use of genetically modified foods. By far, their greatest use has occurred in North America, spurred on by the research and development of GM species.

Europe, by contrast, has been much more hesitant to adopt GM foods, requiring a strict labeling system as well as the reliable separation of GM and non-GM organisms, both at the production level and throughout the whole processing chain. However, the European Union's attempts to block imports of U.S. farm products through its long-standing ban on genetically modified food restrictions was ruled illegal by the World Trade Organization, which stated that the ban violated international trade rules. In March 2010, the European Union quietly gave permission for farmers to grow genetically modified potatoes (developed to produce higher levels of starch, to be used in industries such as paper manufacturing, but not intended for human consumption), on the grounds that it would save energy, water, and chemicals. Even so, six EU member states—Austria, Hungary, France, Greece, Germany, and Luxembourg—continue to ban the cultivation of GM corn in their countries.

Other countries have banned outright the use of genetically modified foods. Zambia, which cut off the flow of GM corn from the UN's World Food Program in 2002, later changed its mind in the face of intensifying famine to allow its importation in 2005. Venezuela announced a total ban on genetically modified seeds in 2004. In India, a moratorium on the cultivation of genetically modified foods was imposed in 2009 after several groups rejected regulatory approval of the cultivation of Bt *brinjal*, a GM eggplant (as the variety of aubergine/eggplant is known in India). In contrast, in November 2009, China granted safety certificates for the domestic production of two varieties of GM rice and one variety of GM corn, with the expectation that commercial production would start in 2013.

Many people believe that pursuing traditional ways of increasing yields is safer and more reliable than genetically engineering plants and animals. On the other side of the question, many scientists argue that the world needs to use GM foods because they can be produced more quickly and are less likely than conventionally bred crops to have unintended consequences. In any case, continued rigorous safety testing of products will need to occur, as well as careful analysis to ensure that the perceived benefits outweigh both the actual and hidden costs of development.

Future Importance of Genetically Modified Foods

It is probably premature to draw any firm conclusion about the future importance of genetically engineered crops. Clearly, not all of the consequences of GM use are fully known. But do the problems that have been identified in the use of GMOs outweigh their benefits of higher yields, lowered and less-toxic pesticide and herbicide usage, and reduced soil erosion as a result of less tilling the ground? Certainly, farmers in the United States do not think so, as they overwhelmingly continue to grow many GM crops. In 2010, a blue-ribbon committee of the

National Academy of Sciences found that crops produced through genetic engineering are, on the whole, beneficial for farmers who plant them. Despite the extra cost of GM seeds, it reports that farmers like these products because they require less labor and fewer chemicals.

As yet, GM seeds have not reached their desired or anticipated potential. Hopes that this new, genetically modified high-technology approach to food production would quickly transform agriculture with higher-yielding crops and fields requiring a lot less spraying and fertilizer have not been realized. Rather, it seems that the path to obtaining most of these highly desirable traits is more complicated and may even be more likely to arise from conventional plant breeding. At the same time, the worst fears of using this technology in terms of health and environmental concerns have not come to pass. Obviously, genetically modified crops are here to stay, even though the jury is still out on their ultimate success.

—Christine Drake

Further Reading

Collier, Paul. "Can Biotech Food Cure World Hunger?" *New York Times*, October 26, 2009.

The Environmental Food Crisis: The Environment's Role in Averting Future Food Crises. Nairobi, Kenya: United Nations Environmental Programme (UNEP), 2009.

Estabrook, Barry. "Supreme Court on Modified Foods: Who Won?" *The Atlantic*, June 22, 2010.

Miller, G. Tyler, and Scott E. Spoolman. "Food, Soil, and Pest Management." *Living in the Environment*. Belmont, CA: Brooks/Cole, 2009, 275–312.

Pickrell, John. "Introduction: GM Organisms." *New Scientist*, September 4, 2006.

U.S. Food and Drug Administration. *The FDA List of Completed Consultations on Bioengineered Foods*. Washington, DC: U.S. Government Printing Office, 2009.

Whitman, Deborah B. "Genetically Modified Foods: Harmful or Helpful?" *ProQuest*, April 2000.

Geotechniques (Remote Sensing and Geographic Information Systems): Tools for Monitoring Change

A well-known axiom holds that the one constant factor that influences human life on Earth is change. Change can have many meanings: to make something different or modify it; to substitute one item, use, or feature for another; to exchange the position or place of an object or use; to improve or make worse by a decision that is manifested by something observable or felt; to become or evolve into something different than there was before; or to transfer from one phase to another, ensuring passage of one phase to another or moving to another phase. Few things on Earth do not experience change; thus change can be detected, quantified, mapped, and analyzed. Change detection tools employed to investigate dynamic systems of energy change, processes, and the events leading to conditions that contribute to hunger and famine include remote sensing, geographic information systems (GIS), and geospatial analysis.

Remote Sensing

Remote sensing is a means of viewing, identifying, and interpreting objects from a distance—hence the often used expression, "sensing remotely." The manner in which objects are captured is through a multispectral scanning instrument housed within a satellite typically orbiting above the Earth, or mounted within aircraft flying at a low altitude. Multispectral sensors measure and store reflected and emitted energy within specific ranges of wavelengths (spectral bands) of the electromagnetic spectrum (EMS). A continuous stream of electronic impulses representing electromagnetic energy is recorded and transformed into discrete digital files representing a number of spectral bands. These bands can be displayed in single- or multiple-band image composites, and they can be further enhanced to bring out specific surface features, atmospheric conditions, or spectral qualities of land and water bodies.

An operating assumption in the most commonly used forms of remote sensing technology is that Earth surface features will exhibit an identifiable spectral response pattern of reflected light and/or emitted heat energy that can be generalized at some level to a surface feature class. One of the applications of digital image processing of multispectral remotely sensed imagery consists of running a systematic array of computer-driven algorithms to discriminate spectral classes, each of which has elements of spectral similarity. Once the digital imagery is classified statistically, it must be further interpreted, analyzed, and associated with specific Earth surface features. It is this process of combining applied science and technology to identify surface conditions that can contribute to an understanding for a variety of process-driven events such as volcanic activity, hurricanes, tornadoes, flooding, mass wasting, land-use changes, and other conditions that can have a severe to catastrophic impact on both the natural and human dimensions of the planet. Remote sensing clearly has applicability to understanding the many variables contributing to conditions leading to drought, crop losses, food shortages, and, ultimately, famine.

A number of variables, observations, and measurements must be taken into account when considering the applicability of remotely sensed imagery and digital image processing to assess potential food shortages and issues of famine. Multispectral remote sensing scanners measure electromagnetic radiation at wavelengths that span both the visible range and ranges beyond visible such as infrared and microwave radiation. An ability to sense, process, and display reflectance and emitting characteristics of wavelengths outside any human's ability to sense with an unaided instrument is critical to understanding likenesses, differences, or uniquenesses for a number of variables that may contribute to crop stress—for example, insect infestations leading to crop disease; local, regional, and seasonal water shortages leading to vegetation stress and drought conditions; or high-intensity rainfall events resulting in excessive water due to widespread flooding, which also contributes to a yearly loss in crop yield. Crop yields affected throughout a longer time period and covering more expansive regions can be impacted by radical changes in climate. Anomalous chemical conditions measured in fields of bare soil may be indicators of pollutants entering from outside sources such as groundwater recharge sites, surface runoff/stream flow, and airborne transport of particulates from any number of point sources.

An ability to classify specific spectral response patterns of surface features using digital image processing applications on remotely sensed imagery provides a means of isolating likenesses, differences, and, perhaps more importantly, surface anomalies that may have a significant impact on crop health and potential seasonal yields. Following an initial classification of regions of interest, field measurements can be recorded at select isolated locations to establish a more precise record for the physical and chemical conditions that may be contributing to weakness in plant anatomy (e.g., biomass, plant density, vigor), crop loss, or other deficiencies identified in the overall spectrum of food production.

The advantage of airborne and satellite remote sensors is that large tracts of the Earth's surface can be scanned and processed at a relatively economical cost per area of land surveyed. This type of evaluation provides a first-order assessment of specific regions that may be stressed and susceptible to drought or other environmental conditions leading to potential food shortages. The processed and classified multispectral imagery is stored as a digital file, and when combined with other spatial data and coupled with a database of field-derived measurements, spatial modeling and geospatial analysis within a GIS can follow.

Geographic Information Systems

Airborne and satellite-derived remotely sensed multispectral digital data provide investigators with imagery that can be enhanced and classified into discrete Earth surface feature classes. Advanced spatial modeling and analytical processing of supporting multivariate data, however, are also undertaken in a GIS. A *geographic information system* is a functional assortment of software that interfaces within a computer for the purpose of addressing spatial issues. The term "geographic" can be broken down into two basic elements: *geo* is Earth-related, and *graphic* means to describe or depict the Earth through maps or drawings. *Information* comprises data with explanation; it may also be defined as data in an applied sense. Information can be word based or graphic. It is derived from data elements through a process of observations, thought, and measurement. *Data* consist of measured phenomena, statistics, or other quantitatively derived values. A *system* is any number of components or parts that, when assembled, function as a unit.

GIS software and computer hardware collectively provide an interface by which spatial objects (objects with a known location on Earth) can be linked to data (a database file, which, in this context, would lists items in tabular format) consisting of relevant information that describes the nature of these spatial objects. The combination of spatial objects and tabular data may be input, stored, retrieved, updated, manipulated, deleted, and analyzed. Queries (questions) may be initiated in a GIS in which only selected segments of a database are combined with specific spatial objects. The GIS can generate a completely new configuration for these spatial objects and their associated database through output in the form of maps, tables, charts, imagery, and other spatial models. The ability to establish spatial models, based on specific information, provides a visual means for identifying trends, likenesses, differences, anomalies, relationships, and physical interactions associated with Earth surface processes and the human conditions that are affected by them. A GIS can serve as an instrument that facilitates activities such as identifying and ranking complex variables by isolating significant processes associated with spatial problems, or as a decision-support system for future planning (e.g., modeling the natural environment, urban–regional planning, or Earth resources planning).

A GIS is, in part, a technology employed to design and construct maps from tabular databases and, in part, an analytical tool focusing on spatial relationships. Maps are a means to convey spatial information. A GIS has certain attributes of a science: It serves as a means of systematically integrating graphical information derived from multiple disciplines. A GIS also functions as a major tool and technique that links the physical and human elements found in a world society. It enables an understanding of place as a physical location identifiable through a geographically referenced (by a grid or coordinate system) set of spatial images and maps that, by extension, establishes spatial boundaries for features of interest. When combined with the human and physical dimensions characteristic of a particular location, a GIS aids in identifying a geographic region and the processes

Spatial Literacy

The means to resolve many of the problems that societies worldwide will face throughout the 21st century will require a person's ability to think, reason, communicate, and analyze a plethora of data. To do so, a person will need to use technical computer programs, coupled with highly developed analytical processing skills that are not incorporated into the traditional curricula. In addition to reading, writing, and arithmetic, the fourth element that is critical to understanding the manner in which the world functions is spatial literacy. We live in a three-dimensional world that operates within a fourth dimension of time. Yet many people have difficulty visualizing complex processes and multivariate phenomena composing the essential foundations that enable us to function. The growing availability of spatially referenced data and advances in computer science, mathematics, spatial statistics, and sophisticated graphical software programs have resulted in the development of research tools, techniques, and ability to analyze massive amounts of data through what has been termed spatial analysis. Moreover, the U.S. Department of Labor has identified geotechnology among the three fastest-growing industries, along with nanotechnology and biotechnology.

—Danny M. Vaughn

at work within a region. The interrelationships of multiple processes (e.g., weather and long-term climate) operating on Earth features produce or prevent changes in our environment throughout time and at varied scales. The consequences of a dynamic Earth have pronounced effects on all inhabitants, and they are best understood when energy exchanges and the processes resulting from them are systematically identified, modeled, and analyzed through the applications of remotely sensed imagery, image processing, and geographic information systems.

Geospatial Analysis

The applicability and resourcefulness with which *geomatics* can address spatial issues is apparent when considering its ability to combine spectrally enhanced and classified remotely sensed imagery with supporting digital and tabular information. Specific computer-driven software algorithms (programs designed to serve a defined function, such as performing a classification) are implemented to generate spatial models that isolate any criteria relevant to a particular issue. The final outputs in the form of maps, charts, graphs, and tabular data are developed within a GIS by a number of processes referred to as *geospatial analysis.*

A GIS links all spatial objects to a grid or geographic coordinate system (e.g., based on latitude and longitude), so that all spatial objects can be accurately positioned in proper geographic relationships and location, distance, and direction can be established. Spatial objects are formed by graphic elements (often called spatial primitives), and are expressed as a series of points, lines, and polygons. Points can represent the specific locations of objects, such as springs (water sources), stream gauging stations (recording point-source locations for stream discharge and velocity), and weather stations (recording local atmospheric conditions). Lines may represent transportation networks such as roads, railroad tracks (transportation networks), topography (elevation changes), or streams (water courses, flow networks). Polygons represent objects that delineate areas, such as fields of row crops, soil-series associations, town and city boundaries, or climate boundaries. Once these spatial primitives/elements are entered into a GIS, they become a spatial database representing "real-world" phenomena that, when coupled with tabular data, supply the basis for geospatial analysis.

Spatial associations such as *connectivity*, *adjacency*, *orientation*, and *containment* among and between graphic elements (representing objects) can be established because all spatial and tabular information has been geographically linked to a coordinate system. Geospatial analysis addresses a number of spatial relationships, such as *content*, *size*, *shape*, *pattern*, *clustering*, *density*, *location*, *distance*, and *direction*, as means of delineating the spatial relationships of naturally occurring soil types, row crops, other vegetation classes, and other elements. Population, religions, per-capita income, and other concepts are considered abstractions because they cannot be directly observed as a specific spatial entity, although they can be quantitatively derived and exhibited by spatial operators. Hence, a GIS provides a dynamic means of blending spatial and tabular data into useful information. It does so by isolating thematic objects—for example, polygons representing soil-series associations, crop types, and arable land; contour lines illustrating rainfall distributions, elevation models, slope models, and aspect models; and point themes, as well as locations, food distribution points, and transportation ports—into layers. When combined, these layers form the content of output maps.

Building upon the isolated classes of surface features (feature classes) identified through remote sensing applications, a digital map delineating the feature classes can be created and used as one of the primary layers for a GIS analysis and spatial modeling. Because all spatial and tabular data are in digital format, the entire system is dynamic, which implies that any thematic content may be manipulated into an infinite array of spatial and tabular configurations. This feat is accomplished through an ability to perform queries in which only selected pieces of a database are selected and combined with only those spatial objects that support the basic thesis of a query.

Identification of selected spatial associations that are applicable to addressing potential food shortages and ensuing conditions of regional famine begins with the creation of a number of layers that reflect known variables contributing or possessing the potential to affect these issues. Data reflecting radical changes in local weather, such as seasonal drought or areas of continuous storm activity, followed by longer-term climate changes, can be mapped to show patterns of isolated weather extremes. Several layers can be overlaid with one another to assess changes over different time periods (time change analysis). Vegetation (crop) stress can be spatially associated with physical and chemical data reflecting anomalous soil conditions that may be altered by such sources as industrial pollutants, ash and pyroclastic fallout from volcanic eruptions, groundwater, and surface-water pollution. Each variable can be input as a separate layer of information that, when displayed on a computer monitor, graphically illustrates any spatial associations such as distinct visual patterns, regions of clustering, and density modeled through the spatial coincidence of the variables. Elevation data may be modeled with weather and climate data, regional settlement patterns, food types, food sources, religious beliefs, and cultural history in an attempt to understand how each of these factors contributes to lifestyle, conditions affecting poverty, and ultimately food shortages. In this scenario, the conditions that could lead to famine are not totally a direct result of naturally occurring processes, but rather are also based on human conditions. Both natural processes (physical and chemical) and human interactions dynamically interact and result in conditions that require an ability to observe, identify, and understand the complex variables. These variables can lead to conditions that threaten the well-being of world societies.

Patterns suggest that something is influencing the spatial distribution of objects. Conversely, a lack of pattern also can be significant: It may suggest an anomaly that is not a function of a normal condition or process. Crop patterns may be overlaid with elevation contours suggesting that crop yield and vigor may be affected by changes in elevation. Density refers to the number of objects taking up an area or region; when this factor is combined with, for example, soils data (physical and chemical variables), the spatial relationships can be isolated and may show something about these two variables that is contributing to plant health or vigor. When a cluster of objects is observed, there is usually something controlling the manner in which they are assembled. By examining these spatial associations, investigators (agencies and private organizations associated with FEWS NET) can identify the most pertinent factors, variables, and conditions that influence local-, regional-, and continental-scale events and may potentially threaten the health, lifestyle, and future existence of a human population.

Geomatics employs multispectral remotely sensed imagery, digital image processing, geographic information systems, cartography, global positional systems, and other contributing disciplines under the umbrella of the mapping sciences. In combination, these disciplines can generate a completely new configuration for spatial objects by isolating only those variables engaged in impacting an event. Geomatics provides a critical set of instruments whose use facilitates a systematic process of collecting, sorting, compiling, and analyzing the most relevant information required for decision making, solving spatial problems, and support planning.

—Danny M. Vaughn

Further Reading

"Educational Resources." January 2011. United States Geological Survey: http://rockyweb.cr.usgs.gov/outreach/index.html.

"The Geographer's Craft." February 2011. www.colorado.edu/geography/gcraft/contents.html.

"The Guide to Geographic Information Systems." January 2011. www.gis.com/.

Kerski, J. "References: GIS and Related Technologies." January 2011. www.ncsu.edu/midlink/gis/joseph_references.htm.

National Center for Geographic Information and Analysis. January 2011. www.ncgia.ucsb.edu/.

Green Revolution

The Green Revolution refers to the development of high-yielding, disease-resistant, nutritional, semi-dwarf varieties of wheat, rice, and corn as part of an effort to resolve hunger issues, improve the rural economy of underdeveloped countries, and reduce the potential of famine. This revolution, which was based on plant research, crop management, and technology transfer issues, began slowly in 1943 and increased dramatically in the late 1970s. The term "Green Revolution" was coined by William Gaud, former USAID Director, who in 1968 compared the life-saving "Green Revolution" to the life-taking Red Revolution in Imperial Russia. Dr. Norman Borlaug is considered by many to be the "father of the Green Revolution." He and his research team are credited with saving the lives of perhaps a billion people worldwide.

A brilliant researcher and a practical agronomist, Borlaug received his PhD in plant pathology and genetics from the University of Minnesota in 1942. During World War II, he was employed by the U.S. government on many research projects designed to solve problems faced with the deployment of American military personnel in both Asia and Europe. In the 1940s, Vice President Henry

Breakthroughs in Increasing the World's Food Supply

Approximately 10,000 years ago, in both the Old and New Worlds, food availability was increased by the domestication of plants and animals. This process achieved through common sense, keen observation, and then detailed knowledge of wild plants and animals and knowledge of how to control reproduction. Many breakthroughs followed. In about 3000 BCE in the Mediterranean rim-land, the plow was developed to soften the seedbed. Slowly plants and animals diffused throughout the known world and human diets improved, particularly after the discovery of the New World.

With a great number of plant and animal options available to them, agriculturalists soon realized that only three ways to expand world food production remained: (1) plowing up more land and increasing the area harvested; (2) increasing food yields on land already under cultivation through better seeds, improved agricultural techniques, and controlled watering or irrigation; and (3) reducing harvest losses, food storage deterioration, and food preparation waste to a minimum. More breakthroughs and much progress were made over time in all aspects of agriculture. The number of hungry in the world and the frequency of famines diminished.

In the last years of World War II, it became apparent to food planners in the United States that vast amounts of food would be needed in war-devastated Europe and Asia after the war ended. Remarkably, a small team of research scientists working in Mexico produced a revolutionary high-yielding, disease-resistant wheat plant and an approach to transferring developed world agricultural knowledge to the underdeveloped world in the 1940s and 1950s. Their work marked the beginning of the Green Revolution. The resulting increased wheat yields, then rice yields, saved millions of lives in the 1960s and 1970s.

—William A. Dando

Wallace, recognizing the socioeconomic potential of Mexico and Mexico's contributions to the war effort, persuaded the Rockefeller Foundation to assist the Mexican government in aspects of agricultural development. With the support of Wallace and the Rockefeller Foundation, a new Mexican government research agency was organized. The Office of Special Studies (OSS) was to be directed by the Rockefeller Foundation and staffed by American and Mexican soil, corn, wheat, and plant pathology scientists. In 1944, Borlaug accepted the position as head of the OSS's cooperative wheat research and production program in Mexico.

OSS's wheat program in Mexico involved research into plant genetics, plant breeding, plant pathology, entomology, agronomy, soil science, and cereal technology. Its goal was to increase wheat production in Mexico. During his 16-year tenure as a research scientist and project director, Borlaug bred a number of high-yielding, disease-resistant, semi-dwarf wheat varieties. To accelerate new wheat variety development time, he established two experimental farms 700 miles apart with 10 degrees difference in latitude and 8,500 feet difference in altitude. By breeding and testing at two locations in two distinctly unrelated agroclimatic regions, he ensured that the varieties that were finally approved did not need to be bred for every agroclimatic region of the world. Thus Borlaug's wheat varieties were able to be grown in many dissimilar farm environments.

To reduce yield losses to disease, Borlaug also identified and transferred disease-resistant genes from traditional wheat plants to the new wheat plants. Still not satisfied, he observed that the wheat varieties he created had tall, thin stems and tended to collapse (or lodge) from the weight of the extra wheat grains, near harvest time. In 1953, he crossed a Japanese semi-dwarf variety, Norin/10 Brevor, with his high-yielding, disease-resistant plants to create new semi-dwarf wheat varieties. The new wheat plants were one-half to two-thirds the height of standard wheat plants.

Borlaug's final product was a revolutionary wheat plant with thick stems and more ears (heads) of grain per plant, which was very high yielding, disease resistant, and adaptable to both tropical and subtropical agroclimatic regions. His varieties of wheat were so successful that in 1963, 95 percent of all Mexican wheat planted consisted of the varieties that Borlaug and his research team had created. In 1963, the Mexican wheat harvest was six times larger than the corresponding harvest in 1944. In the span of just 20 years, Mexico became self-sufficient in wheat and evolved into an exporter of wheat.

Expansion of the Green Revolution

When the subcontinent of India received its independence from Great Britain on August 15, 1947, the area of the former British colony with a predominant Hindu population became the nation of India. The part of the former colony with a predominantly Muslim population formed Pakistan. Unfortunately, the Indian–Pakistan Commission failed to agree on the complete division of the former colony into two distinct nations. The basic division of area did occur eventually. In 1948, approximately 6 million Hindus and Sikhs migrated from West Pakistan and approximately 6.5 million Muslims left India. Civil strife and religious persecution continued, and in the early 1950s approximately 4 million Hindus exited East Pakistan and more than 1 million Muslims exited India.

Long-lasting socioeconomic difficulties and animosities created constant tension and much social disruption in the region. To compound the food-deficiency problems in India, the new country was deprived of the wheat-growing, food-surplus districts, which had become part of West Pakistan. The Indian subcontinent became a battleground for religious groups unwilling to compromise or peacefully resolve divisive issues. Turmoil continued into the 1960s, when a famine occurred and claimed hundreds of thousands of lives (see the "Indian Famines: 1707–1943" entry in Volume 2). The United States sent millions of tons of grain to feed the starving Indian people. In some years, as much as 20 percent of America's wheat harvest was sent as emergency food aid to the starving and dying of India. In 1965, the famine had spread and the mass starvation had become so bad that Borlaug was given permission to introduce tested and successful agrotechniques and immediately distribute the wheat varieties he and his colleagues had developed in Mexico.

Borlaug's wheat varieties grew very well on the Indian subcontinent, with yields reaching levels higher than ever seen before in India and Pakistan. Pakistan's wheat production increased from 4.6 million tons in 1965 to 7.3 million tons in 1970, and the country became self-sufficient in wheat. India's wheat production increased from 12.3 million tons in 1965 to 20.1 million tons in 1970; it also became self-sufficient in wheat. High yields of wheat led to shortages in both countries of labor to harvest crops, trucks and carts to haul the cut wheat to threshing centers, bags to contain the harvested wheat grains, railroad cars to transport harvested and sorted wheat to market, and wheat grain storage facilities.

The Rockefeller and Ford Foundations' directors, spurred by the success Borlaug and his colleagues had achieved in Mexico and by the success of the Green Revolution in Pakistan and India, established the International Rice Research Institute (IRRI) in the Philippines. With the support of the Philippine government and with the assistance of a superb research team of Borlaug's colleagues, a new rice variety was developed in 1966. The new variety, called IR8, required the use of chemical fertilizers, expensive pesticides, and much water, but yields were very high. Yearly rice production in the Philippines increased from 3.7 million tons to 7.7 in less than 20 years. The Philippines became self-sufficient in rice and evolved into a rice exporter. In addition, many other countries in Southeast Asia began planting IR8. When fertilized and well watered, the IR8 variety produced 10 times the yield of traditional rice per acre. Soon IR8 was dubbed the "Miracle Rice." In 1970, a ton of rice in India cost $550; in 2001, it cost less than $200 per ton. India became not only self-sufficient in rice, but also became a rice exporter.

Pleased with the success of the Green Revolution, the foundations that supported Borlaug established the Consultative Group on International Agricultural Research (CGIAR). CGIAR comprised a worldwide network of agricultural research centers charged with expanding the work that began in Mexico. For his contributions to the Green Revolution and his efforts to increase the world's food supply, Borlaug was awarded the Nobel Peace Prize on December 10, 1970.

Core Ingredients for a Green Revolution

The Green Revolution "package" combined modern agricultural technologies utilized by agriculturalists in developed nations with new wheat, rice, and corn varieties. Technologies transferred to aid-requesting, food-deficit developing countries included chemical fertilizers, pesticides, herbicides, insecticides, modern irrigation methods, and recommendations on planting and harvesting. In traditional agriculture, inputs such as fertilizer and biological insect pest control were secured by the farmer or provided by the bounty of nature. Green Revolution inputs required funds to purchase fertilizers, pesticides, herbicides, insecticides, high-yielding seed, and farm mechanization equipment such as water pumps and tractors. Rural credit institutions were established to lend money to pay for the items needed to plant and grow the high-yielding seeds and purchase new farm equipment. The new wheat, rice, and corn seeds also required more water per plant than traditional seeds. To conserve water, new irrigation methods and supplemental irrigation plans were developed.

To ensure the success of the model developed to increase food production in food-deficit developing countries, there also had to be commitment made to all efforts by national, regional, and local governments. Government-maintained or -subsidized infrastructure, including roads, irrigation systems, storage facilities, and electricity, would require upgrading and expansion. Also, the food producers in rural areas would need to commit themselves to trying new ways to grow crops and new equipment to help them with farm work. For a Green Revolution to occur and food production to increase, the core ingredients necessary were a product of applied agricultural research and the sharing of agricultural technologies and information.

Borlaug contended that grain production levels that had required centuries to reach in developed countries such as Europe and North America could be achieved in developing countries within a few decades if the following conditions were met:

1. Government leaders, commercial and industrial decision makers, and rural food producers strongly supported the new input-integrated strategy that combined high-yielding varieties of seed with a "package" of inputs applied to carefully chosen fields supplied with irrigation water.
2. Loans to purchase necessary inputs and gifts of cash, seed, agricultural materials, and machinery were made available or given to farmers by governments, international aid agencies, and foundations.
3. Local, regional, and national grain price policies were set in place that compensated the producers of food products fairly.

4. National and regional agricultural research institutes were established to resolve issues or problems of local concern.
5. A system of state agricultural universities was founded to direct scientific inquiry into long-term agricultural issues.
6. Countries created a well-supported network of "county agents" who would take proven agricultural innovations to the farmer.

Constraints to Change

The Green Revolution was hailed as the beginning of new food abundance for all who lived on Earth. Many believed that this nonviolent revolution would offer sufficient rewards to overcome conservative small-scale farmers' resistance to change in developing countries. Sincere, concerned agricultural development specialists believed that archaic social and defective political institutions had severely constrained the adoption of new crops, new ideas, and new methods. In their minds, the persistence of poverty, hunger, and malnutrition was self-perpetuating. In developing countries, the rewards the farmers were to receive from the new approach to agriculture were perceived as including higher yields, food security, increased farm income, and improved quality of rural life. Supporters of the Green Revolution concept contended that the problems of rural poverty, hunger, undernutrition, malnutrition, and famine would be solved and slowly would lead to improved social institutions.

Unfortunately, the benefits of the Green Revolution rarely were available to small-scale, poor peasant farmers. Only larger-scale farmers with substantial land, financial resources or access to capital, and political power were able to secure the new varieties of seeds, fertilizers, pesticides, and irrigation water. Water was essential, because the new varieties of seeds required more water and, if denied necessary water, yielded about the same amounts of outputs as traditional seeds. Those farmers who did not or could not adopt the Green Revolution "package" and the landless rural laborers found that their income was reduced by the excess grain now on the market and the depressed market price for grain.

Governments of developing nations also learned that the purchase of imported irrigation pumps, new agricultural machinery, fertilizers, pesticides, fuel, and new seeds reduced the nation's financial reserves and led to borrowing capital from developed countries. In turn, great pressure was exerted on their social and physical institutions. Providing food for a nation is not simply equivalent to producing more food; rather, the food must be made available to consumers at prices they can afford. Marketing, farm supplies, and public and private rural services are required. Borlaug realized that much more had to be done to resolve the persistence of world food problems.

Infrastructure improvement is a basic necessity for the resolution of hunger, poverty, and famine issues, which encompass the following concerns:

1. Transportation, such as roads, railroads, airports, pipelines, ports, and canals
2. Communication networks, including telephone, radio, television, and access to computers
3. Credit such as private banks, government loan agencies, and international aid
4. Education and agricultural research, including all levels of formal education and applied research
5. Markets and marketing (i.e., sites and systems)
6. Irrigation and drainage, including agricultural water sources and drainage of waste water
7. National, state/province, regional, and site planning, such as linking goals and objectives

Addressing these facets of the infrastructure requires change over time. Infrastructure improvements generally lead to improvements in the agricultural economy; improvements in the economy, in turn, make new demands on the existing infrastructure. The infrastructure must be well maintained and constantly upgraded to the national, regional, and producer needs.

In many developing countries of the world, national food issues and agricultural growth are affected by international economics, decisions made by international food corporations' management, climate change and weather variations, and the national leader's political agenda. Major policies that discourage agricultural growth and development include the following: (1) governmental control of agriculture; (2) governmental management of the nation's food pricing system; (3) export tax policies; (4) restrictions on farm credit; (5) constraints on the free movement of food products in a country; and (6) unrealistic money exchange rates.

Some specialists in agricultural development contend that the leaders of developing countries have an urban bias and, therefore, tend to neglect rural development and agriculture. These leaders believe that industrial growth will resolve both national food and rural economic issues. Policies of "cheap food" in a country penalize rural food producers and subsidize urban nonfood producers. Most rural dwellers and most food producers have little voice in this national decision making, even though it affects their livelihood. Few national leaders or governments have the vision, the desire, and the commitment to rural dwellers and farmers to change policies in such a way as to increase food costs for urban dwellers.

Assuring rural food producers adequate prices for their produce is critical to any program or plan for agricultural reform.

Lessons Learned from the Green Revolution

Important lessons can be drawn from decisions made by the Rockefeller Foundation, the Ford Foundation, World Bank, and the Food and Agriculture Organization of the United Nations (FAO) to spread the lessons learned in Mexico to other developing nations of the world. For a Green Revolution to be successful, the following components are necessary:

1. A sincere government commitment and strong leadership desire for socioeconomic changes must be present.
2. Assurances must be given that all citizens will share in the benefits in some way.
3. Basic resources—specifically, land and water—must be equitably available to all food producers.
4. Social, economic, and educational institutions should be transformed to serve the needs of all.
5. National, regional, and local policies must support the application of proven new agrotechnologies.
6. Unbiased distribution of financial assistance, aid funds, technical guidance, and on-farm support must be made to all sections of the country.
7. Recommended changes must include means to make labor more productive.
8. Infrastructure and transformed institutions need to serve both small-farm and large-farm agriculture.
9. Social inhibitions and constraints must be identified early in the project and addressed effectively.
10. Rural women must be included in all phases of innovation and development.
11. Nutrition and sound diets should be a priority.
12. Food losses in harvesting, transporting, and marketing must be reduced.
13. Traditional diet food crops should be supported as well as cash crops.
14. Planning to meet the needs of a country must be done at all levels, then integrated into a national plan with obtainable objectives.
15. Traditional food crop and cash crop production increases can be maximized by geographic (area or place) specialization.

16. Local field and site demonstrations need to be provided to stimulate interest in the project.
17. Reserve food storage facilities must be provided to hold food for use in severe crop failures or natural disasters.
18. Feeder or secondary roads in rural areas need to be improved or constructed.
19. Enhanced food crop activities must focus on agricultural intensification (yields per acre) rather than extensification (acres per yield).
20. Subsidized agricultural production inputs should serve as effective encouragement for farmers to adopt new seeds or new agrotechniques.
21. Agricultural subsidies must be reviewed each year to determine their effectiveness and, if necessary, to redirect their focus.

Developing countries also have learned much from the successes and the overall impact of the Green Revolution:

1. Aid requests should be made only to resolve a critical need for assistance.
2. Acceptance of foreign aid may alter many facets of traditional rural life and effect changes in government services.
3. Donor governments, foundations, and agencies in most instances will require that the funds or aid provided be used specifically to resolve critical issues and improve the quality of life for all citizens who live in that country.
4. Aid asked for and received must be in the managerial and technical competence of those who are targeted for help.
5. Both aid regulators and aid givers should take a long-term view and fully understand commitments and expectations.

The Green Revolution had the least impact in countries whose leaders did not have a sincere commitment to integrated rural development and small-farm agricultural programs. Also, lack of national and regional administration skills and a shortage of technical personnel to work with farmers reduced success in some areas.

—William A. Dando

Further Reading

Bickel, L. *Facing Starvation: Norman Borlaug and the Fight Against Hunger.* Pleasantville, NY: Readers' Digest Press, 1974.

Brown, L. *Seeds of Change*. New York: Praeger, 1970.

Clever, H. "The Contradictions of the Green Revolution." *American Economic Review* 62, no. 2 (1972): 177–186.

Dovring, F. *Progress for Food or Food for Progress*. New York: Praeger, 1988.

French, C., J. Moore, C. Kraenzle, and K. Harling. *Survival Strategies for Agricultural Cooperatives*. Ames, IA: Iowa State University Press, 1980.

Ghai, Dharam, Azizur R. Khan, Eddy Lee, and Samir Radwan. *Agrarian Systems and Rural Development*. New York: Holmes & Meier Publishers, 1979, 22–112.

"Green Revolution." wikipedia.org/wiki/Green_Revolution.

Grigg, D. *The World Food Problem: 1950–1980*. Oxford, UK: Basil Blackwell, 1985.

Nabhan, Gary Paul. *Where Our Food Comes from*. Washington, DC: Island Press/Shearwater Books, 2009.

Poleman, Thomas T., and Donald K. Freebairn. *Food, Population, and Employment: The Impact of the Green Revolution*. New York: Praeger, 1973.

Vallianatos, E. *This Land Is Their Land*. Monroe, ME: Common Courage Press, 2006.

Historiography of Food, Hunger, and Famine

Historiography often has been summed up as the "history of history." A "history of history" is an exploration of how history is written; it acknowledges that while there may be certain facts or ideas, these facts and ideas are open to interpretation, with the interpretation varying with the changing times or the author's ideology or worldview. A history can certainly tell us more about the mindset and concerns of the present than it can reveal truths about the past. No truly objective account of human history is possible, as all humans tell the stories that they want to tell and be told. Even what seems to be a straightforward account of a series of events is shaped by the writer's attempts to understand or explain the events. One author/scholar/historian may look at the events and write a progressive narrative that emphasizes the triumph of the human spirit, while another may present a tragic (or declensionist) narrative that emphasizes the horrors of the same events. When it comes to food and hunger issues, all accounts are written by humans, about other humans and human situations, and they can present quite different accounts and understanding. Not only can two scholars approach a single set of events and interpret them in different ways, but they can also disagree on the most basic issue—that is, how they define famine (see the "Famine" entry).

Of utmost importance is the framing—in other words, the ways in which the presentation of a subject shapes the discourse (conversations) surrounding issues. By framing the issue, the writer and medium limit the range of interpretations by the audience, shaping their perception of issues, institutions, and places, and potentially leading them to share the writer's perceptions of the issues and his or her conclusions. Complicating matters further, work on hunger and, in particular, food is pursued by both popular and academic writers, who seek appeal to both popular audiences interested in food and advanced work in food and hunger research. An attempt to introduce the major works on food and hunger and examine the ways writers present different perspectives on these complex topics begins with a brief overview of food writings and scholarship.

The First Cookbook?

Apicius (*Cookery and Dining in Imperial Rome*), from the first century CE, often is cited as one of the "first" cookbooks and attributed to Marcus Gavius Apicius. Existing copies date from about the ninth century. The text is composed of approximately 500 recipes, divided into "books" or chapters, such as "Poultry," "Seafood," and "Vegetables"—not unlike a modern cookbook.

Apicius, an upper-class Roman, was well known for his love of fine food and elaborate banquets, to the point of obsession, according to his peers Pliny and Seneca. Legend has it that Apicius committed suicide by poisoning his last draught of wine rather than live any life but the high life. As time passed, the myth of Apicius evolved so that beside excess, the name Apicius also was associated with good food and its preparation.

It is extremely unlikely that Apicius wrote or even compiled the text attributed to him. The practical nature of the text suggests that it was likely compiled by a number of individuals—slave cooks, freed men, gourmands—late in the Roman Empire. Today, it provides a fascinating glimpse into the diets of the ancient world (albeit the upper classes), especially given that the foods associated with the region today, such as pasta and tomatoes, were not available at that point in history.

—Christina E. Dando

Overview of Work on Food

References to food, an essential element for human life, are found in many of the texts written on this planet, going back as far as the earliest cuneiform tablets (acknowledging that the earliest and most enduring way of sharing recipes was verbally). Found on a Mesopotamian cuneiform tablet is a hymn to the goddess of beer, including a beer recipe, dating to approximately 4,000 years ago. Other found and translated tablets have recipes for cooking birds and meats. Religion texts dating back thousands of years are filled with references to the foods people ate and dietary directions on how people should eat, with food binding people to their faiths as well as setting groups apart. Food and feasting were central to Roman culture: No description of Roman life is complete without an account of the excesses of Roman feasting, including reports of the legendary vomitoria. A text widely acknowledged as the "first" cookbook was Marcus Gaius Apicius's *Cookery and Dining in Imperial Rome*, dating to the first century CE. Roman author Athenaeus of Naucratis, writing in the third century, recorded "ancient" food references in classical literature in *Deipnosophistae* (whose title may be translated as "The Banquet of the Learned" or "Philosophers at Dinner"). Discussions of food also appear in histories and geographies, with recipes cropping up on tablets, walls, and other locations handy to cooks (such as fireplace lintels, as found in Pompeii).

It was not until the rise of academics in the Renaissance that more serious contemplation of food and dining habits, facilitated by the invention of the printing press, were undertaken. Europeans were literally hungry for information, both to feed their curiosity and to properly feed their bodies. Works on food and diet were published throughout Europe and offer a view of Renaissance attitudes on health, food, and consumption. The first English cookery book, the *Forme of Curye* (roughly "The Form of Culinary Recipes"), was published in the 1390s. Many of the works on diet and food draw on the second century CE writings of the Greek physician Galen of Pergamum. For example, Ioannes Bruyerin Campegius's *De Re Cibaria* (1560) uses classical texts to critique the excesses of the age as well as to discuss the eating habits of the known world.

In the 17th and 18th centuries, cookbooks began to be published for audiences of a specific gender. As the notion of cuisine developed (that is, high-quality food), cookbooks were published for male professional chefs that took culinary artistry to high levels. Robert May's 1660 *The Accomplisht Cook, or the Art and Mystery of Cookery* included directions on creating an edible galleon complete with cannon. This perspective stands in sharp contrast to cookbooks written largely by women for female housekeepers, which specifically focused on practicality and economy. This gendered approach to cooking continues even today, with most home-cooking being done by women and most restaurant chefs being male (although more men are cooking at home and more women chefs are becoming known for their cuisine, such as Alice Waters of the restaurant Chez Panisse, in Berkeley, California, or Cat Cora, the only female Iron Chef on the Food Network show of the same name). Julia Child bridged this divide to some extent in the 1960s, when she took lessons learned at the famed Le Cordon Bleu hospitality school in Paris and translated them (literally and figuratively) into recipes that could be made in an American kitchen through her cookbook *Mastering the Art of French Cooking* (co-authored by Simone Beck and Louisette Bertholle, and first published in 1961) and her public television series *The French Chef* (airing from 1963 to 1973).

As social science scholarship developed over the 19th century, scholars in emerging fields such as anthropology began to include food as a topic of research. An early example is Garrick Mallery's paper "Manners and Meats," which was published in an 1888 issue of *American Anthropologist*. In geography, in the wake of World War I and food rationing, J. Russell Smith published *The World's Food Resources* (1919), exploring the spectrum of available food resources, the requirements of food production, and potential yields. The work of historian Lucien Febvre and the French Annales school included food in their studies of regions. The Annales school advocated an approach to long-term historical studies (*la longue durée*) that, in focusing on social and economic history, embraced geography and material culture while attempting to understand the psychology of the times (*mentalité*). The Annales inclusion of food, as part of material culture, its production, and its preparation, is indicative of these scholars' view that to understand a people, one needs to comprehend their daily lives as well as their celebrations and high culture. For example, Fernand Braudel's account of world economic history, *Civilisation matérielle, économie et capitalisme, XVe–XVIIIe siècle* (1967), includes chapters on population and food supply, on "Daily Bread" (wheat, rice, maize), and on food and drink.

By the late 20th century, food studies had "arrived"; that is, they had reached a point where writers and scholars no longer felt the need to justify their interest in the subject. Today, food scholarship is being published that appeals both to an informed public and to a growing academic multidisciplinary field, and that covers the spectrum from diet to food production to food traditions. As a result of the multiple audiences, boundaries are increasingly blurred between writing genres. Works of literature such as Kurt Vonnegut's *Deadeye Dick* (1982) and Laura Esquivel's *Like Water for Chocolate* (1989) include recipes that are tied to the text. Madhur Jaffrey's cookbook *From Curries to Kebabs: Recipes from the Indian Spice Trail* (2003) begins with a discussion of the diffusion of Indian emigrants and the resulting "fusion cuisine" that can be found globally as a result. Academically, the multidisciplinary field of food studies addresses a broad spectrum of food topics: studies on cultural practices and their evolution, on single commodities, on globalization and its impacts, on food history, and on biology and food, to name just a few. The increasing interest in food studies and its presence today as well as historically is not surprising, given its centrality to human existence. As Sidney Mintz and Christine DuBois write:

> In-depth studies of food systems remind us of the pervasive role of food in human life. Next to breathing, eating is perhaps the most essential of all human activities, and one with which much of social life is entwined.

Overview of Work on Famine

Countless references to famine exist throughout recorded history. Religious texts, steles, and historians' accounts document events and provide some sense of what led to the hunger and/or famine. More systematic attempts to understand famine and its causes were really undertaken only as social sciences research began to occur in Europe at the end of the 18th century and beginning of the 19th century, when scholars began to seek explanations of the mechanisms that underlie events. Famine research generally has been undertaken in developed countries that are less vulnerable to hunger and seldom in the countries experiencing famine, with much of the work being concentrated on the Indian subcontinent and Africa. As scholars have sought to understand famine, its definition, causes (or triggers), and possible solutions have shifted, and then shifted yet again.

A significant evolution in our understanding of famine over the past 250 years has occurred regarding its causes. Early understandings of famine linked food shortages and eventual deaths from extreme hunger to "natural causes," whether they entailed variations in rainfall or temperatures or disasters such as flooding or drought. At one point, famines were viewed as synonymous with crop failures produced by abnormal climate conditions. This view of famine as a natural disaster derived from a worldview that was simplistic and somewhat narrow, with humans seen as powerless before nature and/or gods. According to this view, famine was neutral and no one was to blame—no one was responsible, save perhaps rulers or religious leaders who were held accountable by the gods, or perhaps the gods themselves. Famine was something to be endured, not a problem to be solved.

Thomas Malthus viewed famine as a natural check on population growth (see entry on Malthus later in this volume). As the father of modern demography, Malthus was among the first to theorize the relationship between food supply and population. A British clergyman, Malthus suggested, in his treatise *An Essay on the Principle of Population* (1798), that famine would keep population in balance with food supply, and that when population grew faster than food production, famine would "naturally" check the population growth. Malthus's generalizations fail on a number of levels: Famine seldom affects an entire population evenly; food production has increased and continues to increase, through the industrialization of agriculture; and food scarcity is local, not global—where one area or region may have shortages, others have surpluses. Nevertheless, Malthus is considered significant as the first scholar to consider population and the role of famine in population equilibrium.

One of the first systematic attempts to examine famine on a global level was British statistician Cornelius Walford's *The Famines of the World Past and Present* (1879), published by the Royal Statistical Society. Walford created a chronological list of world famines, delineating 350 famines, predominantly in Europe, the Middle East, and India, and their causes if they were known. Walford, while acknowledging Malthus's views, expressed some concerns over a completely "natural hazard" approach, commenting, "The occurrence of famines would appear to me to be likely to result rather from the failure of human means and foresight in many instances than otherwise" (4). Walford instead pointed to both natural and "artificial causes" (within human control, such as war or "defective agriculture").

While the human role in famine causation is increasingly being acknowledged, some subject areas continue to focus on famine as a natural hazard. Several works published in the 1970s linked climate change to famine risk, such as Stephen Schneider's *The Genesis Strategy: Climate and Global Survival* (1977) and Reid Bryson and Thomas Murray's *Climates of Hunger* (1977). Recently, as public awareness has been focused again on climate change in the 21st century, these arguments—that climate change may lead to both greater frequency and a greater extent of famine—have resurfaced.

In what may be one of the most significant works on famine in the 20th century focusing on the human element, Indian economist Amartya Sen's *Poverty and Famines: An Essay on Entitlement and Deprivation* (1981) takes what some have termed an "entitlement approach" to famine. Sen focuses on the distinction between the availability or supply of food and the access to or ownership of food, and an individual's entitlement to food, which may be based on trade, production, sale of labor, or inheritance/transfer. Within this model, a person is reduced to starvation when some change happens to his or her entitlement, whether it is loss of land or loss of ability to labor or rising food prices. Sen writes: "Starvation is the characteristic of some people not *having* enough food to eat. It is not characteristic of there *being* not enough food to eat" (1). Sen's monograph is divided between his development of the entitlement approach and case studies of famine in Bengal, Ethiopia, the Sahel, and Bangladesh. His approach addresses the issues of unequal famine within an area (how in any famine some people have food, whereas others do not) and, in doing so, avoids gross generalizations. Nevertheless, by focusing on the individual, Sen does not address the issue of scale—that is, the point at which individuals without food are recognized as a group of individuals faced with simultaneous food shortages. Sen received the Nobel Prize in Economic Sciences in 1998 for his work on welfare economics, social choice, and poverty.

American geographers William A. Dando and Michael Watts brought their discipline's approach to the topic of famine. Geography, as both a physical and a social science, is concerned with the Earth, its features, its environments, and its peoples. This perspective has a definite advantage when approaching a subject such as famine, whose causes can be physical (climatic) and/or social (political, cultural). Dando's *The Geography of Famine* (1980) takes a global, historical approach to famine, ultimately delineating world famine areas from 4000 BCE to 1980 CE. Through Dando's analysis of 800 famines, he creates a famine typology, identifying five basic famine types (physical, transportation, cultural, political, overpopulation), and recognizing that famines can have different causes, rather than insisting on a single theory to explain famine. The strength of Dando's work lays in his synthesis of tremendous amounts of temporal and spatial information and his emphasis on famine as being "human-made"—that is, ultimately a social problem.

Watts' *Silent Violence: Food, Famine and Peasantry in Northern Nigeria* (1983) focuses on the case study of Hausaland and is based on fieldwork after the Sahel famine of 1972–1974, oral interviews, and archival research. Watts examines the interplay between a marginal landscape and social conditions, arguing that before the British colonized the region, mechanisms were in place that provided relief for small farmers in times of famine, but that after colonialism, this safety net was disrupted, setting the region up for disaster. By considering the region both in times of famine and in times of normal production, Watts examines the specific historic contexts of each situation, from the pre-colonization period through the post-colonization era. The poor are ever vulnerable in Hausaland, but it is the circumstances and effects of famine that shift according to the social and economic conditions.

Like Watts' work, Alex de Waal's *Famine That Kills* (1989) focuses on a specific time and place—in this case, the 1984–1985 famine in Darfur, Sudan. De Waal, a British writer and researcher, begins by critiquing outsiders' (researchers and aid workers) approaches to famine, noting that they often have only a limited view of disasters such as famine. He questions how outsiders (i.e., those far away from famine) define famine. De Waal then uses surveys, interviews, and archival interviews to construct a view of the Darfur famine from the experiences of those who lived through it. If they do not fully comprehend the people and their culture, he argues, outsiders cannot hope to understand the priorities of the people suffering and cannot understand the actions they take, thus limiting the success of any aid program undertaken. De Waal emphasizes the role of diseases leading to famine deaths; notably, as people cluster together in small areas to access resources (water, food), they are at greater risk for health problems such as waterborne diseases. He goes on to critique the relief efforts in the Darfur region, pointing out the different ideologies of relief organizations, and he suggests new strategies for famine relief in the area that would reduce health risks. De Waal's approach is useful for considering not only responses to famine, but responses to natural disasters as a whole.

Finally, Irish economist Cormac Ó Gráda seeks to delineate the changing symptoms of famine, exploring why they happened in the past as compared to today, and how they may be less likely in the future, in his recent book, *Famine: A Short History* (2009). Like Dando, Ó Gráda takes a temporal, spatial approach to famine, covering famines across the globe over the past thousand years, with an emphasis on China and India. From his perspective as an economist, he closely examines, in addition to the context (environment, politics), the prices, supplies, and markets to bring out the complex interplay that results in famine. Ó Gráda makes the argument that while famine has been a threat in the past (although he questions the frequency that other scholars suggest), famines have been less frequent in recent history and will continue to diminish in frequency and fatalities, given global economic growth, the decline of certain totalitarian regimes, and improvements in medical and communication technologies.

Common to virtually all major scholars' approaches is the recognition of the role of human action in famine. A second, perhaps more significant commonality is the hope that their work might lead to greater understanding and ultimately fewer famines. As Cormac Ó Gráda writes, "Writing about famine today is, one hopes, part of the process of making it less likely in the future."

Case Study of a Particular History: The Holodomor

The Ukrainian famine of 1932–1933 provides an example of how the facts of a particular event can be explained in quite different ways (see the "Russian and Soviet Famines: 971–1947" entry in Volume 2). There is no dispute that, between 1932 and 1933, millions of people died from starvation in the Ukraine and that this famine was tied to the efforts to collectivize agriculture as well as poor grain harvests and planned procurements by the Soviets. The severity of the situation led to mass migration to cities and even to cannibalism. Ukrainians refer to this tragic period as *Holodomor* ("death by hunger"). Beyond these basics however, there is much disagreement. Estimates on the number of people who starved to death range from 3 to 10 million, although this estimate depends, in part, on whether those who are estimating focus exclusively on the Ukraine or whether they include other areas that were also extensively affected, such as Kazakhstan. More seriously, the underlying causes of the famine are deeply debated, broken roughly into Ukrainian/pro-Ukrainian and Soviet/pro-Soviet camps.

Ukrainian and pro-Ukrainian writers have designated the Soviet Union as the perpetrator of the famine and have suggested that it be recognized as an act of genocide. They suggest that Soviet officials were attempting to eliminate a nation (the Ukrainians) by issuing excessive grain quotas, requisitioning food from the affected area, and targeting nationalist elements. Probably the most well-known book advocating this perspective is Robert Conquest's *Harvest of Sorrow* (1986), which takes the position that the famine was deliberately inflicted and politically motivated.

Pro-Soviet writers reject the claims of ethnic cleansing and instead focus on broader harvest problems in the Soviet Union beginning in 1931 and resulting in famine in areas with intensive agriculture, not just the Ukraine, although recognizing that the Ukraine was particularly hard hit. While acknowledging the famine, they argue that there is no evidence that Stalin deliberately inflicted the famine. Davies and Wheatcroft's *The Years of Hunger* (2004) is the most prominent book arguing that the famine was not genocide. These authors' focus on a general Soviet picture shifts attention away from the ethnic factor and Stalin's nationalities policy and emphasizes the clash between the Soviet regime and the Ukrainian lower class's resistance to forced collectivization and dehumanizing rapid industrialization.

Since the fall of the Soviet Union in 1991, Soviet-era records have become more readily available to researchers seeking to further our understanding of events such as the Ukrainian famine. Recently, historian David Marples has argued that the increasingly available archival records contain enough evidence to suggest that the Soviets were concerned about Ukrainian nationalism and that, while there may not have been a systematic attempt to eliminate Ukrainians, there are too numerous references to Ukrainian nationalism and too much evidence of the removal of this movement's leaders to be ignored.

Although the Ukrainian famine has been used as an example here, virtually any famine can be approached similarly. For example, work on the Great Irish Famine before the 1960s was tied to Irish nationalism and the extent to which British policy was responsible, with an emphasis on Ireland being wronged. More recent scholarship on the Great Irish Famine has sought to develop an understanding of the complex elements at work in this time and place—the actual conditions, the state of the economy, the actions of individuals, and the role of outside intervention. More scholarship needs to be conducted on all famines using primary sources so that a more nuanced understanding of famine and its triggers can be developed.

A Geography of Famine Scholarship

The geography of famine scholarship has been tilted principally to the West, with scholars in the United Kingdom and the United States dominating research. Why? Some scholars have suggested that there is a certain amount of survivor's guilt or shame on the part of U.K. and U.S. researchers, and that there may be some reluctance by non-Western authors to revisit or reexamine a famine that resulted in the loss of family and friends. Another possibility is that, historically, oral traditions have been important in non-Western societies; thus, when famine hit such societies hard, striking the young and old first, the elderly who were the historians and thus the histories they were preserving were both quickly lost. Both de Waal and Ó Gráda have commented in their writings of the importance of paying attention to oral traditions (when available) and the experiences of those living with starvation and famine.

Other factors may also contribute to the understanding of this geography. In the cases of the United States and Great Britain, periods of extended hunger or even the potential for famine have not threatened their populations for an extremely long time. With current high levels of development and hunger and famine a distant memory in the United States and Great Britain, there is no pressing need for famine research. Famine to U.S. and U.K. scholars primarily involves "others" and affects those "elsewhere." In locations where famine is still a frequent visitor, such as parts of Sudan or Ethiopia, economies are still precarious, populations are living hand to mouth, and governments are in flux. They do not have the luxury of being able to stand back and assess the causes and prevention of famine: They are often too busy worrying where the next meal will come from.

In the case of the Irish famine and even the Ukrainian famine of the 1930s, scholars in Ireland and the Ukraine increasingly engaged in research into these events as time passed. In both cases, researchers now have the safety of the passage of time and an improved quality of life buffering them from the historical events. Nevertheless, these historical famines have come to play an important role in national identity, especially because someone else is assigned blame for the event in each case (for the Irish, it is the British; for the Ukrainians, the Soviets). They become events to rally the people and unify them.

Framing Famine

Writing about the historiography of the Ukrainian famine, David Marples states that "it is the scholarly community that defines and explains events, not governments and public officials or even the United Nations." When considering topics such as food and famine, which can be politically charged, readers need to be aware of the various ways that writers frame and present topics to their audiences. To obtain a well-rounded view of a topic such as famine, several perspectives should be consulted. Readers also need to acknowledge the disjuncture between the location where the research is being done—attempts to understand what leads to hunger, what triggers famine, and what might be done to protect it—and the locations that experience hunger and famine. As de Waal demonstrates, much can be gained by undertaking field research into the experiences of food insecurity and famine and into the needs of the population at risk: Sometimes food alone is not enough. Despite the differences in explanations and approaches to famine, the ultimate aim is what all famine writers/scholars profess at one time or another—never again.

—Christina E. Dando

Further Reading

Albala, K. *Eating Right in the Renaissance*. Berkeley/Los Angeles, CA: University of California Press, 2002.

Albala, K. "History on the Plate: The Current State of Food History." *Historically Speaking* 10, no. 5 (November 2009): 6–8.

Bober, P. *Art, Culture, and Cuisine: Ancient and Medieval Gastronomy.* Chicago/London: University of Chicago Press, 1999.

Cronon, W. "A Place for Stories: Nature, History, and Narrative." *Journal of American History* 78, no. 4 (March 1992): 1347–1376.

Cullen, L. "The Politics of the Famine and of Famine Historiography." In *Comhdáil an Chraoíbhín 1996*. Roscommon, Ireland (1997): 9–31.

Dennis, A. "From Apicius to Gastroporn: Form, Function and Ideology in the History of Cookery Books." *Studies in Popular Culture* 31, no. 1 (Fall 2008): 1–17.

Devereux, S. *Theories of Famine*. New York/London: Harvester Wheatsheaf, 1993.

Edkins, J. *Whose Hunger? Concepts of Famine, Practices of Aid.* Borderlines, Vol. 17. Minneapolis, MN/London: University of Minnesota Press, 2000.

Janmaat, J. "History and National Identity Construction: The Great Famine in Irish and Ukrainian History Textbooks" *History of Education* 35, no. 3 (May 2006): 345–368.

Marples, D. "Debate: Ethnic Issues in the Famine of 1932–1933 in Ukraine." *Europe-Asian Studies* 61, no. 3 (May 2009): 505–518.

Mintz, A., and C. Du Bois. "The Anthropology of Food and Eating." *Annual Review of Anthropology* 31 (2002): 99–119.

Torry, W. "Social Science Research on Famine: A Critical Evaluation." *Human Ecology* 12, no. 3 (1984): 227–252.

Hunger and Starvation

Hunger

"Hunger" is a familiar but misunderstood term used to describe a motivational force to stimulate responses for a very important activity necessary for human survival—that is, a desire to obtain and eat food. This term is used to explain a normal condition when the stomach is empty of food. Hunger is a collection of conscious feelings or sensations manifested in a craving for food, stomach rumbling, and an uncomfortable feeling commonly called "hunger pangs" or "stomach cramps." Pain and cramps are produced by strong contractions of the stomach muscles. The desire to eat or the desire for a specific food is alleviated when a hungry person eats. Once food is ingested into the stomach, the inner walls of the stomach contract rhythmically to reduce food particle size and mix this broken-up food with gastric juices. These stomach contractions push the food mix into the small intestine.

Food that enters the stomach is normally digested in three hours or so, at which point the cravings for food and hunger pangs return. If the body's need for food is not met, the stomach cramps or stomach pangs become more painful and increase in intensity for three or four days, then slowly recede.

Hunger is the sensation of a biological need to sustain growth and life. Hundreds of millions of children and adults throughout the world suffer from hunger. Hunger is a felt, observable, and measurable event.

According to the World Health Organization (WHO), hunger is the world's major health risk. WHO estimates suggest that 2 billion people suffer from chronic hunger and food insecurity. Food insecurity means people do not have enough food to satisfy their cravings, nor do they have quality, nutritionally adequate food. World food insecurity has 10 dimensions:

1. In the developed world, hunger is caused primarily by *poverty* and poor household fiscal management. (In the United States, 36 million households are considered to be food insecure, and 45 percent of the cities are not able to help their citizens meet basic food needs.)
2. In the underdeveloped nations of the world, *population growth*, combined with poverty in areas of overpopulation on marginal agricultural land and in the sprawling urban slums of tropical and subtropical megacities, has created acute food insecurity problems. Approximately 1.4 billion people of the world exist in extreme poverty and more than 924 million people live in urban slums.
3. *Natural disasters*, such as droughts, floods, misuse of agricultural land or land conversion for non-food-producing use, overgrazing, deforestation, and climate change or variations, can all result in a reduction of food produced.
4. *National or international priorities* may have become misplaced and result in food insecurity. These problems may include urban biases in development fund allocation and infrastructure improvement; repression of ethnic, racial, or religious groups by powerful national decision makers; and government

decisions to expend scarce resources for military equipment, ego-building grandiose projects, or opulent leader lifestyles.

5. When *resources are not distributed equitably*, hunger may result. These disparities may include the unavailability of credit or the inflated cost of seeds, equipment, and fertilizers; lack of disease-free drinking water in rural areas; few roads from food producers to urban food markets; inadequate transport equipment or means for crops to be moved to market; limited numbers of agricultural technical support workers to improve agricultural practices and introduce new crops; poorly trained and limited number of rural teachers to improve education and expand knowledge; and few nurses to assist in the maintenance of rural and slum dwellers' health and nutrition practices.
6. *"Cheap food policies" and inflation* may restrict the cost of basic foods paid to food producers so that relatively inexpensive food can be sold to those who live in urban areas. Some governments spend money that they do not have. They print money that has little or no value, which in turn inflates the costs of basic food items. In some urban areas in developing countries, poor urban dwellers spend 80 percent of their income to buy food.
7. *Rising oil and gasoline costs* restrict the ability of the poor or "land-rich, capital-poor farmers" to purchase fuel to run their agricultural equipment and irrigation pumps or buy fertilizers, herbicides, and pesticides produced from oil.
8. *Wars, civil strife, and religious confrontations* reduce food production and destroy stored food. Scorched-earth policies, sieges, looting of property and food, criminal protection rackets imposed on food merchants and food traders, smuggling, and restrictions placed by those directing the actions of civil disobedience on food and medical relief agencies create food shortages, hunger, and famine.
9. *The struggle for basic food-producing resources and land reform* (involving prime agricultural land, grazing pastures, and water) has become extremely violent and has disrupted food production, food preservation and storage, food distribution, and food availability. Most developing countries have transferred, or are in the process of transferring, large land holdings from the ownership of the upper-class minority and multinational ownership to the hands of peasant farmers, and the resistance to land reform is not easily overcome.
10. *Meat and poultry feed* remove grain from the world grain market and increase the price of grain used for food by the poor in developing countries. Funds to purchase imported grains are secured from local cash crops that have not increased in price; cash cropland and infrastructure that could be used for the production of locally needed foods, such as rice, beans, fruits,

War Often Fuels Hunger in Africa

In many African nations, particularly in the Congo, Sudan, Chad, Central African Republic, and Uganda, war and hunger are linked in a cycle of horror, desperation, and death. At the same time, food shortages and food prices have triggered riots and civil disturbances and toppled the governments in Egypt, Libya, and Algeria. Decades of corruption under dictators have wrecked national economies in this region and created great unemployment and underemployment. As a result, many Africans eat only once a day. They cannot purchase more food, which they desperately need, because food is too expensive.

The human costs of the wars in Africa are astonishing. In the Congo, a vast country with a population of more than 50 million, nearly a decade of fighting has left almost 4 million people dead. At least 2.4 million people have become refugees at risk of starvation and death, and hundreds of thousands have perished due to lack of food. In the Sudan, the many-year conflict in the Darfur region has claimed more than 200,000 lives, while thousands of innocent women and children are physically scarred from disease and malnutrition. Also, more than 2 million Sudanese people have been displaced. Massacres and attacks by ethnic militias have forced thousands to face hard choices: flee, fall prey to hunger and disease, or be slaughtered by marauding militiamen. Many farmers have had to abandon their fields of cassava, corn, beans, and sorghum because of the violence. Insecurity on roads has disrupted food aid shipments. Those who have sought refuge in large towns or cities often find that they offer little relief, as these centers provide little food and no jobs.

—William A. Dando

and vegetables, to reduce hunger, instead are directed to cultivation of tea, tobacco, cocoa, and coffee for export to developing countries.

The combined efforts of national governments, the United Nations, religious relief organizations, and private foundations have reduced the number of estimated hungry people in the world today when compared to 25 years ago. Even so, at least 854 million people go hungry every day, and 16,000 children die of hunger-related diseases (one child dies from this cause every 5 seconds).

Starvation: Definition and Description

Starvation is the state of extreme lack of food that induces death by long-continued deprivation of that which the human body needs to maintain life. Use of substitute foods (i.e., "famine foods"), which are usually coarse, unfamiliar, and low in nutritional content, may cause inflammation of portions of the digestive system,

Table 15. Degrees of Starvation (140-pound adult; little or no food available)

Starvation Degree	Weight Loss	Description	Characteristics
1. Moderate	10%	Normal	Little change in body characteristics
2. Severe	20%	Thin	Substantial change in body functions and activities
3. Extreme	30–40%	Very thin	Serious body weakening, pronounced change in activities
4. Unrecoverable	50–60%	Skin and bones	Ill, unable to move, drifts into a coma, death

indigestion, and diarrhea. A starving person's basal metabolism drops, blood pressure declines, pulse rates become slow, skin dries and feels cold to the touch, eyes redden, and bones become brittle. Abnormal accumulation of a watery fluid swells body tissues, the stomach, body cavities, ankles, and feet (see Table 15). At times, a burning sensation in the hands and feet is experienced, and the mouth of a person starving may salivate abnormally. Prolonged starvation results in slow termination of life as a result of the human body's inability to secure necessary life-supporting food.

Starvation's Slow Death Characteristics

After a period of prolonged hunger, the gnawing pain of hunger slowly disappears. It is followed by a sensation of extreme weakness and faintness. The craving for water becomes intense and painful. Heart action becomes rapid, and breathing becomes shallow, slow, and irregular. Constipation is followed by diarrhea. Facial expressions change; eyes stare at objects, look about anxiously, and become glassy. The body slowly appears emaciated. Muscles become soft and shrink by more than one-half their pre-hunger, pre-starvation size. Skin loses color and luster and emits a secretion. Feet and ankles swell. Those starving will eat almost anything. The ability to speak is lost, and the mind becomes more and more listless, irrational, and even idiotic. As starvation intensifies, the eyes sink into the skull and become glassy. The mind becomes further unbalanced. Hallucinations, insomnia, and dreams of food are common.

In its attempt to remain alive, the human body will turn on itself for sustenance. First, the body will burn fatty acids from body fat. Next, it begins to consume lean tissues and muscles. Finally, less important organs are sacrificed to maintain the ability of the body's most important organs to function. Catabolysis is the process of a body breaking down its own muscles and other tissues in an attempt to keep vital body systems, such as the nervous system and heart, functioning.

Starvation causes extreme fatigue and renders the starving person more apathetic over time; he or she becomes too weak to move or fight diseases. Death occurs when the body has nothing left remaining to support life while the starvation victim is in a coma or stupor or suffers a convulsion.

Starvation as Capital Punishment

For thousands of years, starvation has been used to punish criminals for civil infractions and as a death sentence for capital crimes or crimes against the state. Death by starvation included people or families being entombed in a wall, buried alive, chained to the wall of a cave, or marooned on a desolate island. In European societies, starvation was used to rid a family or clan of erring members or perceived political competitors from achieving their goals, and for horrible crimes against society. Examples of this cruel form of execution include the following:

- In 31 CE, Livilla, the niece and daughter-in-law of Emperor Tiberius, was starved to death by her mother for involvement in the murder of her husband and for an adulterous relationship.
- In 33 CE, Agrippina the Elder, a Roman imperial princess, granddaughter of Emperor Augustus, and mother of Caligula, was starved to death on orders of Emperor Tiberius.
- Also in 33 CE, Drusus Caesar, a Roman imperial prince and brother of Emperor Caligula, was starved to death in prison on the orders of Emperor Tiberius.
- In 41 CE, Julia Livilla, a Roman imperial princess and sister of Emperor Caligula, was starved to death on the orders of Emperor Claudius and Empress Messalina.
- In 1317 CE, two brothers of King Berger of Sweden were imprisoned for a coup they staged to overpower the king and died in prison from starvation a few weeks later.
- In 1671 CE, John Trehenban was starved to death while locked in a cage at Castle An Dinas in Cornwall as punishment for the death of two girls.

In prisons, the nutritional value of food served was so low at times that it was insufficient to maintain life. Death by slow starvation was common.

Starvation as a Means of Repression: Stalin and Hitler

Food as a political weapon, hunger to control large numbers of people, and starvation as a means to eliminate political opposition or "self-defined" enemies of the state were facets of Stalin's policies in the Soviet Union and Hitler's policies in Germany and Europe from the 1930s to the 1950s.

Soviet concentration camps were first established to rid the country's society of anti-Soviet elements. The CHEKA, or Soviet secret police, organized camps for members of anti-Soviet political parties, landlords, capitalists, hostages, and select political and ethnic groups. This new "penal system" expanded and grew in scope in the initial period of agricultural collectivization and industrialization. In 1928, 6 million peasant families disappeared. Political terrorism and purges swelled the populations of the prison camps. According to the best estimates, 12 to 15 million women and men were placed in concentration camps. Working and living conditions and food provisions in the camps were appalling. Unlimited working hours, utterly inadequate food rationed according to work norms fulfilled by each prisoner or set based on each so-called crime, and harsh treatment that included beatings and solitary confinement turned helpless prisoners into emaciated human beings. Millions of Soviet citizens starved to death in these camps or gulags. Stalin's despotism, cruelty, and paranoia destroyed productive segments of many ethnic groups in Eastern Europe and the 15 republics of the former Soviet Union.

Hitler, once in power, established an absolute dictatorship in Germany. He issued a decree overriding all guarantees of individual freedom and introduced an intensive campaign of violence and terror. Concentration camps were established to house dissidents and control internal security. In the early 1940s, concentration and work camps were expanded and new extermination camps were established. Personnel at the Auschwitz, Mauthausen, Treblinka, Dachau, Buchenwald, and Bergen-Belsen camps, as well as "extermination squads," began to kill members of dissident groups. Jews were deprived of all human rights, and the Jews of Germany, Poland, and occupied areas of the Soviet Union were most numerous among those killed. In German-occupied Europe, an estimated 6 to 9 million Jews, Slavs, Gypsies, and opponents of Hitler's dictatorship were killed by starvation, gas chambers, shootings, torture, and disease. Such cruelty and barbarianism were indiscriminate, even in Germany. The plans for the extermination of Jews and dissenters were implemented with such efficiency that those who had the ability and will to resist were sapped of their health and strength by prolonged starvation and disease. Although Allied governments knew of Hitler's plans to exterminate select cultural groups, and particularly the Jews, they were so concerned with defeating Hitler and saving their own citizens from harm that they failed to develop and implement effective plans to save the Jews and others from designed starvation, disease, and death.

The Minnesota Starvation Study

During World War II, a ground-breaking clinical study to determine the physiological and psychological effects of starvation and rehabilitation strategies was conducted. The goals of the study were to produce a definitive and accurate descriptive report on human starvation, based on data secured in a laboratory-based and controlled study, and to provide a guide for Allied relief assistance to those starving in Europe and Asia in the post–World War II period. Dr. Ancel Keys was the director of the Minnesota Starvation Study and responsible for all aspects of the project. Keys had worked with the U.S. Army to develop combat rations for troops (K-rations). Thirty-six men were selected from more than 400 volunteers to endure starvation, deprivation, and hardship for the 12-month study. This 12-month study was divided into four distinct phases:

1. Control period (12 weeks)—subjects created a database and received a control diet of 3,200 calories each day.
2. Semi-starvation period (24 weeks)—subjects' diets were reduced to approximately 1,560 calories each day.
3. Restricted rehabilitation period (12 weeks)—staff noted responses to subjects fed four different caloric levels and distinct food supplements.
4. Unrestricted rehabilitation period (8 weeks)—subjects were provided with a diet unrestricted in caloric intake and foods. All were carefully monitored.

Results from this study confirmed that subjects experienced the following effects:

- Became very depressed, experienced hysteria, and emotional swings
- Manifested extreme psychological effects and decline in their ability to comprehend
- Exhibited preoccupation with food
- Withdrew from social interaction and preferred isolation
- Experienced a decrease in basal metabolic rate (i.e., the energy required by a body in a state of rest)
- Reduced body temperature and heart rate
- Suffered from edema (swelling) in the arms, legs, feet, and stomach from fluid accumulations

Recovery of all subjects at the conclusion of the study depended on the diet available at that time and the psychological treatment/care given.

Eradicating Hunger and Starvation

Some members of world society have the strength, the knowledge, and the ability to take leadership roles in addressing local, regional, and national hunger and starvation issues. International economic development and social and relief agencies, in partnership with local decision makers, must expand their efforts to implement sustainable development initiatives and self-help activities. These efforts must be undertaken to ensure that the seeds for long-term economic and social growth take root in communities, and that regions of the world move from endemic poverty to prosperity. Assistance also must be given in the aftermath of human and natural-caused disasters to achieve long-term physical, psychological, and socioeconomic recovery. Vulnerable individuals, families, clans, ethnic groups, and countries require immediate assistance when disaster strikes; those who wish to help the victims in need at the time of a crisis must be prepared, have a mitigation plan formulated, and respond quickly. Confronting the root causes of hunger and starvation requires activities that raise the awareness about the actions of misguided military dictators and for-life national leaders, insensitive religious fundamentalists, and urban minorities composed of landowners, business tycoons, exploitative industrialists, and racial bigots. Governments dominated by such groups typically concentrate national wealth under the control of a few and focus national development investments on nonrural/nonagricultural projects or schemes. Food shortages, hunger, and starvation are manifestations of poverty, unemployment/underemployment, and powerlessness.

—William A. Dando

Further Reading

Barstow, C. *The Eco-foods Guide: What's Good for the Earth Is Good for You.* Gabriola Island, Canada: New Society Publishers, 2002, 227–260.

Bray, G. A. "Hunger." In *Encyclopedia of Food and Culture*, edited by H. Katz and W. W. Weaver. New York: Charles Scribner's Sons, 2003, 219–222.

Dando, William A. *The Geography of Famine*. New York: John Wiley & Sons, 1980, 141–157.

Dolot, M. *Execution by Hunger: The Hidden Holocaust*. New York: W. W. Norton & Company, 1985.

Grigg, D. *The World Food Problem: 1950–1980*. New York: Basil Blackwell, 1985, 2–53.

Halford, J. C. C., A. J. Hill, and J. E. Blundale. "Hunger." In *Encyclopedia of Human Nutrition*, 2nd ed., edited by Benjamin Caballario, Lindsay Allen, and Andrew Prentice. Oxford, UK: Elsevier Academic Press, 2005, 469–474.

Heldke, L. "Food Security: Three Conceptions of Access—Charity, Rights, and Co-responsibility." In *Critical Food Issues: Problems and State-of-the-Art*

Solutions Worldwide, edited by Lynn Walter. Santa Barbara, CA: ABC-CLIO, 2009, 213–225.

Hughes, D. J. *Science and Starvation*. Oxford, UK: Pergamon Press, 1968, 1–157

Keys, A., J. Brozek, A. Henschel, O. Mickelsen, and H. Taylor, eds. *The Biology of Human Starvation*. Minneapolis, MN: University of Minnesota Press, 1950.

Post, J. D. *The Last Great Subsistence Crisis in the Western World*. Baltimore/London: Johns Hopkins University Press, 1977, 109–140.

Tucker, T. *The Great Starvation Experiment: The Heroic Men Who Starved So That Millions Could Live*. New York: Simon & Schuster, Inc., 2006.

Turner, B., ed. *The Statesman's Yearbook*. London: Macmillan, 2009, xv.

Malthus, Thomas Robert

Although there have been countless individuals who have taken "holy orders" or studied in seminaries to become priests or pastors and have influenced society in a negative way, few have had the lasting impact as Reverend Thomas Robert Malthus. It has been postulated that every serious student of food and famine and of population and welfare issues has read Malthus's essays or learned of his concepts. Pre-Malthusian, anti-Malthusian, pro-Malthusian, neo-Malthusian, and Malthusianism are terms understood by almost all concerned about the impact of charity, including that directed toward reducing the number of people suffering from hunger or eliminating famine as a cultural hazard. To many, Malthus misread the passage in 2 Thessalonians 3:10 of the Bible, in which the apostle Paul writes, "He who is unwilling to work, neither will he eat." Malthusianism is perceived as a callous disregard for human welfare, scorn of those in need, reproach, shamefulness, cruelty, and injustice. The use of Malthus's population doctrine to advance the self-interest of individuals, social classes, political dogmas, and governments has resulted in the horrible deaths of millions. To many, his name stands for an insensitive and misguided doctrine of despair. Malthus's writings have long had a dominant influence in economic thought and hunger/famine relief policies in the European Cultural Realm and British colonial possessions.

Thomas Robert Malthus was born on February 14, 1766, the son of a prosperous English landowner, Daniel Malthus. Daniel was a social liberal, a friend of David Hume, and a supporter of the social principles put forward by Jean Jacques Rousseau. An optimist, Daniel Malthus was a firm believer in the good and the perfectibility of humankind. In contrast, his son Thomas was not an optimist and was influenced by Adam Smith. Smith believed in the concept of equilibrium and the "food arithmetic" progression law. He stated that a population equilibrium was reached when the number of deaths equals the number of births. According to Smith, a population limit could be reached in a place or in a country. Malthus, in turn, concluded that the limit set on population growth was food availability.

Thomas was educated primarily at home, entered college in 1793, took holy orders in 1796, and secured a position in a small Surrey parish. He married in 1804 and was appointed professor of Political Economy and History in 1805 at the British East India Company's Haileybury College. Malthus died near Bath,

Are You a Malthusian?

There has been an increase in concern by neo-Malthusians for the perceived threat of food scarcity in a future characterized by climate change and unceasing growth in the world's population. It was Malthus who wrote, in 1798, that food normally would increase in a mathematical ratio and that population would increase in a geometrical proportion. He argued that starvation was a natural event, stating that eventually, if population increases were not reduced, a determination of who would be fed and enabled to live and who would not be fed and slowly die would need to be made. This same theory is being advanced today by a group of thinkers known as neo-Malthusians.

Much of Malthusian philosophy is said to have begun with Adam Smith in 1776 and been refined by John Stuart Mill in 1848. Malthus and his followers did not comprehend the potential of the New World, possible advancements in science and technology, and the creativity of humans. Instead, they believed that everything during their period of history was in a static condition and that there was a set limit to land, labor, capital, and human intelligence. The Malthusians did not envision the Industrial Revolution, the Agricultural Revolution, and a world linked together by communications and trade.

Neo-Malthusians contend that food shortages in the early decades of the 21st century will become so acute that there will be no surplus food, such that food trade will almost completely cease. They believe that there will be a global famine, which could eventually consume a vast segment of humankind.

—*William A. Dando*

England, on December 23, 1834. He was a gentle, kindly, sensitive Christian minister. Thomas wrote his most widely read and discussed publication, an *Essay on Population*, in 1798 while serving as an assistant to the vicar in Surrey.

Malthus lived in a period when England and the world were undergoing tumultuous political and economic changes. The American Revolution, the French Revolution, the "industrial revolution," the Napoleonic Wars, and the "age of steam power and rail transportation" were some of the momentous events or eras occurring during this period. New technologies had led to the invention of steam engines, steam locomotives, spinning jennys (spinning/textile mills), the cotton gin, telegraph, steamboats, and the first college devoted to the creation, invention, and application of new technologies. Science flourished with the "classification of species," modern chemistry, mathematical advances, progress in astronomy, use of electricity, and advances in agriculture, medicine, and sanitation. Serious social changes were occurring during this era, dramatically affecting both individuals and vast segments of societies. Social life was secularized, and "moral or Christian philosophy" gave way to a new concept of a "political economy." The questions of how a subject/citizen relates to the state or the economic system and what role governments and business enterprises have in creating and maintaining a "good life" for all citizens and all employees were hotly debated. Political, social, and economic decisions were being made on the basis of untested principles or alleged "natural laws."

The Dismal Theorem

Malthus's 1798 *Essay on Population* was a controversial work when it was first printed, and it aimed at the concepts articulated by some of the great "social" thinkers of the time. Malthus believed that a natural law controlled the number of people who live on the Earth and that this law had its basis in food supplies. He concluded that only two things could bring an end to uncontrolled population growth: declining fertility (fewer births) or increasing mortality (more deaths than births until equilibrium is reached). His First Law states that "food is necessary to the existence of man" and "passion (sexual desire) between the sexes is necessary and will not change." His Second Law asserts that "the power of population is greater than the power of the Earth to produce food for man." Malthus hypothesized that a population, when not reduced by some means, will grow unchecked and increase at a geometrical ratio: 1, 2, 4, 8, 16, 32, 64, 128, 256, 512. . . . He contended that food availability, in contrast, increases in an arithmetical ratio: 1, 2, 3, 4, 5, 6, 7, 8, 9, 10, . . .

According to Malthus's hypothesis, in the 250 years after the publication of his essay, the ratio of population growth to the increase in food availability would be expected to increase 512 to 10. Because the increase in population could only be commensurate with the increase in the means of subsistence, he insisted, there must be strong and constant checks on population growth. His checks initially were human misery and hunger, which he thought would cause severe distress among the very poor, especially when combined with high infant mortality. Postponed marriage was dismissed by Malthus as unable to limit population growth, as he reasoned that it would not reduce population enough to resolve the problem. He noted that premature deaths, "sickly seasons," epidemics, pestilences, and plagues were positive ways to reduce population. After much thought, he concluded that "famine" was the most effective means for nature to create the equilibrium between population and food.

Charity and the Laws to Help the Poor

Providing a "safety net" for the poor and indigent of England at the end of the 1700s and the first decade of the 1800s was difficult. The laws in place at the time were defective and imperfect. Malthus believed that providing charity and relief to the poor was an inherently unsound policy, resulting in an increase of food costs to all consumers that spread misery to the marginally poor. He asserted that public charity created a class of society that lusted after the dole, married early, had many children, did not seek work, and remained where they were, assured that they would be provided for without working. Malthus contended that the evils of safety nets and "poor" laws were irremediable. Helping those in need of food and subsistence, he said, was a violation of natural laws. According to Malthus, if the "poor" laws never existed, there would have been fewer incidents of severe stress among the poor and the common people would be happier. After a review of the economic data available to him, he wrote that wages are the expressed demand of society for labor, and that low wages mean too many workers are available. In this scenario, he suggested, poor laws prevented the reduction of family sizes and prevented food costs from exercising their natural function of rationing food and forcing the poor to eat the cheapest foods available.

Malthus believed that the extension of relief to the poor in time of dire need increased the cost of food and lowered the value of money held by the rich and those who were not receiving aid. Providing food, medical care, and financial aid, contended Malthus, interfered with the "natural law of population" and destroyed the work ethic and more lives than they saved. Refuting the thoughts of his father, Malthus came to the conclusion that hopes for social happiness for all classes of society are in vain and that poverty, hunger, and disease are the underclass's inescapable lot in life. He exonerated the rich and the privileged from providing for the poorer segments of society, and did not believe that they needed to increase moral restraints or decrease their luxurious living style, extravagant spending, and society-destroying greed.

Malthus and Welfare

Although Malthus was an ordained clergyman and was taught that Jesus Christ said to "feed the hungry, clothe the naked, and give comfort to strangers," he wrote that the distressed poor are redundant and any welfare aid provided by society be minimal. In his judgment, aid to the poor did little to alleviate an individual's plight. Instead, he suggested, society should stigmatize those on relief or facing pauperdom by forcing them to experience a sense of shame. Dependent poverty ought to be held out as disgraceful, regardless of whether those who were on welfare were unable to work because of age or disabilities, because they were laid off from their jobs, or because their place of employment was closed. Malthus also stated that welfare cannot make the poor comfortable without making them increase and multiply. To reduce births by women on welfare, Malthus proposed that children born should receive public assistance for only a year or two. Malthus did not believe that the poor had a "natural right" to support from the rest of society; in his view, the government's role in ministering to the poor should be terminated and replaced by the efforts of private charities.

Malthus was a pessimist who viewed human poverty as a fact of life. He never defined his harsh statements and cruel premises well, nor did he analyze what little factual data he cited with any level of sophistication. Malthus's theories never had the validity that he claimed. Yet, for centuries after his death, his teachings and opinions continued to influence the decision making of many governmental leaders, philanthropists, and political groups. Malthus's pronouncements on population, poverty, and relief were used to justify laissez-faire or non-interference in famine situations, little or no relief or over-concern, and opposition to life-saving socioeconomic reforms. His contentions that the world had a set "carrying capacity" that is maintained by famine, war, and disease were accepted by many who had the power to prevent or temper these horrible events. Forcefully repeating his message in his writings, Malthus stressed that population growth reduces the standard of living for all, is an economic liability, and causes societies to degenerate. He worked to transform what he believed to be the correct thinking into state and national policy and immediate practice, and to discourage traditional forms of charity. His writings and theories did have an immediate influence on English social policy and the social policy implemented by the cadets whom he taught and who went out to rule British India. The most devastating effect of Malthus's theories was to prejudice the British upper and middle classes' attitudes against social policies designed to aid the poor, the hungry, and the starving.

Examples of Malthus's Influence

Lifeboat Ethics

In the 1970s, the questions of who should live, who should die, and who was worth sacrificing were asked because it was believed that large-scale international famines were inevitable. Many were asking for models or guidelines to select the famine-impacted peoples and countries in which aid would make a difference in number of deaths and the quality of life after a famine. Questions were asked about the responsibility or obligations of those who lived in the affluent developed countries relative to the poor, hungry, and starving inhabitants of less affluent developing countries.

As part of this debate, Professor Garret Harden proposed a "lifeboat" approach to world food/population problems, arguing against providing famine relief and in favor of using famine as a population control tool. Harden's arguments against delivery of famine relief to all starving citizens of the world may be summarized metaphorically as a disaster at sea where only two lifeboats manage to be launched from a sinking ship. One of the lifeboats is in good repair and full to the brim with uninjured and healthy survivors, along with emergency food rations for all. The other lifeboat is damaged and full to the brim with injured, sick, and elderly survivors, with no emergency food rations. Those who are able to do so swim toward the lifeboat with food. Should the survivors in the boat with adequate food rations bring aboard the swimmers from the overcrowded lifeboat and share their emergency rations, thereby endangering their own lives?

Harden contended that every life saved diminished the potential of survival for those in the lifeboat where there was food. In turn, many decision makers, politicians, and government policymakers accepted the lifeboat food policy, then expanded it to advocate that rich, food-surplus nations should follow a set of harsh recommendations:

- Not provide food aid or relief to poor developing countries in time of acute hunger, mass starvation, or famine
- Limit immigration and refugees from food-deficit to food-surplus countries
- Reduce food, material, and financial assistance until the famine-impacted, and presumably overpopulated, poor countries have lowered their population to be within the carrying capacity of their countries

Harden and his followers were supported by many neo-Malthusian advocates, and Harden's reasoning was accepted by those in positions of power who were more concerned with their own well-being, profits, and power than with the lives of innocent suffering, starving people. His use of a lifeboat analogy to represent nations of the world, however, was a poor one. A lifeboat actually has the following characteristics:

- A lifeboat is designed to be self-sufficient for a short period of time, whereas countries have long-term ties and are dependent on one another for markets, materials, and support.
- A lifeboat has a definite carrying capacity, whereas the carrying capacity of countries has never been determined and food production can be increased.
- A lifeboat floats on a body of water that can be calm or storm-tossed (a sighting leading to rescue in a calm body of water can occur and rescue can be made quickly; alternatively, in a violent storm, all survivors from the ship, regardless of the condition of their lifeboats, may be lost).

Harden's detractors observed that the cultural environment of the world (as seen in World Wars I and II) can be so destructive that if a world food war or a nuclear war were to be triggered by a food issue, it would impact all life on Earth.

Triage Ethics

In 1967, William and Paul Paddock outlined their efficient method "of food allocation or food aid in times of widespread famine" in a book entitled *Famine—1975: America's Decision: Who Will Survive*. They strongly believed that exploding populations and static food production in food-deficit, developing nations would make large-scale international famines inevitable. In that case, the U.S. government and philanthropic agencies would be required to make the decision of which countries in the world are likely to fall into the following three categories:

- Never solve their food/population problems
- Probably solve their food/population problems if aid is not provided
- Solve their food/population problems and contribute to the advancement of world society if food aid is provided immediately

According to the Paddock brothers, an international survival triage would require the United States to sort out countries into three groups:

- Nations that have a future and should receive U.S. food aid
- Nations that can survive a catastrophic famine without food aid but will require decades to restore normalcy
- Nations without a future whose people must be sacrificed for the well-being of the United States and the world community

The basic concepts of triage were formulated during the trench-warfare slaughters of World War I. Medical units in the front lines separated the wounded or gassed into three categories:

- Those likely to die regardless of the treatment they receive
- Those who probably would recover even if untreated
- Those who could survive only if cared for immediately

With limited medical personnel, medical supplies, and time, only the third group received medical attention. Classifying the wounded into groups for treatment was called "triage."

Triage ethics, as a food policy, was discussed and debated in the 1960s and 1970s at a time when many people were concerned about mounting global food shortages and the maldistribution of food reserves. As a food policy, it found favor in some circles but was not accepted by most scientific and academic scholars. Malthus's ideas and writings were cited as inspiration by the Paddock brothers and others who searched for policies and strategies beyond the framework of traditional values.

The Malthusian, lifeboat, and triage approaches to hunger, starvation, and famine were morally unacceptable to a great segment of American society, who perceived these concepts as violating the most elementary demands of human justice. Numerous examples can be cited of constructive theories, postulations, and means by which the world could reduce hunger, save the number of people who starve to death, and even eliminate famines. For example, believing that drought and natural catastrophes cause crop failures and that humans, by not providing food relief to those who have no food, cause famines, Frances Lappe and Joseph Collins began advocating "food first" in the late 1970s. They proposed six food-first principles:

1. If no demands are placed on a country by outside forces (multinational corporations, grain merchants, and external affluent consumers), every country in the world can feed its citizens.
2. Inequality, prejudices, mismanagement of resources, and greed are the greatest stumbling blocks to socioeconomic development in the developing world.
3. Safeguarding the world's fragile environment and feeding people are complementary goals.
4. World food safety nets and security are not threatened by the hungry masses, but rather by those parties who control opportunities for profit making through the concentration and internationalization of food resources.
5. Agriculture must first provide food for those who grow the food and then for the people who do not and consume food.
6. Elimination of hunger, starvation, and famine can be realized through the redistribution of international, national, or regional control of food-producing resources.

Those who believe that Lappe and Collins' approach to food/famine issues has merit, or at least makes sense in most instances, contend that little evidence supports the supposition that the world today lacks the ability to feed those who inhabit the Earth. More equitable and liberalized agricultural production and trading policies in both developed and developing countries, plus efforts to address the problem of population growth in developing countries, are considered to form the basis for enhanced world food security. All who have studied and researched the causes of hunger, starvation, and famine know that those who have the power and resources also have the power to allocate those resources. The decision makers and the power brokers of the world must direct their attention to worldwide food needs and formulate policies and strategies to resolve this critical issue. Malthus may be dead, but his approaches to food and hunger and his belief that famine is a natural law that checks population imbalances linger.

—*William A. Dando*

Further Reading

Dando, William A. *The Geography of Famine*. New York: John Wiley & Sons, 1980, 193–194.

Fogel, R. *The Escape from Hunger and Premature Death*. Cambridge, UK: Cambridge University Press, 2004.

Harden, G. "Carrying Capacity as an Ethical Concept." In *Lifeboat Ethics*, edited by G. Lucas and T. Ogletree. New York: Harper and Row, 1976.

Harden, G. "Lifeboat Ethics: The Case Against Helping the Poor." *Psychology Today* 8, no. 4 (September 1974), pp. 38-46, 123–124, 126.

Ingram, R. A. *Disquisitions on Population in Which the Principles of the Essay on Population, by the Rev. T. R. Malthus, Are Examined and Refuted*. London: J. Hatchard, 1808, 71.

Lappe, F., J. Collins, and P. Rosset. *World Hunger: Twelve Myths*. New York: Institute for Food and Development Policy and Grove/Atlantic, 1998.

Lappe, F., and J. Collins. "Food First." In *Global Perspective on Ecology*, edited by T. Emmel. Palo Alto, CA: Mayfield Publishing, 1977, 464.

Malthus, T. R. *The Essay on Population*, Vol. 2. London: J. M. Dent & Sons, 1914, 38–46.

Malthus, T. R. *First Essay on Population 1798*. London: Royal Economic Society and Macmillan & Co., 1926, 1–396.

Malthus, T. R. *Population: The First Essay*. Ann Arbor, MI: University of Michigan Press, 1959, 2–13.

Paddock, W., and P. Paddock. *Famine—1975: America's Decision: Who Will Survive*. Boston: Little, Brown, 1967.

Peterson, W. *Malthus*. Cambridge, MA: Harvard University Press, 1979, 1–20.

Smith, K. *The Malthusian Controversy*. London: Routledge & Kegan Paul, 1951, 3–7.

Turner, M. *Malthus and His Time*. New York: St. Martin's Press, 1986, 40–59.

Mapping the Geography of Hunger in the United States

In September 1999, the U.S. Department of Agriculture (USDA) released a report providing the first ever statistically valid and reliable state-level estimates of vulnerability to hunger, referred in the official lexicon as "food insecurity," in the United States. The map of those estimates revealed a distinct spatial pattern, featuring a "hunger belt" running from the Southeast, across the South-Central region, up the West Coast to the Northwest. The lowest rates clustered in the North-Central and Mid-Atlantic regions.

A decade later, a new map was published. Vestiges of the banded pattern from the earlier version were still evident in the revised map. While high rates of food insecurity were still found in the Southeast, South-Central, and Northwest regions, they no longer formed a continuous belt. Moreover, in this more recent map, high rates now were seen in the Midwest.

Taken together, these maps and the data used to draw them represent an important waystation on the political and methodological journey toward understanding the geography of hunger in the United States.

Mapping Hunger: The Politics and Methods of Revealing Hunger's Spatial Patterns

The USDA's publication of state-level estimates of vulnerability to hunger was an important moment in the geographical understanding of hunger in the United States. For decades, researchers, policymakers, health practitioners, and activists had sought and argued over the accuracy and validity of hunger maps. In a Senate subcommittee hearing in 1967, Senator Robert F. Kennedy asked in which parts of the United States malnutrition and hunger were occurring and whether a map existed. One of his senatorial colleagues replied that it would be a good idea to have such a map drawn. The record does not indicate whether the senators were able to produce a map, but that same year an antipoverty activist organization issued a report observing that reliable statistics revealing the geographic distribution of hunger in the United States did not exist (Citizens' Bureau of Inquiry into Hunger and Malnutrition in the United States, 1968). A year later, some members of that organization, which since had become part of a larger team of civil rights activists and doctors called the Citizens' Bureau of Inquiry into Hunger and Malnutrition, produced a controversial map showing the geography of "hunger counties"—that is, places where high poverty and infant mortality rates and a lack of support through food assistance and welfare programs were postulated to produce a greater risk of inadequate food intakes for people living in them. According to the map, hunger was largely a Southern phenomenon. Indeed, all but 15 of the most at-risk hunger counties were located in the South, with four distinct concentrations found in the Mississippi Delta, Georgia/Alabama, South Carolina/North Carolina/Virginia, and Arizona/New Mexico. Elected officials from these areas objected to the Citizens' Bureau's mapmaking, particularly its reliance on what they saw as a dubious measure of hunger.

Almost two decades later, an attempt to reveal the spatial extent and variation of hunger in the United States again emerged at the center of a national political debate over hunger. In 1984, President Ronald Reagan's Task Force on Food Assistance observed that "there are no hard data available to estimate the extent of hunger . . . [Thus there is no] definitive, quantitative proof" that the problem exists (quoted in Bickel et al., 2000, 1). Without an appropriate conceptual basis, researchers and policymakers were left to rely on invalid measures of hunger.

Nevertheless, around that time, a team of doctors and public health academics formed the Physician Task Force on Hunger in America. In its endeavor to reveal a national problem warranting a public response, the Physician Task Force approximated the Citizens' Bureau's technique for measuring hunger and produced a new map of hunger counties, this time using only the poverty and public assistance criteria. By proceeding in this way, the task force intended to identify those places where the incidence of hunger was most likely, as measured by the ratio between the number of people in poverty and the number of people receiving food stamps. While the result was not directly comparable to the hunger counties map from 1968, it is nevertheless noteworthy that the new map showed a spatial pattern of "hunger counties" that contrasted rather strikingly and gave rise to specific critiques. Whereas most of the hunger counties in the Citizens' Bureau map were concentrated in the southern parts of the country, in the Physician Task Force map most of these counties clustered in the Midwest (primarily Missouri) and especially the upper Plains (predominantly North and South Dakota) regions. A third, more dispersed cluster was found in the South-Central/West region, mainly Arizona, New Mexico, and Texas.

As before, various political figures challenged the map. One Congressional representative asked the General Accounting Office (GAO) to determine the accuracy of the "hunger counties" map. The GAO concluded that the data used to draw the map validly measured "risk" related to hunger. Shortly thereafter, in 1991, an academic sociologist, D. McMurry, observed the spatial pattern of hunger counties, particularly their concentration in the country's great food-producing regions, and contended that it revealed conceptual and methodological problems. Hunger is a physiological state that is not, he argued, either necessarily or uniformly related to income (i.e., poverty) and the paucity of public assistance.

In 1999, the map of hunger again found itself in the political spotlight. When the USDA released the first state-level hard data on food insecurity, its report included the map seen in Figure 14, which could be considered the first scientific map of hunger in the United States. Shortly after the release of this data, Texas Governor and presidential candidate George W. Bush was asked about the USDA report, which showed Texas as part of the hunger belt and as having the second highest rate of food insecurity in the country (5 percent of households). Governor Bush acknowledged that he had seen the report and asked where in Texas these hungry children might be. "You'd think the governor would have heard if there are pockets of hunger in Texas" (Yardley, 1999). While this moment of political contention over the hunger map was brief, it nevertheless echoed longstanding questions regarding the geography of hunger in the United States—namely, what does the map of hunger look like and what explains its geographical dimensions?

A geographical account of hunger, like any explanation, presupposes an ontological view of the nature of its existence. In more prosaic terms, to explain the geography of hunger, the condition needs to be mapped. To be mapped, it needs to be measured; and to be measured, it needs to be conceptualized and defined. Thus the question "What is hunger?" is important both methodologically and theoretically. This was the task faced in the early 1990s by a federal working group formed to develop a conceptual basis for the measurement of hunger. Legislation signed into law in 1990 had mandated the establishment of a national nutrition monitoring network involving federal, state, and local governments, the implementation of which entailed the development of a sound national measure of food insecurity and hunger. This tool eventually served as the basis for the Food Security Survey (FSS), administered by the Current Population Survey (CPS), which has been used since 1995 to generate the estimates of household food insecurity in the United States. These estimates are mapped in Figure 14.

The working group drew heavily upon previous research on the conceptualization and measurement of hunger in the United States, as well as upon various efforts to put that research into practice. As a USDA guidebook from 2000 put it, hunger was conceptualized in relation to food security, a condition defined as follows:

> Access by all people [in the household] at all times to enough food for an active, healthy life. Food Security includes at a minimum: (1) the ready availability of nutritionally adequate and safe foods, and (2) an assured ability to acquire acceptable foods in socially acceptable ways (e.g., without resorting to emergency food supplies, scavenging, stealing, or other coping strategies). (Bickel et al., 2000, 6)

Figure 14. Very low food security in the United States

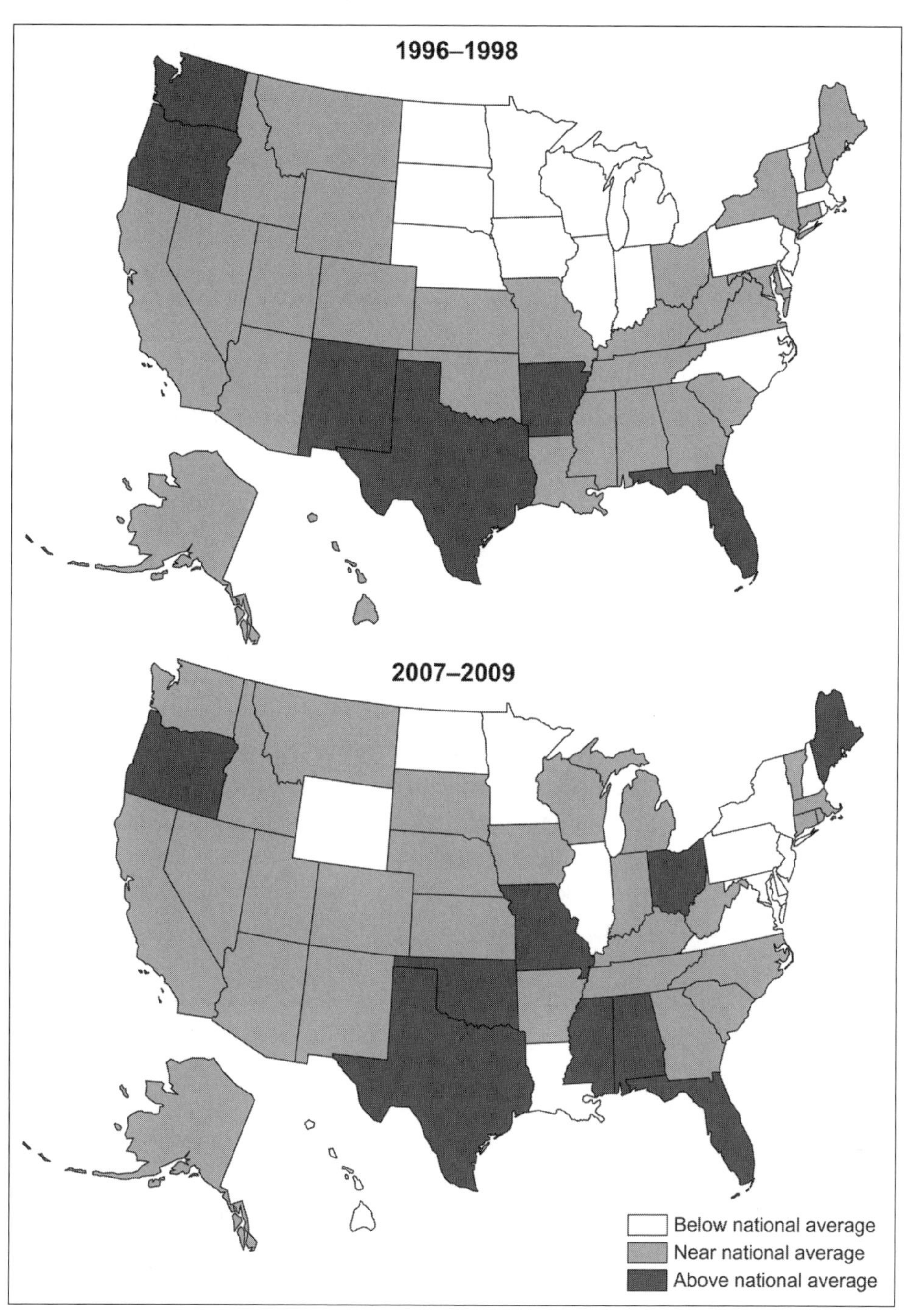

Source: Nord, M., Jemison, K., Bickel, G. *Prevalence of Food Insecurity and Hunger, by State, 1996–1998.* U.S. Department of Agriculture Report; Nord, M., Coleman-Jensen, A., Andrews, M., Carlson, S. *Household Food Security in the United States.* EER-108, U.S. Department of Agriculture, Economic Research Services, November 2010.

The failure of condition (1) or (2) produces a situation of food *in*security. In this framework, hunger is viewed as an extreme stage of food insecurity that will, if it persists, result in an uneasy or painful sensation and, if prolonged, physiological effects caused by an insufficient food intake.

Working with this definition, the food security measurement task force articulated a series of questions that sought to gauge the prevalence of certain conditions, experiences, and behaviors reflecting the food security status of households:

1. Anxiety that the household food budget or food supply may be insufficient to meet basic needs
2. The experience of running out of food, without money to obtain more
3. Perceptions that the food eaten by household members is inadequate in quality or quantity
4. Adjustments to normal food use, substituting fewer and cheaper foods than usual
5. Instances of reduced food intake by adults in the household or consequences of reduced intake such as the physical sensation of hunger or loss of weight
6. Instances of reduced food intake or consequences of reduced intake for children in the household

Specifically, among other questions, respondents were asked whether in the previous month and year their household had run out of money for food, whether adults and children had eaten less or skipped meals as a result of lack of food, and whether children had gone to bed hungry.

The development of the Food Security Survey marked a technical improvement over previous efforts to measure hunger (several of which are discussed later in this entry), which had relied on proxy or indirect indicators, such as the poverty rate and health conditions. The Food Security Survey asks individual respondents direct questions about their own and other household members' food consumption experiences. Responses for each household are statistically processed, allowing the household to be categorized as "food secure," "food insecure with low food security," or "food insecure with very low food security." The latter group includes households whose members have experienced reduced food intake. While it is very likely that people in such households have experienced the physical sensation of hunger, the measurement tool does not technically gauge this phenomenon, leading the USDA to drop the word "hunger" from its lexicon.

In 2009, according to the most recent estimates, 14.7 percent of U.S. households were food insecure. One third of those households, or 5.7 percent of all households in the United States, experienced the most severe condition of very low food security. Both percentages represent the highest estimates of food insecurity since 1995, when the government first collected these data using the Food Security Survey. Both rates have climbed steadily since 1999, when the rates of food insecurity and very low food security were 10.1 percent and 3.0 percent, respectively. A comparison with the 2007 estimates, which were 11.1 percent and 4.1 percent, respectively, reveals the damaging impact of the economic recession on Americans' well-being.

It is useful to make several points regarding the maps of state-level versions of these estimates, such as those seen in Figure 14. First, the USDA does not measure or report data on food insecurity and hunger at any sub-state level, such as at the county level. While the FSS was created to be appropriate and feasible for use in locally designed and conducted food-security surveys, it has been used to generate sub-state-level estimates in only a handful of states and communities so far. Second, state-level prevalence rates cannot be obtained for individual years because, while the overall CPS sample is sufficient in size to produce a reliable national-level estimate, the number of households surveyed within states in a given year is not large enough to create reliable estimates for the states. Thus, to achieve within-state samples of a sufficient size, USDA analysts create pooled samples by combining samples taken in three consecutive years. This means that the hunger and food insecurity maps represent the spatial variation in prevalence rates (or averages) for particular three-year periods. Hence, the geographic representation of hunger generated by these data enables a rough mapping. Even so, from a scientific point of view, the FSS map offers a superior means of revealing the geography of food insecurity and hunger than do previous approaches, such as the Citizens' Bureau and Physician Task Force mappings.

Explaining the Geography of "Very Low Food Security"

The map of "very low food security" has produced two related streams of research. The first seeks to explain the state-to-state variation in these estimates. The surprisingly high rate in Oregon (given its poverty and income levels), which led the state's governor to enact a strategic plan to bring it down, drew specific attention from researchers, such as Taponga in 2004 and Bernell, Weber, and Edwards in 2006. Taponga et al., for example, found a strong association between very low food security and unemployment, high housing costs (as a share of income), and residential mobility. The latter was a key finding that also explained, to a significant degree, the lower-than-expected rate in West Virginia, judging by its poverty

Where Does Hunger Occur in the United States and Why?

Since the 1960s, questions about hunger's geography have played a key role in both political and academic debates. The absence of an accepted map revealed disagreements about the nature of hunger and ways to measure it. In the early 1990s, federal legislation led to the development of a scientific approach to measuring hunger and the collection of "objective" data. The first map created using these data was published in 1999. The most recent map shows the highest rates found in the southern and western regions of the country, while the lowest rates tend to be clustered in the upper Midwest and in the northeastern states. Analyses of these geographic patterns have discovered associations between a higher risk of hunger and high rates of unemployment, low wages, high shelter costs, high prevalence of residential mobility, and gaps in government assistance programs. While these studies have been revealing, they have ignored other forces that are not readily analyzed through statistical analysis, such as power relations and social structures operating across geographic scales to produce places and patterns of hunger.

—Andrew Walter

rate. As the authors explained, residential mobility often suggests various forms of household disruption, such as job loss, divorce and separation, and eviction. Moreover, the move to a new house often distances households from supportive networks of family, friends, and familiar institutions such as churches and local government. Others have corroborated the importance of these findings and have identified a range of additional factors having to do with participation in federal food assistance programs, rates of taxation on low-income households, and wage levels.

The second stream of research seeks to develop a better understanding of the relationship of very low food security to place context. Indeed, from a geographical standpoint, one of the most compelling features of the measurement tool administered via the Food Security Survey is that it conceptually embeds hunger in place. This viewpoint is reflected in the questions asked on the FSS, which compel individuals to consider food deprivation in relation to the local situations of their households—and the web of their needs in relation to the costs of food and other basic necessities, employment, and income-earning opportunities, among others. As Mark Nord, a sociologist and one of the leading food security analysts at the USDA, pointed out in 2000, unlike official poverty estimates, which assume a universal deprivation threshold, the estimates derived from the FSS embody geographically specific circumstances. The food security estimates, therefore, represent information about social and economic contexts that may contribute to hunger. Thus a goal of researchers has been to understand those contexts. For example, based on a comprehensive study, Barfield and Dunifon in 2005 identified a strong association between food insecurity and what they call "the food security infrastructure" in places. The food security infrastructure consists of the combination of social and economic attributes of a particular place that come together to shape the everyday socio-spatial context in which people endeavor to obtain an adequate diet. Those attributes include the availability and accessibility of federal food assistance programs, the tax burden on low-income households, the availability of jobs, wage levels, and networks of social material and emotional support, including family and friends.

While these studies have shed new light on the occurrence of food insecurity and its geographies in the United States, they have tended to neglect the ways in which the everyday struggle to access enough food is structured by the forces of politics (e.g., social movements and ideologically driven policy changes) and the political economy (i.e., the various power relations bound into labor markets and welfare states). As the geographer Michael Watts argued in 2000, the quantities and qualities of food that people are able (or not able) to obtain are "expressions of power" that are "both constituted and reproduced through conflict, negotiation, and struggle." This neglect of social power in explanations of food insecurity and its geographies very likely arises from the effort to "scientize" hunger, as Allen wrote in 2007. In an effort to avoid the politics that infused earlier attempts to define, measure, and map hunger in the United States, scholars and policymakers sought to develop ways of generating hard data that could be analyzed objectively, using quantitative techniques.

While this pursuit has been a success and has advanced the understanding of food insecurity in the United States, it also has tended to narrow the scope of analysis. The emphasis on scientific rigor tends to favor certain units of analysis (discrete individuals, such as households) over others (cultural mechanisms, institutions, structural power, and social relations—for example, class, gender, and race) that are not measurable as quantities yet are required for statistical analysis. Additionally, it has contributed to a knowledge hierarchy whereby nonscientific forms of knowledge about hunger and food insecurity are presumed to be, on the face of things, invalid and merely political or "subjective," especially in the realm of policymaking in particular. In 2007, Allen observed that the need for explanations and critiques to be based on complex, technical analyses to be heard undermines not only advocacy and policymaking but ultimately theoretical work seeking to explain hunger's occurrence as well.

—Andrew Walter

Further Reading

Allen, P. "The Disappearance of Hunger in America." *Gastronomica* 7, no. 3 (2007): 19–23.

Amin, A. *Post-Fordism: A Reader.* New York: Wiley-Blackwell, 1995.

Barfield, J., and R. Dunifon. *State-Level Predictors of Food Insecurity and Hunger Among Households With Children.* Contractor and Cooperator Report No. 13. Washington, DC: Economic Research Service, U.S. Department of Agriculture, 2005.

Barfield, J., R. Dunifon, M. Nord, and S. Carlson. "What Factors Account for State-to-State Differences in Food Security?" *Economic Information Bulletin 20.* Washington, DC: Economic Research Service, U.S. Department of Agriculture, 2005.

Bernell, S. L., B. A. Weber, and M. E. Edwards. "Restricted Opportunities, Personal Choices, Ineffective Policies: What Explains Food Insecurity in Oregon?" *Journal of Agricultural and Resource Economics* 31, no. 2 (2006): 193–211.

Bickel, G., M. Nord, C. Price, W. Hamilton, and J. Cook. *Guide to Measuring Household Food Security: Measuring Food Security in the United States.* Washington, DC: Economic Research Service, U.S. Department of Agriculture, 2000.

Brown, L., and H. F. Pizer. *Living Hungry in America.* New York: Macmillan, 1987.

Citizens' Board of Inquiry into Hunger and Malnutrition in the United States. *Hunger, U.S.A.* Washington, DC: New Community Press, 1968.

Economic Research Service. *Food Security in the United States: Measuring Household Food Security.* Washington, DC: Economic Research Service, U.S. Department of Agriculture, 2010.

Edwards, M. E., B. Weber, and S. Bernell. "Identifying Factors That Influence State-Specific Hunger Rates in the U.S.: A Simple Analytic Method for Understanding a Persistent Problem." *Social Indicators Research* 81 (2007): 579–595.

McMurry, D. "Hunger in Rural America: Myth or Reality? A Reexamination of the Harvard Physician Task Force Report on Hunger in America Using Statistical Data and Field Observations." *Sociological Spectrum* 11 (1991): 1–18.

Nord, M. "Does It Cost Less to Live in Rural Areas? Evidence from New Data on Food Security and Hunger." *Rural Sociology* 65, no. 1 (2000): 104–125.

Nord, M., K. Jemison, and G. Bickel. *Prevalence of Food Insecurity and Hunger, by State, 1996–1998.* Washington, DC: Economic Research Service, U.S. Department of Agriculture, 1999.

Physician Task Force on Hunger in America. *Hunger in America: The Growing Epidemic*. Middletown, CT: Wesleyan University Press, 1985.

Ryu, H. K., and D. J. Slottje. "Analyzing Perceived Hunger Across States in the U.S." *Empirical Economics* 24 (1999): 323–329.

Taponga, J., A. Suter, M. Nord, and M. Leachman. "Explaining Variations in State Hunger Rates." *Family Economics and Nutrition Review* 16, no. 2 (2004): 12–22.

U.S. Congress. *Hunger and Malnutrition in the United States*. Senate Subcommittee on Employment, Manpower, and Poverty, 90th Congress, 2nd Session. Washington, DC: U.S. Government Printing Office, May 23 and 29, 1968/ June 12 and 14, 1968.

Watts, M. "Entitlements or Empowerment? Famine and Starvation in Africa." *Review of African Political Economy* 51 (1991): 9–26.

Watts, M. "The Great Tablecloth: Bread and Butter Politics, and the Political Economy of Food and Poverty." In *The Oxford Handbook of Economic Geography*, edited by G. Clark, M. Gertler, and M. Feldman. Oxford, UK: Oxford University Press, 2000, 195–214.

Watts, M., and H. Bohle. "The Space of Vulnerability: The Causal Structure of Hunger and Famine." *Progress in Human Geography* 17 (1993): 43–67.

Williamson, E. "Some Americans Lack Food, but USDA Won't Call Them Hungry." *The Washington Post*, November 16, 2006, p. A01.

Yardley, J. "Democrats Criticize Bush as Being Out of Touch on Poverty." *The New York Times*, December 24, 1999.

N

New Energy Sources

In the continuing human story, access to energy resources is necessary for the sustainability of societies. In fact, energy is so basic that its manner of utilization is a limiting factor on the complexity of societies. Hunting and gathering societies used human power, while the intense utilization of fossil fuels (coal, petroleum, and natural gas) facilitated the complex, mobile, technological societies of today's developed nations.

Agriculture is the most organized manner that societies have of feeding themselves. Even partial regional failures of agriculture can result in famine. Availability of energy, therefore, is of immense importance in the ability of a country to feed itself. A case in point can be seen in the agricultural development of the United States. In 1900, 41 percent of the U.S. labor force worked in agriculture; by 2000, this share had shrunk to 1.9 percent while readily feeding a population that had grown by a factor of almost 4. This transition was accomplished through mechanization made possible by fossil fuels. On the darker side, one can point to 20th-century Japan's imperative to import resources that are otherwise nowhere available on its modernizing home islands. In particular, the Japanese imperialistic conquest of the Pacific Rim was driven by the country's need for oil. The resulting clash of Japanese territorial interests with those of Europe and the United States led to the bloody battles of the Pacific Theater in World War II.

Solar energy is the basic energy source needed for the production of food, because photosynthesis in plants fixes solar electromagnetic energy into chemical energy. Modern agriculture, however, requires other substantial inputs of energy. Field preparation, planting, spraying herbicides and pesticides, and harvesting are done by internal combustion machines requiring large amounts of fossil fuels. After harvest, energy is used to transport the product to market and, in many cases, add value to the raw agricultural commodity by turning it into something more useful (e.g., processing tomatoes into tomato soup).

Future provision for two classes of power must be considered. The first is specific to the agricultural pursuits on farms and ranches. Energy sources must be mobile in such a way as to allow machines to do their work at the sites where crops are grown and livestock are raised and to transport the results to markets or to processing. The second class of power, which is common to all sectors of society, is central generation of electricity distributed over power lines. This latter class has somewhat more flexibility in that many sources of energy can be harnessed to generate electricity.

At present, there is no viable substitute for the work of fossil-fuel–powered internal combustion engines working on large agricultural fields. For instance, solar panels and battery power output fed to onboard electric motors generally cannot match the performance of internal combustion engines on heavy field equipment over a course of hours.

Other energy inputs to modern agriculture are large and, perhaps, surprising. Prominent among them is the prolific use of fertilizer created by burning natural gas to heat and react with atmospheric nitrogen, thereby fixing the nitrogen into crop-usable anhydrous ammonia. A ton of anhydrous ammonia requires the input of approximately 1,000 cubic meters (33,500 cubic feet) of natural gas. Thus any insecurity in a country's natural gas supply also results in concerns about food production.

When considering the sobering prospect of feeding the more than 10 billion people who might well populate the planet in the year 2100, scientists must examine a wide variety of energy sources that could power the future of food production. Humans will improvise, innovate, and mix and match energy sources in an attempt to ensure food to eat. Changes are under way, but it is not clear which mix of energy resources will make it possible for humans to avoid famine. New energy sources must be found for the future. Many of these fuel sources are not truly new; rather, improving the technologies associated with their use and their percentages in the overall energy portfolio will expand their role dramatically over the remainder of the 21st century. Fossil fuels now predominate in mechanized agriculture, but this situation appears poised to change. With a large number of possibilities, the future of energy sources will be drastically different than that at present and, as an ensemble, will be truly new sources.

Fossil Fuels

Fossil fuels represent solar energy captured over millions of years and stored in chemical form in the remains of plants and animals. Humans are consuming fossil fuels at many thousands of time the rates at which they were deposited. Since the middle of the 19th century, fossil fuels have become heavily intertwined with human progress. The Industrial Revolution and concomitant agricultural advances were made by applying controlled amounts of mechanical energy at the right times and places.

The nations of the world have been unequally blessed with the occurrence of fossil fuels, a situation that is problematic for today's agriculture. The United States extracts much oil from within its borders: It is currently the world's third-leading producer, but uses four times more than that. It also is 12th in the known amount of reserves left in the ground. U.S. production has been declining since 1970, and many energy pundits are convinced that world production has now peaked.

The United States possesses the world's largest natural gas reserves and imports only a small percentage of its total consumption. New extraction methods targeting shale have greatly increased estimated reserves. Natural gas production will peak sometime in the 21st century, but natural gas may well be a bridge to other energy sources until they become economically viable and geographically widespread. Compressed natural gas is already used in cars, trucks, and buses, and there is no technological reason that its use cannot be extended to food production.

Coal was once heavily used in steam-powered agricultural machinery. Because it did not burn as cleanly or have as favorable an energy/weight ratio as gasoline, however, it was supplanted by oil. Nevertheless, coal is plentiful in the United States, with more than 100 years of reserves left at current rates of production, and coal is burned to generate more than 40 percent of the world's electricity.

Today, there is serious global-scale competition for the world's dwindling fossil fuel supplies, and this factor promises to influence the geography of food and famine. In the last two decades, China and India have increased their joint share of fossil fuel consumption, from 10 percent of the world's consumption in 1990 to more than one-third of the current total. As a result, other nations must compete more vigorously for the world's production to maintain their own fossil fuel supplies.

Use of fossil fuels has many environmental drawbacks. Current concerns about global warming caused by carbon dioxide release have been extended to agricultural production. Fossil fuel combustion releases the greenhouse gas carbon dioxide as a by-product. Worldwide, agriculture contributes somewhat less than 10 percent of the anthropogenic carbon dioxide emissions into the atmosphere.

Fossil fuel extraction and transportation also have large environmental consequences at the Earth's surface. The easiest-to-obtain reserves have already been exploited, so that production must now move to places that are notably harder to reach (deeper or geographically remote). Also, the fossil fuels extracted are frequently of lower quality and require increased refinement.

Humankind cannot sustain itself ad infinitum on the present mix of energy sources because of the heavy dependence on shrinking supplies of fossil fuels. This prospect is exacerbated by world population growth and modern lifestyles that demand increasingly more energy use per person.

Storage Batteries and Fuel Cells

In lieu of fossil fuels, the thought of concentrating energy in batteries and fuel cells is quite attractive. The electric battery has been around since Alessandro Volta's experiments of the early 1800s, and this device is useful in low-power operations such as remote controllers and radio receivers. Research and development attention is being given to batteries, but they still cannot power mobile machinery in the same way as gasoline. Unlike with gasoline power, it is impossible to run a wheat combine for a day on batteries, much less fit the batteries onboard to produce enough instantaneous power. Gasoline/battery hybrid vehicles are now being produced in increasing quantities, and it is hoped that the future will bring hybrid vehicles, lessening fossil fuel use by storing some of the energy in batteries.

Fuel cell technology is analogous to battery technology in that a fuel cell generates electricity through a chemical reaction between a fuel and an oxidizing agent. Unlike batteries, fuel cells feature external replacement of the fuel. Many substances can be used as fuel, but universally available hydrogen concentrated and loaded into fuel cells and oxidized with atmospheric oxygen promises an unending fuel supply. Such fuel cells are being developed but, given current progress, powerful fuel cells will not be available for many years.

Solar Energy

The sun provides immense energy to the Earth system, and scientists are attempting to directly tap solar energy to power the equipment used to produce food. Several types of photovoltaic devices are able to capture solar energy and convert it into electrical current. As yet, the technology remains limited in how much energy it can capture per square area. Solar-powered vehicles have delivered only modest performance because they cannot generate much electricity from panels placed on their exteriors. Large solar arrays have been employed and integrated into electric power grids and local agricultural applications. Also, solar energy is geographically uneven and, at night, nonexistent, so it is apparent that any improvements in solar technology will need to be matched with other technologies.

Wind Energy

Wind has been powering grain grinding since the ninth century CE in Persia. Now wind can generate electricity to be used in places far from the windmill. Projects to deploy wind-generated electricity are increasing rapidly, reflecting this technology's ability to deliver large amounts of electricity with virtually zero pollution. In fact,

Large Wind Turbines

Significant advances in wind power generation at the beginning of the 21st century have been made possible by advances in technology and economics. The cost of utility-scale wind power has decreased dramatically, to the point that it has become competitive with other energy sources of electrical generation. The reason for this cost reduction relates to the engineering of these mammoth machines. Now made of fiberglass composites and various polymers, the large blades of wind turbines are light enough to spin effortlessly. Computer control of electric motors point the machine into the wind and pitch the individual blades to capture the wind most efficiently. By placing the blade hub many feet above the surface, a turbine can take advantage of considerably faster winds and produce more energy. These wind turbine behemoths must be financed with huge amounts of upfront capital. Costs are generally more than $1.5 million per installed megawatt of capacity. Larger turbines presently are being designed, and in a few years 10-megawatt turbines will be deployed.

—Stephen J. Stadler

wind power technology is the energy source with today's largest annual increases in worldwide generation capacity. It provides more than 1 percent of the world's electrical generation. The underlying premise is simple: Wind moves large turbine blades, and this kinetic energy turns a crankshaft and energizes an onboard electrical generator. The genius of today's machines is that they capture economies of scale by being massive and tapping stronger winds 150 or more feet above the Earth's surface.

Wind farms are typically placed on rural landscapes and in shallow coastal waters. The loss of crop and grazing land is not a major issue because each turbine and associated access roads take up approximately 0.5 acre; farming and ranching can go on as usual. The world's largest wind farm has a capacity of 780 megawatts, which is on par with the capacity of a large coal-fired generation plant. Wind is spatially variable, so its implementation must be carefully planned. For instance, the Great Plains of the United States has tremendous wind resources but little population, so an improved transmission infrastructure will be needed if the resource is to be fully exploited. The wind does not always blow, but this is not as much trouble as might be imagined if wind power is co-mingled with other sources on large power grids.

Farms and ranches are characterized by both space and electrical demand. Installation of smaller, distributed turbine generation is ongoing at these sites, albeit not at the frenetic pace at which wind farms are being established. The smaller turbines always produce more costly electricity than utility-scale wind farms, but this sort of generation will increase dramatically to aid mechanized food production in remote and lesser developed areas where good transmission is not available. Denmark, for example, already produces more than 25 percent of its electricity by wind power. World wind geography dictates that most countries are not so advantageously positioned with as many favorable wind generation sites. Most estimates suggest that wind power will account for 10 percent of the world's electrical supply by the end of the 21st century. Although wind generation has many positive attributes, by itself it cannot meet the increased world demand for energy.

Water Energy

Hydropower provides power without carbon emissions. The sun fuels the hydrologic cycle, a never-ending exchange of the various phases of water between ocean, atmosphere, and land. For hundreds of years, water has been used to grind grain and, since the late 19th century, dams have been built for hydroelectric generation. The 21st century is likely to see an increase in hydropower exploitation, though with increasing constraints on its use. Most of the viable sites in developed countries already have been used, and environmental pushbacks on issues such as ready fish migration and siltation behind dams will decrease the number of future projects. Clearly, hydropower will not play an increasing role in global energy production. In fact, energy demand in developing countries is likely to outstrip any additions to the supply via hydropower generation.

The future hints at intriguing ways to harness huge amounts of ocean energy. Most of the Earth's surface is covered by ocean water, and the near-shore oceanic environment provides opportunities for the production of electric power, including wind power, which is always greater over water than over the adjacent land. In certain areas, waves driven by consistent winds have led to the development of several types of floating electric generation devices now undergoing testing. In most cases, the vertical motion of waves causes an up-and-down motion in a buoy or hinged float. This motion is used to pressurize a fluid, which then turns a turbine to generate electricity. In some places, these devices are connected to power grids, but much more development is needed to turn ocean energy into a large source of power.

In a quite different approach based on hydropower, osmotic generation can be accomplished at the interface of salt and fresh water where fresh water enters the ocean. In one prototype, a large polymer membrane separates salt water from fresh water. The salinity gradient forces fresh water to move through the membrane via osmosis; seawater is squeezed into a chamber and then circulates to turn an electric turbine. This technology was first deployed as a small demonstration project near Oslo, Norway; the hope is to expand osmotic electric generation to wind-farm-like output as soon as possible. The appeal is obvious: Here is another renewable energy source where the fuel cost is zero.

Biofuel Energy

Biofuel refers to a large number of organic substances burned to create useful energy. Humans have used biofuels for thousands of years—they have burned both wood and peat throughout the ages. Present interest centers on refining organic materials to supplement and replace the fossil fuels used in internal combustion engines. Crops with plentiful sugars such as corn and sugar can be fermented to produce ethanol; many vehicles are already equipped to use as much as 15 percent ethanol.

Although biofuels might appear as renewable and "green," their environmental and social costs can be substantial. Feedstocks must be heated, which requires energy input. In some cases, more energy might be input than is actually derived from the use of the biofuel. Ongoing research is seeking to identify processes to use grasses and wood as feedstocks, thereby making the energy payback equation much more favorable.

Moreover, combustion of biofuels produces carbon dioxide emissions that would otherwise contribute to Earth's greenhouse effect. More directly, growing biofuels creates conflicts with food production. Land used for biofuel production is essentially unavailable for food production. Countries that are able to feed themselves have the freedom of choosing to grow biofuel crops as economic conditions dictate. In contrast, developing countries with limited food supplies can be unintentionally affected by this trend, in that growing biofuel crops in one country may increase the cost of food and food availability elsewhere.

Atomic Energy

Utility-scale electrical production by nuclear fission has been employed successfully since the 1950s. In fission, atomic nuclei from refined uranium are split and produce huge amounts of heat used to create steam for electric turbines. Producing somewhat less than 20 percent of the word's electricity, fission has undergone lethargic growth in the last few decades. Nuclear reactors require large capital investment. Mobile nuclear reactors have been installed on ships and spacecraft, but they are not used in agriculture because of their cost and danger.

Nuclear reactors run on enriched uranium. Common in the Earth's crust, uranium must be found in ore deposits of great enough concentrations to make feasible its refinement to reactor fuel. Three countries—Kazakhstan, Canada, and Australia—account for nearly two-thirds of the world's uranium mining. Most countries with nuclear power generation capabilities must import enriched uranium.

Two specters make the use of fission problematic. The first is that uranium can be refined into fuel for atomic bombs, and the countries that have atomic bomb technology zealously attempt to regulate nuclear materials so that their enemies cannot build bombs. The second is the possibility of accidents with nuclear power generation. The 1986 incident at Chernobyl in the Ukraine killed a few hundred people, caused 300,000 people to be resettled, and sent radioactive materials into the atmosphere over much of the world. Japan is recovering from a major nuclear disaster caused by the combined impact of an earthquake and a tsunami on a coastal nuclear generation facility.

In fusion, two or more atomic nuclei collide to form a single heavy nucleus, a process that is accompanied by the release of massive amounts of energy. Scientists have achieved fusion, but not in the controlled manner that would be required to produce reliable electricity. In addition, this reaction requires huge amounts of energy to force fusion and become self-sustaining. Despite the drawbacks, fusion remains a tantalizing avenue of active research because the feedstock would be the unlimited supply of hydrogen in Earth's atmosphere. The control of fusion in a manner necessary for central electrical generation plants is still decades away.

Conservation of Energy

Energy in the early 21st century is supplied to the users in the world from nonrenewable fuel sources. This situation must eventually change, no matter what governments decide and politics demand. World history has seen many examples of dire social consequences—up to and including war—as energy supplies become inadequate. In the search to secure plentiful energy, there are ways to decrease energy demand and to use the energy sources at hand more efficiently. Those methods are corporately labeled "conservation." The increase in world food demand is certain, but the energy sources needed to meet that demand are not certain. As the world attempts to move to remixed portfolios of energy resources, conservation of energy will be necessary to allow a bridge to the future.

Conservation is an inexpensive manner of decreasing energy use; it is certainly possible that energy usage can be cut 10 or 20 percent without altering the quality of life. Serious concern can already be seen in the European Union, which has plans to cut its primary energy production by 20 percent. Can this reduction be achieved worldwide? The lynchpin of conservation of energy is changes in human behavior, which is difficult to regulate. Given the various energy situations in the almost 200 countries in the world, it is unlikely that conservation of energy will be smoothly implemented in the remainder of the 21st century.

Speculation

As fossil fuel supplies become depleted, world population grows, and energy demand per person increases, the present mix of energy sources will not be able to support a secure world food supply. Therefore, humans must reinvent energy sources in ways leading to maintenance and improvement of humans' quality of life. Improvements in technology are currently occurring, and others are expected to emerge in the future. There are countless numbers of potential energy sources; when they are fully developed and combined, humans could be a step closer to the elimination of famine. This is all well to say, but which strategy should be followed? There is no world government, and energy geography varies markedly by country. A good energy outcome in one country at the end of the 21st century will not guarantee that its neighbors will be secure. Will the world's energy sources be able to keep up with and sustain agricultural output? The world must give new energy sources immediate and consistent attention through the application of monetary and intellectual capital, because humans' future as a species is at stake.

—Stephen J. Stadler

Further Reading

Bagotskiĭ, Vladimir S. *Fuel Cells: Problems and Solutions*. Hoboken, NJ: John Wiley and Sons, 2009.

Hazeltine, Barrett, and Christopher Bull. *Appropriate Technology: Tools, Choices, and Implications*. San Diego, CA: Academic Press, 1999.

Hirsch, Robert L., Roger H. Bezdek, and Robert M. Wendling. *Peaking of World Oil Production: Impacts, Mitigation, and Risk Management*. New York: Novinka Books, 2007.

International Atomic Energy Agency. *Energy, Electricity, and Nuclear Power Estimates for the Period up to 2030*. Vienna: International Atomic Energy Agency, 2008.

Leone, Daniel A., ed. *Is the World Heading Towards an Energy Crisis?* Farmington Hills, MI: Greenhaven Press, 2006.

Lior, Noam. "Energy Resources and Use: The Present Situation and Possible Paths to the Future." *Energy* 33 (2008): 842–857.

Muller-Kraenner, Sascha. *Energy Security: Re-measuring the World.* London: Earth Scan, 2008.

Murray, Raymond LeRoy. *Nuclear Energy: An Introduction to the Concepts, Systems, and Applications of Nuclear Energy*, 6th ed. Burlington, VT: Butterworth-Heinenmann/Elsevier, 2009.

Sternberg, R. "Hydropower: Dimensions of Social and Environmental Coexistence." *Renewable and Sustainable Energy Reviews* 12, no. 6 (2008): 1588–1621.

Smil, Vaclav. *Energy in Nature and Society: General Energetics of Complex Systems*. Cambridge, MA: MIT Press, 2008.

Tertzakian, Peter, and Keith Hollihan. *The End of Energy Obesity: Breaking Today's Energy Addiction for a Prosperous and Secure Tomorrow.* Hoboken, NJ: John Wiley and Sons, 2009.

North American Agroclimatic Regions, 2000–2100 CE

Agriculture is the most important of all primary economic activities; it also is the most dependent upon weather and climate. In the same way that weather and climate help to determine climax vegetation formation, they also set limits for crop and animal production. Agricultural regions and crop–livestock combinations have diffused and shifted through time with climatic oscillations. To ensure a relatively stable food base for humankind in 2100 CE, an understanding by world decision makers of climatic parameters for agriculture and food production in a period of rapid change is critical. Agriculture in general and food production specifically are affected not only by flooding and drought, heat waves and cold waves, and frost and hail, but also by episodic surges of plant and animal diseases.

Agroclimatology

The integrated discipline of agroclimatology developed out of a recognized need in the mid-20th century. It is concerned with climatology in all its scales (macro, meso, and micro) and agriculture in the broad sense (from trees to plants, and from water buffalo to Cornish hens). Agroclimatologists utilize weather and climate data to assist farmers or herdsmen in improving their agricultural practices, increasing productivity, improving the quality of produced food items or technical crops, and adjusting to or tempering natural hazards. They are involved in, conduct research, and provide advice or assistance to agronomists concerned about the following issues: (1) land use, crop selection, and planning (selecting suitable and economically profitable crops and farming systems for a given region); (2) agricultural practices (e.g., dates of planting and harvesting and when to irrigate); (3) crop–weather relationships (plant development—from germination to harvest); and (4) protective measures (against frost, flood, drought, wind damage, and wind-borne insects). Agroclimatologists assist all who grow food on a broad scale by (1) providing long-range agricultural forecasts of significant weather possibilities, (2) conducting regional surveys and defining agroclimatic regions best suited for particular crops or crop combinations, (3) taking micro-climatic and micro-meteorological measurements of plant elements and controls, and (4) studying the response of plant growth to climate elements and controls and the effect of not receiving the necessary physical components on yields, reproduction, and health.

Climate as a Determinant to Crop Distribution

All crops and livestock have optimum and maximum temperature limits for each stage of growth and reproduction. High temperatures are not as destructive as low temperatures if moisture is available; low temperatures and frosts can kill plants. Night-time temperatures are important for the growth processes of many plants. For example, cool night-time temperatures are a requirement for potatoes and sugar beets, whereas corn requires warm nights for maximum development. Every plant requires a set number of heat degree-days above its threshold temperature to reach maturity or produce a crop. For some crops, the length of day is critical to flowering. Soybeans, for example, will flower and reproduce only if the days are shortened to a critical length. Sweet potatoes do best with a natural day length of 11 to 13 hours. Onions require a long day to set bulbs. In the absence of irrigation, rainfall determines crop distribution, dates of planting, and harvesting times. For upland rice, a monthly rainfall of at least 4 inches is necessary. Some crops cannot tolerate excessive rainfall, whereas others are very susceptible to wind damage.

Weather-related pests and diseases also set limits on crop distributions, planting dates, and harvest times. Changes in humidity, wind movement, or formation of dew can spread disease or insect pests and can cause yield-reducing injury to plants. Potato blight, Colorado potato beetle, desert locusts, and Japanese beetle are examples of the close relationship that exists between climate, disease, and insect pests. Every climatic region has set moisture and temperature limits, and every climatic region has natural hazards, such as frost, hail, drought, floods, and strong winds, that can wreak havoc despite a food producer's best efforts.

Water, in all its forms, plays a fundamental role in the growth and yield of all food crops and livestock. Heavy rains and hail can damage food plants, and flooding can kill plants and animals alike. Excessive rain and high humidity can stimulate destructive insect and disease numbers and lower food quality. Droughts can result in plants wilting and dying, and livestock can perish from lack of water. Every food crop and every animal have specific water needs, and these needs are increased or decreased according to the temperature and the food source's stage of development.

The main processes in food plant survival are photosynthesis and photoperiodism. In photosynthesis, the visible rays of the sun are most effective. Photoperiodism—the response to hours of sunlight that enables plants to react to seasonal change—optimizes plant growth and yields. For instance, plants of the tropics and subtropics are generally short-day types, whereas plants of the mid- and high latitudes generally are long-day types and do not do well if the length of the day is reduced.

The Climatic Base for Food Production in North America

The North American continent contains an all-encompassing variety of climatic types. Every major climatic type found on the surface of the Earth is present somewhere in North America. This is an important factor in the ability of agriculturalists to produce so much high-quality and inexpensive food and such a variety of foodstuffs. Table 16 provides insight into the origins of food plants and the locations where these food plants were first found, domesticated, and became a part of a distinct regional diet.

Tropical rainforest (Af) climates (3 percent of total area) are found in the Caribbean coastal plains of Central America, from Panama to southern Mexico, the eastern half of Hispaniola, Jamaica, portions of Puerto Rico, and the Lesser Antilles. Rainfall in these locations averages between 50 and 140 inches per year. The tropical savannah (Aw) climate (4 percent of total area), with moderately heavy rainfall and a well-defined dry season, is found on the Pacific coast, from central Panama northward to 22° N latitude, on the coast of Mexico, Cuba, southwestern Hispaniola, most of the Yucatan, and the southern tip of Florida. Mean annual precipitation in the Aw climate is similar to that in the Af climate, but the Aw climate has a distinct, very dry winter season.

The BS and BW climates of North America extend from 21° N to 54° N in western Canada, and from the 100° meridian westward to the Pacific coast in Mexico and to the Sierra Nevada and Cascades in the United States and Canada. The BW climates (4 percent of total area) extend from northern Mexico to the rain

Table 16. Climatic Regions of Food Staples Origin (World)

Climatic Symbol	Descriptive Term	Food Plant
Af/Am	Tropical rainforest/ monsoon	Rice, yams, bananas, plantains, taro, sugar cane, cocoa, sweet potatoes, manioc (cassava), squash
Aw	Tropical savannah	Millet, sorghum, cucumbers, citrus fruits, corn, pineapple, yams, coconut, peanuts, pumpkins
BS	Steppe	Millet, sorghum, peas, beans, sunflowers, chickpeas, apples, cherries, garlic, carrots, coffee, melons, watermelon
Cs	Mediterranean	Wheat, barley, oats, pears, grapes, cherries, lentils, figs, walnuts, almonds, olives, lettuce, peas
Ca	Humid subtropical	Rice, yams, citrus fruits, mulberries, tea
Cb	West Coast marine	Rye, sugar beets, oats
Da	Humid continental (warm summers)	Buckwheat, soybeans, legumes, millet
Db	Humid continental (cool summers)	Cabbage, onions, wild rice, maple sugar, cranberries
H	Highland	Wheat, barley, buckwheat, millet, corn, potatoes, lima beans, peanuts

Note: Many food plants have multiple regions of origin or are of different common food plant species.
Source: William A. Dando.

shadow area of extreme southwestern Canada. Summer days are extremely hot and dry in the hot deserts (BWh) of Baja California, northwestern Mexico, southeastern California, southwestern Arizona, and parts of northern California. Average annual rainfall is as low as 3 inches and as high as 20 inches or more. The cold mid-latitude desert climates (BWk) of the Great Basin and Columbian Plateau physiographic regions of the United States and Canada have distinct continental temperature characteristics. Precipitation in the North American cold desert climates totals anywhere from a low of about 4 inches to a high of 10 inches.

The humid subtropical climates of North America are found on the east coast of the United States (Caf, or humid subtropical with warm summers), on the southwest coast of the United States (Cs, or Mediterranean/dry summer subtropical), and on the northwest coast of the United States and western Canada (Cb, or marine West Coast). The warm, moist Caf climate of the United States (approximately 9 percent of total area) extends from the 100° meridian in central Texas, eastern Oklahoma, and Kansas, and eastward to the Atlantic coast. Average annual rainfall ranges from 40 to 60 inches. A large portion of California has a climate characterized by warm to hot summers, mild winters with temperatures seldom falling below 32°F, absence of rain in summer months, and moderate rain in winter. It is classified as the Cs, or Mediterranean/dry summer subtropical, climate (about 1 percent of North

America). A very distinctive climatic type, much of the Cs climate receives less than 20 inches of precipitation per year.

North of the Mediterranean climatic region of California, a cool and humid marine climate of the Pacific Ocean littoral stretches to the southern coastal area of eastern Alaska. The mild, wet winter and cool summer of the Cb, or marine West Coast climatic, region (nearly 2 percent of the continent) extend from about 38° to 60° N latitude, in a narrow coastal belt (only 40 to 100 miles wide). Winter average monthly temperatures in this region exceed 32°F. Average annual precipitation ranges from about 40 inches to more than 200 inches.

The Daf, or humid continental with warm summers, climatic region (6 percent of total area) of the northeastern United States is noted for its cold winters and warm summers. Average annual precipitation ranges from 20 to 60 inches. Summers are hot, with temperatures in July averaging 76°F to 82°F or more, and the January average is between 20°F and 30°F over most of the region. The Dbf, or humid continental with cool summers, climate (6 percent of North America) extends from the BSk climatic boundary in the Dakotas and western Canada to the Atlantic Coast of New England and Nova Scotia. Winters are generally snowy and severe, and weather here is characterized by rapid changes in temperature and precipitation. Temperatures can drop to –40°F or less in winter and 100°F or more in summer. Average annual rainfall ranges from less than 18 inches to in excess of 35 inches.

The Dc, or subarctic taiga climatic, region and the Dd, or subpolar willow-aspen, climatic region are cold and dry; at least one month out of the year has a mean temperature of 50°F or higher. The northern boundary of these climatic regions is the northern limit of forests. This region covers more than 32 percent of the North American continent. There is great contrast in summer and winter insulation, from almost continuous daylight in June and July to the long dark nights of winter. As a result of daylight/night duration differences, the Dc and Dd climates have the largest annual temperature ranges in the world. Precipitation ranges mostly from between 10 and 20 inches.

The ET, or tundra, climate (11 percent of the total area) includes all of North America north of the 50° isotherm for the warmest month. Average annual monthly temperatures range from –45°F in February to 49°F in July. Precipitation over the greater part of the ET climatic region ranges between 10 and 20 inches. The H, or highland, climate (about 8 percent) of North America reflects the significance of altitude and exposure in causing differences in climates of similar latitudes. There is no single highland type of climate; rather, an almost endless variety of climates, subclimates, and microclimates exist within a mountain complex.

Table 17 (a modified Köppen classification system base) summarizes actual and conjectured changes in these climatic regions in North America.

Table 17. Current and Postulated Climatic Areas of North America, 2000 and 2100 CE (Modified Köppen System)

Climatic Type	2000 Actual	2100 Conjectured	Percent Change
Af (tropical rainforest)	3%	1%	–2%
Aw (tropical savannah)	4%	9%	5%
BW (desert)	4%	8%	4%
BS (steppe)	14%	12%	–2%
Cf (mild, moist)	9%	10%	1%
Cs (Mediterranean)	1%	1%	0%
Cb (mild, wet winter; cool summer)	2%	1%	–1%
Daf (humid, cold winter; warm summer)	6%	7%	1%
Dbf (humid, cold winter; cool summer)	6%	13%	7%
Dc, Dd (subarctic)	32%	26%	–6%
ET (tundra)	11%	8%	–3%
H (highland)	8%	4%	–4%
Total	100%	100%	

Potential Impact of Climatic Change

Attempting to understand the potential effects of climate change and weather variations is complicated by two factors: (1) the great diversity of existing climatic regions and subregions and (2) the critical uncertainties in long-term weather/climate predictions. Nevertheless, the following postulations can be made regarding the impact of climate change on agriculture in 2100 CE:

- Mean temperature will increase by a minimum of 3°F to 6°F.
- Minimum temperatures will increase dramatically.
- Current northern subarctic (ET, Dc, and Dd climates), tundra, and taiga wetland ecosystems will be reduced in size by as much as 60 percent.
- Forested areas in northern Canada will increase in size.
- Permafrost thawing in Alaska and Canada will shift the southern limit of permanently frozen ground, at minimum, 350 miles.
- Arid lands (BW desert and BS steppe climates) will expand in size.
- Water resources available for agriculture in the southern Great Plains will be reduced.
- Violent rainstorms and tornadoes will be more frequent.
- Droughts in all marginal climates will be more frequent.
- Ice in the Arctic Ocean will be reduced by 85 percent.

Global Warming and Future Food Production in North America

Currently, the world has the technological capabilities to deal with many future agricultural contingencies related to climate change and regional food crop losses. Efficiency, production techniques, and distribution of world foods have improved tremendously in the past 50 years. Nevertheless, to maintain a famine-proof world food balance, the world needs the high-quality and inexpensive surplus foods produced in North America. Millions of world citizens depend on the United States and Canada as sources of commercial food commodities in normal times and to provide food aid in times of food shortages or famine.

Fluctuations in weather and climate—from wet to dry, cold to warm, and windy to calm—add uncertainty or risk to world food supplies and have the greatest impact when they occur in food-surplus, food-exporting nations. North America's agroclimatic regions are shifting, and these changes in climates will inevitably affect food production. In the 21st century, average temperatures are projected to rise, at a minimum, 3°F to 6°F. The higher temperatures could cause intense lingering droughts, inducing rural crop failures and urban heatstrokes. Warmer and wetter weather will amplify insect-borne diseases, while arid lands will expand. Conflicts between states and provinces over scarce water may erupt, as warmer ocean waters deplete the bounty of fish in offshore fishing banks. Wildfires will be very frequent. Sea level will rise, and ocean saltwater incursions into aquifers will contaminate drinking water sources. No individual in North America can escape the effects of climate change in the 21st century.

—William A. Dando

- Precipitation in the A, C, and D climates (tropical savannah, humid subtropical, and humid continental regions) will increase by at least 3 percent.
- All crop patterns will shift northward and higher in mountain areas.
- The Great Lakes' water level will decline.
- Sea level will rise, at minimum, 15 inches.
- Increased carbon dioxide in the atmosphere will elevate wheat and soybean yields by approximately 30 percent.

Regionalizing North American Agriculture

In 2000, eight major classes of farming areas were distinguished in North America, which are commonly referred to as major agricultural regions. The proportion of farm income in a region that is derived from a given commodity or crop–livestock combination determines, to a large extent, the type into which a region is classified. These broad regions are shown on the map of major agricultural regions of North

Figure 15. Mapping agricultural regions of North America

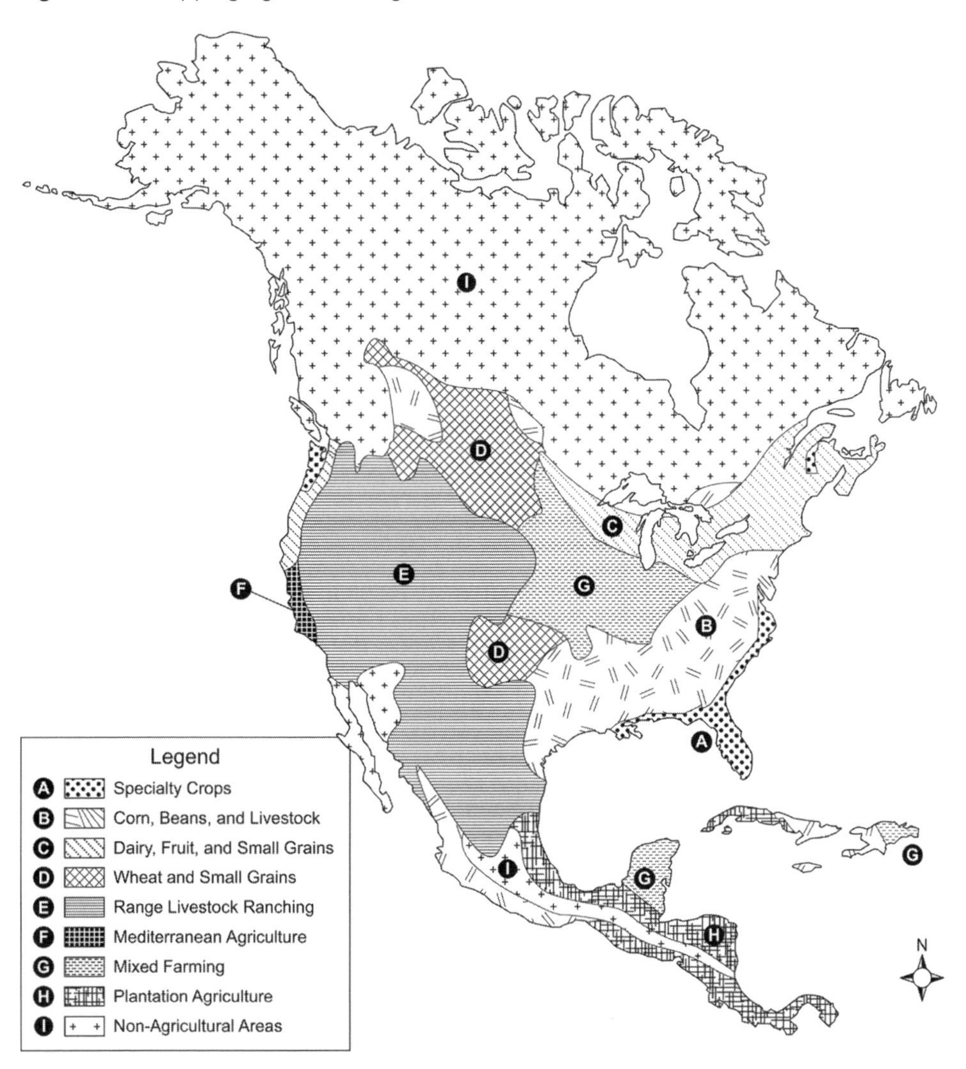

Source: William A. Dando.

America in 2100 CE (see Figure 15). Although major agricultural regions, when delimited, seem to end abruptly at their borders, the outer fringes are transitional zones that shade into contiguous regions. Each major type of farming region has a core in which the climate, landform, and location combine such that agriculturalists there employ a crop–livestock specific technology, factor into a region-specific economy, and evolve a distinctive social life. The common economic activities in which rural dwellers engage produce broadly similar interests, attitudes, values, and life patterns. The core area of each major region is, in general, homogenous in use of technology, economic structures, and social attitudes.

In 1960, the U.S. Department of Agriculture developed eight labels for farming regions in North America: (1) Corn Belt; (2) Cotton Belt; (3) Dairy; (4) Wheat; (5) Range-Livestock; (6) General and Self-Sufficing; (7) Western Specialty Crops; and (8) All Other Areas. The terms used on the 1960-era map became the standard terms used on most maps of agriculture in North America. In the past 50 years, however, agriculturalists in North America have become more knowledgeable, more efficient, and more cognizant of the global market for their products. The U.S. Department of Agriculture (USDA) maps of 2000 reflect the evolution of American agriculture and the impact of agro-economics on the rural landscape. In 2100, the map of agricultural regions will reflect the impact of climate change and increased world demand for food.

Conjectured Map of Major Agricultural Regions of North America, 2100

North America, the "food basket" of the world, will, in 2100, experience dramatic shifts in agricultural regions and crop–livestock combinations in those regions. A synthesis of USDA maps and maps produced by geographers who specialize in agriculture, plus intergovernmental agencies' postulations on climate change, serve as the basis of the conjectured map of major agricultural regions of North America in 2100 (see Figure 16):

- An enlarged "Specialty Crops" agricultural region, extending from eastern Mexico along the Gulf Coast and north to Connecticut, will produce not only fruits and vegetables but also house plants, ornamentals, and new tropical food items.
- An expanded "Corn, Beans, and Livestock" agricultural region in the American Southeast will increase its yields and numbers of livestock produced as a result of more rainfall and higher plant growth stimulating carbon dioxide amounts in the atmosphere.
- A slightly reduced "Dairy, Fruit, and Small Grains" agricultural region will shift northward and eastward; the Canadian portion of this region will experience the greatest increase in size and agricultural productivity.
- An enlarged "Wheat and Small Grains" region will shift westward and northward into north-central Canada. Both wheat and soybean crop yields will increase, though the number of droughts, tornadoes, and blizzards will also increase.

Figure 16. Conjectured map of major agricultural regions of North America

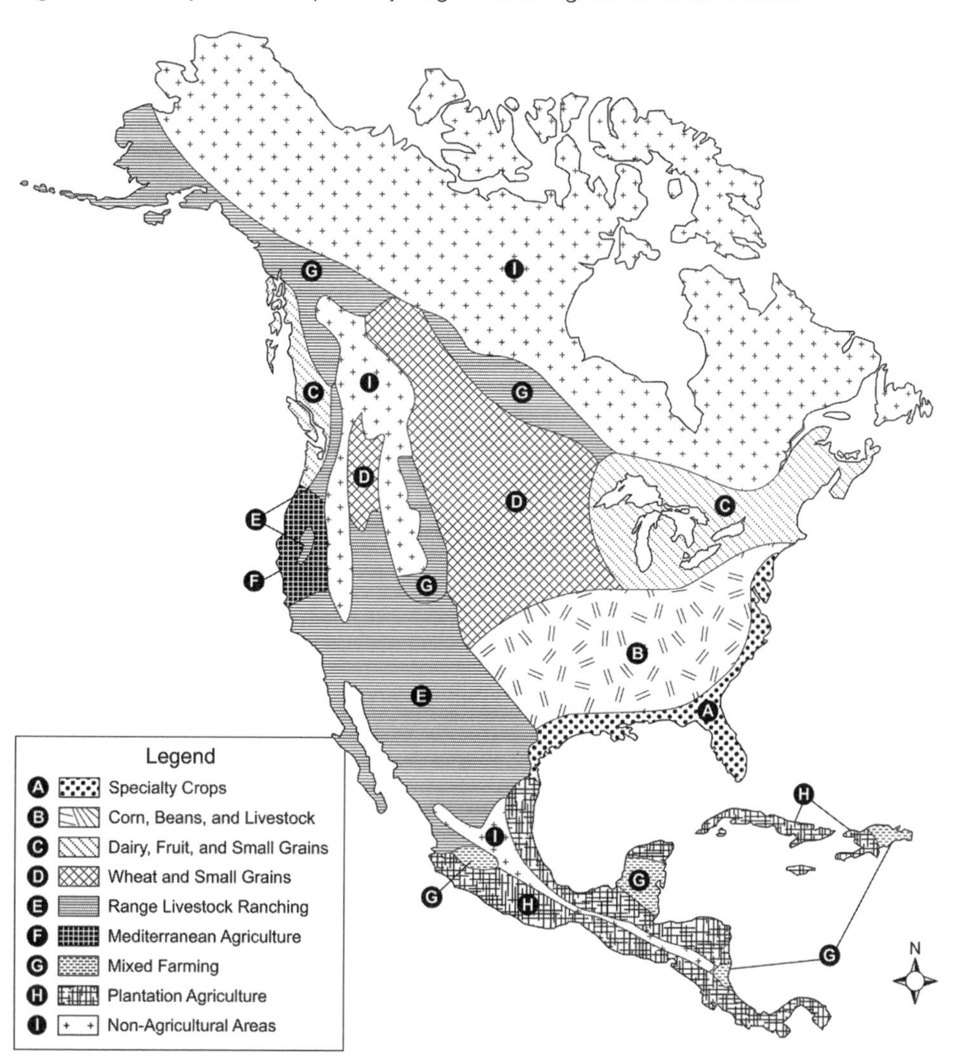

Source: William A. Dando.

- An expanded "Range Livestock Ranching" region and "Non-Agricultural Areas" will become drier and hotter in the year 2100. The critical problem will be a shortage of water. Also, massive dust storms and violent infrequent rains will negatively affect both livestock and grasslands.
- A reduced area of "Mediterranean Agriculture" will experience a reduction of winter precipitation, an increase in night-time temperatures, and higher day-time temperatures. Irrigated agriculture will be restricted to very high-yielding

fruits and vegetables. Water will be a major factor in agricultural yields, quality of product, and productivity.

- A general reduction will occur in the area of "Mixed Farming" in the United States and Middle America. The greatest increase in mixed farming will take place in the northern limits of agricultural production in Canada and Alaska.
- An enlarged "Plantation Agriculture" region will affect subsistence agriculture in all forms in areas from mid-Mexico to Panama. Climate change will force a change in crops produced and careful use of a reduced amount of precipitation received each year. Irrigation of crops that offer high value per acre will offer new opportunities for some, but force many independent farmers to become day laborers on foreign-owned plantations.

A New Era for American Agriculture

Change in world climates and world weather patterns will affect not only North American food production, but also North American food trade patterns. Currently, North America is blessed with an abundant supply of high-quality, inexpensive food. In many ways, North American agriculture has set the standards for the world. There has been little increase in the demand for basic foods in North America. Instead, the greatest increase in food demands in the past 50 years has occurred in non-industrial or developing countries. In 2000, rapid population growth was a major factor in the increased demands for food. In 2100, the demands for more foodstuffs will double those seen in 2000. The greatest negative impact of climate change in 2100 will occur in the countries of the world with the least resources to develop new farmland or make existing farmland more productive. The impact of conjectured climate change on North American agriculture, in general, is small and may be positive. Shifts will and are being made in the location of major agricultural regions. Local impacts may be great, but continent-wide impacts will not diminish North America's status as the "food basket" of the world in 2100. In fact, climate change and variation, combined with increased world population and affluence, will usher in a new era in North American agriculture—an era of innovation, adjustment, and prosperity.

—William A. Dando and Bharath Ganesh-Babu

Further Reading

Chang, J. "Agroclimatology." In *The Encyclopedia of Climatology*, edited by J. Oliver and R. Fairbridge. New York: Van Nostrand Reinhold, 1987, 16–22.

Corcoran, W. T., and E. Johnson. "Climate of North America." In *The Encyclopedia of World Climatology*, edited by J. Oliver. Dordrecht, Netherlands: Springer, 2004, 525–534.

Dando, William A., and Bharath Ganesh-Babu. "Climates of the Future: Case Studies of the Middle East and North America." In *Climate Change and Variation: A Primer for Teachers*, edited by William A. Dando. Washington, DC: National Council for Geographic Education, 2007, 31–48.

Dando, William A., and Bharath Ganesh-Babu. "Mapping the Shifting Agroclimatic Parameters and Agricultural Regions of North America: A.D. 2100." Paper presented at the Annual Meeting of the National Council for Geographic Education, Detroit, MI, October 9, 2008.

Nordhaus, W. D. "Climate Change's Global Warming Economics." *Science* 294, no. 5545 (2001): 1283–1284.

Nuttonson, M. "Ecological Crop Geography of the Ukraine and Agroclimatic Analogues in North America." *Institute of Crop Ecology* 1 (1947): 1–24.

Watson, R. T., et al. *The Regional Impacts of Climate Change: An Assessment in Vulnerability*. Cambridge, UK: Cambridge University Press, 1998.

Nutrition

Consumed foods and beverages provide the nutrients needed for the biochemical activities of the body. Each body cell is a miniature biochemical factory that uses basic components of food, known as nutrients, to create energy, heat, and water needed by the body. When needed nutrients are missing, the body temporally uses a backup system to compensate for variations in the diet. If a deficiency continues for a long period of time, the system no longer functions as designed and undesirable consequences occur. Depending on when this happens during the life cycle, the changes can be either minor or catastrophic.

Importance of Nutrition

Human biological systems are complex and varied in nature. Genetic background, previous diet environments, stress, and other factors influence the degree of response to diet. Nutrition is important to all. Food and the nutrients in foods power the human body through periods of growth, maintenance, and repair. Eating a variety of foods ensures that essential nutrients are consumed.

During periods of limited food, nutrient imbalances occur. The most vulnerable population groups are pregnant women, fetuses, infants, and children. A fetus grows extremely quickly. If needed nutrients are not present during this critical growth period, the body structures might not be properly developed. For example, the closure of fetal neural tubes requires folate, a B-vitamin. If this nutrient is not present, the neural tubes do not close, and a cleft palate or spinal bifida occurs. A pregnant woman with iodine deficiency may miscarry or the infant may be born with diminished brain development, mental retardation, learning disabilities, and problems in movement, speech, or hearing.

Chronic hunger and the nutritional imbalances that accompany it lay the foundation for chronic diseases, birth defects, impaired mental function, and an overall reduced quality of life. The impact on the community at large is a diminished healthy workforce, which in turn affects the economic resources of the affected community.

Nutrients in Food

The primary need of the body is for energy. Macronutrients provide energy, contribute to cell and body structure, and regulate body processes. The amount of energy in food is measured in kilocalories (kcal). One kilocalorie is the amount of energy required to raise the temperature of one liter of water by one degree Centigrade at sea level.

The nutrients in foods are classified as either macronutrients or micronutrients. Macronutrients are needed in large quantities, whereas micronutrients are needed in small amounts. Macronutrients include carbohydrates, proteins, lipids (fats), and water. Micronutrients include vitamins and minerals.

Macronutrients

Carbohydrates

Carbohydrates function as an energy (kilocalorie) source, regulate metabolic processes, and contribute to cell and body structure. Food sources of this macronutrient include grains, vegetables, fruits, and dairy products. The carbohydrates in foods take the form of sugars, starches, and fiber. Sugars are simple carbohydrates that require little to no further breakdown before they can be absorbed by the body. They are found in fruits, vegetables, honey, syrups, cornstarch, table sugar, and milk. Starches and fiber are complex carbohydrates. Starches are absorbed by the body after digestion. Fiber cannot be digested, so it cannot be absorbed. Fiber is needed to prevent constipation and to lower blood cholesterol and blood sugar.

The digestion of carbohydrates ultimately liberates glucose, the simplest form of carbohydrate that serves as an essential body fuel, especially for the brain.

If carbohydrate sources are missing or reduced in the diet, the body will use energy supplies within the body itself. Glycogen (stored carbohydrate) from the liver, protein from muscle, fatty acids from adipose tissue, and structural protein all serve as energy sources in such a case. If a person does not eat, then the body gets its energy supply from itself.

Protein

Protein is an important component of body structure. It also provides energy and regulates body processes. Proteins are composed of essential and non-essential amino acids. Essential amino acids come from food sources, as the body cannot manufacture them. Non-essential amino acids can be produced by the body. Making body proteins from amino acids is an expensive, energy-requiring process, however. Tissue synthesis takes place when adequate energy and protein are available. The energy level of a diet is a critical factor for protein synthesis, especially for pregnant and nursing women, infants, and children.

Foods with a high biologic protein content contain the essential amino acids; these foods include eggs, meat, milk, and soy. Legumes, vegetables, and grains are incomplete protein foods; they can be combined with one another or with animal meats to add in the missing essential amino acids to make a complete protein.

Lipids

Lipids, also known as fats, provide energy and body structure; they also regulate body processes. Lipids are composed of fatty acids. Some of these fatty acids are considered essential, whereas others are non-essential. Essential fatty acids must be provided in the diet, as the body does not synthesize them. Essential fatty acids, such as linoleic acid (omega-6 family) and alpha-linolenic acid (omega-3 family), are used to synthesize eicosanoids. Eicosanoids are involved in immunity, inflammation, and central nervous system communication messengers. Excessive carbohydrates, beyond what the body can easily metabolize, are stored as body fat. Lipids are found in oils, shortening, beef tallow, lard, butter, cream, olives, and nuts.

Water

Water is the most critical of all nutrients. Individuals will die sooner from water deprivation than from starvation. The human body contains more water than any other compound. Approximately 70 percent of the fat-free body is water. Water is an essential component of all cell structures. It is a medium in which the chemical reactions of cellular metabolism take place. Water moves nutrients around the body, participates in biochemical reactions, influences cognition, and maintains body temperature. Body water is in balance when the water consumed is equivalent to the water excreted. The daily estimate for water needed is one milliliter of water per kilocalorie intake. Excessive water loss, such as occurs with diarrhea or prolonged periods of vomiting, results in dehydration and loss of electrolyte minerals from the body. Conversely, water intoxication occurs when the body water is replenished without electrolytes. Daily total water needs are available from ordinary intakes of food, drinking water, and water produced by the body's biochemical reactions (metabolic water).

Micronutrients

Vitamins

Vitamins are classified into fat-soluble and water-soluble vitamins. Fat-soluble vitamins include vitamins A, D, E, and K; these vitamins are absorbed through the gastrointestinal tract in the company of dietary fats. Unused fat-soluble vitamins are stored by the body. Both vitamins A and D function like hormones. Vitamin A regulates cell and tissue growth and differentiation, whereas vitamin D regulates calcium metabolism. Vitamin E functions as an antioxidant that protects the body cells from the damaging effects of oxidation. Antioxidants are essential for the proper functioning of the immune system.

Water-soluble vitamins dissolve easily in water, so that excess amounts are excreted in the urine. Because they are not stored in the body, these vitamins must be provided as part of the daily diet. Water-soluble vitamins include the B-complex vitamins (thiamine, niacin, riboflavin, pantothenic acid, biotin, vitamin B_6 folate, vitamin B_{12}) and vitamin C. The B-complex vitamins function as coenzymes in metabolic processes. Vitamin C functions as an antioxidant. Vitamin K and biotin are synthesized by gut bacteria. The body can synthesize some vitamins from consumed precursors. For example, vitamin A can be created from beta carotene and niacin; niacin can be derived from tryptophan, an amino acid.

Some vitamin-like compounds are needed and produced by the body. Ubiquinone (coenzyme Q_{10} [CoQ_{10}]) is necessary for the basic functioning of cells. Lipoic acid is an antioxidant. Choline is usually grouped with the B-complex vitamins; it is a constituent of acetylcholine, a hormone necessary for the transmission of nerve impulses. Inositol also helps transmit nerve impulses.

Minerals

Minerals are essential for the body to stay healthy. They have various functions, including building bones, making hormones, and regulating the heartbeat. They are cofactors for metabolic reactions.

Minerals are classified into macrominerals and microminerals (also known as trace minerals). Macrominerals are needed by the body in larger quantities, whereas microminerals are needed in small amounts. Macrominerals include calcium, phosphorus, potassium, sodium, chloride, magnesium, and sulfur. Iron, zinc, selenium, iodine, copper, manganese, fluoride, molybdenum, boron, nickel, chromium, and vanadium are trace minerals.

Among the essential macrominerals, calcium plays a role in blood clotting, nerve impulse transmission, muscle contraction, and bone formation. Sodium is important in fluid balance and nerve impulse transmission. Potassium is needed for muscle contraction, nerve impulse transmission, and fluid balance. Phosphorous functions in bone formation and is necessary for energy production; it is also a component of deoxyribonucleic acid (DNA) and ribonucleic acid (RNA) proteins. Magnesium plays a role in the metabolic process and nerve conduction. Also included in the microminerals category is iron—a component of hemoglobin and myoglobin, which are oxygen carriers for the body. Zinc, which serves as a cofactor in more than 400 metabolic reactions, is involved in gene regulation and the immune system. Selenium is an antioxidant. Iodine is needed for thyroid hormone production. Copper helps release energy and plays a role in melanin, collagen, elastin, and red blood cell production. Manganese is a factor in cartilage production and in antioxidant enzyme systems. Fluoride is part of bone and tooth structure. Molybdenum functions in protein synthesis, metabolism, and growth. The physiological roles of nickel and boron, which are considered ultra-trace minerals, remain under study. Chromium functions in glucose metabolism, while vanadium may increase insulin sensitivity.

Consequences of Nutritional Inadequacies

The effects of nutritional inadequacies can range from subtle to profound, depending on when the diet insufficiencies occur during the human life cycle. More than 1.2 billion people in the world do not have enough to eat. This number is larger than the combined population of the United States, Canada, and the European Union. In the United States, 5 percent of newborns have intrauterine growth retardation due to malnutrition during the mother's pregnancy.

The impact of famine on human pregnancies and those individuals' surviving children during World War II is still being felt today in the Netherlands. During the winter of 1944–1945, a food embargo was imposed on the Netherlands by occupying German authorities. A famine in Amsterdam began in mid-September 1944 and ended on April 29, 1945, with the first Allied food drop. An estimated 100,000 Amsterdamers died of disease and starvation (see the "Dutch Famine: 1944–1945" entry in Volume 2). Public heath care records were maintained during

Nutrition Axioms

An axiom is a statement taken to be true without proof. As a self-evident statement, it can become accepted as a well-established rule or law. In ancient times, every scientific, medical, or mathematical question or research project began from an assumption or tested common notion. Early mathematicians began their research by defining "postulates" and "common notions" or as-required assumptions. In modern times, the terms "postulate" and "axiom" are synonymous. To many, an axiom is akin to a "hypothesis." Many nutrition axioms exist, including the following:

"Let food be thy medicine, thy medicine shall be thy food."—Hippocrates

"Leave your drugs in the chemist's pot if you can heal the patient with food."—Hippocrates

"When diet is wrong, medicine is of no use. When diet is correct, medicine is of no need."—Ancient Ayurvedic Proverb

"A man too busy to take care of his health is like a mechanic too busy to take care of his tools."—Spanish Proverb

"He who takes medicine and neglects diet wastes the skills of the physician."—Chinese Proverb

"Let nothing which can be treated by diet be treated by other means."—Maimonides

"The doctor of the future will no longer treat the human frame with drugs, but rather will cure and prevent disease with nutrition."—Thomas Edison

"The first wealth is health."—Emerson

"To eat is a necessity, but to eat intelligently is an art."—La Rochefoucauld

"The belly is the giver of genius."—Persius

"A hungry stomach rarely despises common food."—Horace

"He who has health has hope, and he who has hope has everything."—Arabian Proverb

—Sara A. Blackburn and Kausar F. Siddiqi

this period, and birth weights and the amount of food available were documented. From these records, it is possible to determine the level of malnutrition during the stages of pregnancy in Dutch women during this famine.

The Dutch famine records suggest that individuals who were exposed to famine during the first trimester of pregnancy had mutations in their DNA sequence, affecting gene expression. Genes carry the blueprints to make proteins in the cell. The DNA sequence of a gene is transcribed into RNA, which is then translated into the sequence of a protein. Methyl groups are needed for production of protein. The bulk of human DNA is methylated and function is impaired when methylation is reduced. The essential amino acid, methionine, is needed for methylation; animal proteins represent food sources of methionine. DNA combines with histone (a protein) to form chromatin and organize into chromosomes that carry the genes. Persons exposed to famine in the early part of gestation had lower methylation rates than individuals exposed in the late gestation period. An inadequate diet during pregnancy is thought to cause a disruption in gene expression by altering methylation of DNA and modifying histone. Individuals exposed to malnutrition in utero had higher rates of chronic diseases, metabolic disorders, insulin resistance, vascular disease, obstructive airway disease, morbidity, and mortality in later life.

Retrospective research suggests greater complications in obstetrical outcomes occur during times of famine, war, and poor environmental conditions. Maternal nutrient and energy insufficiencies during the first trimester of pregnancy increase the risk of hypertension, obesity, and changes in lipid profile. Nutrient insufficiency during the second trimester is associated with an increased risk of glucose intolerance.

Critical growth periods in human brain development occur when the brain tissues and cells are rapidly growing and changing. Alterations in the availability of nutrients during these growth periods appear to have long-lasting effects on neuroanatomy, neurochemistry, neurotransmitters, and neurometabolism. Evidence suggests that genes are altered, affecting protein synthesis in the brain. These effects may account for reduced brain size and overall mental function. Aggressive nutritional intervention at critical growth periods (premature infant, postnatal period, and weaning) may reduce the severity of the consequences of undernutrition.

The type of malnutrition is related directly to the nutrients missing in the diet. While some individuals may have adequate energy, the needed levels of protein, fat, and other nutrients are severely restricted. Nutrient deficiencies observed in malnourished infants and children include vitamin A, iron, zinc, folate, and iodine. These nutrients are known to collaborate with each other. Thus vitamin A deficiency alters iron status, leading to an iron overload in tissues. Many children with a chronic infectious disease, such as measles, have iron-deficiency anemia, which may in turn be a nutritional adaptation to the disease. The body sequesters iron so that it is not available to the infectious agent. When iron levels return to a normal level, the infection intensifies, increasing mortality rates. In areas where a person's diet lacks variety, micronutrient deficiencies are most prevalent, and the most vulnerable groups are women and children. Eating a wide variety of foods that contain an array of micronutrients is important to preserve nutritional health, an overall good quality of life, and a sense of well-being (see Tables 18 and 19).

—Sara A. Blackburn and Kausar F. Siddiqi

Table 18. Food Sources and Functions of Major Nutrients

Nutrient	Food Sources	Functions in the Body
Protein	Meat, fish, poultry, eggs, dried beans and peas, milk, yogurt, cheese, tofu	Provides energy (4 kcal/g); growth and maintenance of body tissues; building blocks for enzymes
Carbohydrates	Breads, cereals, pasta, rice, potatoes, corn, green peas, dried beans and peas, yellow squash, fruits, honey, sugars	Provide energy (4 kcal/g); major energy source.
Fats	Butter, oils, margarine, shortening, cream cheese, sour cream, olives, avocado	Provide energy (9 kcal/g); part of every cell in the body; transport fat-soluble vitamins (A, D, E, and K); insulate the body
Water	Beverages, fruits, vegetables	Part of every cell in the body; transports nutrients; removes waste products; maintains body temperature
Calcium	Milk, yogurt, cheese, salmon, and sardines with bones, tofu, spinach, collard greens, figs, almonds	Builds and maintains strong bones and teeth; helps lower blood pressure; reduces risk of certain cancers
Iron	Liver, beef, veal, lamb, pork, oysters, dried beans and peas, spinach, molasses, raisins	Part of the blood cells that carries oxygen throughout the body; resists infection; helps the body convert food to energy
Potassium	Milk, bananas, plantains, oranges, lychee, mango, spinach, kale, prunes, dried beans, potato, tomato, honeydew, cantaloupe	Helps maintain normal blood pressure; balances fluid distribution throughout the body; needed for muscle contraction and relaxation, and nerve transmission; regulates the heartbeat
Vitamin A	Liver, carrots, sweet potatoes, pumpkin, winter squash, bok-choy, spinach, collard greens, kale, broccoli, tomatoes, kumquats, mango, apricots, cantaloupe	Helps with night vision, bone growth, healthy skin, and resistance to infection
Vitamin D	Milk, fatty fish, eggs, liver, butter	Helps with calcium absorption; helps maintain healthy muscles and nervous system; helps the body resist infection
Vitamin E	Vegetable oils, margarine, nuts, seeds, dark green leafy vegetables	Antioxidant; helps the body resist infection
Vitamin K	Dark green leafy vegetables, brussel sprouts, cabbage	Needed for blood clotting
Thiamin (vitamin B_1)	Organ meats, pork, wheat germ, dried beans and peas, nuts, oranges, oysters, green peas, whole grains	Needed for normal functioning of all body cells, especially nerves; necessary for heart and nervous system

Table 18. (Continued)

Nutrient	Food Sources	Functions in the Body
Riboflavin (vitamin B_2)	Milk, yogurt, cottage cheese, liver, beef, veal, lamb, pork	Carbohydrate metabolism and tissue repair; helps with normal vision and prevents light sensitivity in the eyes; helps maintain skin; needed for growth and production of red blood cells; helps in the production and regulation of some hormones
Niacin (vitamin B_3)	Liver, beef, lamb, veal, pork, fish, nuts	Needed for metabolism and absorption of carbohydrates; helps keep a healthy skin, nervous system and gastrointestinal system; needed to form blood cells; helps with production of some hormones
Vitamin C	Oranges, grapefruit, kumquat, kiwi, lychee, pomegranates, cantaloupe, papaya, mango, strawberries, broccoli, brussel sprouts, bell peppers	Antioxidant; helps heal wounds; helps nervous system function normally; helps the body to absorb iron

Source: Adapted from USDA Food Pyramid: http://www.mypyramid.gov/index.html.

Table 19. Major Nutrients in Food

Foods	Major Nutrients
Grains (bread, cereal, rice, and pasta)	Carbohydrates, fiber, several B vitamins (thiamin, riboflavin, niacin, and folate), and minerals (iron, magnesium, and selenium)
Fruits (such as apples, oranges, grapes, kiwi, lychee, papaya, and strawberries)	Carbohydrates, potassium, fiber, vitamin C, and folate (folic acid)
Meat and meat substitutes (such as meat, poultry, fish, dried beans and peas, eggs, tofu, textured vegetable protein, seeds, and nuts)	Protein, essential fatty acids (flax, walnuts), B vitamins (niacin, thiamin, riboflavin, and vitamin B_6), vitamin E (sunflower seeds, almonds, hazelnuts), iron, zinc, and magnesium
Vegetables (such as potatoes, tomatoes, green beans, carrots, spinach, and squash)	Fiber, potassium, vitamin A, folate (folic acid), vitamin C, and vitamin E
Milk (such as milk, yogurt, and cheese)	Protein, calcium, potassium, riboflavin, and vitamin D
Fats and oils (such as oils, shortening, and butter)	Energy (kilocalories) and essential fatty acids
Sugars and sweets (such as cane sugar, beet sugar, honey, and syrups)	Energy (kilocalories)

Source: Adapted from USDA Food Pyramid: http://www.mypyramid.gov/index.html.

Further Reading

Brown, Judith, Janet Isaacs, Bea Krinke, and Ellen Lechtenberg, *Nutrition Through the Life Cycle*, 4th ed. Belmont, CA: Wadsworth, 2010.

Duyff, Roberta. *ADA Complete Food and Nutrition Guide*, 3rd ed. Chicago: American Dietetic Association, 2006.

Heijmans, B. T., E. W. Tobi, A. D. Stein, H. Putter, G. J. Blauw, E. S. Susser, P. E. Slagboom, and L. H. Lumey. "Persistent Epigenetic Differences Associated with Prenatal Exposure to Famine in Humans." *Proceedings of the National Academy of Science of the USA* 105 (2008): 17046–17049.

MyPyramid.com. Accessed August 2010. U.S. Department of Agriculture: www.mypyramid.gov.

Popkin, B. M., K. E. D'Anci, and I. H. Rosenberg. "Water, Hydration, and Health." *Nutrition Reviews* 68 (2010): 439–458.

Roseboom, T., and R. Painter. "The Dutch Famine and Its Long-Term Consequences for Adult Health." *Early Human Development* 82 (2006): 485–491.

"Vitamins and Minerals." Accessed August 2010. www.rd411.com.

Organic Agriculture

Organic agriculture entails production of food and fiber without the use of chemical pesticides and fertilizers. Grown in a manner that enhances the ecological balance of the natural system, organic products have become a booming segment of the world food industry. Even during and after the recent global food crisis and the ongoing global economic crisis, the world's appetite for certified organic produce substantially outpaced the conventional food market. Statistics from the Organic Trade Association show that in the United States, organic food sales grew more than twice as fast as conventional food sales in 2009, quadrupling this segment's market share to nearly 4 percent since 2000. In the nonfood sector, demand was even stronger, with organic sales increasing almost 10 percent in 2009, compared to a small decrease for the conventional nonfood sector.

In describing organic agriculture, this entry refers specifically to certified organic production and consumption. This clarification is important, as many farmers in developing countries are deemed de facto organic, meaning they have not applied chemical pesticides and fertilizers for many years, if ever. These farmers, however, do not receive the organic price premium that is available after farmers traverse the costly organic certification process. Furthermore, because organic agriculture is an active practice, many traditional slash-and-burn and shifting cultivation practices lacking external inputs run counter to its long-term goals of improving the fertility of soils and health of ecosystems. This holistic approach is captured by organic agriculture's four principles of health, ecology, fairness, and care, as envisioned by the International Federation of Organic Agriculture Movements (IFOAM), the global representative body of organic agriculture.

No matter the specific motivations for purchasing organic food, which range from taste and freshness to food safety and social justice issues, the consumption boom has contributed to the evolution of organic agriculture from a niche market to part of the mainstream in the last decade. This shift is most vividly illustrated by a shopping trip to its two most prominent players, located at the extremes of the retail spectrum: Wal-Mart and Whole Foods. Although alternative initiatives to produce and consume food locally and outside the networks of these mainstream retailers, such as community-supported agriculture (CSA; see the entry by Greenwood and Leichenko earlier in this volume) and farmers markets, are gaining in popularity, their share of the overall organics market remains small.

Looking at the global picture, Europeans and Americans are the main consumers of organic products, accounting for more than 95 percent of all consumption. Due to limited local supplies and relatively higher prices, the clustering of global organic consumption has resulted in a worldwide boom in organic export production. Involving millions of smallholder farmers in the poorest regions of the world, this organic revolution holds vast promise as a means to reduce poverty and increase food security in a sustainable and equitable manner.

The Shortcomings of the Green and Biotechnology Revolution

Modern agricultural methods have been at the center of a long-standing interdisciplinary debate surrounding their effectiveness and sustainability. Conventional wisdom holds that the provisioning of improved high-yielding varieties for food crops in Mexico and India dramatically increased food production. For his efforts to increase food supplies and avert mass starvation, Norman Borlaug (the so-called father of the Green Revolution) received the Nobel Peace Prize in 1980 (see the "Green Revolution" entry). What remains contested is the ability to extend the positive results from the combined usage of improved seeds with chemical fertilizer, pesticides, and herbicides into the 21st century. According to its critics, rates of return using high-yielding varieties have stagnated, with exhausted soils needing higher (and costlier) synthetic fertilizer inputs to make up for the long-term loss in soil fertility.

Biotechnology promises to address these main drawbacks of the Green Revolution. For example, by implanting *Bacillus thuringiensis* (Bt) genes in cotton, pesticide usage is reduced because the main predator, the cotton bollworm, dies upon feeding from the cotton plant. This example often serves as the departing point for two opposing views on biotechnology.

Proponents of this approach argue that this innovation offers higher yields for the farmers in conjunction with reduced pesticide expenditures. This latter point is especially illustrative of the ecological potential of biotechnology, given that large amounts of chemical pesticides in the world are used in cotton production. These inputs include such pesticides that contain Bt protein derived from bacterial culture, which have been used as an organic pesticide for more than 50 years.

Opponents declare that this technological solution, once again, fails at improving either the economic situation of the farmer or the key issue of soil fertility in the long run. The main economic critique centers on the proprietary nature of this new technology—that is, the genetically modified (GM) seed, which creates a strong dependency on the seed supplier. This fear has been attenuated in the aftermath of the global food crisis, as GM seed prices (as a percentage of operating costs) have increased to levels well above the prices of their conventional and organic counterparts. From an ecological standpoint, critics assert that the simplification of using herbicide-tolerant or insect-resistant GM seeds reduces farm-level biodiversity, displaces older seed varieties, and, in the case of Bt cotton, has spurred the rise of previously minor pests, potentially necessitating an increase in long-term pesticide usage.

In developing countries, the diffusion of GM seeds has been slowed by inadequate institutional settings (i.e., weak intellectual property rights protections and seed delivery systems) and lack of research spending on adapting varieties to their local environments. Especially in Africa, where the majority of agriculturalists are small-scale farmers, the labor-saving methods introduced by the Green and Biotechnology Revolutions are of lesser importance, compared to the impact these developments have had on their counterparts in developed countries. Such technology-intensive modern agricultural methods are also capital intensive. Depending on access to credit, better-educated, wealthier, and larger farmers have tended to be among the first adopters of these technologies, often at the cost of smaller, less-educated, and poorer farmers. Even in cases where improved seeds are accessible, farmers can struggle to repay the inputs-related debts during unsuccessful cropping years. The increased number of suicides among Indian cotton farmers provides a tragic example of the consequences of credit-based farming during drought years. In sum, even though these improved technologies greatly increased global food production and enabled a shift toward industrialization in the developed world, they have a mixed record in alleviating poverty and reducing inequality among the poorer nations.

Impacts in the Developing World

As reflected in the recent findings from the International Assessment of Agricultural Knowledge, Science and Technology for Development (IAASTD), organized by the World Bank and the Food and Agriculture Organization (FAO) of the United Nations (UN), organic agriculture has moved to center stage in developing countries. The leading edge for this shift is located in sub-Saharan Africa. Accounting for only 1 to 3 percent of all global land under organic production, certified African organic farmers make up almost 20 to 25 percent of all such farmers across the world, according to Willer and Kilcher. Compared to the outcomes of previous agricultural revolutions, the ecological and health benefits of organic production are well known. For example, organic farmers make use of regular manual scouting of the fields and targeted application of botanical pesticides. These methods are not only better for one's health, as they reduce the incidence of pesticide poisoning, but also more environmentally sound, as they do not produce any groundwater pollution. Organic farmers also minimize off-farm inputs, such as use of chemical fertilizers. In developing countries, these inputs are often replaced with locally available fertilizers, such as farmyard manures and composting strategies. Integrated with crop rotation systems and intercropping of nitrogen-fixing leguminous crops, these comparatively labor- and knowledge-intensive organic methods not only result in increased soil fertility and biodiversity, but also reduce negative ecological externalities, such as fertilizer runoff.

While the ecological benefits of organic production are well established, there are long-standing debates surrounding the economic viability of increased organic farming and its impact on food security. Most prominently, two related studies—one conducted by Pretty et al. in 2006 and another carried out by Badgley et al. in 2007—showed that organic agriculture can feed the entire world and achieve yields at least comparable to (if not even higher than) conventional farming. The findings from these studies, however, have come under fire from two angles. First, researchers find that relying on organic manure and nitrogen-fixing crops alone will be insufficient to make up for large-scale soil fertility losses. Second, due to the lack of funding for comparative studies on organic versus conventional versus biotech yields, some organic yield figures for developing countries are self-reported and selective. The Badgley et al. study has been targeted with especially heavy criticism, as the researchers broadened their definition of organic agriculture beyond a singular focus on certified organics and ignored the possibility that even higher yields might be achieved by combining biodynamic farming with GM seeds or other technologies.

While these debates highlight the tensions present between these two camps in their claim for scientific proof for their paradigms, they also serve as reminders of how important it is to understand the different assumptions and localities. For example, significant differences can exist between geographical regions and the farming systems used as a baseline for comparison, with organic agriculture seeming more favorable in developing countries, where farmers are converting from traditional production, as opposed to in industrial agriculture.

Three examples related to coffee, soybeans, and cotton showcase this diversity in organic production. In Central and South America, coffee is the most important permanent organic crop. Because a lot of organic coffee is grown through cooperatives, producers are often simultaneously double-certified for Fair Trade and organic markets, thereby increasing their marketing opportunities. Coffee production also entails fewer processing steps from harvesting to consumption, easing the process of organic certification. Conversion to organic coffee production is fairly

Challenges to Organic Farmers

As of spring 2011, a new food and fiber commodity boom was under way, and cotton prices had reached all-time highs. While these upsurges in commodity prices likely foreshadow new struggles for the urban poor, organic farmers and traders might be expected to rejoice in the trends. Yet cashing in on skyrocketing prices has been anything but automatic. Ongoing research by Samuel Ledermann in Tanzania indicates that lack of savings and knowledge about market trends have meant that many organic farmers are selling their crops too early to benefit from rising prices. For organic traders, the struggle to secure certified commodities has increased with each price hike for conventional cotton, because the profit margin on the fixed organic premium is reduced and many organic farmers are selling their cotton to conventional traders (also known as side-selling). On the retail side, rising prices also are likely to slow demand for organic products. Wal-Mart, for example, has recently signaled its intention to scale back its organic selection, revealing limits to passing higher commodity prices on to Western consumers.

—Samuel T. Ledermann and Robin Leichenko

straightforward, as it is a perennial crop planted on small landholdings, which simplifies the labor-intensive monitoring needed for successful organic production (i.e., scouting of pests or application of fertilizer, composting, or mulching).

According to the GMO Compass database, 77 percent of soybeans globally are grown with GM seeds. It is in this context that organic production becomes a viable option for satisfying the rising demands made by environmentally conscious consumers in developed countries. While soybeans also are grown organically in the United States, incidents of cross-contamination with GM soybeans and difficulties in meeting organic demands have opened up markets in countries that rely on non-GM soybeans, such as China. This crop thus illustrates the fears of decertification within the organic industry due to cross-contamination, and highlights the central role of trust within the industry and among consumers.

In sub-Saharan Africa, one of the most prominent export crops is cotton. While West Africa is one of the world's largest cotton-exporting regions, the largest organic cotton production is based in Tanzania, East Africa. Because this crop is grown on larger landholdings, providing timely extension service to the farmers is important. Export production is organized via an innovative vertically integrated value chain that has as its foundation bioRe Tanzania, which contracts with cotton farmers and provides them with social and agricultural services; bioRe Tanzania's partner in this venture is the world's leading organic cotton trader, Swiss-based Remei. Compared to organic coffee, increasing the market share for organic cotton has proceeded more slowly, as price differences are often larger than for organic food, fashion tastes are more diverse, and raw cotton requires numerous processing steps prior to consumption.

Cotton also illustrates the challenges facing the organic sector from GM seeds. In Africa, only farmers in South Africa and (recently) Burkina Faso use GM cotton seeds; in contrast, in India (one of the world's largest producers and importers of both organic and conventional cotton), acres planted with GM cotton seeds have skyrocketed from less than 1 percent in 2002 to nearly 80 percent by 2008, according to the GMO Compass database. This simultaneous boom in GM and organic production has resulted in cross-contamination, and incidents of faulty certification of organic cotton. Although these developments illustrate that organic consumption can be a viable opportunity in a market saturated with GM products, its production is likely to be messy and involve costly monitoring.

Amid all these latest controversies, it is important not to lose sight of the improvements that organic agriculture has wrought in terms of food security and poverty alleviation. First, engaging in intercropping and crop rotations improves soil fertility. When these practices are combined with strategies to combat soil erosion, these lands can better deal with extreme weather phenomena of both droughts and floods, as outlined by Niggli in 2009. Compared to conventional soils, organically managed soils have been shown to have better water retention and infiltration capacities, with the latter being as much as 40 percent higher in a long-term trial. Besides the improved ecological factors, intercropping and crop rotations increase the food security and lessen the vulnerability of the farmers. This effect can take the form of either obtaining another income source by selling the leguminous crops or increasing food self-sufficiency altogether by planting a greater diversity of crops. Finally, these labor- and knowledge-intensive methods are able to decrease poverty in a non-capital-intensive matter via the price premium that organic farmers are receiving, which averages 15 percent over conventional products' prices.

This lack of emphasis on capital is especially relevant, given the previous agricultural revolutions' capital-intensive nature. Organic farmers remain vulnerable to lower yields during the one- to three-year conversion period required to practice certified organic agriculture. In a conducive institutional setting, the organic revolution can reach both small and large-scale, wealthier and poorer, and more and less educated farmers. Women, who play a central role in improving food security, are specifically targeted and are—at least through increased household incomes—beneficiaries from organic projects. Even neighboring farmers who are not engaging in certified organic production can benefit by the transfer of knowledge, which has the potential to significantly reduce inequality and increase the organic movement's impacts beyond the higher price premium received. While landless laborers and those who only rent land on an annual basis are certainly at a disadvantage, they can benefit directly from the increasing labor demands and resultant higher wages, as well as indirectly from greater local food availability.

Pathways for a Sustainable Organic Revolution

The boom in organic production in the developing world can be sustained only if the increasing yields also are met with sustained consumption. Although the current economic crisis has had a weaker impact on the organic market, some signs indicate that the price premium available for organic products is eroding, with oversupply emerging in some markets (e.g., organic cotton). Consequently, there is a constant need to reevaluate and innovate when it comes to organic products, both for the end market and at the site of production. In short, organic producers face several long-term challenges to sustaining the boom's momentum and contributing to increased food security.

First, to combat the erosion between organic and conventional prices, organic farmers need to consider strategies to receive a higher share of the consumer price. In addition to strengthening international markets by achieving a meaningful agreement at the World Trade Organization's Doha meeting, these efforts include developing local and regional organic markets in developing countries and supporting innovative value chains. Examples include empowering small-scale farmers as shareholders of agribusinesses or paying the organic premium directly from the point of consumption to the farmer.

A second challenge is to expand awareness of the ecological benefits of organic agriculture, including adaptation to and mitigation of climate change. By focusing on improving soil fertility with locally available inputs, organic production provides farmers with flexibility to adapt to drought and other climate change-related stresses, while also increasing food security when facing other types of economic shocks. Organic initiatives can be combined with the use of biogas stoves and other techniques, such as agroforestry, to market carbon-neutral products and increase soil sequestration. Drawing on conservation agriculture, such as reduced tillage, also can increase resiliency and enhance adaptation to global climate change.

The final challenge is also the most hotly contested and potentially crucial. As shown with the cases of contamination, the production, certification, and marketing of organic crops can become difficult in cases where competition exists at the farm level from GM seeds. Second- and third-generation GM crops that not only reduce pesticide use but also provide nutritional advances, such as vitamin or mineral deficiencies or drought tolerance, are likely to become of greater importance. From a moral standpoint, as Paarlberg (2008) argues, it becomes increasingly more difficult to justify banning these GM seed varieties from reaching the farmers in developing countries.

As long as differences between organic and biotechnology-based agriculture are small in yield, organic agriculture will find a positive adoption rate from local farmers. However, in an environment where GM seeds outcompete organic seeds significantly in both yield and profitability, monitoring becomes a herculean task, as many farmers are tempted—justifiably—by GMs' promise. While this battle once again highlights the need for increased funding for extension service and organic seed research, it also opens the door for considering an integrated agricultural strategy with GM seeds. Such an approach is illustrated by the recently launched Better Cotton Initiative (BCI), which includes such diverse players as Adidas, IKEA, and WWF; the BCI focuses on reducing pesticide and water usage, but also accepts the use of GM seeds. Efforts along these lines could integrate improved organic practices of soil management and extend them beyond the certified organic niche. As proposed by Ammann in 2008, such practices would strengthen integrated farming, soil fertility, and ultimately food security.

—Samuel T. Ledermann and Robin Leichenko

Further Reading

Ammann, Klaus. "Integrated Farming: Why Organic Farmers Should Use Transgenic Crops." *New Biotechnology* 25 (2008): 101–107.

Avery, Alex. " 'Organic Abundance' Report: Fatally Flawed." *Renewable Agriculture and Food Systems* 22 (2007): 321–323.

Badgley, Catherine, Jeremy Moghtader, Eileen Quintero, Emily Zakem, M. Jahi Chappell, Katia Avilés-Vázquez, Andrea Samulon, and Ivette Perfecto. "Organic Agriculture and the Global Food Supply." *Renewable Agriculture and Food Systems* 22 (2007): 86–108.

Benbrook, Charles. "The Magnitude and Impacts of the Biotech and Organic Seed Price Premium." December 2009. The Organic Center: www.organic-center.org.

Bolwig, Simon, Moses Odeke, and Peter Gibbon. "Household Food Security Effects of Certified Organic Production in Tropical Africa: A Gendered Analysis." 2007. www.epopa.info.

GMO Compass. "Genetically Modified Plants: Global Cultivation Area Cotton/Soybeans." July 2010. www.gmo-compass.org.

International Federation of Organic Agriculture Movements (IFOAM). "Definition of Organic Agriculture." 2010. www.ifoam.org/growing_organic/definitions/.

Kirchmann, Holger, Lars Bergström, Thomas Kätterer, Olof Andrén, and Rune Andersson. "Can Organic Crop Production Feed the World?" In *Organic Crop Production: Ambitions and Limitations*, edited by Holger Kirchmann and Lars Bergström. Netherlands: Springer, 2008, 39–72.

Lu, Yanhui, Kongming Wu, Yujing Jian, Bing Xia, Ping Li, Hongqiang Feng, Kris A.G. Wyckhuys, and Yuyuan Guo. "Mirid Bug Outbreaks in Multiple Crops Correlated with Wide-Scale Adoption of Bt Cotton in China." *Science* 328 (2010): 1151–1154.

McIntyre, Beverly D., Hans R. Herren, Judi Wakhungu, and Robert T. Watson. "Agriculture at a Crossroads: Global Report." International Assessment of Agricultural Knowledge, Science and Technology for Development (IAASTD), 2009. www.agassessment.org.

Niggli, Urs. "Organic Agriculture: A Productive Means of Low-Carbon and High Biodiversity Food Production." In *Trade and Environment Review 2009/2010*, edited by Ulrich Hoffmann and Darlan F. Marti. UNCTAD, 2010, 112–118.

Organic Trade Association (OTA). "US Organic Product Sales Reach $26.6 Billion in 2009." April 2010. Organic Newsroom: www.organicnewsroom.com.

Paarlberg, Robert. *Starved for Science: How Biotechnology Is Being Kept Out of Africa*. Cambridge, MA: Harvard University Press, 2008.

Pretty, J. N., A. D. Noble, D. Bossio, J. Dixon, R. E. Hine, F. W. T. Penning De Vries, and J. I. L. Morison. "Resource-Conserving Agriculture Increases Yields in Developing Countries." *Environmental Science and Technology* 40 (2006):1114–1119.

Ronald, Pamela C., and Raoul W. Adamchak. *Tomorrow's Table: Organic Farming, Genetics and the Future of Food*. New York: Oxford University Press, 2008.

Willer, Helga, and Lars Kilcher, eds. *The World of Organic Agriculture: Statistics and Emerging Trends 2009*. Bonn: IFOAM; Frick: FiBL; Geneva: ITC, 2009.

P

Protein and Protein Deficiency

A balanced diet requires both macronutrients (needed in large quantities) and micronutrients (needed in small quantities). Chief among the macronutrients are proteins and carbohydrates. Proteins are needed to build and repair tissue; carbohydrates (sugars and starch) are needed for energy. Especially when under stress, however, the body also uses proteins for energy. A third category of macronutrients constitutes fats and oils. As a percentage of total daily calorie intake, a normal diet might be proportioned as follows: 50 percent carbohydrates, 25 percent proteins and oils, and 25 percent fats. In a "low-carb" diet, protein might account for 40 percent of a day's caloric intake (see the "Deficiency Diseases" entry).

Proteins

The word "protein" is from the Greek *prota*, meaning "first in importance." Proteins are made up of amino acid molecules arranged in linear chains. They are nothing more than carbon, hydrogen, oxygen, and nitrogen atoms, variously arranged. Proteins were first described by Dutch chemist G. J. Mulder in the 1830s. Although at least 100,000 different proteins are present in the human body, they are made up by combining only 20 standard amino acids. Half of these 20 amino acids can be manufactured by the human body. The other half must be ingested; they are called essential amino acids (e.g., lysine, tryptophan), because they are essential in the diet. Some authorities count 8 essential amino acids. The disagreement arises because children seem not to be able to manufacture as many different amino acids as adults; their list of essential amino acids is therefore longer.

A complete protein contains all of the essential amino acids in the right proportions. An incomplete protein lacks one or more essential amino acids. Digestion breaks protein down into amino acids, which then become available for building and repairing cells in every organ system from bones and muscles to skin and blood. Amino acids account for approximately 20 percent of the body's weight (water accounts for approximately 60 percent).

The first amino acid was discovered in 1806 in France. It was named asparagine because it had been isolated from experiments on asparagus. The last amino acid was discovered in 1935 in the United States; it was named threonine. In 2009, the simplest amino acid, glycine, was found in a comet by NASA's Stardust spacecraft, suggesting to some scientists (and disputed by others) that the essential building blocks of life are not limited to the Earth's biosphere.

The body does not store proteins the way it stores fat. Some protein is broken down and excreted in urine, feces, and sweat; more is flaked from the skin or is used to build hair and nails. Therefore, a continuous dietary supply of complete proteins is necessary. Mother's milk is an excellent source of proteins and their building blocks, amino acids. After weaning, complete proteins must be found elsewhere. In general, animal-derived foods are excellent sources of essential amino acids, including meat, poultry, fish, milk, eggs. Animal proteins, except for gelatin, are complete proteins, but they are often delivered with other potentially health-harming constituents, most notably fat. The only plant source of complete proteins is the soybean, which is what makes soy products so valuable. Tofu ("bean curds" in English) is one of those products. It originated in China and is a common element of the diet in all of eastern Asia, from Japan to Indonesia. Soy "milk" (preferably fortified with calcium) is increasing in popularity in North America as a complete-protein alternative to cow's milk.

Beans (navy, black, pinto, kidney) and other legumes, plus grains, seeds, and nuts, are also excellent sources of protein, but none of them individually delivers all essential amino acids. By combining these foods (not necessarily at the same meal), the body may have access to essential amino acids through a plant-based diet alone. Various combinations have become part of traditional foodways around the world. In Latin America, beans and corn (a legume and a grain) are staples of the diet. In the Caribbean and parts of Africa, this combination becomes beans and rice. Throughout the Middle East, hummus (a chick pea paste with sesame) is combined with pita bread, and snack foods consist of high-protein almonds, pistachios, and sunflower seeds. In Japan, miso soup (with bean paste, often made from soybeans) is served with rice. In South Asia, lentils are made into dal, a thick stew, which is served with rice. Peanut butter on whole-wheat bread provides a popular complete protein in the United States. These legume–grain combinations often are garnished with vegetables for micronutrients, as well as to add taste and color. In such dietary contexts, less meat is needed to supply protein, thus freeing up agricultural resources to feed more people.

The Recommended Daily Allowance (RDA) of protein in the adult diet is 0.36 gram per pound (0.8 gram per kilogram) of body weight. As an alternative to the RDA, however, the Food and Nutrition Board of the National Research Council has adopted Dietary Reference Intakes (DRIs) of protein for various age and

Table 20. Dietary Reference Intakes of Protein

Stage in the Life Cycle		Grams of Protein per Day
Infants		9–11
Ages 1–3		13
Ages 4–8		19
Ages 9–13		34
Ages 14–18	Males	52
	Females	46
Ages 19–70	Males	56
	Females	46
Pregnant/lactating women		71

Source: National Research Council, 2010.

gender cohorts. They are presented in Table 20 as guidelines for various stages in the life cycle. The average protein content of selected foods is presented in Table 21, which indicates some of the most common sources of protein in the diet, such as meat, fish, eggs, dairy products, and grains.

Malnutrition

In quantity and quality, proteins and carbohydrates are critical to bodily health. Too much or too little of any nutrient can result in *malnutrition*, which literally means "bad nourishment." The World Health Organization defines malnutrition as "the cellular imbalance between the supply of nutrients and energy and the body's demand for them to [e]nsure growth, maintenance, and specific functions."

Table 21. Approximate Protein Content of Selected Foods

Food	Protein Content
Beef and pork	7.5 grams per ounce
Chicken and fish	7.5 grams per ounce
Eggs	7 grams per large egg
Milk	8 grams per cup
Cheese (hard)	10 grams per ounce
Cheese (medium)	7 grams per ounce
Cheese (soft)	6 grams per ounce
Soybeans	10 grams per ounce
Peanuts	7 grams per ounce
Lentils	6.5 grams per ounce
Red beans	6 grams per ounce
Pumpkin seeds	2 grams per ounce

For most of human history, malnutrition was equated with undernourishment. People could not get enough food or could not get the proper food. Over the course of the late 20th century, however, malnutrition came to include overnourishment, commonly manifested as being overweight or obese. Undernourishment is sometimes called hypoalimentation ("under-feeding") and overnourishment is sometimes called hyperalimentation ("over-feeding").

Malnutrition is probably the greatest worldwide threat to public health. It is manifested in developing countries as protein-energy malnutrition (PEM), which arises from macronutrient deficiencies, particularly in childhood. PEM is associated with endemic poverty and acute disasters (natural and political) that precipitate famine and refugee movements. Clinically, it is manifested in four forms: (1) kwashiorkor, (2) marasmus, (3) marasmic kwashiorkor, and (4) catabolysis. Undernourishment is significant, however, even before it becomes evident clinically. The Food and Agriculture Organization has reported that 1 billion people are malnourished worldwide, and that malnutrition affects perhaps one-fourth of the world's children. Infants and children represent the most vulnerable demographic because of the high demands for protein and energy made by their growing bodies. In the United States and other developed countries, clinical PEM is extremely rare. When it is diagnosed, it is usually associated with severe infant and child neglect. Among adults in developed countries, undernourishment also may be associated with eating disorders or diseases that impede the body's ability to digest or absorb nutrients. Institutionalized elderly populations are especially vulnerable to subclinical protein malnutrition.

At the opposite end of the scale, macronutrient overload has become a widespread public health problem in developed countries of the world. With affluence, food becomes more available, advertising ups sales, and more people look for "cheap eats" on the run, which usually means indulging in fats, refined carbohydrates, and salt. Affluence is also a concomitant outcome of urbanization, and urban lifestyles usually mean less physical activity. Even in developing countries, overweight and obesity are becoming increasingly common in cities, as the forces of globalization shape diets. Women may be especially vulnerable to this effect in countries such as Egypt, where traditional cultural norms encourage a home-based lifestyle without regular exercise. On the basis of these trends, some authorities recognize a "nutrition transition" (analogous to the epidemiological transition) in which nutritional disorders initially are accounted for by undernourishment and deficiency diseases. As urbanization and development proceed, overweight and obesity emerge as the predominant nutritional health threats. The United States leads the world in the nutrition transition, followed by Germany and the United Kingdom in the European Union.

Kwashiorkor

Kwashiorkor is protein malnutrition; it is a manifestation of PEM. The name was first attached to the disease by a Jamaican pediatrician in the 1930s. Kwashiorkor is caused by a diet of barely adequate calories and inadequate proteins. Micronutrient deficiencies (vitamins and minerals) usually compound the disease. The defining symptoms are distended abdomens, widespread swelling (edema), listlessness, skin lesions, and depigmentation of the hair, which often takes on a red cast. In need of protein, the body begins to draw essential amino acids from the muscles. As the abdominal muscles are weakened, the internal organs fall outward, producing the "big bellies" that look so incongruous on starving children. Enlarged, fatty livers merely add to the distentions. Measles often precedes the onslaught of kwashiorkor.

Kwashiorkor is a disease of childhood, rather than of infancy. It carries a Ghanaian name that means "the disease of weaning" and often is associated with the birth of the next child. While being breastfed, an infant gets a full supply of essential nutrients (given a healthy mother). In a poor country, however, an infant often is weaned to a diet of gruels and starches or sugar water. Coffee lighteners (compounds of sugars and vegetable oils with no protein) also have been used as a substitute for breastmilk or baby formula, because they are inexpensive and widely marketed by commercial venders. In some countries of Africa and the Pacific, infants are weaned to dietary staples such as yams, cassava, or green bananas (as opposed to whole grains), none of which are good sources of protein. As a result, proteins may not be part of a child's diet at all. With barely adequate calories and a shortage of protein, kwashiorkor manifests itself, generally when the child is between 18 months and 5 years of age. Growth and intellectual development are stunted. Kwashiorkor may be fatal, but death often results from secondary infections that take advantage of a weakened body and a suppressed immune system. Kwashiorkor is almost unknown in the United States.

Treatment of kwashiorkor itself is relatively simple and inexpensive. It begins with rehydration and continues with the administration of simple sugars and fat, then proteins. Vitamins and minerals also must be included in the diet. If this disease is treated too late, however, affected children may never reach their full height and weight. Public health authorities in the developing world generally take a two-pronged approach to kwashiorkor prevention: (1) educating mothers and (2) making balanced diets (proteins, carbohydrates, fats, trace elements) available and affordable. Mothers who see their infants survive and thrive on a balanced diet are more likely to understand that they do not need a large family to assure that at least a few children will survive. In this respect, treating all forms of protein-energy malnutrition among infants and children encourages families to reduce fertility.

Marasmus

Marasmus, another form of PEM, is more common than kwashiorkor. It is caused by a deficiency of both proteins and carbohydrates, aggravated by insufficient vitamins and minerals; in essence, it can be envisioned as starvation. The defining symptoms are low body weight, tissue and muscle wasting, and emaciation. Marasmic children may be less than 60 to 80 percent of normal body weight, basically "skin and bones." In fact, the word *marasmus* (first used in the 1600s) is derived from a Greek root that means "a wasting away." In the absence of nutrition, body systems begin to shut down, beginning with the liver, heart, and digestive tract. Death eventually ensues.

In contrast to kwashiorkor, marasmus is not accompanied by edema (swelling) or depigmentation of the hair, but it is often associated with weaning. Like all PEM disorders, marasmus is associated with a palette of infectious diseases (e.g., gastroenteritis), many of which further impair the body's ability to absorb nutrients. In developing countries, marasmus is most common during the first year of life in regions where there is an acute shortage of food. Sub-Saharan Africa and South Asia, for example, are the regions of widespread prevalence. In the United States, marasmus is extremely rare among children, but more common among elderly adults.

Other Forms of Protein-Energy Malnutrition

When the symptoms of marasmus and kwashiorkor are both present, the condition is called marasmic kwashiorkor (marasmic KW). Children in this category are severely underweight (like children with marasmus) but also suffer from edema (like children with kwashiorkor).

The most severe form of PEM is catabolysis, the end stage of starvation. It begins when the body reaches into its last reserves of fat and begins to break down its own tissue to stay alive.

In reality, it is difficult to distinguish among the four forms of protein-energy malnutrition. A more widely used classification system simply divides macrodeficiencies into mild, moderate, and severe undernutrition based on weight, body mass, and serum protein. In an effort to isolate these diseases from those of overnutrition, the term protein-energy undernutrition (PEU) is being more commonly used as a replacement for PEM.

—Donald J. Zeigler

Further Reading

Bassett, Thomas J., and Alex Winter-Nelson. *The Atlas of World Hunger*. Chicago: University of Chicago Press, 2010.

Gardner, Gary T. *Underfed and Overfed: The Global Epidemic of Malnutrition*. Washington, DC: Worldwatch Institute, 2000.

Haerens, Margaret. *Malnutrition*. Farmington Hills, MI: Greenhaven Press, 2009.

Jacobsen, Kathryn H. *Introduction to Global Health*. Sudbury, MA: Jones and Bartlett, 2008.

Remote Sensing of Food Production

In the 21st century, assessing the potential of land to provide food to support life often employs remote sensing and geographic information systems (GIS) approaches, supplemented by ground observations and ground measurements. A single date of remote sensing analysis of an area, which is a snapshot in time, can help provide a reasonably accurate estimate of food production when it is analyzed at an appropriate stage of growth. For example, the vegetation vigor of an area planted in corn in the midwestern United States in May might indicate a high future yield, based on remote sensing and field observations in July, but it may or may not result in a good harvest in late September. Multiple dates of data acquisition are often needed to estimate the productivity of land to produce food over the course of a growing season or year, and assessment of longer-term productivity associated with famines always requires more frequent analysis.

Estimation of Food Production Over Large Areas

Estimation of food production over large areas, predicted and actual, commonly relies on an integration of remote sensing, climate analysis, field observations and measurements, soil conditions, modeling software, and other physical, cultural, and economic factors. Remote sensing is the major tool that is used to estimate actual or potential food production over areas, large and small, at a given time for an acceptable cost. Data acquisition and analysis of remote sensing to estimate food production exist and are used at a variety of scales and costs. However, remote sensing's ability to positively intervene in reducing or increasing the magnitude of food production losses or gains may or may not be possible, depending on site-specific circumstances. For example, crops and pastures in areas requiring a certain level of precipitation that fails to come at critical times due to occasional severe droughts—be it in areas of the western Corn Belt in the United States or in generally very humid areas in the Amazon Basin of Brazil—will have reduced food production. Remote sensing can help estimate this production, but it cannot alleviate production losses during a drought, except possibly by prompting farmers to initiate irrigation at most appropriate times if such an option is available.

Nevertheless, huge amounts of food are saved each year in the United States and other areas of the world where remote sensing and related technologies are effectively applied, often with the support of GIS and global positioning system (GPS) satellites. A majority of farmers engaged in corn, soybeans, and wheat production in the United States assess their fields using remote sensing at least once a year, checking on infestations, whether they involve insect, mold, rust, or other. If the interpreted remotely sensed data indicate a problem exists, it usually is found in time to halt widespread expansion of the disease. Records are kept of field disease patterns, year after year, in modern farming, and changes in these patterns provide the information needed to implement strategies to reduce or stop the degree of crop infestation and loss of food production early in a crop cycle. Seeds that incorporate special genes to make the plants resistant to most of the common infestations are increasingly being used, with the locations selected to plant these resistant seeds determined, in large part, through analysis of remote sensing data. Field changes from period 1 to period 2 and beyond are considered when implementing protective strategies. Many fields that were historically disease-prone and offered no or low yields can be made very productive through the use of genetically engineered seeds, for example (see the "Genetically Modified Foods" entry).

Analysis of a single remote sensing image can be beneficial in increasing food production and helping alleviate hunger. By comparison, remotely sensed images taken on multiple dates for use in change detection are needed to fully analyze and model strategies that are required to address food production problems and solutions to those problems, including elements of famine. To understand how large-scale crop losses in the 21st century can be detected through the use of remote sensing–based change detection, it is prudent to gain insights into selected remote sensing products and techniques that are already successfully used for determining food production changes or crop losses.

Sensors and Scales in Remote Sensing of Food Production

Sensors that are useful in collecting spectral data, and that can be employed to identify Earth features relevant to food production (e.g., field crops, planted pasture, native grasslands, tree crops, soils, and water), are available on various types of Earth-orbiting satellites. The spatial resolution or pixel size (minimum area on the ground from which data are collected) available with these sensors ranges from 0.4 meter to 500 meters. The cost of data, which ranges from free to thousands of dollars for small areas; the number of spectral (light reflectance) bands of data collected; spectral band suitability or effectiveness for food production estimation use; availability and frequency of data collection for a given area; and the image processing techniques used for analysis are all variables that need to be addressed to ensure that the system acquires the food production information most relevant to problems associated with hunger and famine. The list of satellite sensors of value for food production studies is long, but only three examples of the most widely used sensors representing three scales (high, medium, and low resolution) are presented here for illustrative purposes.

MODIS (Moderate-Resolution Imaging Spectroradiometer): Low Resolution

MODIS features free NASA 36-band data with world coverage every 1 to 2 days. The most useful bands for food production have spatial resolution of 250 m (two bands) and 500 m (five bands), collectively providing spectral data segments of the blue, green, red, near-infrared, and mid/short-wave infrared parts of the electromagnetic spectrum. An example of a 500-m resolution MODIS scene showing an Amazon area is provided in Figure 17. The Amazon rainforest at the top left in the figure is shown as almost black, but is becoming fragmented by areas of

Figure 17. Amazon region of Brazil in natural color (blue, green, and red light combined), captured by MODIS Satellite and displayed as a gray-scale image. (Courtesy of NASA/MODIS)

deforestation, which appear as lighter gray tones amid the darker tones of original forest. Rivers are shown in white, as are forest fire smoke-induced clouds. The lighter gray tone represents some combination of pasture, crop, and young successional forests (e.g., tree/grass mix).

The advantages of using MODIS include that it is free, covers very large areas or scenes, and collects data from an area of interest at least once every two days. As is evident in Figure 17, however, a spatial (ground) resolution of 500 m does not permit detailed analysis of most features needed for use in food production studies. Nevertheless, MODIS can show broad changes in the landscape, such as major deforestation with crops and pasture replacing forests, as well as areas that are returning to forests (afforestation). Small fields and the types of crops and pasture cannot be specifically identified, but major crop or pasture replacement of forest equates to an increase in food production (crops and cattle) in the area. Also, analysis of MODIS data can identify large changes in water bodies, which can help assess the intensity of drought or famine. Using MODIS, it is easy to get the number of images needed for change detection, but difficult to get the appropriate change perspective detail needed to help model detailed increases or decreases in food production or changes in surface water availability.

Landsat: Medium Resolution

Numerous Earth resources satellites collect or have collected multispectral data in the 10-m to 30-m spatial resolution range, including Landsat, IRS, Spot, and Aster. These satellites collectively have been among the most used in the world to collect Earth resources data, such as for water, crops, soils, minerals, plant disease, or urban analysis, and offer a level of detail vastly better than that provided by MODIS data. The image shown in Figure 18 depicts Landsat multispectral data whose spatial resolution is 30 m for its six light reflectance bands (blue through mid/short-wave infrared), which are used for classification of Earth features. Landsat thermal infrared (emitted heat energy) is also available at 60-m or 120-m resolution, as is a 15-m broad panchromatic band in the most recent Landsat-ETM satellite.

A Landsat image provides a level of detail suitable for making some types of regional or subregional estimates of food production. As is evident in Figure 18, such items as roads, individual crop fields, soil plots, small water bodies, big buildings, coal stripmines, rivers, forests, airports, and residential, urban, and other features can be identified visually or by computer. However, a 15-m or 30-m resolution image will not show features such as an individual tree unless it is very large. It cannot identify an individual corn plant, and it will not easily show a family swimming pool and many other features if they are smaller than 30 or so meters. A medium-resolution sensor still can provide very important details about

Figure 18. Landsat TM Satellite 30-meter resolution three-band color infrared composite (green, red, and near-infrared light combined), Terre Haute, Indiana, area (August, 1994). (Courtesy of NASA/Indiana State University, Department of Earth and Environmental Systems Landsat Archives)

features related to food production, but the improvement in resolution compared with MODIS is also accompanied by an increase in cost of data acquisition and analysis, the area of data collected per scene is less, and there are longer periods between data collection.

The value of medium-resolution sensors for food production has been widely recognized for more than 35 years, and it will continue to be important given that MODIS, with some important exceptions, has limited use for detailed studies of food production because of its low resolution. Also, the high-resolution sensors are too expensive to use for large-area analysis. Selected data from some types of medium-resolution sensors, such as Landsat, can now be obtained for free or at low cost, but most still require purchase. As in the case with any resolution of remotely sensed data, a multiple-image change detection product is needed to maximize the contribution that remote sensing and related technologies can make relative to addressing food production and famine issues.

Figure 19. A 1.0-meter resolution panchromatic IKONOS image of Washington, D.C. (AP Photo/Satellite Imagery)

IKONOS: High Resolution

Several high-resolution multispectral and panchromatic sensors have become available for use during the last decade or so, such as IKONOS and Geoeye/Quickbird. The spatial resolution of these sensors range anywhere between 0.4 m to 4.0 m, depending on the sensor and whether the data are collected as a single extra-high-resolution broadband (panchromatic) or as several narrow single bands (multispectral). For example, in the case of IKONOS, its black-and-white broadband version offers 1.0 m resolution, while its multispectral (multiple narrow bands) version has four narrow bands with 4.0-m resolution.

IKONOS is currently one of the most commonly used high-resolution sensors; an example of an image produced by this sensor is shown in Figure 19. At this resolution it is evident that Earth feature detail has been greatly increased. Individual cars, trees, branches on trees, very small features in fields, narrow paths, small buildings, and other features can be identified in its 1.0-m resolution panchromatic (black-and-white) version. Buildings such as the Hirshhorn Museum (light tone circular building), the National Air and Space Museum, and the former NASA building across the street are readily visible in Figure 19. In this figure, the trees and shadows are shown in black; buildings, cars, and some roads appear white; and grassy fields are in medium gray tones.

Many more different Earth features can be identified using IKONOS than using Landsat or other medium-resolution sensors. At the same time, IKONOS and its relatives have some difficulties related to large-area analysis of food production and physical analysis of famine conditions. IKONOS resolution is very good, but its data cost per acre often limits its use as an important supportive ground truth for medium-resolution sensors when large-area analyses are required. For some purposes, the higher level of detail of IKONOS data increases the complexity of analysis.

Figure 20. Color composite aerial digital photo using three bands of digital data (blue, green and red), in Indiana. (www.indiana.edu/~gisdata/Indiana University-Bloomington)

Airborne Sensors

Standard digital cameras flown on aircraft such as that shown in Figure 20, usually at altitudes from 1,000 to 50,000 feet, can provide very high-resolution data from lower altitudes and even good medium- to high-resolution data from higher altitudes. In this figure of a rural residential area in Indiana, cars can be seen on an interstate highway. Details of residential features are shown in light tones; forests are shown in very dark tones; and agricultural fields most commonly shown in medium gray tones. Airborne hyperspectral cameras can also collect data at submeter or lower resolutions. In this 2.0-m resolution grayscale version of a three-band color composite image of South Padre Island in south Texas, built-up areas are shown as white; water, as black; mangrove forest, as nearly black; grass and trees, as medium gray; and roads, as light gray (Figure 21).

Resolution of data acquired by airborne digital cameras or spectroradiometers—a type of large and costly digital camera that often collects more than 200 very

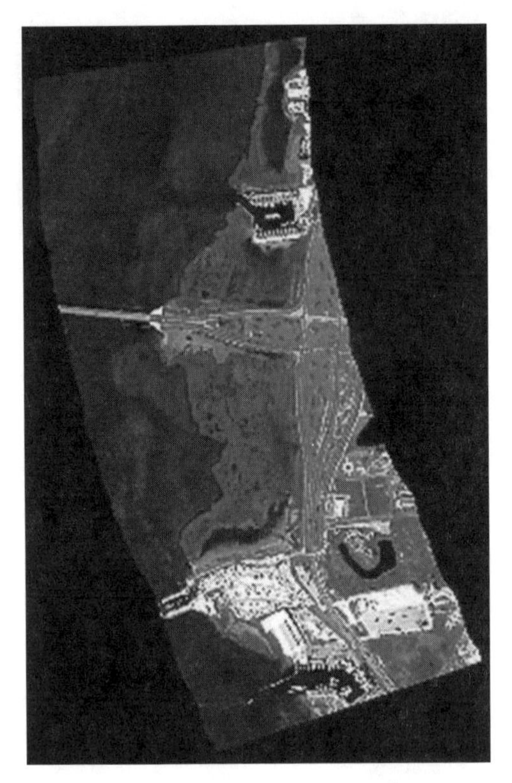

Figure 21. A 2.0-meter resolution near-infrared color composite image of South Padre Island, Lower Rio Grande Valley, Texas, using three bands of hyperspectral data (green, red, and near infrared). (Original data collected, acquired, and processed by Indiana State University, Department of Earth and Environmental Systems using an AISA+ hyperspectral sensor)

narrow bands of spectral data, reflected or emitted from Earth features from low-flying aircraft—usually can be more detailed than the images produced by most Earth resources satellites. The best commercial high-resolution satellite-acquired imagery has a resolution of approximately 0.4 m, though some sensors on low-flying aircraft can achieve a spatial resolution of approximately 0.1 m (3.5 inches). Depending on the nature of food production assessment, airborne sensors collect most of their data in the 0.2-m to 10.0-m spatial resolution range. With many important exceptions, high-resolution satellite imagery increasingly is taking over many of the roles that high-resolution airborne cameras traditionally filled. However, when special purposes, cost considerations, and research uses become important, airborne-acquired imagery relevant to food production will continue to be used. Satellite data with high or very nearly high spatial resolution can replicate the resolution of most airborne-acquired data. Thus selection of airborne versus satellite technology for a given use, such as food production, will depend on the advantages or disadvantages of the technology, such as its cost, frequency of data acquisition, number of spectral bands of data acquired, and available analyst expertise.

Change Detection Basics Using Remote Sensing Data

The variety of remotely sensed data available to address food production and famine issues is very large and increasing, as are the tools and approaches used to analyze these data in ways that can help predict current and future food productivity. The results can describe the temporal or time dimensions of food production and processes that contribute to the spatial nature of famine events. Currently, relatively accurate predictions of world corn, wheat, soybean, and other major crop yields are obtained by integrating remotely sensed data, climate data, plant disease data, field sample observations, and other relevant data using analytical techniques such as statistical/spatial modeling, remote sensing, and GIS. Inherent to success in identifying food production parameters is following the health of a food source from inception through harvest during a growing season.

In many cases, droughts, plant diseases, fires, war, floods, early frosts, and other factors may affect food production during the course of a growing season, which changes the outlook for the food source from time to time. To be able to estimate the potential production of a crop or pasture over a season requires knowledge about events that may affect yields, positively or negatively, so as to plan for food distribution in response to deficiencies or surpluses. For example, many famines are associated, at least in part, not only with actual reduction in food supply, but also with poor distribution of what food supply exists. In this scenario, actions need to be taken to conserve the existing food, as well as to make arrangements with food surplus areas not affected by acute food problems. To be best prepared for seasonal or multiple-year food production events, it is necessary to incorporate temporal elements into analyses. How many different time periods need to be studied depends on the nature of the food production–related study. Tracing the progress of a corn crop likely will require analysis conducted over several dates using remotely sensed data, because detecting and following a plant disease, starting in its early stages, cannot be performed effectively using remote sensing data gathered on only a single date per season. In contrast, in the huge Amazon Basin, perhaps a remote sensing analysis carried out every two or three years might be sufficient to identify basic changes in agricultural and other vegetation forms, as current interest focuses on deforestation issues as well as on the amount of new land opening up for grazing and crops.

At least 25 different approaches to change detection using remote sensing data have been described. These approaches can be divided into several categories and illustrated by change examples from the Amazon Basin in Brazil. Change detection approaches basically focus on visual interpretation and computer-based interpretation, which may contain elements of visual interpretation depending on the exact approach used. The great majority of change detection estimates of features related to food production are now based primarily on use of computer algorithms, analyzing various types of satellite data. However, visual interpretations of airborne and satellite data are still used for some small-area analysis and to help provide "ground truths" for many types of satellite analyses. For example, the side-by-side visual interpretation in Example 1 later in this entry (see Figure 22) primarily uses variations of techniques that are more than 80 years old, while

Figure 22. & 23. Example 1: Visual interpretation of change in the Altamira area of the Brazilian Amazon: Landsat color images, August 1985 (top) and August 1991 (bottom). (Original Landsat Data Courtesy of NASA, Anthropological Center for Training and Research on Global Environmental Change [ACT] LBA Ecology Archives, Indiana University–Bloomington. Processing conducted by Indiana State University, Department of Earth and Environmental Systems.)

Examples 2 and 3 (see Figure 23) are computer-intensive techniques that offer many advantages in providing Earth feature change information. Example 2 almost exclusively focuses on using remote sensing and image processing algorithms (programs) to classify Earth feature data and provide time perspectives needed to identify changes. A variation of Example 2 is shown in Example 3, which uses remote sensing for initial classification of features related to food production, but then relies on GIS computer programs to organize classified data into Earth feature change information. Many other variations of change detection also use computer analysis.

Analysis Method 1: Side-by-Side Image Visual Interpretation

In this approach, two images (preferably three-band color composite images) are placed side by side and visually interpreted to identify the areas, direction, and types of Earth feature change. If each of the images has been visually interpreted and classified into Earth features, then the direction and type of change can be identified by comparing the two images—for example, Area A changed from forest to pasture between Dates 1 and 2. If each of the images has not been classified or interpreted prior to assessing change detection, then areas of change and direction of change can be identified. To determine the exact nature of changes requires additional visual image interpretation. For instance, assuming Area A had a change in reflectance pattern that appeared as dark red on an image on Date 1 and then bright red on Date 2, this difference would indicate that major sunlight reflectance change occurred between the two dates. An interpreter could readily note that a change occurred, but still would need to be able to determine which kinds of features were represented by dark red and bright red on the images, based on knowledge of light interactions with Earth features, to accurately extract the most useful change information.

This approach to change detection often uses inexpensive data and requires little or no computer processing, but it is difficult to provide detailed quantitative change information. Moreover, this approach is impractical to use for detailed large-area change detection, and it is not digital, making it difficult to integrate the results with computer software. Visual image interpretation has value in broadly assessing aspects of change detection, particularly for small areas. An explanation of the use of this technique in an Amazon Basin study area is provided in the "Examples of Change Detection" section later in this entry.

Analysis Method 2: Unsupervised Classifications

The first approach discussed relies on simple visual image interpretation techniques, but it has problems in providing detailed Earth feature change information in a format such as "from Feature A on Date 1 to Feature B on Date 2," particularly over large areas with complex land-use or land-cover patterns. This inability to be as specific about types of changes in a quantitative form limits applications, although for some purposes, the information developed satisfies needs.

Analysis Method 2 illustrates a more comprehensive computer use in change detection, whereby detailed Earth feature change information in a "from Feature A on Date 1 to Feature B on Date 2" format can be developed. The procedures using comprehensive change detection through overlaying multiple-date unsupervised classification are too complex to discuss here. Essentially, raw multiple-band digital data representing reflectance from several spectral bands with two or more dates of data are overlaid or merged by computer, classified into Earth features, and then analyzed to show changes. Unsupervised or cluster classification is a commonly used remote sensing approach that automatically groups pixels with similar reflectance patterns into spectral classes, each of which usually can be identified as an Earth feature class or subclass based on an analyst's interpretation of two or more date spectral pattern statistics. An Amazon Basin change detection example using this method is provided in the "Examples of Change Detection" section.

Analysis Method 3: Supervised Classifications

A final product that produces results similar to those obtained with unsupervised classification is to categorize individual areas of interest that are assessed on different dates using supervised classification. Supervised classification uses known representative training spectral samples of features of interest to train a computer to classify pixels with similarly known spectral patterns derived from field observations and/or through detailed high-resolution image analysis. Each individual date of Earth feature classification is merged in some manner; for example, it may be used as input into a GIS, to show feature change details.

Examples of Change Detection

An example of change detection (1985 and 1991) using Landsat three-band color composite images in a study area of the Amazon Basin of Brazil, with selected implications, illustrates the visual interpretation approach. An example of using a comprehensive computer analysis approach to change detection (1985, 1988, 1991) in the same Amazon study area, using six bands of Landsat, is also provided. The third example illustrates how a GIS can be used to identify Earth feature changes.

Example 1: Visual Interpretation of Change in the Altamira Area of the Brazilian Amazon, 1985 and 1991

Example 1 consists of two Landsat images of a small area in the Amazon Basin acquired six years apart. The area shown is 14.3 km by 12.0 km in size (110,000 acres), but the full Landsat scene itself is approximately 185 km by 185 km. These images are displayed in a three-band color composite format that replicates a color near-infrared photo by merging light (sun) reflectance from Earth features from the following Landsat bands: near-infrared light (0.76- to 0.90-micrometer wavelength); red light (0.63- to 69-micrometer wavelength); and green light (0.53- to 60-micrometer wavelength). The Example 1 images can be visually interpreted to identify changes, knowing that analysis of variations in reflectance patterns can be employed to identify changes in land use and land cover that can be roughly associated with aspects of food production changes in the area between 1985 and1991. The more an analyst knows about an area, the more accurate the identification and interpretation of changes on the images from 1985 to 1991 will be. The information acquired related to food production changes often will be general, yet still valuable in addressing food production issues. A visual interpretation effort to acquire quantitative data is limited, often inaccurate, time consuming, and a difficult method to use if the area of analysis is large and complex.

Figure 24. Three-date (1985, 1988, 1991) land-use and land-cover changes in an Altamira Brazil study area. (Original Landsat data courtesy of NASA. Data processing and analysis conducted by Paul Mausel [Indiana State University]. Image published in Mausel, P., Y. Wu, Y. Li, E. Moran, and E. Brondizio, "Spectral Identification of Successional Stages Following Deforestation in the Amazon," *Geocarto International*, 8 (1993): 1–11.)

In the Example 1 Landsat images (Figure 24), which are grayscale versions of three-band color infrared composites, changes in grayscale tone and texture are easily seen on both dates, but require knowledge about what each tone/texture pattern means to provide useful information. In the grayscale version of Figure 24, water is indicated by black, rainforest as dark gray, crop or pasture as light gray, and some type of successional forest (original forest cut and new trees growing) as medium gray. The two Amazon color near-infrared images provide greater differentiation of features in the study area, and all black-and-white (grayscale) images were originally in color (see the note at the end of this entry). In the color version of Figure 22, a dense tropical forest (rainforest) is shown as dark red with a rough texture; crop, pasture, or early stages of forest regrowth/forest succession, as bright red/pink; more advanced forest succession, as somewhat dark red with an indication of some roughness (texture); water, as black (or sometimes dark green or blue); and variations of soil and other bare features (e.g., roads or buildings), as light blue or light green.

Using visual interpretation to determine broad changes in land use and land cover is possible, applying an approach whereby the same area, identified on two or more images but on different dates, is compared and the changes noted. In the two black-and-white images in Example 1 are two small areas in the northwest part, identified as FIELD 1 and FIELD 2. FIELD 1 in 1985 is shown as dark, highly textured gray area that can be interpreted as dense rainforest; the identical area in 1991 is shown as light gray with little roughness or texture that can be interpreted as a crop or dense pasture field. From 1985 to 1991, FIELD 1 was deforested and converted into a dense crop or pasture use with a much higher food production potential than the rainforest. In 1985, FIELD 2 (the rectangle on the northwestern edge of the image) was bare soil (shown in light gray) and likely to have been in crop or pasture shortly after the date of image acquisition. In 1991, this same area appears as a medium gray with small patches of light gray that could be interpreted as crop or pasture with some trees reestablishing themselves. From a food productivity perspective, FIELD 2 in 1991 probably had no increase or even a slight loss in food production compared with several years earlier, as some parts of the field were turning back into forest. From a food productivity perspective, FIELD 1 had much higher food production in 1991 than it had in 1985.

Considering the full study area, dense forest significantly declined between 1985 and 1991 and was replaced by crops and pasture initially, thereby significantly increasing food production in the area. Another trend that could be identified through interpretation of images created on two or more dates was that many of the more productive pasture and crop areas were becoming less productive (from a food production perspective) due to regrowth of successional forests (smaller trees, often accompanied by some grasses). However, dense forest continued to be cut or burned in this area to provide new lands for crop and pasture; thus this area, post-1991, continued to have increased food production overall. Secondary succession forests also were being deforested to increase the acreage in food production, and more intensive use of technology added to food production to help support an increasing population in the study area.

Example 2: Multiple-Date Overlay of Remotely Sensed Data Using Merged Unsupervised Classifications—Comprehensive Change Detection in an Altamira Area of the Brazilian Amazon, 1985, 1988, and 1991

Example 2 (Figure 23) demonstrates a three-date (1985, 1988, 1991) change detection. The small 9.6 km by 6.0 km study area (37,000 acres) is a subarea of the images shown in Example 1, but the analysis uses data collected over three dates for change detection instead of on two dates for a smaller area. In Example 2, remote sensing techniques were used for data display, data analysis (unsupervised classification), and change detection.

The Earth feature classes that can be accurately identified in this study area through analysis of Landsat multispectral reflectance data include dense, moist, mature original forest (rainforest); bare soil; water; dense crop or pasture; early secondary succession (SS; crop or pasture with significant growth of small trees); and more advanced secondary succession (forest succession in which larger trees fully cover the ground). From a food production perspective, the class crops or pasture provide a majority of the food that is consumed or exported from this study area, with tree crops contributing to the total production. Prior to the early 1970s, this study area was almost fully forested, with only a few people living off the forest products. After the trans-Amazon highway passed through this area, the government supported immigration of poor farmers from other areas of Brazil, providing them with free small parcels of land. It also allowed wealthier citizens to obtain larger plots of land for large-scale ranching.

The study area shown is a change detection image representing three-date change, illustrating two computer approaches (Examples 2 and 3). The study area was almost 100 percent covered with dense original forest in the late 1960s, but was covered with only 35 percent of its original forest in 1991. Over 20 to 25 years, almost 65 percent of the forests were cut or burned, which has created deforestation problems. However, viewed from a food production perspective, this deforestation has resulted in additions to the food supply from the area. It now supports a large number of people and, in most years, produces a surplus of food (e.g., corn, sugar, cattle, spices, soybeans, and tree crops) that is sent to other areas in Brazil needing additional food. The Amazon is relatively drought-free, but a severe drought covering large portions of the Amazon Basin took place in 2005. In the past, severe droughts occurred about every decade or two.

The three-date change image developed in Example 2 used remote sensing only to identify changes in a rapidly changing part of the Amazon during the mid-1980s to early 1990s. The categories of change require some interpretation on the ground, conducted by local officials, to translate changes in terms of economic value, yields per unit of area, or some other way of expressing the amount of food available to support a population. The 1985–1991 change figure (in which acres or hectares of change are known for each category on the legend) provides a visual and statistical database that officials in the field can use to convert to food production and note food production changes.

The changes that can occur from one time period to another can follow several paths. Immediately after deforestation, land goes into crop or pasture, both of which are associated with a relatively high level of food production. After a few years, cropland may be converted to pasture or, if it was initially pasture, to cropland for several years. Thus it is common for recently deforested land to have relatively intensive food production in the form of crops and or pasture for six years or more.

Figure 23 is a complex classification identifying many change class possibilities. Thus it is difficult to interpret in its black-and-white form, though its color version can also be accessed (see the note at the end of this entry). In the color version, red areas on the change figure show where original forests were very recently cut or burned. Hence, those areas in 1991 were covered in either crops or pasture and provided a relatively high level of food production. The areas in yellow on the figure were deforested for 4 to 6 years, but were likely used for either crops or pasture in 1991. Often, after several years, fields of crops and pasture are allowed to revert back to successional forests—a practice that helps to replenish soil fertility—or, in some cases, are deforested to meet farmer pressure to expand crop or pasture acreage. The first stage of forest succession is for small trees and shrubs to establish themselves, albeit mixed with grasses; thus some modest food production in the form of cattle grazing exists. The areas in light purple (slow SS) are likely a mixture of young trees, pasture grasses, and small plots of cropland with possible modest food production. The areas in dark blue were formerly used for crops or pasture, but by 1991 were fully forested, although not as dense as the original forest. Little crop or pasture exists in these areas, but some tree crops might add to food production. Once forest succession is well established, it is not uncommon to cut or burn a secondary succession forest to use the area for crops or pasture again.

Example 3: Multiple-Date Overlay of Classified Remotely Sensed Data Using a Geographic Information System for Comprehensive Change Detection

The change detection image shown in Example 2 could have been created using three dates of classification, followed by use of GIS technology, or it could identify and display changes in a manner different than that developed in Example 2. Example 3 focuses on providing insights into how a GIS can be used to provide temporal perspectives to multiple-date classification of Earth features that affect food production. Although most GIS change detection methods require classified remotely sensed imagery as inputs, the approaches available vary greatly in their complexity. At the most basic level, binary (0's and 1's) analysis can be used to indicate change (or lack thereof) in one category of land cover. More complicated applications of these change mapping techniques include outputs similar to that seen with Example 3, which shows the change over time in multiple land-cover categories. Such categorical maps are commonplace; they are particularly intuitive displays of changes that might affect food production. The application of GIS is not limited to categorical comparisons, however, and probabilistic analyses of change are becoming more widely accepted. These change detection maps indicate the likelihood that a change will be observed, giving the map user a better sense of the certainty that the change has been or will be observed.

GIS allows for other forms of change detection analyses as well, including those focusing on areas changed rather than just locations. One of the most straightforward GIS tools to track change across multiple categories (e.g., a forested area converted to an agricultural field one year and then the same agricultural

Table 22. Example of Change Matrix Showing Area Converted Between 2000 and 2010

From (2000)	Forest	Agriculture	Barren	Total (2010)
Forest	0	57 acres	128 acres	185 acres
Agriculture	69 acres	0	49 acres	118 acres
Barren	8 acres	15 acres	0	23 acres
Total (2000)	**77 acres**	**72 acres**	**177 acres**	

field converted to residential housing the next year) is the cross-tabulation or "change matrix." An example of a change matrix is presented in Table 22.

The change matrix is a useful GIS output because it assesses both the direction of change (e.g., the fact that agricultural area increased between 2000 and 2010) and the type of change (e.g., the fact that increases in agriculture came at the cost of forest and barren land categories). The "0" values along the diagonal indicate that land use cannot convert into itself. In interpreting the change matrix, one also can see the total area affected by changes between categories by looking at the total columns. Beyond mapping, GIS data sets and tools can illuminate more than just the location or amount of change. A GIS can help characterize the changes observed by remote sensing.

Value of Remotely Sensed Data

Remote sensing and associated technologies such as GIS are now considered crucial elements in estimating various types of food production throughout the world, and their accuracy and importance are increasing as technology advances. Medium-resolution Earth resources satellites are collecting spectral (e.g., light reflectance) data that, when analyzed and supported by ground observations and high-resolution satellite data, provide local, regional, and world estimates of food production. Satellites are providing the most data relevant to food production estimates, but airborne digital camera data also play an important role in food production estimation in many areas. This information is vital to identifying annual food deficits and surpluses worldwide and to supplying data that can be modeled to help identify and alleviate areas of hunger and famine.

Satellite and airborne sensors collecting data and modern technologies in remote sensing, GIS, spatial modeling methods, and GPS are being used more widely around the world. Not only do they provide accurate estimates of major crop and other agricultural yields, but they also provide temporal (time) perspectives associated with food production. Change detection is being employed to increase food

Science Education and Research Opportunities for Students: Frontiers in Science Education

Science in practice in the research community is usually interdisciplinary, requiring integration of multiple knowledge bases, tools, and techniques to address science problems and solve them. The K–12 science education community has not yet fully adopted interdisciplinary and related systems approaches to many sciences as understood and practiced by scientists worldwide. With some exceptions (e.g., the Global System Science group at the University of California-Berkeley), publishers provide science textbooks that are primarily topic based, offering only modest interdisciplinary or systems content. Recognizing this shortcoming, numerous education associations have urged greater use of interdisciplinary/system approaches in many types of science education. For example, spatial literacy approaches to learning recently received support from the National Research Council and other associations, which recommended its use as an approach that should be implemented in many types of science teaching. Spatial literacy focuses on understanding spatial relationships, determining how geographic space is represented, and reasoning and decision making about spatial concepts, using modern geotechniques.

SEROS, a not-for-profit organization, focuses on environmental and other sciences with major spatial elements and employs interdisciplinary, systems, and spatial literacy approaches to learning, particularly for grades 6–14 (approximately ages 12 through 20) students. An example of one of its recent projects was "Learning Environmental and Related Sciences Through Student Hands-on Research." Among the universities with participants engaged in SEROS activities are Indiana State University, University of New Hampshire, University of California–Berkeley, and Western Carolina University.

—Paul Mausel, Dennis Skelton, and Stephen Aldrich

production, identify areas of food shortage, and facilitate the efficiency of food distribution to help alleviate extreme hunger. For many food production activities, data collection and analysis on two or more dates, primarily using satellite and airborne sensor data and interpreted by computer or visual methods applying modern geotechniques, are required. For example, large areas of food crops are saved each year when a crop disease is caught early in a growing season when there is time to halt and monitor the spread of the infestation problem. Measures that prevent unnecessary losses of food production include acquiring and analyzing multiple dates of remotely sensed data every year.

Remote sensing techniques that classify data and associated techniques such as GIS are now well established and used worldwide to estimate food production and help analyze processes that have the potential to influence hunger and famine. The amount of available remote sensing data is huge and growing, while costs of its use for food production applications are declining. The value of multiple-date temporal analysis of remotely sensed data is obvious, as many food production applications and other uses are greatly increasing, thereby strengthening the role geotechniques will play in the foreseeable future in supporting important aspects of hunger and famine-related issues.

Note: The images in Figures 17, 18, 20, 21, 22, and 23 can be accessed in color by going to www.seros.us/moodle/ and clicking on "Remote Sensing of Food Production Color Images."

Paul Mausel, Dennis Skelton, and Stephen Aldrich

Further Reading

Cardwell, Shannell. "MODIS Web." Accessed January 2011. modis.gsfc.nasa.gov/.

Copenhaver, Ken. *Development of a Monitoring System for the Proactive Detection of Insect Pest Resistance to Transgenic Pesticidal Crops Using Geospatial Technologies.* EPA/NASA Technical Report. Chicago: Energy Resources Center, University of Illinois at Chicago, 2010.

"Hyperspectral Imaging." Accessed January 2011. www.microimages.com/getstart/pdf/hyprspec.pdf.

Irons, James. "The Landsat Program." Accessed March 2011. landsat.gsfc.nasa.gov.

Jensen, John R. *Remote Sensing of the Environment: An Earth Resource Perspective.* Upper Saddle River, NJ: Prentice-Hall, 2006.

Lu, D., M. Batistella, P. Mausel, and E. Moran. "Mapping and Monitoring Land Degradation Risks in the Western Brazilian Amazon Using Multitemporal Landsat^{TM}/ETM+ Images." *Land Degradation and Development* 18 (2007): 41–54.

Lu, Dengsheng, Paul Mausel, Eduardo Brondizio, and Emilio Moran. "Change Detection Techniques." *International Journal of Remote Sensing* 25, no. 12 (2007): 2365–2407.

Mausel, P., D. Lu, and N. Dias. "Remote Sensing of Amazon Deforestation and Vegetation Regrowth (Succession): Inputs to Climate Change Research." In *The Global Climate System: Pattern, Process and Teleconnections*, edited by Howard Bridgman and John Oliver. Cambridge, UK: Cambridge University Press, 2006, 79–95.

Mausel, P., Y. Wu, Y. Li, E. Moran, and E. Brondizio, "Spectral Identification of Successional Stages Following Deforestation in the Amazon." *Geocarto International* 8 (1993): 1–11.

Satellite Imaging Corporation. "IKONOS." Accessed February 2011. www.satimagingcorp.com/satellite-sensors/ikonos.html.

Wardlow, B. D., and S. L. Egbert. "Large-Area Crop Mapping Using Time-Series MODIS 250 M NDVI Data: An Assessment for the U.S. Central Great Plains." *Remote Sensing of Environment* 112, no. 3 (2008): 1096–1116.

S

Soil

Soil is an unconsolidated deposit, created through natural weathering and erosion processes, consisting of mineral or organic material that functions as a natural medium for the growth of rooted, land-based plants. Soil displays genetic and environmental factors influenced by climate (water and temperature effects), liquid and gases, macroorganisms and microorganisms, parent material, and topographic relief, all of which change over time. Naturally occurring soils form distinguishable horizons or layers based on zones of accumulation (creating an illuvial horizon), zones of loss (creating an eluvial horizon), and the transfer and transformation of organic and inorganic matter over thousands of years. The movement of minerals and organic material occurs when they are placed in suspension or solution, due to the presence of water, and they penetrate through a porous and permeable unconsolidated medium under the influence of gravity.

The upper limit of soil is generally defined as a boundary between the top soil layer and the lower atmosphere in contact with it. A surface is not considered to possess soil if it is under water, and if it cannot sustain living rooted plants or other organic material that has not begun to change into a state of decomposition (chemical breakdown). The lower boundary of soil is more difficult to measure from a non-soil. Because soil is derived from parent material that differs in its biological, chemical, physical, and morphological properties, the characteristics of the lower soil layer generally will grade to bedrock or unconsolidated glacial deposits, sediment from river systems, or wind-blown sand. Parent material is devoid of animals, roots, or other indicators of biological activity. Living matter must be present for the material to be classified as soil.

Matrix for Sustaining the World Food Supply

Ninety percent of the food supply generated throughout the world comes from only 15 plant and 8 animal species. Seventy percent of the genetic diversity of agricultural (crop) plants has been lost over the past century. The ability to promote agricultural growth in the future will require an understanding for best farming practices, including intelligent land use and natural resource management.

Soils vary geographically. Their physical and chemical properties can change in a matter of a few yards. Natural land cover has been defined as a biophysical state of the Earth's surface, which includes the overlying vegetation, soil layers, and water status. Natural land cover is a good indicator of the general conditions of the soil. The soils of the humid tropics are typically deep and severely weathered. They lack major nutrients and tend to be overlain by dense, tropical forests. The subtropics can have cyclical periods of dry and wet conditions with shallower soil profiles, less intense weathering, and more available nutrients derived from the weathered minerals broken down from bedrock parent material. These weather conditions promote moderately dense, drought-resistant grasses and xerophytic trees as primary, naturally occurring vegetation. Soil profiles are much thinner and poorly developed in higher latitudes with cold-resistant plants that tend to be seasonal, developing during the short, warmer summer period.

Climate, latitude, and altitude are significant factors in determining the length of a growing season and, by extension, the types of crops that can be grown in a region. While people tend to settle in a region for many reasons, the health and sustainability of a society ultimately will be predicated on the presence of an adequate supply of food, water, and shelter. If the land is nonproductive or underproductive, then the major source of food must be provided by outside sources. The cost of importing food can be offset only if a country's economic base can provide goods and services that can be sold or traded in kind to sustain the population's basic needs. If land-use policies violate the natural environmental conditions that have been established for a particular species of vegetation and crops, then living conditions will be severely strained. As an example, many tropical environments have been stripped of their natural forests to develop land for agriculture. The ability of these tropical soils to replenish depleted minerals and nutrients cannot keep up with the nutrient demands of non-indigenous crops. As a result, large tracts of land are becoming sterile, nonproductive, and exposed to mass wasting and erosion during rainy seasons, drought, and periods of sustained high wind.

Issues Requiring Mitigation

Next to population growth, soil erosion due to natural processes, poor land use, and corrupt management practices are among the highest-priority environmental problems facing the world in the 21st century. Cornell University's David Pimentel has estimated that the United States is losing soil 10 times faster than the natural replenishment rate. China and India are losing soil 30 to 40 times faster than this rate. Over the past 40 years, 30 percent of the world's arable land has become unproductive.

Poorly managed agricultural practices, such as overgrazing, have significantly contributed to soil degradation. In arid climates, this practice often has led to desertification. Desertification can be caused by overgrazing, a breakdown of soil structure, a loss of fertility due to the introduction of salts into the soil layers (salinization), poor soil water management, forest removal leading to mass erosion, and climate change (see the "Desertification" entry). Overgrazing promotes soil compaction, leading to a loss of porosity and permeability. Water is unable to infiltrate the soil layers, but rather flows off as surface runoff. This process causes rapid erosion and soil loss. Excessive grazing of cattle removes grasses at a rate that cannot be sustained, resulting in destabilization of the soil structure and erosion. In addition, preferred perennial pasture vegetation is replaced with less desirable annual species that do not provide the structural stability for soil to withstand the processes of erosion, driven by wind and water. In the North African Sahel, for example, farmers plant trees as barriers to reduce wind speed and loss of topsoil.

Drought can accelerate conditions conducive to desertification, but land abuse and misuse are the principal triggers of this phenomenon. Population growth leads to greater demands for crop yields. Higher expectations are set forth, and the land is pressed beyond its ability to recover. Overuse of the soil can lead to crustal development over the top layer, which promotes excessive runoff, development of deep gullies, and massive soil losses. Once the top layer of soil is removed, the more nutrient-rich illuvial B horizons will follow. Drying occurs after the soil layers are exposed to the wind, with additional erosion removing valuable topsoil. Encroachment of sand and silt follows, such that large regions gradually become covered with sterile, infertile deposits. The 2010 United Nations Conference on Desertification has ranked the hazards of desertification throughout the world, based on whether a region's agricultural productivity has declined to any great extent. If there is no significant decrease in agricultural productivity, there is less than a 10 percent likelihood of desertification developing. A moderate reduction in agricultural productivity results in a 10 to 25 percent risk of desertification; a high drop in agricultural productivity leads to a 25 to 50 percent chance of desertification; and a very high drop in agricultural productivity is associated with a greater than 50 percent likelihood of desertification.

Africa and Asia have experienced the most desertification. To date, three-fourths of the drylands of Africa have been degraded. The greatest impact on population is currently occurring throughout pockets of Asia, including massive sand dune migration throughout Syria, erosion along the mountain slopes in Nepal, and deforestation practices leading to overgrazing on the highlands of Laos in Southeast Asia. The northern Mediterranean region has suffered from poor agricultural practices leading to salinization; this condition, which is affected by drought, forest fires, occasional flooding, and hot, desiccating winds, diminishes the soil's ability to maintain fertility. Poor land-use practices also have damaged soils in Central and Eastern Europe and along the coast of the Adriatic Sea. Poor irrigation practices, resulting from rapid aquifer depletions, have resulted in unsustainable water demand throughout the Caribbean and Latin American countries. Irrigation management issues have created problems with salinization and chemical contamination, poisoning valuable soil and groundwater sources. A more than 25 percent poverty rate throughout these regions and lack of efficient farming skills have further contributed to the problem. This trend can be reversed by exposure to better land-use practices, education, and introduction of better farming equipment and supplies.

An ingenious technique to enhance soil fertility and productivity is practiced by the farmers of Burkina Faso, a small landlocked country in western Africa. They dig small pits in the fields and fill them with manure. This action attracts termites that digest the manure and spread their excrement throughout the soil, providing a natural fertilizer. The burrowing nature of termites creates a network of channels that aerates the soil and provides pathways for water to infiltrate.

The Amazon Basin throughout Brazil possesses some of the most fertile soil on the planet, often referred to as the "Black Indian Earth" or *terra preta*. These human-impacted soils contain large amounts of charcoal that were added to the soil during colonial times by the indigenous people as they employed "slash and char" management techniques. *Terre preta* soils have an ability to effectively collect carbon, nutrients, and water, to the point that they contain three times more phosphorus and nitrogen than surrounding soils. These human-influenced soils can provide a significant food-producing resource that is sustainable and serves as an alternative method for restoring a carbon balance.

In developing countries, nearly 80 to 90 percent of fresh water is used for agriculture. While approximately 17 percent of all cropland is irrigated in these regions, it produces only 30 to 40 percent of the world's food. Sixty percent of the world's irrigated land is in Asia and is used primarily to grow rice. Although improved seeds and hybrids, fertilizers, and irrigation practices have increased crop yields throughout Asia and other emerging countries, they also have enhanced water-intensive farming that is more dependent upon chemicals and fertilizers, releasing nitrous oxides that are degrading the world's soils. Intensive plowing releases carbon dioxide, which in turn breaks down the structure of the soil, thereby reducing its water-holding capacity and weakening its ability to fix carbon dioxide, nitrogen, and other nutrients. Where water is in limited supply, rainwater harvesting and drip-feed irrigation systems can mean the difference between a productive crop yield and starvation in such arid regions as South Asia, the Near East, and North Africa. Stone fences often are built around a cultivated field to capture rainwater, a practice that helps to prevent rapid runoff and ensuing

Impact of Sea Level Rise on Soils

Bangladesh was predominately settled and developed on a massive delta formed by sediments deposited from the joint discharge of the Ganges and Brahmaputra rivers system. The low-lying plains created from these deposits feature some of the most fertile alluvial soils in the world. Unfortunately, the majority of the country is less than 39.4 feet above the mean sea level. A rise of just 3.28 feet (1 meter) will result in a loss of approximately 10 percent of the country's land. Currently, 30 to 50 percent of the cultivable land along Bangladesh's coastal margins resides in a fallow state. Soil wetness, water stagnancy, tidal surges (directed by tropical cyclones/hurricanes), periodic and unpredictable drought conditions, increased salinity due to sea level impingement on low-lying coastal regions, insect infestations, and crop diseases have contributed to a 20 to 40 percent reduction in crop yields in the coastal districts of Khulna, Bagerhat, Satkhira, Barisal, Bhola, Barguna, Pirojpur, Patuakhali, Cox's Bazar, and Noakhali. The effects of sea level rise in Bangladesh are clear; now one must consider the impact these conditions will have on other coastal agricultural regions throughout the world as sea level continues to rise.

—Danny M. Vaughn

soil losses during high-intensity rainfall events. Seeds migrate within these enclosures, germinate, and grow a variety of shrubs and bushes, providing added protection for crops.

The basic means for mitigating these problems can be addressed, in part, through conservation agriculture. Soil conservation restores a soil's ability to sequester carbon, increases its water-holding capacity, retains nutrients through reduced plowing (no-till farming), minimizing the need for agrochemical compounds, and protects the soil from excessive rain, sun, and wind. Crop rotation will increase the variety of soil plant matter and the water and nutrient-holding capacity. The Food and Agriculture Organization of the United Nations reports that farmers who apply modern conservation practices use 30 to 40 percent less energy than current traditional farming practices.

Taxonomic Classifications: What Constitutes Prime Farmland

Many forms of soil classifications exist throughout the world. Most are local, regional, national, or continent specific. A standard system of soil classification, based on agricultural chemistry, biology, geology, and morphology, was developed as a unified concept by the end of the 19th century. More contemporary studies report soil classification as an aggregate of interrelated physical, chemical, and

Table 23. Major Soil Orders

Soil Order	Description
Alfisols	Alfisols occur under semi-arid to moist climates. Soils form by leaching clay minerals and other materials from the surface layer and are deposited into the subsoil. They form under a forest and vegetative cover and result in good production for most crops. Alfisols make up about 10% of the Earth's ice-free land surface.
Andisols	Andisols occur in cool regions with moderate to high precipitation. Soils develop from the weathering of volcanic materials with high water and nutrient-holding capacity. Andisols make up about 1% of the Earth's ice-free land surface.
Aridisols	Aridisols occur in arid (desert) regions. These soils are too dry to sustain growth of mesophytic plants. The dry climate restricts weathering processes, which results in an accumulation of gypsum, salt, calcium carbonate, and other materials. Aridisols make up about 12% of the Earth's ice-free land surface.
Entisols	Entisols occur in regions of recently deposited parent material (dunes, steep slopes, floodplains) due to rapid erosion and deposition. The rate of soil development lags behind weathering processes, resulting in a weakly developed pedogenic horizon. Endisols make up about 16% of the Earth's ice-free land surface.
Gelisols	Gelisols occur at higher latitudes and elevations, and have permafrost near the soil surface with evidence of frost churning or ice segregation throughout the upper horizon. Gelisols make up about 9% of the Earth's ice-free land surface.
Histosols	Histosols occur mostly in saturated environments, although some are freely drained. The soils are rich with organic material and derived from the decomposition of plant remains within forest litter, water, and mosses. Histosols make up about 1% of the Earth's ice-free land surface.

Source: Natural Resources Conservation Service of the U.S. Department of Agriculture. Modified from http://soils.usda.gov/technical/soil_orders/.

biological processes. Quantitative studies have provided a detailed data set, describing the physics, chemistry, mineralogy, and biology of various soil profiles as they have developed historically throughout specific environments.

The world distribution of soil is identified based on five factors: living organisms, parent material, climate, topography (local relief), and time. The soil taxonomy published by the Natural Resources Conservation Service of the U.S. Department of Agriculture (USDA) includes 12 orders; Tables 23 and 24 briefly outline the fundamental characteristics of the soil orders.

The USDA describes prime farmland as having the best combination of physical and chemical properties that will produce the highest yield and healthiest product as a source for food, animal feed, and fiber from specific vegetation types. The basic criteria of optimal soil quality, moisture content and availability, growing season,

Table 24. Soil Orders

Soil Order	Description
Inceptisols	Inceptisols occur under semi-humid to humid environments from a moderate but rapid degree of weathering processes on parent material. They form quickly and do not possess measurable amounts of clay, iron, aluminum, or organic matter. Inceptisols make up about 17% of the Earth's ice-free land surface.
Mollisols	Mollisols form under grasslands with climates characterized by a pronounced moisture deficit. The surface is dark in tone, resulting from a relatively high organic content, and very fertile. Mollisols make up about 7% of the Earth's ice-free land surface.
Oxisols	Oxisols occur throughout the tropics and subtropics and are highly weathered. The horizons tend to be indistinct, and possess inactive minerals such as quartz, kaolinite, and iron oxides with low fertility. These soils have an inability to retain additions of lime and fertilizers. Oxisols make up about 8% of the Earth's ice-free land surface.
Spodosols	Spodosols develop from coarse-textured deposits underlying coniferous forests in humid climatic regions. They tend to be infertile and acidic with a gray (quartz rich) upper horizon overlying a reddish brown to black subsoil due to eluviation of minerals and organic matter. Spodosols make up about 4% of the Earth's ice-free land surface.
Ultisols	Utisols form under humid conditions from intense weathering and leaching processes resulting in clay-rich (quartz, kaolinite, and iron oxides) subsoil. The soil is acidic and has a low capacity to retain additions of lime and fertilizers. Ultisols make up about 8% of the Earth's ice-free land surface.
Vertisols	Vertisols occur under varied climates and are characterized by expandable clay minerals that undergo changes in volume with the addition of water. Swelling of the clay results in very little leaching and transmission of water to the lower layers, although the soils are high in natural fertility. Vertisols make up about 2% of the Earth's ice-free land surface.

Source: The Natural Resources Conservation Service of the U.S. Department of Agriculture. Modified from http://soils.usda.gov/technical/soil_orders/.

and location must be met to warrant this label. A mixture of minerals, organic matter, and soil moisture must exist and be capable of being replenished by natural processes or human-induced processes involving fertilizers and growth stimulation. Water availability is critical and must be supplied in a timely and dependable manner through precipitation events or irrigation systems.

Other conditions necessary for good crop yields include moderate temperatures, a lengthy growing season, an acceptable level of soil acidity or alkalinity (crop dependent), limited amounts of sodium or other salt content, and a minimum of rocks within the soil layers. Air and water must be able to infiltrate the soil layers, so the soil's permeability and porosity must combine to form a soil that is not compacted or possessing layers of hardpan and other obstructions. A prime soil is not susceptible to flooding, excessive droughts, mass wasting, or other conditions that would significantly change its physical and chemical composition. Other important considerations are topography, which should not be overly steep, excessively undulating (rolling), or located at too high of an elevation or latitude. Higher elevations and higher latitudes will result in shorter growing seasons, extreme temperature variations, and an uneven seasonal distribution of precipitation, with extremes in terms of both water excess and shortages being common. Likewise, those areas are characterized by a propensity toward mass wasting processes that remove valuable top soil due to high-wind speeds, driven by polar jet streams in higher latitudes, and a closer proximity to subtropical jet streams at higher elevations within the mid-latitude belts.

Each geographic region is unique in its ability to sustain agriculture. When assessing an area's prospects, it is important to understand which types of crops can be successfully grown, given the physical and chemical conditions of the soil, its ability to replenish necessary nutrients, and a climate conducive for production. Soil is a critical natural resource that plays a major role as a host medium promoting agricultural yields directly associated with food production. Soil is considered a primary matrix necessary to sustain the world food supply for the 21st century.

—Danny M. Vaughn

Further Reading

Finfacts Team. "Global Food Production Will Have to Increase 70% for Additional 2.3 Billion People by 2050." September 24, 2009. www.finfacts.ie/irishfinancenews/article_1017966.shtml.

Ghanem, H., K. Stamoulis, and M. Smulders. "The State of Food Insecurity in the World 2008: High Food Prices and Food Security—Threats and Opportunities." 2008. ftp://ftp.fao.org/docrep/fao/011/i0291e/i0291e00a.pdf.

Kotkin, J. *The Next Hundred Million: America in 2050*. New York: Penguin Press, 2010.

McGourty, C. "Global Crisis to Strike by 2030." British Broadcasting Company, 2009. http://news.bbc.co.uk/2/hi/uk_news/7951838.stm.

Media Centre. "One-Sixth of Humanity Undernourished—More Than Ever Before." Food and Agriculture Organization of the United Nations, June 19, 2009. www.fao.org/news/story/en/item/20568/icode/.

Molden, D. "Water for Life: A Comprehensive Assessment of Water Management in Agriculture." International Water Management Institute, 2007. www.iwmi.cgiar.org/assessment/files_new/synthesis/Summary_SynthesisBook.pdf/.

Soil Survey Staff. "Soil Survey Manual: Chapter One." In *Soil Taxonomy: A Basic Systems of Soils Classification for Making and Interpreting Soil Surveys*. Natural Resources Conservation Service, U.S. Department of Agriculture, 1999. http://soils.usda.gov/technical/manual/contents/chapter1.html.

Soil Survey Staff. "The Twelve Orders of Soil Taxonomy." In *Soil Taxonomy: A Basic Systems of Soils Classification for Making and Interpreting Soil Surveys*. Natural Resources Conservation Service, U.S. Department of Agriculture, 1999. http://soils.usda.gov/technical/classification/taxonomy/.

Tulloch, J., ed. "Agriculture and Climate: Saving the Soil." 2008. http://knowledge.allianz.com/en/globalissues/safety_health/food_water/agriculture_sustainable_farming.html.

"United Nations Conference on Desertification 2010." April 3, 2010. http://europa.eu/legislation_summaries/development/sectoral_development_policies/r12523_en.htm.

Verheye, W. H. "Land Cover, Land Use and Global Change." In *Encyclopedia of Life Support Systems*, Vol. I, edited by W. H. Verheye. Washington, DC: National Science Foundation.

Water, Energy, Health, Agriculture and Biodiversity (WEHAB) Working Group. "A Framework for Action on Biodiversity and Ecosystem Management." In *World Development Report 2008: Agriculture for Development*, World Summit. World Bank, 2008. http://siteresources.worldbank.org/INTWDR2008/Resources/2795087-1192111580172/WDROver2008-ENG.pdf.

Sustainable Food Production

Sustainability of food production has been a topic of concern since the early 1800s, when Thomas Malthus addressed the issue with respect to feeding the growing global population.This topic has been discussed in great detail by academics for more than 200 years. Clive A. Edwards and his team, in their 1989 book *Sustainable Agricultural Systems*, outlined a series of recommendations that policymakers were advised to consider regarding the sustainability of food production.It was not until 1993, however, with the establishment of the U.S. President's Council on Sustainable Development and its task force on Sustainable Agriculture, that serious policy talks began to develop surrounding sustainable food production.

Sustainability defies definition, but generally refers to practices that are able to continue indefinitely without exhausting the resources and degrading the environment that support the practices. Most human activities on the planet do not meet this criterion. Instead, practices that are initiated continue for some time until one or more critical resources become unavailable or too costly. At that time, the activity stops. Examples of this linear progression are many, such as those described in Jared M. Diamond's *Collapse: How Societies Choose to Fail or Succeed*. Usually the intensity of the activity increases because of growing pressure on its critical resources caused by population increase. At the same time, serious environment degradation occurs in the process.

Problems of Sustainability

Agriculture has successfully produced enough food to feed the planet's inhabitants up until this time. Although hunger still occurs in certain parts of the planet, William A. Dando writes that this phenomenon is not due to an overall lack of food, but rather the lack of distribution of the existing food to those places. Innovations in agriculture over time have enabled food production and its potential to far exceed the human demand for food.

When the second Agricultural Revolution in the 19th century introduced the use of animal fertilizer, crop rotation, and crop genetics, the Malthusian worries were overcome. In the 20th century, hybridization, the Green Revolution, mechanization, chemical fertilizers, pesticides, and herbicides enabled mass production of food for the planet. Today, in the 21st century, with the onset of genetic engineering, agriculture has made possible a vast food potential.

Concerns About Sustainable Food Production

Modern-day farming's heavy dependence on fossil fuels to power farm machinery and produce chemical fertilizers, pesticides, and herbicides is not sustainable. Fossil fuels are limited, and it is only a matter of time before petroleum-based agriculture will collapse. Much of the infrastructure of modern agriculture relies on this expensive technology. To overcome this collapse, either other technology will need to be discovered or agriculture in other parts of the planet will need to be developed, using different strategies. Dependable agricultural solutions must be dispersed over the planet and not disempowered by the overwhelming agro-economies of the scale of developed countries.

A Case Study: Traditional African Agriculture

Traditional African agriculture has been able to sustainably feed its populations in the past. Although famines and much loss of life have marked the history of

Africa, the African people have survived. As an evaluation of African agriculture shows, there are major cultural differences between definitions of sustainable food production. The level of sustainability in sub-Saharan Africa has not been deemed acceptable by North American and European standards, and much intervention has taken place to "improve" the lives of those who live in sub-Saharan Africa. Since the last half of the 20th century, major efforts have been undertaken to enhance development in Africa by investing money, donating food, introducing new agricultural systems, and advancing medical care.

The donation of food by the developed countries has had disruptive results on existing African farming systems. While the food is going to hungry people, the effort has undermined the prices and profitability of local African farm production. Capital investment in agriculture, although welcomed gratefully, has had short-term impacts that do little to empower small farmers to engage with the larger commercial system. Frequently, foreign aid arrives in the form of high technology that may function for a short time, yet there is no allocation for long-term maintenance; thus, when repairs are needed, no parts are available. Developed nations are ill equipped to provide appropriate technology necessary in the sub-Saharan context.

Sub-Saharan Africa is capable of producing its own food and meeting the challenge of feeding its growing populations, but not in the context of the large-scale commercial farming undertaken in Western nations. An African solution is possible if Africans can limit the influence of Western agricultural models in their countries. At the same time, they must empower their own farmers with price supports while imposing import duties on foreign food, provide infrastructure in the form of roads and markets, and develop and protect agro-industries. Agricultural extension services could then be successfully implemented, not to promote established approved practices, but rather to disperse knowledge of practices that have proved successful in context of the nearby local environments.

A disheartening example of agro-disempowerment occurred in Mozambique in 2004–2005. A developing broiler industry was successfully providing chickens to the Maputo community. Suddenly containers of frozen chickens began arriving from Brazil, which were sold for half the price of the Mozambican chickens. The government offered no protection to its local chicken farmers and their self-developed infrastructure. The infant broiler industry collapsed overnight.

Sub-Saharan Africa has repeatedly experienced food shortages because of lack of governmental support, poor agricultural planning, political instability, and inadequate infrastructure. Local African governments have looked to wealthier countries for ways to improve their agriculture but have not looked within to discover their own innate agriculture potentials.

The Developed Country's Model

The world sees agriculture in North America and Western Europe as the model for the underdeveloped world because of its present success in producing huge surpluses that have been used to feed the world. Western agencies routinely spend millions of dollars trying to implement their agricultural models in tropical Africa. Africans sit by and allow these investments, which put at least some money into their economies and perhaps even more into the pockets of the politicians. Inevitably, implementation of the Western agricultural models entails costly inputs that can be sustained only as long as the foreign aid continues. Few of these efforts survive over the long term.

African governments typically ignore their own small farmers and regard them as inefficient and not worthy of investment because most of these farmers contribute very little to their national economies and certainly do not attract extravagant grants from the West. Having been marginalized by their governments, the small farmers in turn find no other alternative except to engage in survival agriculture that can feed their families. Usually these farmers produce some surpluses and are able to sell small amounts to support special needs such as paying school fees for their children and purchasing a few Western items such as medicines and second-hand clothing. Nevertheless, most of the surplus crops are not sold; they can be found rotting in the fields because of the lack of infrastructure, the poor quality of rural roads, the lack of means to transport them to markets, and the absence of cottage agro-industries to preserve them.

Temperate agriculture in the Northern and Southern Hemispheres was successful originally because of cold winters that killed weeds and insects. In the spring, the farmers planted their seeds before any of the weeds, insects, and other yield-reducing pests had emerged. This successful start gave temperate agriculture credibility around the world. Western farmers also found that monoculture was extremely profitable once they began using heavy applications of fossil fuels and chemicals to enhance growth and to kill pests and weeds. Although not sustainable in the long run, these practices became part of the developed world's agriculture model. Today, developed countries continue to protect this model by supporting their farmers with many government programs.

As a profit-driven industry, agriculture in developed countries has become "big business," operating on economies of scale and generating surpluses large enough to feed the world twice over. These surpluses are exported around the world, driving prices below the level where the small farms in developing countries can compete. As they become further marginalized, millions of these farms retreat into their subsistence modes of operation. According to the Population Reference Bureau, approximately half the world's population and two-thirds of those who live in Africa engage in semi-subsistence agriculture.

Farming in the Tropics

In tropical agriculture, there is no winter to kill the higher diversity of pests. Western colonialists and later food corporations, with their advanced technologies, costly fossil fuel needs, and high chemical inputs, have been very successful in the cultivation of high-demand specialty crops that can be grown only in the tropics. Coffee, tea, bananas, cocoa, sugarcane, pineapple, cashews, and a variety of other foods are examples. These crops are organized largely in plantation format, employ cheap local labor, and contribute to the national income. Very few of these products are consumed locally, and very little of the value added in the processing contributes to the local economy.

Approximately 65 percent of Africans live in rural areas are marginalized on the periphery of small urban-based African economies. It appears that if you can do nothing else, you farm. With limited capital, marginal agricultural land, little water, and minimal farm equipment, millions of Africans attempt to produce enough food to supply the needs of their families—a type of farming known as subsistence farming, Many Africans have been left to their own devices to discover, through trial and error, how to feed their families on the leftover land. Many of their agricultural practices had been in place long before plantation farming was introduced and long before the colonial masters arrived. Subsistence farming has been the standard way of life for centuries, if not millennia. For most Africans, this tradition means continuing the way that father did, and as grandfather before him.

Africans have had to adapt to their natural environment—one where many things survive naturally, without the disease and pest controls that a winter provides. Monoculture is not possible because the expense of various inputs, such as fuel, chemicals, and technology, is insurmountable if not unthinkable. Traditional farmers are totally dependent on themselves and their community when crop failures occur. However, total crop failures are rare in most of Africa: If one crop falters, the other crops will likely succeed in producing the necessary food. Polyculture is Africans' crop insurance; there are no government bailouts and there are no price supports.

The African Solution

The high biodiversity that is the bane of monoculturists is a blessing to those who engage in polyculture. The solution is to match the high biodiversity of the natural tropical environments with the high biodiversity of local agriculture. Here, many crops are grown together, side by side, planted in the same hole, arranged vertically in three-dimensional space. Tall trees tower over shorter trees. They, in turn, shade the crops growing closer to or on the ground, each nurturing the other through various symbiotic relationships.

Figure 25. Cacana (Cucurbitaceous) more formally known as *Momordica balsamina*, is an invasive plant that grows spontaneously following land clearing in Mozambique. Locals find that it makes a nutritious salad. (Photo by Frederick L. Bein)

One plant provides nitrogen for another, while sturdier plants protect delicate plants from the hot tropical sun. One plant may repel the insects that plague a neighboring plant, while another provides support for a climbing plant. The system maximizes the capture of solar energy. Every space is filled with plants. Useful wild plants fit into the mixture, providing spices, medicines, and artisan materials. Invasive plants or weeds have little chance to survive because their niches are claimed by producing plants. When weeds do occur, farmers find uses for them. Weed dishes add balance to many meals (see Figure 25).

Tropical Biodiversity

Biodiversity is greater in the humid tropics because environmental limitations are less constrictive than in the temperate climatic regions of the world. Because winter kill does not exist and prolonged drought is infrequent in areas where most Africans live, the natural biodiversity of pests such as weeds, insects, and diseases are much more expansive. Such factors are extremely difficult and expensive to eliminate in the tropics. However, the natural predators of the weeds, pests, and

Figure 26. The ratio of insect predators to insect pests in Papua New Guinea garden plots

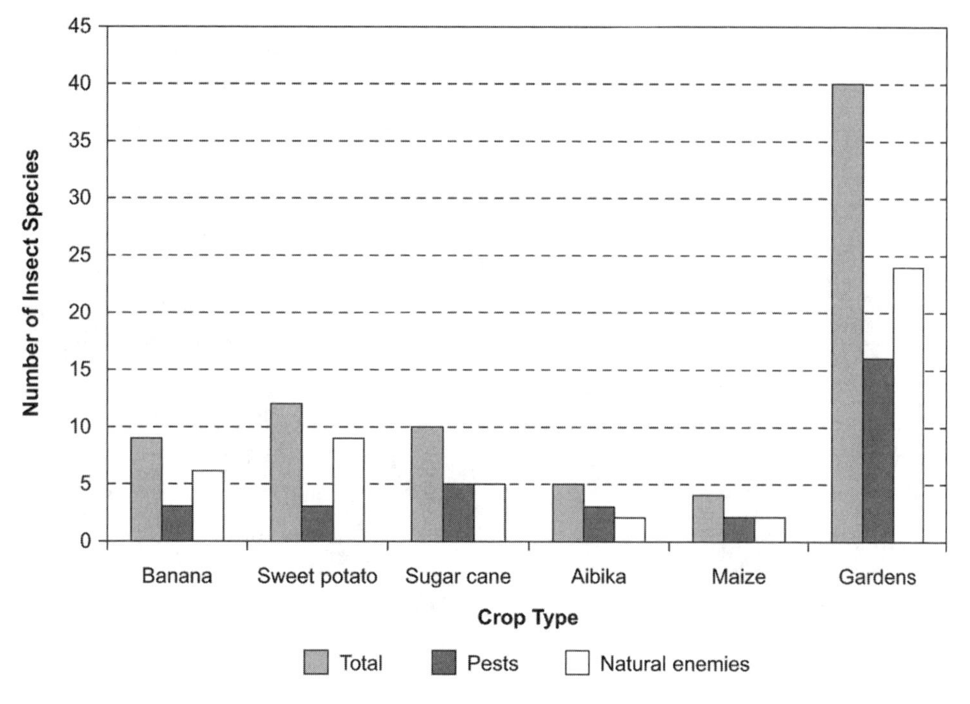

Source: Frederick L. Bein.

diseases also are more diverse and react to control their numbers. The higher biodiversity reduces the pest and disease problems that would have been stimulated by monoculture. It is the biodiversity itself that controls (not eliminates) pests. A study done by K. S. Powel in Papua New Guinea in 1998 showed that insect pest species outnumbered their insect predator species outnumbered the insect pest by 50 percent (see Figure 26).

Genetic Biodiversity in Traditional Seed Stock

The Nuba Mountain people of Sudan in the 1970s maintained their ancient sorghum grain stock by continually planting and replanting collected seed every year. They did not accept the hybrid seeds that were promoted in the western Sudan, even though the hybrids produced two to three times their yields. When asked how they survived the devastating drought of the early 1970s, they responded, "What drought?" In Figure 27, unthreshed sorghum grain heads of many different colors and seed sizes can be seen. This apparent biodiversity in seed color and size is matched with much unseen biodiversity that exhibits resistance to drought, plant diseases, flooding, bird invasions, and many unknown crop hazards.

Figure 27. Genetic biodiversity is evident in the color and size of the sorghum seed stock maintained by the Nuba Mountain people. (Photo by Frederick L. Bein)

Conservation of Labor

Mechanization is impossible in traditional African agriculture, as it would be destructive to the large mixture of crops grown together. This is not a problem in the foreseeable future, as Africa has a surplus of cheap labor, most of which consists of free family labor. On traditional farms in Africa, there is never a time when farmers have nothing to do. Because of the diversity of useful plants, many types of agricultural activities create a high demand for human labor. Animals for farming tasks are used by some families and are best suited for this type of farming. Also, on these farms, only the tasks essential to the farming process are performed, leaving little time for neatness. Jobs such as clearing large dead trees or removing termite mounds are left undone because the trees provide nesting places for insectivores that help control some of the pests and the termites can supplement diets.

Controlled burning is undertaken to clear the land—a practice that saves labor and also helps reduce pests. Dead trees may remain in place for some years while the small controlled burns eat away at the wood. In addition, pieces of the wood are collected for fuel when necessary.

Most of the farmers' labor is involved with growing annual and biannual crops. By having good ground-cover crops, they have little need for weeding. Land

Figure 28. Study areas in Africa

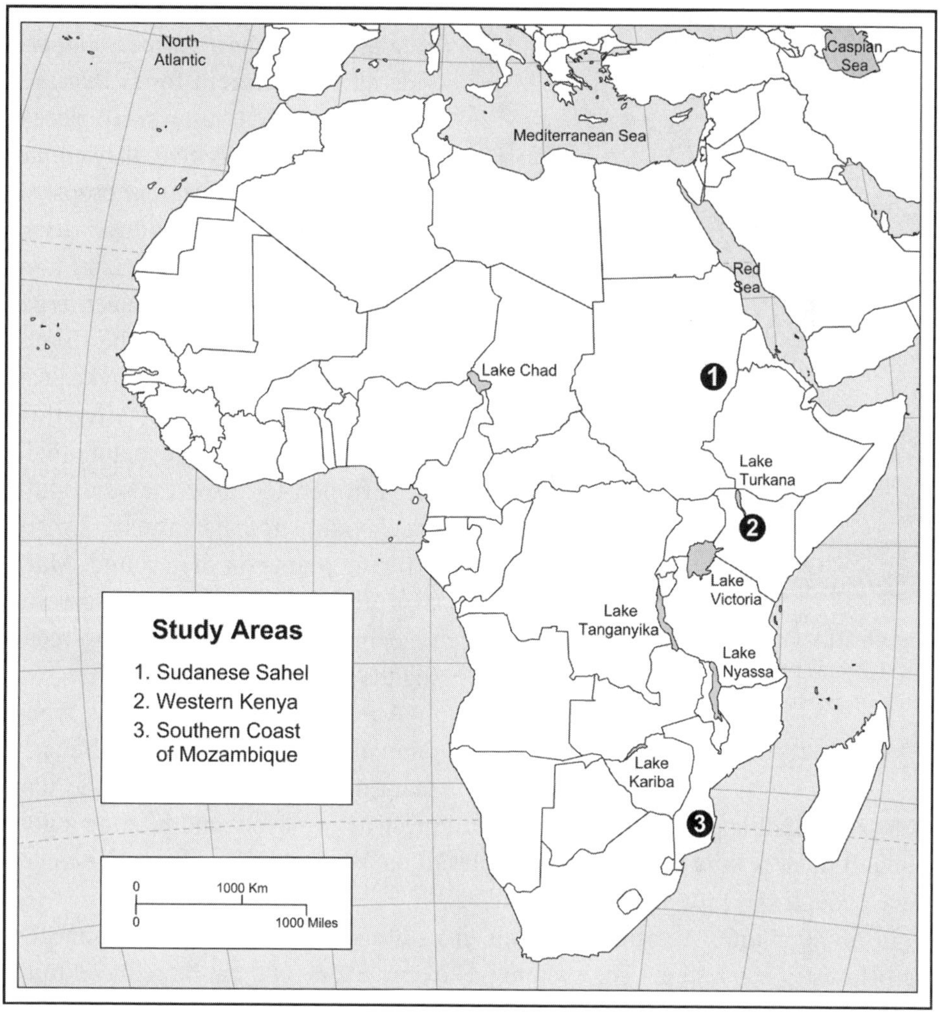

Source: Frederick L. Bein.

preparation can be arduous, particularly if the soil being prepared has been fallow for a long time. Here, the use of controlled burning reduces labor requirements.

Survival Strategies

African farmers have discovered many survival strategies over the course of hundreds and thousands of years. These strategies or adaptations differ in each region depending on the local climate, soils, ecology, farm size, and cultural traditions (see Figure 28).

Figure 29. Coconut trees tower over papaya and banana plants, which shade leafy vegetables on the ground. (Photo by Frederick L. Bein)

Generally speaking, the smaller the farm size, the more biodiverse the cropping systems. These farmers can provide all the different foods that their families need from just a small piece of land. To achieve this goal, they employ a wide mix of the different crops, yet achieve higher total productivity on their farms. In western Kenya, the farmers plant longer-cycle crops such as cassava and pineapple under the shorter-cycle maize. The pineapple and cassava grow slowly at first, but after three months the leaves of the mature maize are stripped to allow the sunlight to reach and nourish those slower-growing plants on the ground. Maize ears are left on the stalk and harvested on demand over several months, reducing the need for storage.

Another Kenyan strategy, where annual rainfall is highly variable, is to plant maize and sorghum in the same hole. If rainfall is plentiful, the maize will grow aggressively and provide a good yield. It there is little rain, the maize will fail but the sorghum will survive to produce enough of a crop to feed the family.

In Mozambique, a variety of crops grow at different heights in a four-story agro-forestry ecosystem. The system has many layers, but can be reduced to the four dominant ones described here:

1. The highest (fourth) story is occupied by well-spaced coconut palms. These palms, when fully grown, generally protrude above any other trees. Their products include copra, palm wine, roofing, brooms, construction beams, coconut milk, and many other items.
2. The third story contains both cultivated and wild trees. These trees include introduced domestic trees such as the mango, cashew, and citrus and many native wild trees and bushes. Generally, the native species are useful in providing many different products, including fruit, material for construction or artifacts, fiber, medicines, and firewood. This layer varies in height: Some trees, such as the mango and the indigenous waterberry, grow almost as high

as the coconuts, whereas others, such as the citrus and guava, grow to only12 feet or so.

3. The second story is occupied by short-lived plants that grow upright off the ground, including cassava, maize, sorghum, and bananas.
4. The first story covers the ground and includes peanuts, cowpeas, pumpkins, wild ground cover such as cacana, and a variety of vegetables (see Figure 29).

Figure 30. Farmers harvest cassava in Inhambani Province, Mozambique. (Photo by Frederick L. Bein)

Plant life cycles and environmental micro-zones add further agricultural diversity to this high-precipitation coastal area. Plant life cycles vary as crops mature throughout the year. Some crops, such as cassava (see Figure 30), papaya, bananas, and coconuts, can be harvested at any time during the year. In contrast, the annual crops and tree fruits must be collected at the end of their specific growing seasons. The life span of the different food plants (see Table 25) offers another element of diversity, as some (e.g., coconut trees) live for decades and others survive for only a few months.

Traditional Food Storage Systems

The storage of food after harvest is as important as the cropping system. This is especially true when particular staples need to be ready for processing. As

Table 25. Crop Life Cycles in Four-Storied Agriculture, Inhambane, Mozambique

Cycle Time	Crops
Long	Palms and fruit trees
Medium	Bananas and papaya
1–2 years	Cassava, pineapple, sugar cane
Annual	Maize, cowpea, peanuts, vegetables

Figure 31. Maize being fumigated by kitchen smoke in Limpopo District in Mozambique. (Photo by Frederick L. Bein)

previously mentioned, dried maize is left connected to the stalks where it is grown on very small farms in western Kenya. Other storage examples that effectively maintain the quality of food throughout the year include these two examples:

- Storing maize in "corn cribs" over the farmer's household kitchen, as in the Limpopo District in Mozambique, is part of a long tradition. Ears of maize in these corn cribs undergo fumigation from the smoke flowing upward from the charcoal cooking stoves in the kitchens. This tradition eliminates many of the insect and fungus pests that would otherwise destroy much of the food stored (see Figure 31).
- Farmers in the Sudan have learned that they can store their sorghum and millet in pits under the residual black cracking clays where they remain dry. When wet, the clay swells and prevents any downward percolation of water below 6 inches. Sorghum can be kept for years this way, providing a steady supply of food during drought years (see Figure 32). Pest problems are minimal if these granaries are constructed in open areas away from habitations where rodents are more prevalent. Because of the aridity of the region, insect problems are few.

Figure 32. Sorghum grain being stored in clay pits. The clay soil expands when wet and keeps the sorghum dry underneath. Chaff from the sorghum harvest is added as an insulator layer to separate the grainy soil particles from the grain. (Photo by Frederick L. Bein)

Sustainable Agriculture and Population Growth

Because traditional agriculture is labor intensive, children are viewed as assets to work the farms and birth rates are extremely high. High population growth can challenge the sustainability of the system. Currently, birth rates in sub-Saharan Africa average 36 per 1,000; death rates are approximately 13 per 1,000. At these rates, the area's population could double in 29 years. Such a growth rate—the highest in the world—poses a serious challenge to the sustainability of this traditional agricultural system.

An African Model for Food Self-Sufficiency

Traditional agriculture in sub-Saharan Africa has developed over thousands of years. This peasant technology is truly dependable and sustainable. While their techniques maintain the local peoples in a subsistence mode, it is possible that commercialization can be achieved if local governments provide assistance with development of roads, markets, subsidies, and agro-industries. There is little use of fossil fuel or other chemical inputs, which contrasts dramatically with the agricultural practices used in developed countries.

Rural–urban migration is proceeding rapidly, but the slums of the African cities offer little improvement in a person's or family's lifestyle. If governments decided to give priority to providing infrastructure, financial support, and guidance to their agrarian communities, the migration to the cities could be reduced. A reduction in rural–urban migration would enable city planners and administrators to more humanely absorb those immigrants who do come. At the same time, national self-sufficiency in food production and food products for export could be achieved.

—Frederick L. Bein

Further Reading

Bein, F. L. "Response to Drought in the Sahel." *Journal of Soil and Water Conservation* 35, no. 3 (1980): 121–123.

Bein, F. L., and Christopher Hill. "Four Storey Agriculture in the District of Massinga, Province of Inhambane, Along the Mozambican Coast." *Focus on Geography* 53, no. 4 (2009): 62–65.

Bein, F. L., and Gilbert Nauru. "Traditional Agriculture: Indigenous Crops, Production and Utilization in Kenya." Paper presented at the annual meeting of the Association of American Geographers, Las Vegas, NV, March 2009.

Bonar, James, ed. *Malthus and His Work*. New York: Augustus M. Kelley, 1966.

Dando, William A. *The Geography of Famine*. London: Arnold, 1980.

Diamond, Jared M. *Collapse: How Societies Choose to Fail or Succeed*. New York: Viking Press, 2005.

Dover, M. J., and L. M. Talbot. *To Feed the Earth: Agro-ecology for Sustainable Development*. Washington, DC: World Resources Institute, 2006.

Edwards, Clive A., Rattan Lai, Patrick Madden, Robert H. Miller, and Gar House, eds. *Sustainable Agricultural Systems*. Soil and Water Conservation Society, 1989.

Powel, Kevin S. "Pest and Natural Enemy Survey of Kamiali Garden Area." In *Biodiversity Inventory of the Kamiali Wildlife Management Area*, edited by F. L. Bein. Lei, Papua New Guinea: Papua New Guinea University of Technology, 1997.

T

Teaching Concepts

In an attempt to support the teaching of the concepts of famine and chronic hunger (also known as chronic undernourishment or food insecurity) and to teach about food and soils, this entry provides some resource suggestions. The suggested resources enable an educator or researcher to become more knowledgeable about the topics and concepts. They are meant as a starting point, however, and are not the only possibilities available. When studying or researching the concepts of famine and chronic hunger, a person must be cognizant of a resource's funding sources, political affiliations, and other aspects that may skew the source's perspective. Teaching students and others to become more careful about resource sources and their information is a key component of 21st-century thinking skills. Empowering students and others with definitions of key terms and concepts also becomes a starting point for instruction and a foundation upon which further research can take place.

Food

Food is the substance eaten to provide nutritional support for the body or consumed for pleasure. What do most people think when the word "food" is mentioned? A snack? A five-course meal? Dessert? Junk food? To some people, food equates to being satisfied, eliminating the pangs of hunger, healthy diets, eating whatever and whenever desired, be it at home or in a restaurant. Most Americans do not think of hunger on a daily basis. They do not think about the lack of access to food but of diverse foods, healthy foods, foods to satisfy heart, mind, soul, and body. Food production, food distribution, money to purchase food, seasons, weather, climate, soils, poor political leadership (or mis-governance), war, civil unrest, and extreme natural hazards, such as an earthquake or flood—all are factors that contribute toward the local, state, national, continental, and global access to food. Enough food is produced throughout the world for all, but it is not adequately distributed throughout the world.

What does it mean to a person to be hungry? Hunger may be a feeling of emptiness when a person skips breakfast. It may be the sensation of "starving" after a physically demanding day of activity. To millions of people around the world, chronic hunger (consistently hungry) is a part of their lives, a daily challenge. Often, it means eating only one or two "meals" a day for months and years. Meals with little or no nutritional value lead to a weakened immune system and an underdeveloped body and mind. To most Americans, a meal for the very poor may be just one food item: a potato, a mango, mashed tuber roots, grains, or milk. Most chronic hunger could be eradicated. Famine occurs when large numbers of people are dying from acute malnutrition—literally wasting away. Famines occur when there is a drastic and widespread shortage of food. The causes of famine are usually existing poverty and chronic hunger, combined with climatic shock (drought or flooding), civil unrest, political conflict, or unwise governmental decisions.

In 2010, nearly 9 million children younger than the age of 5 died needlessly, more than half from hunger-related causes (see the "Famine" entry). One can imagine 25 jumbo jets filled with children, crashing into the ground on a daily basis. Chronic hunger/undernourishment affects more than 800 million people in the world and is the leading contributor to global deaths. Many of these deaths have very little to do with starvation; instead, lack of access to food and, therefore, to consistent nutrition contributes to susceptibility to common illnesses and other factors that may arise, such as diarrhea/dysentery, malaria, measles, and famine. A consistently hungry body is a greatly weakened system.

More than 1.4 billion people currently live below the international poverty line in the developing world. The international poverty line, as defined by the World Bank in 2005, is the equivalent of earning $1.25 (U.S. dollars) per day. The concept of poverty can be viewed from many definitions: absolute poverty, relative poverty, national poverty, overstating poverty, and understating poverty. Due to this lack of economic viability (earned income), more than 1 billion people in the developing world consume less than the minimum amount of calories essential for sound health and growth. When a person must spend time and energy securing food, less time is available for work and earning income.

Chronic Undernourishment and Hunger

The *Freedom from Hunger* website provides helpful statistics and explanations for use in the classroom and stresses that chronic undernutrition and malnutrition have very little to do with food shortages. Global food supplies exceed demand, yet global food distribution is not efficient or equitable for many reasons: lack of infrastructure, poor political leadership, corruption, conflict, spoiled natural resources, and lack of empowerment. According to several sources, five key factors contribute to world hunger:

1. *Poverty.* Poor people do not have the resources (i.e., land, tools, money) needed to grow or purchase food on a consistent basis.
2. *Armed conflict.* War disrupts agricultural production, and governments often spend more money on arms than on social programs.
3. *Environmental overload.* Over-consumption by wealthier nations and rapid population growth in poorer nations strain natural resources and distribution of resources, making it more difficult for poor people to feed themselves.
4. *Discrimination.* Lack of access to education, credit, and employment is often the result of racial, gender, or ethnic discrimination.
5. *Lack of power.* People who do not have ability or means to protect their own interests go hungry. The full impact of this lack of power falls mostly on children, women, and the elderly.

Chronic hunger tends to affect generations of people within a geographic region. The regions most intensely affected by chronic hunger are sub-Saharan Africa and South Asia. On the World Bank website (http://devdata.worldbank.org/atlas-mdg), *Millennium Development Goals* (MDG), Goal 1—Eradicating Poverty and Hunger, provides an online atlas with two valuable maps focusing on malnutrition and poverty. Effects of chronic hunger include the following problems (taken directly from the *Freedom from Hunger* website):

1. *High infant mortality rates.* Malnourished women are more likely to be sick, have small babies, and die at a younger age. Results produce a high infant mortality rate, and a higher child mortality rate equates to higher birth rates.
2. *Vulnerability to common illnesses.* More than 2 million children die every year from dehydration caused by diarrhea.
3. *Increased risk of infection.* Malnourishment causes a weakened immune system, making a child more vulnerable to infection. Loss of appetite due to illness further weakens the hungry child.
4. *Acute vulnerability in times of disaster.* Crop failure, floods, epidemics, earthquakes, locusts, and typhoons lead to devastation and almost certain death to those people already living on the edge of survival.
5. *Impediments to development.* Chronic hunger deprives children of essential proteins, micronutrients, and fatty acids they need to grow. Physical and mental developments are stunted due to consistent deprivations.

Studying Hunger and Human Environment Mismanagement Issues

The study of patterns and relationships between and within physical systems and human systems forms the discipline known as geography. A classic example would be to compare two remotely sensed images or aerial photographs of the Aral Sea in the former Soviet Union, specifically an image prior to 1973 and then an image in 1987 or later. Great changes will be immediately seen. Why do you think the Aral Sea diminished in size? How will this change affect the local physical and human environments? How will it impact the region—both human and physical systems? What is the relationship between the shrinking of the Aral Sea and food, famine, hunger, and poverty? Why should this issue be of importance to others around the globe? Can the situation be rectified? Can we apply solutions from similar situations throughout the globe?

Visual imagery can be a powerful tool when used honestly. The difficulty for educators and students, and the general public, lies in their ability to discern trustworthy data. Both casual users of scientific information and experts alike face this problem. To encourage critical review of data, teachers can instruct their students that not all data are reliable and note that some data are designed to fool the observer. In the 21st century, students will need to become savvy and skilled in the art of data acquisition and data management/manipulation, which includes visual imagery, to solve problems and plan for the future.

—Kathleen Lamb Kozenski and Melissa Martin

6. *Impediments to economic growth.* Studies show that underweight children spend fewer years in school, which has a measurable impact on their earning potential as adults.

In conjunction with the two World Bank MDG maps that focus on malnutrition and poverty, the U.S. Central Intelligence Agency's *World Factbook* (www.cia.gov/library/publications/the-world-factbook/index.html) provides volumes of information about places around the world, as well as maps, flags, and photographs. This website can be searched by country, region, or keyword. Searching the site with the term "poverty" produces an alphabetical listing of countries. Beside each country listed is a statistical percentage of those residents living below the global poverty line. For example, 80 percent of Chad's citizens live below the global poverty threshold.

Poverty and Famine

A bibliography of academic articles and books available via the *World Food Habits and Bibliography* can be found at www.lilt.ilstu.edu/rtdirks.FAMINE.html. Many of the suggestions found at this website were prepared for research and postsecondary academic purposes and may not be appropriate for kindergarten through 12th-grade classroom applications. Even so, some of the suggestions may aid the educator or researcher in gaining a better perspective on teaching the topics of food and famine. For example, *The Geography of Famine*, by the editor of this book, Dr. William A. Dando (1980), provides a good background on the history of famines from a geographic perspective. Many of the suggested articles and books are written through a very definitive perspective, which offers another lens through which students and others can view food, famine, and hunger.

Without consistent, reliable access to diverse food products, tens of thousands of children will die monthly, due directly and indirectly to chronic hunger in the form of malnutrition. The world moves forward without the contributions from these children. Famines can be mitigated through planning and visionary thinking—a task that could be accomplished if every hungry or malnourished person had access to nourishing food. Nations of this world must work toward eliminating hunger at the local and international levels through various outreach organizations, such as the national and international Bread for the World and the local food pantries.

—Kathleen Lamb Kozenski and Melissa Martin

Further Reading

Bread for the World. www.bread.org. A nonprofit Christian citizens' organization, dedicated to eradicating hunger locally and internationally.

Freedom from Hunger: 1995–2010. 1644 DaVinci Court, Davis, CA 95618; 1.800.708.2555; fax 530.758.6241; info@freedomfromhunger.org; www.freedomfromhunger.org.

"Global Data Monitoring Information System." World Bank, Millennium Development Goals. http://devdata.worldbank.org.

"Hunger and Famine." Save the Children, 2010. www.savethechildren.org/programs/hunger-malnutrition/hunger-and-faminehtml.

Overview: Understanding, Measuring, Overcoming Poverty. Washington, DC: World Bank, 2010.

Statistics by Area/Child Survival and Health. Rome: UNICEF, November 2009.

The State of Food Insecurity in the World: Economic Crisis—Impacts and Lessons Learned. Rome: Food and Agriculture Organizations of the United Nations, 2009.

U.S. Census Bureau. www.census.gov. Local, state, and national data; Fact Finder; educator support and classroom curriculum for all grade levels (Advanced Placement high school and postsecondary geography and history courses may need enrichments); maps; and more are available at the Census Bureau site.

Webster's Online Dictionary. www.merriam-webster.com.

Teaching Data Sources

Teaching, presenting, or researching about food, famine, hunger, and poverty can be a very daunting and intimidating task, but today's students must understand the 21st century world as it currently exists and as it can exist in the future. Students, today and in the future, will use a variety of resources to access and analyze data, to solve problems, and to plan for the future. The following information provides a starting point for an educator, student, parent, resource specialist, or citizen wishing to better understand the topics of food, famine, hunger, and poverty:

Bread for the World regularly updates a *Bread Blog* about hunger issues internationally and domestically, noting links to new articles and research. Bread for the World also includes short videos from persons involved with the leadership of Hunger 2009/Global Development: Charting a New Course. At the same link, accessible and downloadable maps and data are available. Its website addresses are www.breadblog.org and www.hungerreport.org/2009.

TED Talks provides a forum where people can introduce and discuss ideas that they believe will make life better for everyone. A few discussions focusing on agriculture (food, production, and distribution) and climate (which affects agriculture via soil, precipitation, temperatures, and other factors) are available. Some of the presentations may be inappropriate for certain age groups; teachers should watch the talk prior to classroom use.

Crops for the Future (www.cropsforthefuture.org) has created a website that connects to very diverse articles, blogs, tags, upcoming agricultural events, species, projects, experts, and additional information that can be read through (probably by a high school student or adult) to look for a variety of thoughts, perspectives, and opinions. The idea of the development of crops that are strongly favorable to a specific area or region has been around since humans settled into an agricultural lifestyle millennia ago. Much of the site connects to in-depth research and opinions, but it also enables students to understand the variations in perspectives and thoughts. Most people, ultimately, want the best for humanity, yet the process through which a person arrives at "the best" can vary greatly, depending on his or her perspective.

Teaching Hunger Sources

The best resources for teaching about food, famine, hunger, or poverty are local, state, or national specialists. For example, a teacher might invite a geographic information systems (GIS) specialist to the classroom to discuss local, state, national, and global poverty indicators through data. Data represent a very powerful tool, but viewing data through a geographic lens (spatially) enables the user to better understand any patterns and relationships that the data may convey. In other words, the user gains a better sense of space via physical and human geography. Students and educators become temporary geographers as they delve into the analysis of data through the various patterns and relationships. Specialists on food, famine, hunger, or poverty may be found at local food pantries, in postsecondary departments of geography/sociology/health and nutrition/anthropology, within various city and state agencies, and as members of local religious organizations that provide outreach programs to those in need. Teachers can fruitfully take advantage of the diverse and dynamic resources at their disposal.

—Kathleen Lamb Kozenski and Melissa Martin

Purdue Agriculture Connections (newsletter) provides agricultural research and agricultural news updates. For example, Gebisa Ejeta, Distinguished Professor of Agronomy from Purdue University, recently received the 2009 World Food Prize Laureate award for his development of varieties of sorghum (a staple food for millions of people throughout Africa) that are resistant to drought and to a certain parasitic weed. As a result of years of research and collaboration with hundreds of individuals and farmers, sorghum varieties now can be grown in larger yields (production) with the virtual elimination of the parasitic weed. Hundreds of students around the globe helped with this research over decades. Researchers and teachers can access the *Purdue Agriculture Connections* (http://www.agriculture.purdue.edu/connections) and other schools of agriculture or agronomy newsletters to read about additional research.

Freedom from Hunger is a nonprofit, nongovernmental organization, established in 1946 to provide food assistance around the world, and it has evolved since that time to confront issues faced by rural females in the poorest countries. Its assistance, in the form of financial credit and savings, health protection, and education, is intended to lead to a sustainable, self-help end to hunger and poverty. On the group's website (www.freedomfromhunger.org), a variety of short videos (Multimedia Highlights) can be found about people from around the world who are involved with Freedom from Hunger, either as advocates or as recipients. Also, a photo gallery highlights people in the various countries with which Freedom from Hunger is involved. Information about hunger and poverty aids students' perception of what hunger and poverty truly mean.

The **United Nations World Food Programme—Fighting Hunger** addresses the first of the Millennium Development Goals, which is to eradicate extreme poverty and hunger. Found on the website are downloadable lesson plans, including "The Hunger Obstacle Course," "The Hunger Tree," "Stone Soup," and much more, videos, statistics, and information about many of the world's poorest countries, with a focus on local obstacles to food security. The teacher, researcher, or student should type in the website address, http://www.wfp.org, and follow the "for students and teachers" link.

The World Bank (www.worldbank.org) provides data and information by way of the Global Data Monitoring Information System. Data can be viewed via bar charts, line graphs, maps, and other graphic visuals. In addition, information through photographs, articles, and short videos can be obtained. Data and information can be found by world region. In the search engine, the researcher should type in "World Bank and Hunger" or type in "Global Data Monitoring Information System."

Documentaries and motion pictures can provide learning experiences for the teacher, researcher, and students. Understanding food, hunger, famine, and poverty begins with understanding soil. For example, a student-friendly video about soil, *Dirt! The Movie*, may be purchased at http://dirtthemovie.org. A link is provided to downloadable classroom applications about dirt and soil.

National Geographic Society's website and publications are rich sources of information about food, hunger, famine, and poverty. In a period of climate change and increasing world temperatures, water becomes more critical in world food production. The significance of water to the 21st century's world society is emphasized at various websites, including *National Geographic Magazine*, where hundreds of photographs highlighting drought, desertification, barren landscapes, water, poverty, and agriculture are available. A puzzle-generator, articles, short video stories, and more are available for student use. *National Geographic* magazine is an affordable resource that can be archived for classroom use. During March 2010, a special issue devoted to water was published. Free, downloadable, digital editions are available at www.ngm.nationalgeographic.com.

The **National Geographic Xpeditions** website (lesson plans for classroom applications) hosts a lesson entitled "Drought, Famine, and Geographic Diversity," about the famine in Ethiopia. This online lesson links to photographs, maps, and articles supporting the implementation of the materials in the classroom.

NASA Earth Observatory or **Visible Earth** provides satellite imagery for diverse categories on a yearly basis, including "Crops and Drought," "Floods," "Fires," and others. Satellite images are a good geo-tool that empowers students in making personal connections to a place. For instance, using satellite imagery to study the current food crisis/famine in Niger, the second poorest country in the world, demonstrates that 21st-century geospatial technologies can provide early detection, solutions, and plans to rectify the problems.

World Economic Forum (www.weforum.org; under "Business Alliance Against Chronic Hunger") provides an interactive Map of Hunger. Following the internal links via "Initiatives, Agriculture and Food Security," and "Interactive Map/Hunger" provides a quick view and summarized information about hunger and hotspots around the globe.

The FOOD Museum Online (www.foodmuseum.com/issuehunger.html) gives an interesting visual and written story of various foods. The teacher, researcher, or student can follow the "Exhibits" internal link to view information about certain foods (listed alphabetically). Additional information about foods, places, and histories is available as well, such as facts, blogs, issues, and ask an expert. Students can research a food that they have never heard of or have never tasted.

Famine Foods (www.hort.purdue.edu/newcrop/FamineFoods/ff_home.html), a website developed by Robert L. Freedman, provides a database of nontraditional food sources that may be utilized in times of famine. Thinking innovatively and creatively may provide unexpected food sources and contributors during times of famine. Many of the ideas have not been researched, but the possibilities enable students to envision a future free of famine.

The History Place—Irish Potato Famine website (www.historyplace.com/worldhistory/famine/index.html) gives a historical perspective on the Irish potato famine. The site provides summaries of segments of the Irish potato famine of the mid-1800s. Historic visuals from newspapers are available on the site via the "Author/Bibliography" link. The teacher, researcher, or student can analyze the differing perspectives of the famine (nation).

The **Visual Thesaurus** (www.visualthesaurus.com) can be used to create diagrams of intersecting words for key terms that revolve around the issue of food and famine—for example, famine, hunger, poverty, food, soil, dirt, water, employment. A subscription is encouraged, but a 14-day free trial is an option. The *Visual Thesaurus* is a great tool to empower students with understanding the interconnectedness of words and meanings.

—Kathleen Lamb Kozenski and Melissa Martin

Further Reading

Bassett, Thomas J., and Alex Winter-Nelson. *The Atlas of World Hunger*. Chicago, IL: University of Chicago Press, 2010.

Beckmann, David, et al. *Hunger Report 2011 Our Common Interest: Ending Hunger and Malnutrition.* Washington, DC: Bread for the World Institute, 2011.

Dando, William, Mary Snow, Richard Snow, Melissa Martin, and Kathleen Kozenski, eds. *Climate Change and Variation: Lesson Plans, Data Sets, and Activities. Volume II.* Washington, DC: Pathways in Geography Series No. 37, 2008.

Hubbard, Jim. Lives *Turned Upside Down: Homeless Children in Their Own Words and Photographs.* New York: Aladdin Paperbacks/Simon & Schuster for Young Readers, 2007.

Jensen, Eric. *Teaching with Poverty in Mind: What Being Poor Does to Kid's Brains and What Schools Can Do About It.* Alexandria, VA: Association for Supervision and Curriculum Development, 2009.

Kogan, Felix, Alfred Powell, and Oleg Fedorov. *Use of Satellite and In-Situ Data to Improve Sustainability.* New York: Springer, 2010.

National Geographic Student Atlas of the World. Washington, DC: National Geographic Children's Books, 2009.

Parkinson, Claire L. *Earth from Above: Using Color-Coded Satellite Images to Examine the Global Environment.* Herndon, VA: University Science Books, 1997.

Shaw, D. John. *The World's Largest Humanitarian Agency: The Transformation of the UN World Food Programme and of Food Aid.* New York: Palgrave Macmillan, 2011.

Warner, Carl. *Carl Warner's Food Landscapes.* New York: Abrams Image, 2010.

Teaching Definitions

The two-volume book set *Food and Famine in the 21st Century* is designed to provide all who have interest and concern about food, hunger, and famine with the basic definitions of common terms, insights into current issues, and prospects for the future. Entries in Volume 1 and chapters in Volume 2 provide more insights and are more up-to-date than those found in any other sources. Definitions are offered for all of the basic terms, as food, hunger, and famine have been concerns of humans since the beginning of recorded history. Also, those concerned about food, hunger, and famine come from different cultures, various academic disciplines, diverse religions, different political leanings, and various time periods. A meaningful student introduction to food, hunger, and famine topics and issues is to ask them to define the basic terms and then have the students compare these terms to those found in this two-volume book set.

Starting with a common foundation enables a classroom educator to have a base from which to begin teaching about food, hunger, and famine. The most noted international source reference regarding related definitions or descriptions is the United Nations' Food and Agriculture Organization (FAO) publications. Also, *Wikipedia* has similar definitions, as does *Webster's Dictionary*, though Wikipedia has the advantage of descriptions and cross-referencing. Educators and students must realize that, on occasion, *Wikipedia* has incorrect or inadequate information. Luckily, the site enjoys receiving additional information from students. Students or participants can update, modify, improve, and rewrite the following definitions:

1. Food—Any substance or material eaten to provide nutritional support for the body (*Wikipedia*).
2. Hunger—A feeling experienced when one has a need (not desire, which is appetite) to eat food (*Wikipedia*).
3. Famine—Any widespread scarcity of food that may apply to any faunal species, which phenomenon is usually accompanied by regional malnutrition, starvation, epidemic, and increased mortality (*Wikipedia*).
4. Chronic hunger—Constant hunger leading to undernourishment and malnutrition.
5. Endemic deprivation—Means the same as chronic hunger.
6. Food insecurity—Ranges from a temporary, food-insecure situation to full-scale famine (*Wikipedia*).
7. Food security—Exists when all people, at all times, have physical and economic access to sufficient, safe, and nutritious food to meet their dietary needs and food preferences for an active and healthy life (*United Nation's Food and Agriculture Organization).*
8. Food security—Means access by all members of a household at all times to enough food for an active, healthy life. Food security includes, at a minimum, (a) the ready availability of nutritionally adequate and safe foods, and (b) an assured ability to acquire acceptable foods in socially acceptable ways—that is, without resorting to emergency food supplies, scavenging, stealing, or other coping strategies (*United States Department of Agriculture*).
9. Malnutrition—A general term for a condition caused by improper diet or nutrition (*Wikipedia*).
10. Poverty—The state of one who lacks a usual or socially acceptable amount of money or material possessions (*Merriam-Webster's Online Dictionary*).

Understanding and Experiencing Hunger

For many students, understanding hunger is difficult—that is, the idea of hunger versus the actualities of hunger. One Indiana educator teaches her middle school students about hunger by reading a storybook to the class, through researching global current events highlighting Asia and Africa and hunger, and then spending a day going hungry. After reading the book about a young person's life in hunger and conducting the research on hunger, her students experience hunger themselves. For one day, with permission from their parents and guardians, the students eat a small breakfast, watery soup and bread (no butter or jelly) for lunch, and a light meal for dinner at home. The next day, students record in a journal their physical, mental, and emotional feelings about being hungry. In this way, the lessons of the storybook and the current events research become more real for the students. Afterward, the students choose a local community service project to help the hungry. People are never too young to make the world a better place!

—Kathleen Lamb Kozenski and Melissa Martin

11. Scarcity—Insufficient supply of something (*Encarta World English Dictionary*).
12. Starvation—A state of exhaustion of the body caused by lack of food. This state may precede death (*Wikipedia*).

Food, Hunger, and Famine Perceptions

Once a common ground of terminology is shared, a more engaging discussion about student perceptions of food, hunger, and famine can occur. Local, state, national, and international issues related to the theme of food, hunger, and famine are often in the news, which make the topic relevant to most students of all grade levels and presenters, depending on the audience. Hunger and famine usually begin with some level of food scarcity and poverty. Understanding the concept of poverty enables students to truly grasp the nature of doing without (surviving), and food is at the heart of survival. It is the task of an educator to facilitate student learning, not to impart his or her perspectives onto the students. An educator must remember to discuss with students the need to be cognizant of the reality that a particular person's or organization's perspectives may not be based in fact or that their interpretation of fact may be different than another's interpretations—an exciting prospect, as students can read a diverse range of sources that may present diverse opinions or thoughts. Students can gain their own (beginning) perspective by reading on the topic, engaging in classroom discussions, and reviewing colleagues' perspectives.

Suggested "hands-on" learning activities include asking students or participants to do the following:

- Research a local pantry (or soup kitchen) or outreach center's needs, including families served per week or month, intake of foods, and most usable foods; work at the pantry; assist the pantry by hosting a school event; interview the coordinator of the facility; and write an article about the pantry/soup kitchen and publish it in the local newspaper or the school's web page; create an advertising campaign with posters, a website, or video publicity, while collaborating with the facility.
- The Atlanta Community Food Bank has created a series of six, downloadable curricula modules, each representing an aspect of poverty and hunger in the United States, and each well thought out and supplemented with resource materials from *Hunger 101* (www.acfb.org/projects/hunger_101/curricula). Created for people from ages preschool through adult, the modules are designed to inform people about poverty, hunger, and action. Specifically for elementary, middle, and high school students, *The Community Food Game* views poverty and hunger and barriers to food security, and *Stories' Windows: Representations of Hunger and Poverty in Literature* presents brief quotes from books highlighting poverty and hunger. Complete lessons, book lists, resources, and a glossary are provided via the *Hunger 101* materials.
- Read books about poverty, homelessness, soup kitchens/food pantries, hunger, and famine. Create Venn diagrams about poverty, homelessness, and hunger or about poverty, famine, and food. A few trade-book and academic-book suggestions include the following:
  - *Uncle Willie and the Soup Kitchen*, by Dyanne DiSalvo-Ryan, New York, Morrow Junior Books, 1997. Level: elementary school.
  - *The Lady in the Box*, by Ann McGovern and illustrated by Marni Backer, Madison, CT, Turtle Books, 1999. Level: elementary school.
  - *A Shelter in Our Car*, by Monica Gunning and illustrated by Elaine Pedlar, San Francisco, Children's Book Press, 2004. Level: elementary school.
  - *Fly Away Home*, by Eve Bunting and illustrated by Ronald Himler, San Anselmo, CA, Sandpiper, 1993. Level: elementary school.
  - *Sidewalk Story*, by Sharon Bell Mathis, London, Puffin Books, 1986. Level: upper elementary school and middle school.
  - *Homeless Bird*, by Gloria Whelan, New York, Harper Collins, 2001. Level: upper elementary school, middle school, and high school.

- *Breath, Eyes, Memory*, by Edwidge Danticat, New York, Vintage, reprint edition, 1998. Level: middle school and high school.
- *Angela's Ashes*, by Frank McCourt, New York, Scribner, 1999. Level: well-read middle school and high school.
- *Gap Creek: A Story of a Marriage*, by Robert Morgan, Clearwater, FL, Touchstone, first Scribner edition, 2000. Level: high school and adult.
- *Native Son*, by Richard Wright, New York, Perennial Library, 1966. Level: high school and adult.
- *The Grapes of Wrath*, by John Steinbeck, New York, Viking Press, 9th edition, 1939. Level: high school and adult.
- *The Geography of Famine*, by William A. Dando, New York, Wiley, 1980. Level: high school and adults.
- *The Geography of Hunger*, by Josue de Castro, Boston, Little, Brown, 1952. Level: high school and adult.
- *World Hunger: 12 Myths*, edited by Frances Moore Lappe, Joseph Collins, and Peter Rosset, London, Earthscan, 2nd edition, 1998. Level: adult.
- "Twelve Myths About Hunger" (an abbreviated article, based on the book *World Hunger: 12 Myths*) via Share International Archives. Level: high school and adult.
- *Hunger in History: Food Shortage, Poverty, and Deprivation*, edited by Lucille F. Newman, Hoboken, NJ, Wiley-Blackwell, 1995. A collection of 16 articles addressing the topic given in the title. Level: high school Advanced Placement and adult.
- *Famine: A Short History*, by Cormac O Grada, Princeton, NJ, Princeton University Press, 2009. An optimistic view that famines will decline in the future. Level: adult.
- *Outgrowing the Earth: The Food Security Challenge in an Age of Falling Water Tables and Rising Temperatures*, by Lester Brown, New York, W. W. Norton & Company, 2005. A futuristic view of food, water, and basic human needs (an opposing perspective from the book, *Famine: A Short History*). Level: adult.

- Conduct a food–human body analysis by investigating the food intake a person needs at various ages (based on national and international average height and weight). Research the average intake of student-age humans in the United States, compared to their counterparts in other countries: Canada, Guatemala, Brazil, Germany, Switzerland, Niger, the Sudan, Mozambique, Malaysia, and so on. Determine the types of foods grown in the various countries explored and the foods that are eaten by the local peoples. Look into the foods exported by these countries as well. Which agricultural products do these countries export? Where, in the country, are these products grown? Are the same foods exported as are eaten by the local people? Is more food exported than is available to the local people? Why or why not? Research the country's infrastructure, especially internal transportation. Does infrastructure affect distribution of local agricultural products? Read information about regional climate issues within the past 10 years. Analyze the information to answer these questions: Why do local people not receive (or receive) enough of the locally grown agricultural products? Are there regional differences?

- Participate in the Population Reference Bureau's simulation, *Life on the Edge: Land Use, Food Supply, and Economics in a Small African Village*. This is an excellent activity that introduces the concepts of nomadic lifestyles, stationary lifestyles, food, water, subsistence, food assistance, migration, births and deaths, and choices. Educators must spend approximately 20 minutes prior to class or event to prepare the activity. Laminating the simulation cards enables the activity to be used each year (or each semester). Visit the website at www.iupui.edu/~geni, follow the Lesson Plans link and scroll to "Life. . . ." This file should be saved to a hard drive for future use.
- Compare and contrast various country statistics via the Data Sheets at the *Population Reference Bureau* (PRB) website. The Data Sheets are updated annually for each country in the world and are a reliable resource for classroom use. The PRB site also provides educators with lesson plans, including guides that focus on a theme, short articles, videos, population pyramids, and glossaries.
- Construct "Population Pyramids" for developed and underdeveloped countries (e.g., Norway, United States, Mozambique, Chad, Armenia). Map the countries chosen. Compare and contrast the information available via the PRB and the CIA's *World Factbook*. Students are to create a mathematical equation for defining poverty. In small groups, they will choose statistical information categories (e.g., literacy, births, deaths, number of citizens younger than the age of 15) that they believe are vital to defining poverty. They must test their equation with two underdeveloped and two developed countries. Students must write a four-page rationale about their equation and their sample tests. Members of each group should share their equation and their rationale with their classroom colleagues. The class should then choose one underdeveloped country and provide outreach to that country in some capacity: write to a school and send needed supplies; locate a visiting doctor, dentist, or optometrist to aid in gathering information and making personal connections with the students; develop a letter regarding the needs of the country, based on available information, and send the letter to the United Nations or another globally significant group; or produce a YouTube rap/song about the plight of the country. (This idea from Orville Schlatter, a creative retired educator from East Allen Community Schools, Fort Wayne, Indiana.)

- Create "Country Recipes" based on underdeveloped places. Example:

  Ghana, Warrior King: Mix a bit of equatorial Africa with more than 23 million citizens with a pinch of the total GDP $38 million per citizen; add 3.5 percent water within 92,000 square miles and a dash of Accra, the capital; combine 28 percent living below poverty line; slowly stir in an average rainfall of 43 inches in the north and 83 inches in the southeast; mix well to gain independence in 1957; and bake at the yearly average temperature of 80°F. When done, you will find a tropical climate (drier to the north) with more than 100 different ethnic groups speaking English and local languages.

(Idea from Janis Coffman, Bellmont Middle School, Decatur, Indiana.)

- Engage students in "The Hunger Obstacle Course" by the *World Food Programme* (www.wfp.org), a simulation designed to overcome the challenges of extreme hunger and poverty. Created for grades 7–9.
- Examine Heifer International's Educator link on its website, which leads to free, downloadable lesson plans and additional materials. Within the educator materials is a guide, entitled *GET IT!: Global Education To Improve Tomorrow*, for students in grades 6–8. The guide has students analyze hunger in their community and around the world from the perspective of an investigative journalist. If downloading a copy of the *GET IT!* materials is not possible, order one by calling 1.800.422.0474.
- Review the *World Food Programme*'s Students and Teachers link on its website. Within that link is a connection to "Classroom Activities," under which are listed several stimulating lesson plans that include math, economics, language arts, social studies, critical thinking, collaborative learning, and problem-solving activities for students in grades 4–9. Read through the descriptions to determine which program and classroom applications are appropriate for your situation.

So many possibilities exist when teaching or presenting about food, hunger, and famine. The teacher or researcher should begin with a local, regional, or national issue of interest, draw upon the many available resources, and create a learning experience that will prepare all involved to make the world a better place.

—Kathleen Lamb Kozenski and Melissa Martin

Further Reading

Coyle, L. Patrick. *The World Encyclopedia of Food*. New York: Facts on File, 1982.

Dando, William A., and Caroline Z. Dando. *A Reference Guide to World Hunger.* Hillside, NJ: Enslow Publishers, 1991.

DiSalvo-Ryan, Dyanne. *Uncle Willie and the Soup Kitchen.* New York: Morrow Junior Books, 1997.

Kaye, Cathryn Berger. *A Kid's Guide to Hunger and Homelessness: How to Take Action!* Minneapolis, MN: Free Spirit Publishing, 2007.

Lappe, Frances Moore, Joseph Collins, Peter Rosset, and Luis Esparza, eds. *World Hunger: 12 Myths*, 2nd ed. New York: Grove Press, 1998.

Markandaya, Kamala. *Nectar in a Sieve.* New York: Penguin Group, 1982.

Merriam-Webster's Dictionary and Thesaurus. Chicago, IL: Encyclopaedia Britannica Company, 2007.

Park, Linda Sue. *A Long Walk to Water: Based on a True Story.* New York: Clarion Books/Houghton Mifflin Harcourt, 2010.

Pfeffer, Susan. *Life as We Knew It.* New York: Scholastic, 2006.

Williams, Karen Lynn, Khadra Mohammad, and Doug Chayka. *Four Feet Two Sandals.* Grand Rapids, MI: Eerdmans Books for Young Readers, 2007.

U

Unbounded Food Trade: A Deterrent to Hunger and Famine

World food prices are increasing at an alarming rate. In the United States, food prices at the wholesale level experienced the highest increase in 36 years during February 2011. Weather shifts and perturbations account for part of the food price increases, but demands for higher-quality foods and conversion of food grains to energy sources also contributed to this rise in prices, which is occurring throughout the world. While Americans are noticing higher prices at their grocery stores, global prices for corn, wheat, soybeans, coffee, and other commodities have increased sharply. Increased costs of animal feed have resulted in corresponding increases in the cost of eggs, poultry, beef, and milk. Demands for basic grains and high-quality fruits, vegetables, meats, and animal products by China, India, and other new world economic powers also have been exacerbated by weather and water issues. On a worldwide basis, these more expensive foods are affecting household budgets. Because consumers have less money to spend on nonfood items, there is less money available to help local and national economies grow and create new jobs. World food prices, according to the United Nations, are the highest since the organization began keeping records in 1990.

Current challenges to world food security focus on food costs and changes in family diets related to the increasing food prices. Challenges in the future will be compounded by a world population that has been projected to increase from approximately 6 billion in 2011 to possibly 9.3 billion in 2050. Moreover, the world is undergoing rapid urbanization, coupled with a decline in rural food producers. Between 2007 and 2050, the world's urban population is anticipated to increase from 3 billion to 6 or more billion. The vast majority of the world's population—at a minimum, 70 percent—will live in urban areas by 2050. A growing world urban population, with increased incomes and changing food preferences, will require more food to be produced locally and more food to be imported from food-surplus, food-exporting nations. The United Nations projects that the world's annual grain production will need to increase from 2 billion tons in 2011 to 3 billion tons in 2050, and annual meat production from 200 million tons to 470 million tons in 2050, to feed the Earth's human population. Self-sufficiency in meeting the domestic population's food needs is almost impossible for most countries in the world.

Thus countries will be forced to rely more heavily on food trade (i.e., food imports) to satisfy the needs of their people.

Transportation Availability and Reliability

Food exporters generally begin the process of shipping their products by recognizing a food need, then comparing and examining competing forms of transportation in terms of the existing equipment's capabilities to deliver the food commodity to its intended destination without damaging the quality of the item. In addition, they must consider the cost of transportation; the time required to reach the destination; the destination infrastructure, such as ports, storage, and processing facilities; and the availability of roads and railroad connections from ports to move their products to local and regional markets. Finally, food exporters must consider the length of the haul, the nature of the route, traffic density and port unloading time, the weight of the product being exported, and the ease of loading and unloading shipments. The most spectacular long hauls are across oceans. The world average round trip of oil supertankers is approximately 10,000 miles. The average round trip of a grain hauler is more than 6,000 miles. Loss of time in loading, pickup, and delivery outweighs any advantage, except for extremely high-value commodities, of sending these goods by air cargo transport.

Some transportation routes are direct—specifically, large ocean-going merchant ships, large cargo airliners, and large-diameter pipelines. Costs and time expended by unit trains or "dead-headed" freight trains are favorable if the rail mainlines are between major metropolitan areas or private rail-siding operations. Africa is the one continent with the least connected transportation system. It has only a fragment of a rail network and lacks continuous linked navigable waterways to supplement highway and rail routes.

Ocean- and sea-going vessels have great advantages for very heavy consignments, and specifically for grain. Railroads have an advantage over trucks for very heavy overland consignments. Costs per ton carried on ships, river-tows, or trains decline as the load weight increases and as fuel, labor, and terminal costs are spread over greater tonnage. The optimal weight of a food consignment shipped by truck varies according to the carrying capacity of the truck and conditions of the roads. The range of carrying capacities of inland waterway craft is greater than that of either trains or road vehicles. Air freight generally consists of articles of high value, small size, and low weight.

The relationship between the weight of a food commodity and its bulk is termed *loadability*. Road vehicle and freight car loadabilities are measured based on the maximum axle load or the maximum axle weight a road or highway can withstand. A pipeline must be completely full all the time, from one end to the other. River barges are very economical for transporting heavy and bulky agricultural commodities from areas of production to distant markets along rivers or canals or to ports for overseas shipment. Overall, transport costs for export agricultural products generally are cheaper by water than by any other form of transportation. Supertankers and giant ocean-going grain carriers offer the cheapest freight rates of all. Containerization has reduced port expenses, commodity losses, cargo-handling labor, and time for shipping some higher-value agricultural products or processed foods.

Global Food Markets

Long-term consumer spending trends in the food markets of foreign countries indicate a decline in family budget expenditure shares devoted to staples, such as rice and wheat, and increasing budgetary allocations to higher-value food items, such as meat, dairy products, fruits, and high-quality vegetables. Food trade, in some nations of the world, is now categorized by level of processing:

1. Traditional low-value bulk, such as wheat, rice, and corn (little or no processing)
2. Horticultural high-value products, such as fresh fruits and vegetables (ready to eat and perishable; no processing)
3. Semi-processed products, such as flour and oils (requiring quality and safety security; some processing)
4. Processed food products, such as pastas and prepared meats (require processing and, at times, restricted by domestic trade policies)

Trends in shifting dietary demands have increased trade in high-value food products and changed food marketing and the food trade globally. Consider the following examples:

- Food suppliers and retailers where the demands are greatest have responded by importing foodstuffs around the world, modifying the way they display and sell imported products, and modernizing retail formats to meet the new consumer needs.
- Major multinational retail corporations have constructed supermarkets in developing nations where dietary shifts are occurring; the largest of these retailers now accounts for more than 30 percent of world supermarket annual sales.

- National tariff impediments and restrictive trade regulations have led many major multicultural retail corporations to invest in local food processing companies in host countries.

U.S. Trade Expansion and World Food Supply and Demand Factors

With frontage on both the Atlantic and Pacific Oceans, the United States has a favorable position for direct trade contact with Europe, Africa, and eastern and northern South America on the one hand, and with western South America, the islands of the Pacific Ocean, and Asia on the other hand. Its size and agroclimatic diversity mean that the United States is able to cultivate a great variety of agricultural products, which in turn provide it with a favorable position in both internal and international trade.

World War I brought American foodstuff exports to an unprecedentedly high level, and the ratio of exports to imports in America became enormous. The post–World War I boom continued to stimulate trade until 1929, when it then declined because of both the Great Depression and a rise in nationalism that resulted in the imposition of many artificial trade barriers. World War II and the period afterward were a time of rebuilding in many parts of the world. The United States became the "supply house" of the world. It fed and clothed a hungry and bankrupt world, and supplied equipment and money for the reconstruction of war-devastated areas. The ongoing growth in foreign trade since this era has been the result of American technological infrastructure and agricultural equipment improvements, quality products, and automation. Also, investments of U.S. capital in foreign country enterprises, foreign aid, maintenance of armed forces overseas, and reductions of restrictive barriers to world trade have combined to create high demand for American agricultural products on a global scale.

Supply and demand of agricultural products worldwide and U.S. trade are greatly affected by the following factors:

- Changes in world, regional, and country's population, economic growth, and incomes, because these factors alter food demands
- Global and national food reserves, changing exchange rates, government subsidies, and protectionist policies
- American agricultural productivity and surpluses, because U.S. farmers and food merchants rely heavily on export markets to sustain revenues and commodity prices
- American consumers' demands for foreign foodstuffs (imports have increased steadily, though exports exceed imports by a large margin)

- Reciprocal trade that provides expanded American food variety, stabilizes food prices, and enables trading partners to purchase American surplus commodity products from sales and profits

Global Hunger and Food Security Initiatives

Following the world food price crisis of 2007–2008, the United States and other nations unveiled a new food security plan. The international community pledged $16.5 billion and the United States pledged $3.5 billion to global food security. In 2009, the United States released a new plan, the "Global Hunger and Food Security Initiative," that stressed the importance of international and national collaborations with various partners. Acknowledged in this initiative was the key role that trade and market expansion played in enhancing world food security, including expanded market information, improving post-harvest market infrastructure, and agribusiness growth. In May 2010, the Barack Obama administration released the *Feed the Future Guide*, which outlines the strategic approach the United States plans to use in its quest for world food security.

International trade capacity-building, the number one priority, seeks to lower the costs incurred by a developing country desiring to engage in international trade. Specifically, it addresses the development and construction of roads, ports, railroads, airports, storage facilities, water and sanitation systems, and power generation. According to an all-continent international survey, only 40 percent of rural Africans live within a mile or so of an all-season road, and African farmers have developed only 10 million acres of new irrigated fields in the past 10 years. Moreover, the total power generation capacity of all 48 sub-Saharan African countries is about the same as that of Spain. In developing countries, transportation inefficiency and post-harvest losses claim a minimum of 15 percent and a maximum of 50 percent of what is produced. Trade capacity-building assistance also addresses the obstacles developing countries encounter in placing their agricultural products on the world market. Assistance plans are being funded to enhance a developing country's competitiveness in food trade, encourage effective trade policy reforms, and promote technical training programs to enable all to take advantage of new trade opportunities.

Regional integration of infrastructure, the number two priority, is expected to lead to improved agricultural productivity, expansion of products' sales and regional trade, higher farm incomes, a more stable food supply, and increased food security. Trade among developing countries is expanding, and income growth has created fast-growing markets for food and agricultural commodities. Integrated regional exchange of food products increases food availability, decreases food shortages, and lessens food price volatility. Current barriers to regional trade are often higher than those associated with international trade. For example, African trade tariffs for regional imports average 20 percent; for international imports from the developed world, they average 5 percent. As part of regional integration, nations are called upon to resolve food security issues, reduce trade barriers, enhance collective cross-border agricultural research, improve the transfer of agricultural innovations and technology, and provide access to shared market information.

International trade reform, the number three priority, attempts to mitigate the actions of unfair developed nations' "merchants of food" who control the international trade system. This near-monopoly works against the involvement of developing countries' food producers and, according to many UN members and administrators, weakens world food security. Increasing agricultural exports from developing countries augments local incomes, alleviates rural poverty to some extent, and strengthens the capacity of poorer countries to amass capital for improvement of local agriculture. Today, the "tariff escalation" policies of developed countries place lower import tariffs on raw materials from developing nations and higher tariffs on processed or value-added goods. Also, government export subsidies are given to agribusinesses that export food from many developed nations. Subsidies on food exports distort food prices, discourage investments in agriculture, and undermine developing nations' ability to price compete. During the 2007–2008 food crisis, many countries imposed price controls and food export restrictions. China, India, and Vietnam restricted rice exports, and Argentina, Kazakhstan, and Russia limited wheat exports. These restrictions contributed to worldwide food price increases and led to a panic buying of staple food commodities. International trade reform is clearly necessary, as the current agricultural trade rules favor food exporters.

The United States' trade preferences, biofuel conversion policy, and domestic education of citizens to the benefits of agricultural exports must be reexamined to strengthen global food security. Trade preferences, outlined specifically in the 1976 Generalized System of Preferences (GSP), give Americans duty-free access to 4,800 products from 131 countries and territories. These policies have been successful in assisting developing countries to gain access to markets in the United States, but many have not been granted preferences. Another issue that must be reviewed is the subsidies offered by the U.S. government to biofuel producers. As energy prices rise, demands for biofuels are increasing—a trend that has prompted farmers to shift their crop patterns to corn or other biofuel raw materials. It is believed that increased grain demand for conversion into biofuels led to a 30 percent increase in the average price for a bushel of corn or soybeans in recent years. Using cereal grains for conversion into biofuels has far-reaching implications for food prices. Finally, the U.S. government should inform its citizens of the benefits and importance of unbounded food trade. Exports account for 10 to 13 percent of the total U.S. gross domestic product, and they make a major contribution to reducing the size of the annual budget deficit. Exports support employment, and increasing trade with other nations creates jobs in America.

A Snapshot of the Volatile Global Food Market

Until the early 20th century, exports of agricultural products were basic to the economic development of the United States. Contributing greatly to the formation of the country's investment capital pool were the huge exports of cotton, tobacco, wheat, and processed food items such as flour. Exports of agricultural products declined somewhat in the period between 1929 and the early 1940s, when the United States and the world experienced and were recovering from the Great Depression. During the 1940s and early 1950s, the United States again became a significant exporter of foodstuffs, and by the late 1950s, it became an importer as well.

A series of natural and political factors led to world food demands increasing in the 1960s and 1970s; the United States responded to this need, and exports of foodstuffs increased. Total U.S. agricultural exports in 1980 exceeded $41 billion; in 1990, they were approximately $40 billion; in 2000, they topped $51 billion; and by 2007, they reached $72 billion. The highest percentage of exports, 36 percent, went to East Asia. Worldwide agricultural progress in some areas initiated by the Green Revolution increased competition for markets. Commodity prices rose steadily through 2000, but then dropped suddenly in 2008. A severe drought in the American Southwest and the Southeast in the first decade of the 21st century reduced foodstuffs available for export and caused much farm hardship. To assist farmers who suffered from a decline in commodity prices and drought, the government provided billions of dollars in agricultural subsidies.

Markets for foodstuff exports vary by decade, year, and month, and, in some cases, on a daily basis. A trend observed since 2000 has been the declining relative importance of both agricultural and processed commodities in the composition of the United States' export trade. Exports of both categories of products declined from 18 percent of all exports in 1980 to 8 percent in 2001.

Agriculture is among the most highly protected industries in most countries. "Food first" advocates in many countries of the world have convinced national decision makers to protect local food production in case of war, revolution, blockades, crop failures, and natural disasters. "Country first" advocates fear foreign dominance. High import tariffs and trade restrictions hurt American farmers who depend on exporting a large share of their crops.

The U.S. Department of Agriculture's Agricultural Marketing Survey does an excellent public service of publishing monthly reports on world agriculture supply and demand. A brief snapshot of the volatile nature of commodity trading would include statements such as the following:

1. Wheat consumption is projected to be lower in the world in 2011, with the biggest change being a 1.5 million ton reduction in Russian wheat imports. Despite a major drought, Russia appears to be meeting its internal wheat needs, and the government's wheat export ban helps maintain domestic supplies.
2. Corn production in the world is projected to decline in 2011 because of smaller harvests in Mexico and India, with the threat of a drop in harvests in China. Corn demands internationally will be strong, both to replace lost crops and for biofuel use.
3. Rice production, consumption, and exports are projected to be less in 2011 due to a smaller rice crop in India. Drier than normal weather in the eastern and northern Indian rice-growing regions is expected to decrease average yields. Rice production increases in Argentina and Brazil are based on an expected increase in harvested areas. Global consumption of rice has been lower, primarily due to reductions in India and China.
4. Livestock, poultry, and dairy product exports from the United States will be higher, basically because of demands from Asia. Despite higher production forecasts, prices for livestock, poultry, and dairy products will increase in 2011. A strong export market and improving domestic demand in a time of relatively tight meat supplies will result in higher prices, specifically for cattle, hogs, broilers, and turkeys. International demands and improving domestic demands for milk, cheese, and butter will raise the store prices consumers pay worldwide.

Global Diversity and Regional Specialization

Even with the technological advances that mark the modern world, food production and food demand depend on fundamental factors, such as geographic location, climate, soils, power/energy resources, raw materials, culture, and markets. Under modern conditions, production of agricultural commodities tends to be highly specialized. Certain areas have optimal natural conditions for only a small range of agricultural products, while other commodities can be produced more efficiently somewhere else. If people desire to have a diverse diet, composed of a wide range of foodstuff options, they must arrange to exchange their local surpluses for food products produced elsewhere. The local self-sufficiency that characterized the world in its pre-industrial and pre-urban society days is not possible in the 21st century with a world population of 7 billion. High standards of living are inseparably linked with low-cost food production and efficient means for the exchange of foodstuffs. The world's physical infrastructure for such an exchange has been developed to a remarkable degree. The foundation for greater foodstuff production, exchange, and communications has been laid to support further progress—and at a rate more rapid than ever. Unbounded food trade is a deterrent to hunger and famine, and a stimulus for a world of human progress and peace.

—*William A. Dando*

Further Reading

Bennett, M. K. *The World's Food.* New York: Harper & Brothers, 1954.

Connor, J. M., and W. A. Schiek. *Food Processing: An Industrial Powerhouse in Transition.* New York: John Wiley & Sons, 1997.

Enzer, S., R. Drobnick, and S. Alter. *Neither Feast nor Famine: Food Conditions to the Year 2000.* Toronto: Lexington Books, 1978.

Global Food Markets. Washington, DC: Economic Research Service, U.S. Department of Agriculture, 2009.

Hebebrand, C., and K. Wedding. *The Role of Markets and Trade in Food Security.* Washington, DC: Center for Strategic and International Studies, 2010.

Hopkins, R. F., and D. J. Puchala. *Global Food Interdependence: Challenge to American Foreign Policy.* New York: Columbia University Press, 1980.

Johnson, D. G. *Trade and Agriculture.* New York: John Wiley & Sons, Inc., 1950.

U.S. Agricultural Trade. Washington, DC: Economic Research Service, U.S. Department of Agriculture, 2009.

World Agricultural Supply and Demand Estimates. Washington, DC: Economic Research Service, Foreign Agricultural Service, 2011.

U.S. Farm Machinery Industry in a Global Market

Genetic modification of common food crops has led to increasing yields over the last 100 or more years. In the last 10 years, the rate of annual increase of average yields, especially for corn, has continued steadily upward, owing to the use of

Figure 33. Yield of corn used for grain in the United States

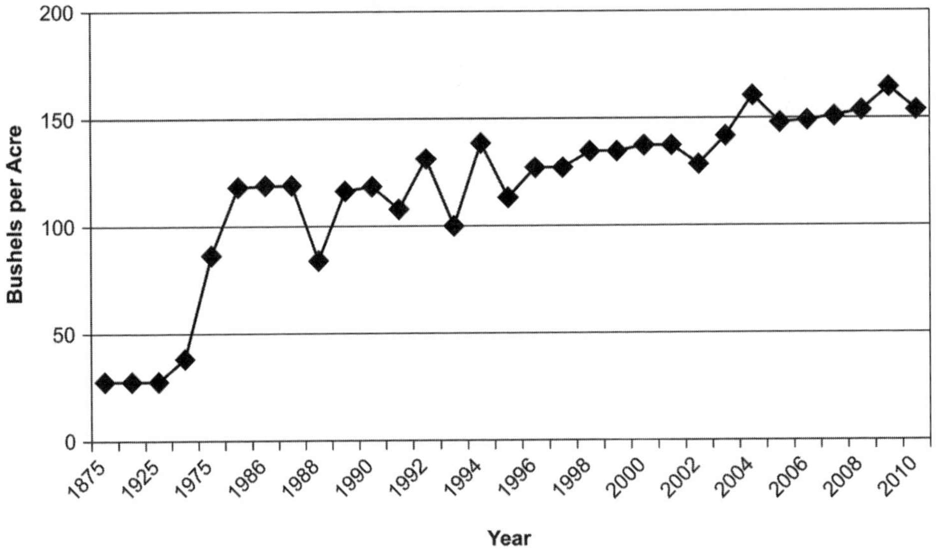

Source: Dawn M. Drake. Data from U.S. Department of Agriculture, 2010.

biotechnology to introduce new traits into existing crops (see Figure 33). In addition to the increased yields, there has been an increase in the acreage used for planting corn in the last 20 years (see Figure 34). These increased yields and acreage are most efficiently managed through the use of mechanized farm machinery, such as high-horsepower tractors and combination harvester-threshers, commonly referred to as combines (see Figure 35).

The United States is among the largest producers of grain crops for food uses in the world. To maintain that productivity despite shrinking agricultural labor forces, U.S. agriculture will require mechanization using tractors and combines. Farmers, even on small operations, increasingly manage their land with large tractors (see Figure 36) and harvest with combines (see Figure 37). The ownership of combines has been on the rise since 1997 (see Figure 38). In addition, U.S. producers with large acreage to manage are increasing the number of tractors with greater than 100 horsepower (75 kilowatts) that they own (see Figure 39). Even livestock farmers have increased their level of mechanization, allowing them to produce larger bales of straw for bedding (see Figure 40).

As the population of the world continues to grow (see Figure 41), other countries will also need to increase their level of mechanization to meet increasing food

Figure 34. Increases in acres of corn planted in the United States

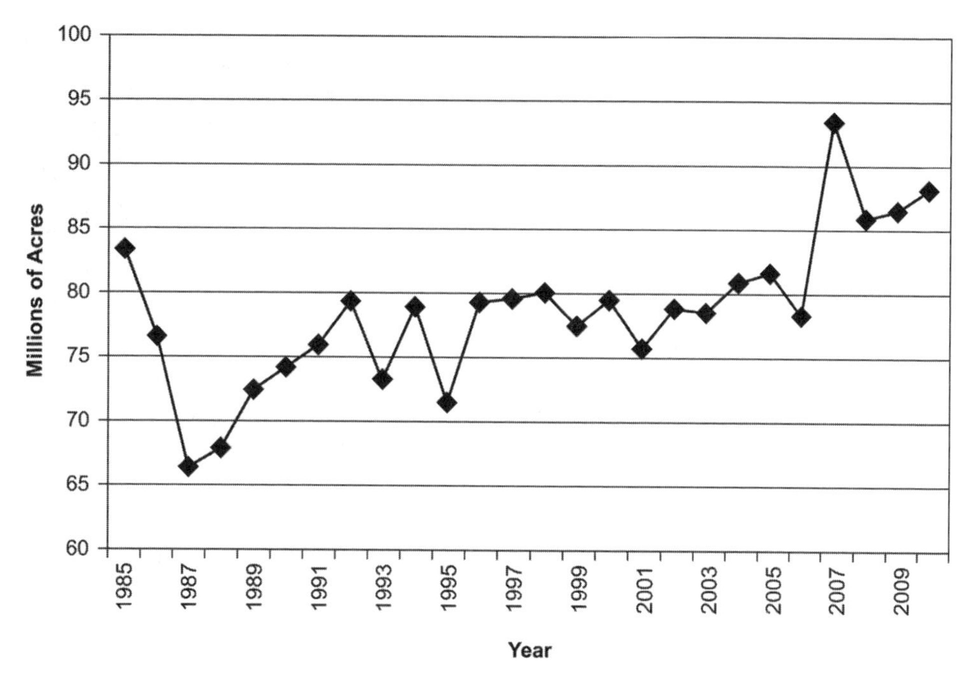

Source: Dawn M. Drake. Data from U.S. Department of Agriculture, 2010.

Figure 35. A Gleaner combine and Case International tractor on display at a farm machinery trade show. (Photo by Dawn M. Drake)

Figure 36. Planting soybean seeds on a farm in eastern Pennsylvania. (Photo by Dawn M. Drake)

Figure 37. Harvesting wheat with a combine in eastern Pennsylvania. (Photo by Dawn M. Drake)

Figure 38. Change in combine ownership from 1997 to 2007

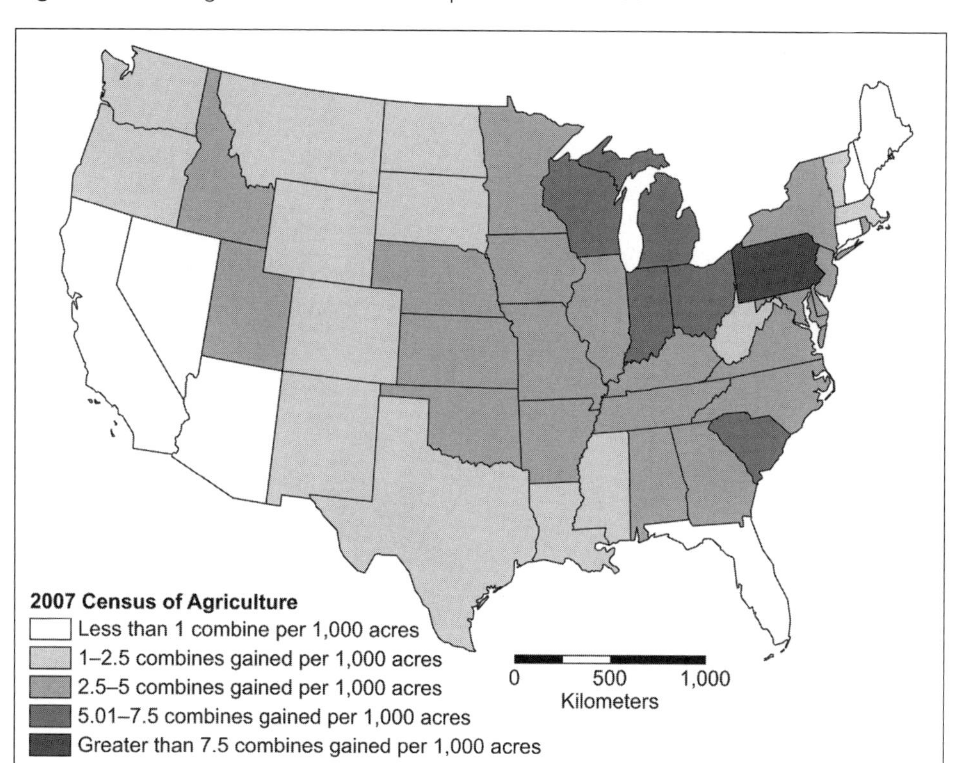

Source: Dawn M. Drake. Data from U.S. Census of Agriculture, 2007.

needs. New growth areas in population, such as India and China, likewise represent emerging markets for the sale of farm machinery. If farm machinery manufacturers can keep pace with demands from a growing population and penetrate into previously untapped markets, their future profit margins will continue to rise, even as they help to prevent food shortages in the world.

A Snapshot of the Farm Machinery Industry

The farm machinery industry is a vast and complicated sector of the manufacturing economy that affects countries throughout the world. The North American Industrial Classification System defines the farm machinery industry as including firms that produce everything from tractors and combines to turf grass tools and feed grinders. For this reason, the size of firms in this sector can range from *Fortune* 500 companies producing a full line of equipment to small family-run operations

Figure 39. Ownership of tractors, greater than 100 horsepower

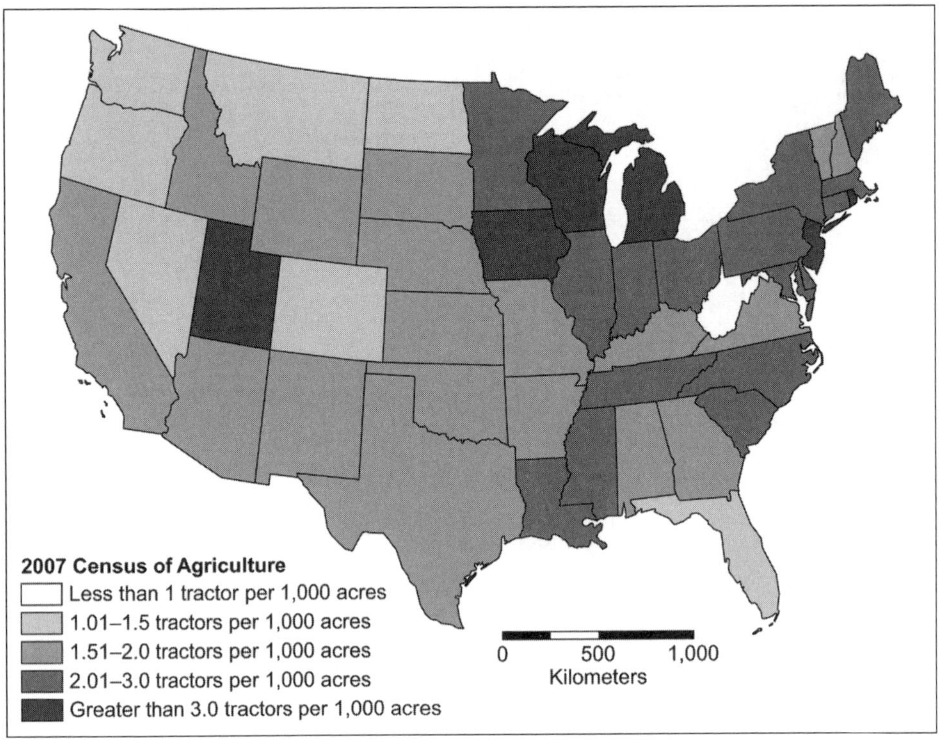

2007 Census of Agriculture
Less than 1 tractor per 1,000 acres
1.01–1.5 tractors per 1,000 acres
1.51–2.0 tractors per 1,000 acres
2.01–3.0 tractors per 1,000 acres
Greater than 3.0 tractors per 1,000 acres
0 500 1,000
Kilometers

Source: Dawn M. Drake. Data from U.S. Census of Agriculture, 2007.

that manufacture only a few units a year. There are literally thousands of firms in the sector. That said, approximately 43 percent of farm machinery sold in the world in a given year is produced by one of three U.S. firms—Deere and Company, also known as John Deere (JD); Case New Holland (CNH); and Allis-Gleaner Corporation (AGCO). This "Big Three" of farm machinery dominate worldwide sales and will be the leaders in bringing mechanization to emerging markets to help feed growing populations.

The largest farm machinery firm in the world is John Deere, headquartered in Moline, Illinois. The firm, which is ranked 87 on the *Fortune* 500 list and 307 on the Global *Fortune* 500 list, produces a full line of farm machinery, including high-horsepower tractors, combines, forage equipment, and irrigation tools (see Figure 42). It controls 22 percent of the global market share and, until recently, was the worldwide leader in tractor sales. John Deere is probably one of the most recognizable brands in farm machinery; it produces equipment that is sold and serviced throughout the world.

Figure 40. A tractor produces 8-foot bales of straw for use on livestock operations in eastern Pennsylvania. (Photo by Dawn M. Drake)

The second-largest farm machinery firm in the world is Case New Holland, which is a wholly owned subsidiary of the Fiat Corporation. Its worldwide headquarters is located in Amsterdam, but all executive decisions are made from one of two North American headquarters in Racine, Wisconsin, and New Holland, Pennsylvania. The firm markets its farm machinery products under two major brand names—Case International Agriculture and New Holland Agriculture (see Figure 43), both of which are recognizable and sold the world over. Fiat is listed as number 64 on the Global *Fortune* 500, with no less than 20 percent of its annual revenues coming from CNH. It controls 11 percent of global farm machinery sales, including market dominance in the area of hay tools.

The third-largest farm machinery firm in the world is Allis-Gleaner Corporation, headquartered in Duluth, Georgia. While the name AGCO may be unfamiliar, some of the many brands that it produces are more likely to be recognized. They include Massey-Ferguson, Gleaner, White Planters, and Challenger, which was the agricultural division of the well-known Caterpillar construction company, prior to its sale to AGCO in 2002 (see Figure 44). Among AGCO's other brands are RoGator, Sunflower, SpraCoupe, Fendt, Hesston, and TerraGator. Approximately 10 percent of the world market share in farm machinery is controlled by

Figure 41. World population growth, 1950–2050

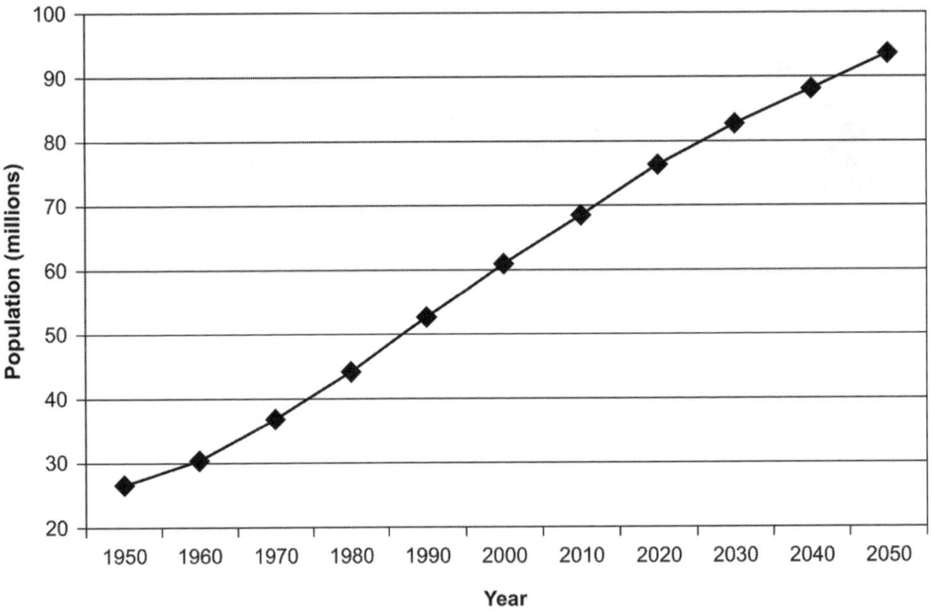

Source: Dawn M. Drake. Data from U.S. Department of Agriculture, 2010.

AGCO, and it is the market leader in South America and Europe. It is listed in the top 300 of *Fortune* 500 companies in the United States and is continuously growing. In 2006, AGCO was among the top 10 performing stocks on the New York Stock Exchange, an honor it shared with companies such as Amazon.com and GameStop. In 2008, it was named one of America's most trustworthy companies, along with Lowe's and Heinz, largely due to its commitment to environmental stewardship and its open accounting practices. As the leader in technology to meet emissions standards for diesel farm equipment, AGCO has continued to expand its global reach and competitiveness among U.S. and international firms.

These three firms are not the only ones that produce farm machinery in the world. They are, however, the largest. In addition to the Big Three, other full-line manufacturers compete internationally for sales in farm machinery. The fourth-largest farm machinery firm in the world is India-based Mahindra and Mahindra, which in 2010 unseated JD as the world's leading seller of tractors (see Figure 45). Mahindra sells mostly compact tractors in the United States, but does market larger tractors and combines for sale in Asia. Currently, it controls approximately 8 percent of the global market share. Other full-line manufacturers include European firms SAME-Deutz Fahr (fifth largest in the world) and CLAAS,

Figure 42. John Deere machinery on display at various trade shows. (Photo by Dawn M. Drake)

Figure 43. A Case International combine (left) and a New Holland tractor (right) on display at the 2010 Farm Progress Show in Boone, Iowa. (Photo by Dawn M. Drake)

Figure 44. AGCO products, including (from top left clockwise) Challenger, Gleaner, Hesston, and Massey Ferguson machinery, on display at trade shows. (Photo by Dawn M. Drake)

both of which produce full lines of equipment, albeit at much smaller scales than any of the U.S. manufacturers. In short, while there are international competitors to the Big Three U.S. firms, only Mahindra and Mahindra has the potential, in the near future, to surpass any of them for control of global market share.

Changing Trajectories in the Farm Machinery Industry

For much of the recent history of the U.S. farm machinery industry, the fortunes of the sector closely tracked those of U.S. automakers. The largest outbreak of bankruptcies and closures of farm machinery firms in the United States, save during the highly volatile early years of the industry, occurred in the late 1970s and 1980s. This was around the same time that Detroit automakers were being forced to innovate and improve the quality of their own products in the face of stiffer competition from international automakers. Chrysler faced a bankruptcy scare and subsequent reorganization. Just after Chrysler reorganized, the Allis-Chalmers

Autonomous Tractors

It was not that long ago that the notion of using a tractor instead of a horse to do farm work was considered a novel idea. Now, the next wave of technology from farm machinery manufacturers stands ready to make farmers themselves virtually obsolete. GPS satellites have been increasing the precision with which a farmer can guide a tractor or combine through a field, in some cases allowing the tractor to drive itself with the operator steering just to make turns at the end of rows. New advances by AGCO, through a partnership with Topcon, are making tractors even more autonomous, with the operator needed only to monitor an onboard computer. It uses the same technology as the automated Prius that Google has been testing throughout the United States. Case New Holland has taken this technology a step further, allowing the navigation system of a combine to control the steering of a tractor and grain-cart driving alongside it, making unloading safer and more efficient. Autonomous tractors could solve the problem faced by countries with declining agricultural labor forces.

—Dawn M. Drake

Corporation (today, part of AGCO) and International Harvester (today, part of CNH) also filed for bankruptcy and were subsequently reorganized. Over most of the last 30 years, this trend has continued, with the fortunes of the farm machinery sector tracking closely the upswings and downturns of the automobile industry.

The pattern has not held in recent years, however, as the farm machinery industry has proved more stable than the automobile sector in the most recent economic downturn. In 2008, when most of the Detroit automakers were again facing bankruptcy and seeking government bailouts, the farm machinery industry was in a much better position fiscally. Part of what drove the 2008 economic downturn was record-high grain prices, which translated into higher consumer food prices. While higher food prices lead consumers to limit their discretionary spending on higher-order goods such as automobiles, the higher grain prices that drive food prices translate into increased income for farmers. Farmers often spend that increased income on profit maximization in the form of new seed hybrids, increased inputs such as fertilizer and pesticides, and more modern and technologically advanced implements, tractors, and combines. This trend, in turn, leads to more profits for farm machinery manufacturers. During 2008, JD and CNH both suffered minor layoffs in their facilities, mostly related to other commodities, such as steel and rubber, being at record-high prices as well. At the same time, AGCO opened three new assembly facilities in the United States and hired 600 workers, funded largely by the increased profits that the automobile industry did not enjoy.

Figure 45. Mahindra and Mahindra tractors on display at Ag Connect Expo in Orlando, Florida, in 2010. (Photo by Dawn M. Drake)

Commodity prices were not the only thing that allowed the U.S. farm machinery industry to remain solvent in the 2008 economic downturn. Recent U.S. Environmental Protection Agency standards have mandated a decrease in the level of emissions produced by diesel-powered farm machinery. Some state and local governments have provided incentives to farmers to purchase tractors and combines with the cleaner-running engine technology. The leader in emissions research and development is AGCO, whose Sisu engines are among the cleanest running in the industry. This technological advancement also helped the sector remain solvent in the face of the 2008 economic downturn that negatively affected the automobile industry.

The ability of the farm machinery industry to remain stable and competitive in the face of a global economic recession is a positive sign for the long-term sustainability of the sector, not just in the United States, but globally. Given the fact that farm machinery will continue to be important as world population grows over the next 20 years, it seems likely that the industry will continue to expand into new markets to help feed the people of the world.

International Penetration of the U.S. Farm Machinery Industry

Given the growing population of the world and the fact that the fastest-growing regions are those outside the traditional markets of the U.S. farm machinery industry, the Big Three have recognized the need to expand their operations abroad to feed the world (see Figure 46). Today, the U.S. farm machinery industry has manufacturing plants, sales offices, and parts distribution centers on every continent, save Antarctica. These facilities provide the machines and support necessary to feed growing populations in emerging markets such as India and China. The Big Three have all entered into joint ventures with Chinese manufacturers to produce cheap tractors that are well suited to the needs of farmers managing small acreages

Figure 46. CNH advertising that recognizes the role of farm machinery in a growing world. (Photo by Dawn M. Drake)

in East Asia. Additionally, AGCO has recently opened two new facilities in Russia to provide tractors to Central Asia. Collectively, these moves demonstrate the increasing penetration of the industry to feed the world.

As a part of this increased market and manufacturing penetration by U.S. farm machinery firms, there has been an equal increase in the global focus on the part of JD, CNH, and AGCO. This increase has been greatest on the part of AGCO, which has shifted much of its production into South America and Europe, maintaining only three manufacturing plants and three assembly facilities in the United States, today. Its dominance in Brazil, coupled with its plants in Russia, India, and China, demonstrates AGCO's increasing penetration into the emerging BRIC economies, which will be able to help feed a growing world population in the future.

Currently, many of the hunger situations in the world are a matter of distribution problems, not production shortages. However, as the world population continues to grow, it will be necessary to increase production on the existing farm acreage to adequately feed more people. Mechanization is necessary, as the number of people working in agriculture is decreasing. In the future, increased yields will be best managed by mechanization, given the diminishing labor force. The farm machinery industry is a key component in fighting future hunger and famine, led by three U.S. firms. John Deere, CNH, and AGCO, as well as Mahindra and Mahindra, will all need to produce low-cost yet efficient farm machinery for sale in developing economies, as well as modern high-horsepower machinery for sale in economically advanced parts of the world. Meeting these increasing demands for farm machinery will allow farmers, in turn, to meet the world's growing food demands. The population of the world will continue to increase and state-of-the art farm machinery, accompanied by increased food production, will reduce the possibilities of famine in the world.

—Dawn M. Drake

Further Reading

AGCO Corporate. "AGCO: Challenger, Fendt, Massey Ferguson, Valtra Tractors and Farm Equipment." Accessed January 2011. www.agcocorp.com/.

Case New Holland. "Case IH Agricultural Equipment." Accessed January 2011. www.caseih.com/.

Case New Holland. "New Holland." Accessed January 2011. www.newholland.com/Pages/index.html.

Deere & Company. "The John Deere Home Page." Accessed January 2011. www.deere.com/.

Gerstner, John, ed. *Genuine Value: The John Deere Journey.* Moline, IL: Deere & Company, 2000.

Gibbard, Stuart. *The Ford Tractor Story Part Two: Basildon to New Holland 1964 to 1999.* East Yorks/Ipswich: Japonica Press/Old Pond Publishing, 1999.

Grooms, Lynn. "Got Food? Can Crop Production Keep Pace with the Growing Global Population?" *Farm Industry News* 43, no. 5 (2010): 34–38.

IBISWorld. *IBISWorld Industry Report: Tractors and Agricultural Machinery Manufacturing in the US.* Santa Monica: IBIS World, 2009.

Mahindra USA. "History and Information About the Company. Accessed October 2009. www.mahindrausa.com/company.php.

Marsh, Barbara. *A Corporate Tragedy: The Agony of International Harvester Company.* Garden City: Doubleday & Company, 1985.

McMahon, Karen. "AGCO Shifts Gears." *Farm Industry News* 43, no. 9 (2010): 22, 24.

Petersen, Walter F. *An Industrial Heritage: Allis Chalmers Corporation.* Milwaukee, WI: Milwaukee County Historical Society, 1978.

Stonehouse, Tom, and Eldon Brumbaugh. *JI Case Agricultural and Construction Equipment 1956–1994*, Vol. 2. St. Joseph, MO: American Society of Agricultural Engineers, 1996.

World Agricultural Systems

An agricultural system is the arrangement and relationships of environmental, political, and socioeconomic factors that merge to form a specialized food-producing region in a particular segment of the world or place. The world's diverse regional patterns of food plants and livestock, once identified and regionalized, provide an excellent means to gain insights into why plants are grown where they are and why animals are reared in specific areas. Regionalizing the diverse combination of factors that are included in agricultural systems facilitates the determination of the quantity and variety of food produced and local and regional food needs. The unity of circumstances—a combination of physical, historical, economic, social, and political factors that contribute to the problems and the successes of agriculture anywhere—has a space and time perspective. Systems of agriculture generally change slowly, but social, political, and economic factors may force changes to occur more rapidly.

Societal-Based World Agricultural Systems

To bring order and to understand the infinite varieties of human activities to produce food require that they be classified into a few broad categories. One attempt to do so sought to divide world agricultural systems into three divisions based on social organization and group dynamics—that is, tribal systems, traditional systems, and modern systems.

Tribal systems are composed of small numbers of generally related members linked together in efforts to reduce food insecurity and dietary uncertainty. Survival demands group solidarity. Mutual aid is common, especially in sharing food surpluses. Rules of custom prescribe rights and obligations. A severe barrier to economic improvement and societal advancement is the denial to the "fruits" of one's labor. Even so, tribal societies do change.

Traditional systems are composed of large numbers of like-minded and similarly educated people whose status is determined largely by birth into a family, clan, or caste. Wealth is determined by land and/or cattle owned and, in some cases, by family size. Traditional systems usually are prescientific, and productive power largely depends on the physical labor of men, women, and children and the power of work animals. A traditional peasant farmer, whose family survival equation leaves no margin for error, is hesitant to try new ways of farming, new technology, or new crops, because the loss of a crop or of cattle could lead to the family starving. Surplus food items are produced and sold to merchants, government agencies, and urban residents in farm markets. Many rural residents are landless and rent or sharecrop land with urban landlords. Those landlords who reside in cities believe that living in rural areas is unacceptable (cultural wastelands) and that agricultural work is beneath their dignity.

Modern systems are composed of agriculturalists who are well educated. In all aspects of their farming and ranching activities, they intensively use scientific findings and new technologies to produce high-quality food products in vast amounts. These commercial farmers and ranchers sell what they produce for local, national, and international consumption. Personal competence and hard work, rather than inherited social position, determine financial rewards and status in society. Decisions made by a modern agriculturalist daily on his or her farm, without government direction or controlling absentee elites, permit quick responses to natural hazards or market trends.

Land use in tribal agricultural systems involves gathering, hunting, herding, fishing, shifting cultivation, and hand tillage of small fields. Land use in traditional agricultural systems involves intensive cultivation with traditional plows and hand tools, extensive cultivation with animal-drawn and small tractor-pulled implements, supplementary feeding of livestock, and pastoral nomadism. Land use in modern agricultural systems, employing state-of-the-art mechanized farm equipment and livestock-rearing devices, includes mixed crop and livestock rearing, dairy farming, intensive and extensive crop farming, plantations, industrial farming, and extensive livestock ranching.

This attempt to differentiate agricultural systems does not distinguish food-producing areas of the world on the basis of a crop or commodity. It also does not distinguish food-producing areas of the world on the basis of production methods such as irrigation, plow, or hoe tillage. Finally, it does not distinguish food-producing areas of the world based on the purpose for which food is produced, such as subsistence, limited surplus sales to local markets, or large-scale commercial sales.

Ecological-Based World Agricultural Systems

Some agricultural scientists contend that the world pattern of food production is, for the most part, the result of ecological factors. They believe that combinations of climate and soil, with the influence of economic and cultural factors, determine which crops or animals are reared, where these activities take place, and why particular foods are produced in a place or region. They classify food-producing areas in the world on the basis of cropping practices and believe that only four basic approaches exist, based on the type of cultivation involved: (1) permanent grasslands; (2) grasslands rotated with cultivated crops; (3) perennial tree or vine crops; and (4) annual crops.

Permanent grasslands agricultural systems are based on the annual growth of natural grasses and plants. The types of natural vegetation used to support food animal growth vary according to climate and soils. The carrying capacity of the rangeland depends on the animal's type, size, and nutrition needs. Permanent agricultural systems support limited numbers of people. Animals and animal products are traded or sold to producers of plant foods and to urban dwellers.

Grasslands rotated with cultivated crops agricultural systems are based on the alternation of crops with grass or other forage crops. Normally, this form of agriculture occurs where the climate is more moist and the soils are more fertile than in areas of permanent grassland. Crop agriculture is threatened by droughts and by weather variations. Grasses sown are selected carefully to resist natural hazards and enrich the soil when they are plowed under and the fields planted in an annual crop. Both the crops and grasses planted must have low plant growth and survival demands for water and great tolerance for weather variations.

Perennial tree and vine agricultural systems are characterized by no-till or limited-till cropping. This system is represented by such agricultural activities as orchards, vineyards, and berry farms, plus banana, tea, and coffee plantations. The return from the original plantings may take years to be realized, because perennial crops of this type require time to reach a profitable bearing age. The original capital investment, in most cases, is high. The products produced in this agricultural system are in great demand, however, and the returns from the investment of land, labor, and capital are substantial. Products are sold off of the agricultural unit and basic foods for farm families and employees are purchased or bartered.

Annual crop agricultural systems focus on cultivated cash crops, such as wheat, corn, rice, oats, rye, barley, and soybeans, and including vegetables. All crops are planted annually in freshly tilled or no-till fields. The heavy emphasis on cereals in this agricultural system reflects the ease of applying modern farming techniques and state-of-the-art equipment to the type of crop that is sown and the ease of mechanical harvesting, transportation, and storage of the crop that is grown. In the tropical areas of the world where storage is a problem and the growing season long, farming units tend to emphasize ever-bearing plants and year-round sequences of food crops. It is estimated that 75 percent of the energy and protein needs of humans is met, one way or the other, by cultivated grains. Products from this agricultural system dominate the world's food commodity trade.

At one time, a great number of agricultural specialists believed that the dominant factor in agriculture and where food is grown is climate. They contended that a combination of environmental conditions set the limits for any crop or domestic animal. Combining optimal growth or habitat conditions with population pressures, level of technology, and cultural traditions, a dominant form or system of agriculture in a place or region emerges. Based on optimal habitat, climate, and cultural circumstances, an agricultural system can be devised by synthesizing the following factors: (1) crop and livestock associations; (2) agricultural methods employed; (3) intensity of utilization of land, labor, and capital; (4) use of items produced specifically for on-farm consumption or for sale; and (5) ensemble of structures used in farming operations. Agricultural systems based on climate include agricultural products of the following regions: (1) humid tropical climate regions; (2) dry tropical climate regions; (3) Mediterranean climate regions; (4) humid subtropical climate regions; (5) mild humid continental climatic regions; (6) semi-arid and arid intermediate climatic regions; and (7) cold forest and tundra climates.

Intermixed-Based Agricultural Systems

Of great importance in the 20th century was the role of government policy in determining the basic agricultural systems in countries of the world. In the United States, federal programs and subsidies exert a major influence on what, where, and when agricultural commodities are grown or animals raised. In the Soviet Union, Eastern Europe, China, and Cuba, the Stalin model for agriculture was adopted and collective decision making significantly affected agricultural systems. Governmental control of water and irrigation facilities, flood control projects, insurance programs, agricultural research, agricultural specialists placed in minor civil divisions to aid farmers, and infrastructure construction and maintenance are examples of the ways governments affect the decision making of food producers. Government policies are very important in determining the regional composition and the success or failure of politically based agricultural systems.

An example of intermixed-based agricultural systems occurs in tropical agroclimatic regions where primitive migrating, gathering, shifting cultivation, and plantation commercial agriculture exist side by side. A framework of intermixed world agricultural systems, with its limitations and need for refining, encompasses the following elements:

I. Primitive Migrating Systems
   A. Gathering
   B. Nomadic Herding
   C. Shifting Cultivation
II. Subsistence Systems
   A. Primitive Sedentary
   B. Intensive Sedentary
   C. Non-Subsistence Mediterranean
III. Commercial Systems
   A. Plantations
   B. Grain Farming
   C. Vegetable and Fruit Farming
   D. Mixed Crop and Livestock Farming
   E. Dairy Farming
   F. Livestock Ranching

This descriptive outline is provided to guide further study, as space does not permit detailed description of the character and interworking of each system. Government-directed agricultural systems in the 21st century, however, will determine food availability in many parts of the world.

Economic-Based World Agricultural Systems

Economic development and progress are basic to social evolution and human survival. Ideas, institutions, law, politics, and artistic expression inevitably change with the transformations that occur in a group's economic foundation. Altruism, food aid, religious devotion, patriotism, and regard for other human races or cultural groups are products of economic conditions and the effects of socioeconomic conditions on the human mind. The economic status of individuals and the economic condition of a nation influence consumption of food and agricultural systems. Through education and technology, humans have learned much about nature and natural processes. They have learned to work with nature and to modify natural processes. They have gained, in some instances, some modicum of control over nature. Technological advances and peaceful progress have enabled humans to create economic-based world agricultural systems that elevate the quality of life of all involved in these systems. If they do not, they are replaced.

Primitive subsistence systems involve individuals engaging in the following activities: (1) hunting, fishing, and gathering; (2) nomadic grazing; (3) shifting and slash-and-burn; (4) primitive sedentary; and (5) near-subsistence sedentary. The concept of sharing the talent and property of individuals began with primitive societies. Individuals could not survive the vicissitudes of nature and the actions of other human groups. Survival dictated that all members of a group, a tribe, or clan work and defend what they had as a concerted unit. Goals and aspirations of individual members were perceived as subordinate to those of the group and those of the collective will. All members of the society shared aspects of hunting, gathering, and later food-growing activities, and all received the food they needed to survive. As numbers of family members survived and as clans or tribes grew in size, and as they grew socially and technologically, the basic survival and defense concept was modified. Groups bonded together and their expanded community of like-minded agriculturalists became the smallest units of local governments. Leaders and decision makers were initially the heads of families and later the village elders.

Feudal agricultural systems evolved when all land in a delimited area was the property of a lord, and the vassals, serfs, or peasants who dwelled on the lord's land owed him specific services and personal loyalty. Food producers in this manorial or seigniorial system (i.e., unfree peasantry) were controlled by a landlord who had police, judicial, fiscal, and other power functions over them. The landed aristocracy had almost unlimited power over dependent peasants and flourished in a closed agricultural economy. Eventually, the feudal landowners or class developed into nobility tied to the king or queen, with large numbers of unfree peasantry in their services. The feudal lord's responsibility was to maintain tranquility and ensure food security for his peasants, and the peasant's duty was to produce foodstuffs for his family, for the manorial lord, and for the lord's sale to others. Famines were infrequent in primitive subsistence agricultural systems, but very common in feudal agricultural systems.

Capitalistic agricultural systems emerged slowly from feudal agricultural systems. This type of system is characterized by a food producer owning his or her own land, producing agricultural commodities for family consumption, and, most importantly, producing goods for cash sale. In a capitalistic agricultural system, decisions concerning food production are made by individual farmers or business people and are intended to yield a profit. This system attained its highest form in the 1800s, after which internal and external forces slowly resulted in an increasing range of farm or ranch decisions being made or directed by governments. To many students of economic systems, exploitive classical or frontier capitalism has evolved into benevolent capitalism in the United States and many democratic countries of the world. Productive use of surplus capital (social surplus) invested into land improvement, infrastructure, new agricultural crops, technologically advanced equipment, and agricultural research, rather than building pyramids or cathedrals, enabled the capitalistic agricultural systems to replace feudal and primitive subsistence agricultural systems. Benevolent capitalistic agricultural systems, with all their faults and problems, led the world from an age of scarcity and ubiquitous famines to an age of prosperity, potential abundance, and limited numbers of region-specific famines.

Socialist agricultural systems have their basis in primitive agricultural systems but emerged in European society during the 1800s to include any visionary system seeking political or social perfection. European philosophers and thinkers sought means to cure the evils of exploitive capitalism, poverty, hunger, and famine. Other socioeconomic thinkers and writers advocated establishment of communal societies and nations in which everyone would be absolutely equal and no one owned private property. Concomitantly, various Christian groups made deliberate attempts to revive the structure of the first Christian community established in Jerusalem during the first century (Acts 2–4). Modern communes, collectives, and cooperatives have as their antecedents the basic concepts, the model developed, and the writings of those who wished to create ideal societies without hunger and famine.

A modern agricultural collective is composed of individuals bound together by political, religious, or social principles of centralized decision making and group control of all means of agricultural production. Activities of a collective are designed to enhance the whole collective, not individual members. The organization is characterized by uniformity of actions in response to a natural hazard, a threat, or a group goal. The government-owned and government-planned collective farms of China, nations of the former Soviet Union, Central Europe, and Cuba can trace their heritage back to the earlier experimental idealistic communities in Western Europe and North America.

The Stalin model for agricultural development, conceived in the Soviet Union during the late 1920s, was designed to extend governmental control over all aspects of food production. All land was nationalized first. Land in the Soviet Union, prior to its dissolution in 1991, could not be owned by anyone except the state. Three means of agricultural production were created:

1. The **collective farm** (*kolkhoz*) initially encompassed as much as 16,000 acres of land. Approximately 600 families worked on each farm. Although the land belonged to the state, all property on it belonged to the members of the collective, including houses, farm machinery, and means of transport. Usually a collective grew diverse crops best suited to the soils and climate and also raised livestock. Collectives were self-governed, but decisions were not

Doing Away with a Class of Small Farmers

Communist leaders in the Soviet Union in the late 1920s were constantly haunted by the fear of "capitalistic encirclement" and a possible direct attack by one or more capitalistic powers. To guard against this prospect, they determined that the Soviet Union should build a powerful armament and munitions industry as fast as possible. Gosplan, the national planning agency, was given the task of preparing a comprehensive plan for the development of Russian industry. This vast five-year plan required equally vast sums of money for its implementation. The immense capital investment required could be obtained in only two ways: (1) reducing the living standards of the people and (2) using the resources and production of the farmer peasants. As part of the program, all assets of the peasants were taken from them, and a program of forcible collectivization of agriculture resulted in a horrible famine. In the early 1930s, starvation and accompanying diseases claimed millions of lives throughout the richest agricultural region of the Soviet Union, particularly the Ukraine. Russia became a police state where citizens were held under a yoke of discipline that was harsh, unfair, and cruel. The Russian people endured great tribulation and travails, and they had few rights and no freedom.

Stalin, the leader of the Soviet Union during this era, used a slogan of "the building of socialism in a single country" to convince the people to sacrifice for Russia. To work for Russia was to the people, in essence, working for their own benefit. Stalin harnessed the love Russians had for their country and aroused in them patriotic feelings to transform Russia from a predominantly agrarian country to an industrialized nation. In doing so, Stalin did away with a class of peasants, the *kulaks* and small farmers, and he created problems with food production within Russia that persist to this day.

—*William A. Dando*

made or accepted unless approved by the state and in accord with a Five-Year Plan. A planned quota committed a collective to sell to the state a certain amount of produce for a plan-determined amount of money. If a collective produced more than planned, it sold its excesses to the state. Profits from sale of products were distributed among members of the farm. If the harvests were bad, a collective farmer would receive little remuneration for the year's work. He or she would suffer periods of deprivation or starvation.

2. The **state farm** (*sovkhoz*) was normally two or three times the size of a collective farm and employed many more families. It ordinarily specialized in one product (e.g., grain, animal husbandry, orchards, vegetables, or poultry) and was highly mechanized. All property on a state farm belonged to the

state. Farm workers were paid a salary that did not depend on the harvest or the quality of the goods produced. A state farm worker was an employee of the state; he or she was eligible for pensions under the national social security system. State farms received much more financial support from the state than did collective farms.

3. The **private plot** was a small parcel of land, ½ to 1¼ acres, depending on the quality of the land and the location of the farm. Both collective and state farm workers were permitted to use private plots for growing or raising food products, primarily for family consumption. In some years, at minimum, 30 percent of all vegetables, fruits, milk, eggs, and meat products came from private plots. In 1986, these plots fed and produced 15 million cows, 13 million pigs, 32 million sheep and goats, and hundreds of millions of poultry. Surplus items from private plots were sold to the state at times, but more so in farmer's markets within urban centers. For years, the Communist Party and state planners worked hard to abolish private plots but could not because the nation needed the food they produced.

Rural and Urban Cooperative-Based Agricultural Systems

Whereas most socialist agricultural systems are rural entities, cooperatives based on the West European and American models are both rural and urban organizations. Cooperatives are primarily business enterprises jointly owned by their members and operated without profit for the benefit of the membership. Many forms of cooperatives exist—marketing, consumer, producer, and service, for example. Some sell goods produced by their membership, some buy to resell to members, some provide loans, and some perform services (e.g., financial, medical, telephone, or rural electricity). A common set of goals and objectives link cooperatives. These organizations are established as a means for like-minded people to pool their talents and resources, share the fruits of their labor, and create a safe social environment without antagonistic competition and interpersonal strife.

World Agricultural Systems in the 21st Century

World agricultural systems will undergo radical changes in the 21st century. Population increases in the billions; climatic change ramifications; shortages of water; new and improved crop, animal, and poultry varieties and types; and government involvement will require that agricultural systems adjust to the times and increase food production. The demand for foodstuffs will stimulate development of new agricultural implements, application of new technology, methods of food processing and storage, and transportation delivery modes to consumers. Constant enhancement in foodstuff quality, nutritional value, and yields per unit area will be essential if the world is to be spared devastating urban and calamitous national famines. Many old systems will disappear as new agricultural systems emerge.

—William A. Dando

Further Reading

Amelia, M. "The (Not So) Mysterious Resilience of Russia's Agricultural Collectives." *World Bank's Transition Newsletter*, November/December 1991.

Anderson, J. *A Geography of Agriculture*. Dubuque, IA: Wm. C. Brown, 1970.

Cone, C. A., and Andrea Myhre. "Community-Supported Agriculture: A Sustainable Alternative to Industrial Agriculture?" *Human Organization* 59, no. 2 (2000): 187.

Dando, W. A., and C. Z. Dando. *A Reference Guide to World Hunger*. Aldershot, UK: Enslow, 1991.

Dando, W. A., and J. D. Schlichting. *Soviet Agriculture Today: Insights, Analysis, and Commentary*. Washington, DC: National Council for Soviet and East European Research, 1987.

Fourier, C. *Theory of Social Organization*. New York: C. P. Somerby, 1876.

Grigg, D. B. *The Agricultural Systems of the World: An Evolutionary Approach*. Cambridge, UK: Cambridge University Press, 1974.

Higbee, E. *American Agriculture*. New York: John Wiley & Sons, 1958.

Lockwood, G. B. *The New Harmony Movement*. New York: D. Appleton and Company, 1905.

World Food Day

World Food Day (WFD) is an event originally established in 1979 to honor the date of the founding of the Food and Agriculture Organization (FAO) of the United Nations in 1945. Celebrated in unique ways around the world, an interactive telecast to 150 countries is the highlight of the event. Dr. Pal Romany, the Hungarian Minister of Agriculture and Food, suggested the idea of an annual worldwide event on October 16 each year to raise awareness of the issues that create poverty and hunger. The theme of WFD varies each year, and each theme has paved the way to organize events in those countries involved:

1. 1981 "Food Comes First"
2. 1982 "Food Comes First"
3. 1983 "Food Security"
4. 1984 "Women in Agriculture"
5. 1985 "Rural Poverty"
6. 1986 "Fishermen and Fishing Communities"
7. 1987 "Small Farmers"
8. 1988 "Rural Youth"
9. 1989 "Food and the Environment"
10. 1990 "Food for the Future"
11. 1991 "Trees of Life"
12. 1992 "Food and Nutrition"
13. 1993 "Harvesting Nature's Diversity"
14. 1994 "Water for Life"
15. 1995 "Food for All"
16. 1996 "Fighting Hunger and Malnutrition"
17. 1997 "Investing in Food Security"
18. 1998 "Women Feed the World"
19. 1999 "Youth Against Hunger"
20. 2000 "A Millennium Free from Hunger"
21. 2001 "Fight Hunger to Reduce Poverty"
22. 2002 "Water: Source of Food Security"
23. 2003 "Working Together for International Alliances Against Hunger"
24. 2004 "Biodiversity for Food Security"
25. 2005 "Agriculture and Intercultural Dialogue"
26. 2006 "Investing in Agriculture for Food Security"
27. 2007 "The Right to Food"
28. 2008 "World Food Security: The Challenge of Climate Change and Bioenergy"
29. 2009 "Achieving Food Security in Times of Crisis"
30. 2010 "United Against Hunger"
31. 2011 "Food Prices—From Crisis to Stability"

On October 16, 1984, the editor of this book, a professor in the Department of Geography at the University of North Dakota at the time, was the National Moderator for the World Food Day video-teleconference transmitted to more than 150 countries from the campus of George Washington University, in Washington, D.C. He had just returned from a year in Hong Kong and China, researching Chinese famines of the past.

A worldwide event, World Food Day is designed to increase hunger and poverty awareness, understanding of basic issues, and informed year-round actions to alleviate hunger. In the United States, WFD activities and events are sponsored by 450 national, private, and voluntary organizations. Planning for each year's theme, events, and activities is at three levels: international, national, and local. International pre-event planning begins at the FAO's headquarters in Rome, Italy. National planning and event coordination is done in a small corner office on K Street NW in Washington, D.C., by Patricia Young. For more than 30 years, she has been the U.S. leader of World Food Day. A volunteer, Young has spent almost half of her life fighting hunger issues. At the local level, concerned citizens, professionals, nonprofit and for-profit organizations, universities and colleges, and the media organize special events, interactive local teleconferences, presentations, and food-raising events. Activities and events are designed to fulfill the following purposes:

1. Provide briefings on hunger and poverty issues for the media
2. Work with schools, colleges, and universities to plan talks and seminars
3. Coordinate projects on community food needs and security
4. Raise funds for local and/or international projects
5. Seek policy commitments from public officials and candidates for public office
6. Increase networking by bringing people, ideas, and resources together
7. Involve people in year-round service and support
8. Make WFD the "annual meeting" for hunger activists

Examples of Events

Millions of world citizens are involved in WFD activities and events. The challenge is to coordinate planning on all levels, thereby increasing the impact of everyone's work. On World Food Day, October 16, 2009, Secretary of State Hillary Rodham Clinton reaffirmed the United States' commitment to combat global hunger. She noted that more than 1 billion people, representing one-sixth of the world's population, suffer from chronic hunger. The United States supports activities that fight hunger and poverty through sustainable agricultural development. It is committed to ensuring that food is available to all and that those suffering from hunger have the resources to purchase food. The Secretary of State noted that food security in food deficit areas is also a facet of America's national security, as well as the national security of all countries of the world. In 2009, Clinton announced that the major industrialized nations of the world had pledged more than $22 billion over the next three years to stimulate agriculture-led economic growth.

WFD activities, plans, and events can have short-term and long-term impacts, such as the following:

1. Governmental agencies, ministries, universities and colleges, international agencies, and NGOs have organized WFD meetings, exhibits, and symposia.

2. Newspapers, radio stations, and television broadcasting systems have reported on press conferences, lectures, and events.
3. Film stars, television celebrities, and sports stand-outs have contributed funds and helped highlight food security issues.
4. Awards and special recognition have been given to those who have done much to resolve hunger issues at national or local levels.
5. Popes John Paul II and Benedict XVI have made public pronouncements to food producers, food processors, and food consumers on WFD.
6. Presidents, vice-presidents, legislative leaders, and prime ministers have spoken, dedicated agricultural research stations or buildings, and worked in farm fields to provide symbolic support to food producers and food/hunger/poverty activists.
7. State, province, parish, and other political entities of nations have organized awareness campaigns, forums on nutrition issues, festivals, and exhibitions.
8. Farmers', fishermen's, and food distributors' workshops have been organized and human nutrition sessions scheduled.
9. Exhibitions of indigenous food products have been held, along with agricultural fairs that include opportunities for scientists, administrators, food producers, and food consumers to exchange views and experiences.
10. National leaders and FAO administrators have issued WFD proclamations and statements and even hosted "conference calls" to discuss the theme of the food year.

World Food Day's Role in FAO's International Efforts

World Food Day is a grassroots event that supports the overall mission of the FAO. This UN organization acts as a neutral forum where all nations of the world can meet as equals to discuss food and hunger issues, negotiate agreements between differing groups, and debate international policy. FAO is also a source of data, information, and knowledge. Since its founding in 1945, FAO has focused on developing countries making the transition to modernize and improve food production. Special attention has been given to rural areas in developing countries, where approximately 70 percent of the world's poor and hungry people live. FAO's activities comprise four main thrusts:

1. *Information gatherer*—collects, analyzes, and disseminates data that aid agricultural development; publishes newsletters, reports, and books; distributes magazines; and creates numerous CD-ROMs.

2. *Sharing policy expertise*—helps nations to devise agricultural policies, plan activities, draft legislation, and create national strategies for rural development and hunger alleviation.
3. *Provider of meeting places*—encourages policymakers and experts to convene in Rome or in FAO field offices to discuss issues and problems and build common understanding.
4. *Brings knowledge to food producers*—funds thousands of field projects throughout the world, designed to resolve local issues and develop models that can be used elsewhere; provides technical support if needed; and in crisis situations, works with other humanitarian agencies to protect lives.

The preamble to the FAO constitution states that ensuring humanity's freedom from hunger is one of the basic purposes of the organization. Human access and rights to food are special concerns to FAO and, therefore, to WFD. President Franklin D. Roosevelt's "Four Freedoms" address in January 1941 was the seed that grew into the United Nations General Assembly's adoption of the Universal Declaration of Human Rights; freedom from want was one of those rights.

World Food Day's Role in the United Nations and National Responses to Emergencies and Economic Crises

Food emergencies can be triggered by any number of natural causes, such as hurricanes, floods, droughts, or earthquakes, or any number of human-induced causes, such as policy decisions, civil strife, economic crises, and war. In most cases, rural populations in the developing world are the hardest hit. Acute hunger, starvation, and famine routinely claim lives on the small-scale farms and rural areas where 70 percent of the world's hungry live and work. Responding to a food emergency or crisis requires exceptional external assistance. WFD, in its programs and connections with people, organizations, and government decision makers, helps marshal public support for FAO's and other relief agencies' activities. For-profit and nonprofit companies and organizations have a particularly important role to play as partners with FAO in interventions to protect lives and restore agriculture-based livelihoods. Nongovernmental organizations often facilitate the distribution of life-saving food to the hungry and of seeds, tools, and fertilizers to affected farmers. WFD publicizes emergency interventions and assists FAO through its briefings to the media, fundraising, and networking activities that bring people together to resolve a problem at a place. FAO's emergency interventions are designed to prevent food emergencies and food crises caused by economic factors from deteriorating further. WFD's ties to grassroots activities and concerned citizens can stimulate relief and survival strategies of those affected.

Achieving food security in times of worldwide economic crises is difficult and costly. In the world recession that began in 2007, trillions of dollars were spent to revive and stimulate the economies in the world's richest countries. Governments bailed out banks, financial institutions, and industries, but the poor and the hungry received very little to ease their suffering. WFD, in contrast, reaches out to all segments of society and all socioeconomic classes to arouse public opinion so as to protect the most vulnerable from hunger.

In mid-2008, the FAO food price index rose by 52 percent relative to the mid-2007 index. Global cereal prices were 63 percent higher than they were in 2005, according to the International Monetary Fund. Factors that caused the food crisis during these years included the following:

- Low agricultural productivity
- High population growth rates in food-insecure countries
- Rural water shortages
- Land tenure issues
- Floods
- Droughts
- Reduced investments in agricultural research

After the food price crisis came the global economic downturn. Thus the poor—both urban and rural—were forced to face two simultaneous crises.

The theme of World Food Day 2009, "Achieving Food Security in Times of Crisis," permitted hunger and famine specialists to present their insights on this issue. All those involved reflected on the numbers affected and the suffering. It was repeated again and again that the world has the know-how to resolve most hunger issues. Organizers of WFD events in 2009 urged all to work together to make sure hunger is recognized as a serious worldwide problem. The 2010 World Food Day theme was "United Against Hunger," and the 2011 theme was "Food Prices—From Crisis to Stability."

World food prices dropped in 2010, largely due to weaker international prices of sugar, dairy products, and cereals, while the prices for meat and oilseed remained stable. However, a review of FAO statistics for seven areas of the world that were impacted severely by the food crises of 2007–2008 revealed the following developments:

1. In East Africa, cereal prices fell in 2009 and 2010 but remained above the level of 2007, while sorghum and millet prices were three times higher.

2. In West Africa, cereal prices increased in 2010, particularly for imported rice.
3. In South Africa, the price of corn, the population's main food staple, has declined.
4. In Asia, rice and wheat prices remained 20 to 70 percent higher than in 2008.
5. Haiti's increases in food prices added to concerns about food security (the price of rice was 50 percent higher and that of corn was 30 percent higher than prior to the earthquake).
6. In Central America, the price of corn, the region's staple food, declined sharply in 2009.
7. South American prices for wheat and rice declined, with the exception of wheat in Argentina (increased 15 percent) and rice in Colombia (increased 25 percent).

WFD publications, videos, films, graphs, and graphics distributed or used at WFD events provide statistics, data, and the current "geography of hunger" maps to keep the public informed.

The Warrior for the Hungry Masses

Patricia Young is a warrior working on behalf of the hungry masses. Now 86 years old, she works tirelessly, coordinating WFD events and activities. She solicits pledges from supporters of WFD, encourages individuals and organizations to schedule WFD programs, and finalizes arrangements for the annual WFD teleconferences. Young recruits speakers and authorities on topics of hunger and famine to become involved and sends emails and telephone reminders when needed. A convincing speaker and a master in the use of the telephone to get things done, she has stimulated discussions of rural poverty, women in agriculture, food and nutrition, water, malnutrition, biodiversity, agricultural research, bioenergy, food security, and more. A superb manager of time and resources, Young coordinates her vast outreach program, involving national, private, and volunteer organizations, with military precision and gentle determination.

Young's website reports that, following an estimated increase of 105 million hungry people in 2009, there are now 1.2 billion malnourished people in the world. Those who are most adversely affected are the poorest of the poor, according to Young. To her, WFD does increase awareness and by doing so, it increases the power of public influence and synthesizes resolution of the hunger problem. Young has built WFD into a vast network of connected groups around the United

States. From a small enterprise, led by a greater-than-life woman, a very successful force for human welfare has emerged. Primarily, the success of WFD is based on her warm, sensitive, and caring personality, the force of her will, her boundless energy, and her commitment.

In 1980, the FAO asked Young to become the United States' coordinator of World Food Day. She keeps a rigorous schedule and returns home to Scranton, Pennsylvania, and her husband every other weekend. Although she is a woman who commands attention and is respected throughout the world for her work, Young has an aversion to being in the spotlight. Compelled by a strong work ethic and a deep commitment to the hungry and poor of the world, Patricia Young is a superb example of what a dedicated American can do to help others.

—William A. Dando

Further Reading

"Achieving Food Security in Times of Crisis." In *World Food Day/TeleFood 2009 Note*. Rome: FAO, 2009, 1–2.

Boustany, Nora. "Warrior for the Hungry Masses." Washington Post Foreign Service, *Washington Post*, Saturday, October 25, 2008, p. A12.

Conversations with Pat Young, U.S. Coordinator of World Food Day, beginning in 1984 and continuing until 2010, experience as a state and local coordinator of WFD activities spanning more than 25 years.

Eide, A. "The Human Right to Adequate Food and Freedom from Hunger." In *The Right to Food in Theory and Practice*. Rome: FAO, 1998, 1–5.

Emergencies. Rome: FAO, 2008, 1–2.

http://en.wikipedia.org/wiki/World_Food_Daywww.fao.org/about/en/

www.state.gov/secretary/rm/2009a/10/130628.htmwww.worldfooddayusa.org/CMS/2953.aspx

World Population and Demographic Projections to 2050

At present, world population growth is in the process of transitioning from its peak annual rate of growth of 2.0 percent in the latter half of the 1960s to a period of slower growth as a result of declines in birth rates in many countries. As of 2011, the world population growth rate stood at 1.2 percent, with approximately 83 million people added to the total each year.

Table 26. World Population by Region, 1900, 1950, and 2010 (millions)

	1900 Millions	Percent	1950 Millions	Percent	2010 Millions	Percent
World	1,650	100	2,529	100	6,892	100
Developed countries	539	33	812	32	1,237	18
Developing countries	1,111	67	1,717	68	5,656	82
Africa	133	8	227	9	1,030	15
Asia and Oceania	904	55	1,323	52	4,040	59
Latin America/ Caribbean	74	4	167	7	585	8

The four regions of Africa, Asia, Latin America/Caribbean, and Oceania are classified by the United Nations as *less developed*. The *more developed* countries are the remaining regions, Europe and North America (excluding Mexico). The UN does not routinely classify individual countries as more or less developed, but it does make three exceptions: Australia and New Zealand in Oceania and Japan in Asia are grouped with the more developed countries.

The 20th century was a period of unprecedented population change in at least two significant ways. First, in 1900, world population stood at 1.6 billion, a figure reached only after tens of thousands of years of human history (see Table 26). At the end of that century, in 2000, those two numbers had reversed and population had grown to 6.1 billion. The sixth billion was reached in 1999 and in record time, just 12 years after the fifth billion. The seventh billion was reached in October 2011. Second, virtually all population growth now takes place in the developing countries of Africa, Asia, Latin America/Caribbean, and Oceania. Both of these developments have had a profound effect on world demography.

In 1900, the population of the developed countries was 539 million, 33 percent of the world total, and that of the developing countries was 1.1 billion, or 67 percent of the world total. In 2010, the total population of the developed countries was 1.2 billion, 18 percent of the world total, and that of the developing countries was 5.7 billion, or 82 percent of the world total. Comparing the current state of world population dynamics in the developed countries as a whole, on an annual basis births outnumber deaths by only 2 million, with a ratio of births to deaths of 1.2. In Europe, however, the annual numbers of births and deaths have nearly drawn equal. In the developing countries, births outnumber deaths by 81 million, with a ratio of births to deaths of 2.8. As a result, 98 percent of the *natural increase* (births minus deaths) takes place in the developing countries. Additionally, nearly 3 million immigrants, on a net basis, move from developing to developed countries so that, overall, two-thirds of population growth in the developed countries is now due to immigration.

The World Is Now Experiencing Its Fastest Population Growth

The world's total population exceeded the 7 billion mark in the latter part of 2011, only 12 years after reaching the 6 billion mark. And that sixth billion was reached in 1999, only 12 years after the fifth billion. Population projections that assume birthrates will decline without interruption for the next few decades suggest that the eighth billion will be reached 13 years after the seventh—but a more realistic prospect is that the eighth billion will arrive in even *less* time than the seventh. No matter which scenario plays out, 3 billion people will have been added to the Earth in 25 years or less. This increase exceeds total world population in 1950 of 2.5 billion—a figure that had taken all of human history to reach. Current population growth is different from the past, however. Today, most growth occurs in the poorest countries and in the poorest areas of countries. It represents one of the greatest challenges humankind has ever faced.

—*Carl Haub*

The remarkable change in population growth and distribution in the 20th century was due primarily to rapid decreases in death rates in developing countries, especially after 1950. Public health measures, such as cleaner water and improved disposal of sewage, along with medical advances, such as immunization programs and other preventive measures such as malaria spraying, have had an unprecedented effect on death rates. These measures had taken centuries to emerge in the developed countries but were "exported" to developing countries in a relatively short period.

World Population Projections

It is typical, and quite understandable, to ask what world population size is projected for the future by organizations such as the United Nations Population Division and the U.S. Bureau of the Census. Only the *results* of projections are frequently cited in articles and media coverage; if the organization produces multiple scenarios, such as high and low series, those are usually ignored. When inquiring about projections, it is the *assumptions* from which they result that should first be investigated. Projections are the result of, first, the base population used as a starting point, and, second, the assumptions about future trends of birth rates (fertility), death rates (mortality), and migration. Unfortunately, these assumptions—like the alternative scenarios—are nearly always ignored in media coverage. A discussion of the factors influencing projections is in order.

First, the confidence currently placed in our knowledge of the *current* size is quite high. Nearly all countries have taken a recent census and most have taken a series of censuses, greatly enhancing their analytical value. Both organizations producing population estimates, the United Nations Population Division and the U.S. Bureau of the Census, agree that world population stood at 6.9 billion in mid-2010.

To project population assumptions must be made. Of the three mentioned previously, fertility frequently has the greatest impact. One must consider, for example, projections for China, Germany, India, Russia, and Zambia. The measure used to quantify the "birth rate" is the *total fertility rate* (TFR). The TFR is the average number of children a woman would have if the rate of childbearing of a particular year were to remain constant. China has a TFR of 1.5, and informed speculation holds that it will lift its "one child" policy in the not too distant future. If so, will China's TFR rise and to what level? In India, the TFR now stands at 2.6, but this national rate was largely the result of TFR decline in states with higher levels of education, of serious attention given to family planning programs, and overall of better governance. Doubt about India's population future now rests in the heavily populated and comparatively illiterate "Hindi Belt" states of the north, whose TFRs still range from 3.5 to 4.0. States in India are very independent, often with their own political parties, and the country's northern states are known for less effective governance.

An important issue in India and in China is their governments' attempts to put an end to the sex-selective abortion of female fetuses. The normal worldwide biological sex ratio at birth is 105 male births per 100 female births. In India, there are about 111 male births per 100 female births; the corresponding rate is 119 male births per 100 female births in China. In some states of India, such as Haryana and Punjab, the male to female birth rate ratio is 120 to 100. Even if the practice of sex-selective abortion can be stopped, the preference for sons is likely to persist, with a resultant upward pressure on the TFR as couples ensure themselves of at least one son.

Germany now has a TFR of 1.4 and has had such a low figure, or lower, for at least 30 years. Despite recent government steps to give greater support to families, the TFR has shown no increase. Russia's TFR reached a low point of 1.2 in 1999, prompting an emergency government program to encourage childbearing through generous payments for second and subsequent births. The Russian TFR rose to 1.5 in 2009 and may approach 1.6 in 2010, suggesting that the government program did have some success.

Finally, Zambia is a good example of an African country whose prospects for fertility decline are quite poor. Surveys have shown that the TFR in Zambia has remained rather constant at about 6.2 since the 1990s, after a decline from 7.2 in 1980. Other African countries that had some initial success in reducing fertility, such as Ghana and Kenya, have shown a decline in the TFR from very high levels to a medium level of about 4.5 when a slowdown in fertility decline evidenced itself. Data suggest that the decline in TFR has now resumed in Ghana.

The U.S. Population Reference Bureau's knowledge of demographic trends in developing countries, most of which do not have effective registration of births and deaths, has been greatly enhanced by a series of nationally representative surveys. These efforts began with the World Fertility Surveys in the 1970s, and now continue with the current Demographic and Health Surveys (DHS) and other similar survey programs.

The role of fertility in developing countries cannot be overstated, as it will result in the greatest variation in future world population size. In terms of the assumptions one might make when running a projection, there are a number of essential questions one must address. When will the TFR decline in a serious way in countries, primarily in Africa, where it is currently high and stationary? Is the two-child family really in the future for all developing countries? The two-child family represents *replacement-level* fertility, whereby a couple simply replaces itself, rather than increasing the size of each generation. When that level is reached, a TFR of about 2.1, population growth will eventually cease, but not for some time due to population *momentum*. Population momentum is the continuation of population growth for generations after reaching a low level of fertility. In Africa, for example, the population younger than age 15 typically accounts for 45 percent of the total population, resulting in a huge pool of new potential parents for a long time to come. In India, with a TFR of 2.6, the percentage of the population younger than age 15 is 32 percent; in Pakistan, the same measures are 4.0 and 38 percent; in Guatemala, 4.4 and 42 percent; in Egypt, 3.0 and 33 percent; and in Indonesia, 2.4 and 28 percent. By comparison, in Europe overall, the TFR is 1.6 and 16 percent of the total population is younger than age 15.

The second of the three components of population change will have significantly less impact on population growth around the world, with the possible exception of Africa. In many parts of the developing world, life expectancy at birth has risen to moderately high levels and tends to increase only slowly in the contemporary era. The basic summary measure of mortality—life expectancy at birth—stands at 67 years in developing countries compared to 77 years in their developed counterparts. As a result, the contribution of improvements in life expectancy to future population growth in developing countries, unlike its very large contribution in the 20th century, will not be a major driving factor. A striking exception to this general principle is likely to occur in Africa—especially in sub-Saharan Africa (SSA), where life expectancy at birth averages only 52 years. This leaves considerable room for improvement in mortality conditions, which in turn places upward pressure on the population growth rate in the absence of parallel fertility decline. In addition, while the infant mortality rate (IMR)—calculated as annual deaths of infants younger than age 1 per 1,000 births—has declined dramatically from 174 in the early 1950s, it is still rather high at 81. The under age five mortality rate (U5MR—the same basic measure as the IMR) is about 140, indicating that approximately 14 percent of children die before age five. Continued improvements in death rates among young children have an important effect on life tables and life expectancy at birth.

An additional factor, the prevalence of the human immunodeficiency virus (HIV) and acquired immune deficiency syndrome (AIDS), has changed mortality assumptions for developing countries, particularly for SSA. While it was never true that HIV would "wipe out" Africa or cause its population to decrease in size, HIV has had and will have a devastating effect on some countries. Currently, approximately 4 percent of males and 6 percent of females are estimated to have HIV/AIDS in SSA as a whole. For most SSA countries, the level of HIV prevalence is comparatively low, except in parts of eastern Africa and all of southern Africa. Four countries in eastern Africa have HIV prevalences between 10 percent and 18 percent, with higher rates occurring among females than among males: Malawi, Mozambique, Zambia, and Zimbabwe. In southern Africa, all five countries have HIV rates ranging from 12 to 32 percent, again with higher rates among females than among males. The highest rates are found in Swaziland—20 percent among males and 32 percent among females. The nine countries just cited total 122 million in population, 14 percent of SSA's population. In many countries, the HIV epidemic is believed to have peaked, and slowly spreading use of antiretroviral therapy (ART) has begun to make a real contribution to greater longevity for HIV-positive people. The UN Population Division now projects that there will be 583 million AIDS deaths from 2005 to 2020, down from the 610 million deaths projected 2 years earlier; the majority of these deaths will occur in SSA.

Assumptions regarding the third component of population change, migration, are particularly difficult to explicate. Migration depends on determinants such as "push" factors, those compelling migrants to seek a better life elsewhere or to escape political persecution, civil violence, or wars. "Pull" factors are largely limited to economic influences, such as the knowledge or belief that migration will result in higher income and a higher standard of living. Another push factor is the ability to send cash remittances home to improve life among relatives in the country of origin. The migration that results is often a consequence of actual economic conditions in the country, which typically change, and the effectiveness of immigration policy and laws in the country of destination. In recent years, the global recession has acted to reduce migration and even compel return migration to the home countries. There also appears to be a growing resistance in Western countries to cultural diversity when the proportion of foreign-born reaches higher than comfortable levels. Very recently, Prime Minister David Cameron of the United Kingdom and Chancellor Angela Merkel of Germany have publicly stated that multiculturalism has failed in their countries—that is, immigrants are not

Figure 47. World population, 1950-2050, United Nations medium variant

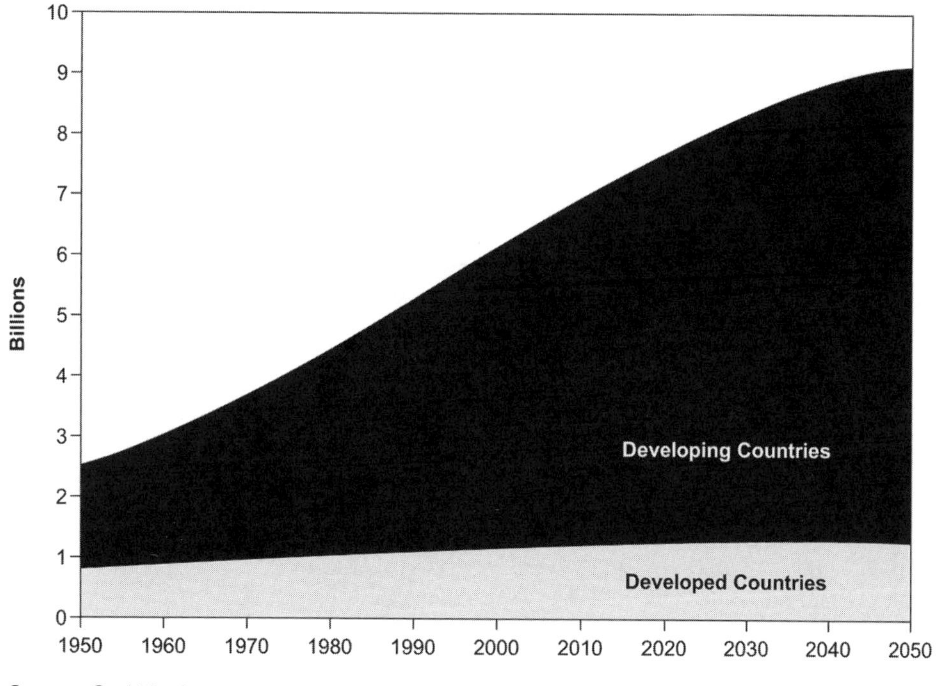

Source: Carl Haub.

assimilating as hoped. The rise of international terrorism, particularly "home grown terrorism," has certainly made an important contribution to changing attitudes regarding migration.

Results of Projections

The preceding discussion was intended to emphasize the fundamental importance of considering the assumptions underlying projections and the country settings for which assumptions for the future must be made. The following discussion of projection results and their assumptions makes primary reference to the biennial series of the UN Population Division, the last of which was issued in the spring of 2008.

Figure 47 illustrates the general *pattern* of population growth expected by 2050, as illustrated by the UN Medium Variant projection, one of a number of projections. This pattern has remained largely unchanged for many years, given that the developed countries have long since completed their *demographic transition*.

Figure 48. Developed countries by age and sex, 2010

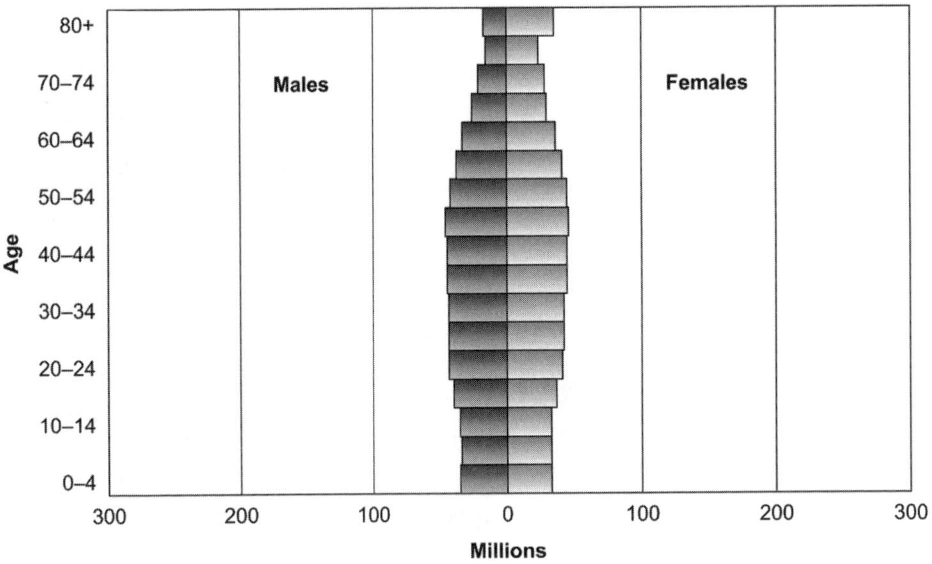

Source: Carl Haub.

The demographic transition is the model of demographic change that underlies all population projections. This model simply holds that, over time, both birth and death rates will decline so that, ultimately, the number of children women bear is historically small and life expectancy reaches new highs. But that is all the model is useful for; it cannot adequately describe the precise manner in which such change is currently taking place, because there have been significant differences in the form of transition between developed and developing countries. As described earlier, the transition took place over several centuries in the developed countries. Although it has been happening more quickly in the developing countries, the pace is very uneven among them. This effect is clearly shown in the two population "pyramids" in Figures 48 and 49. Low birth rates, for many years, have greatly reduced the number of young people at the base of the pyramid who will be the parents of tomorrow. In the developing countries, the broad-based pyramid shows that a large amount of potential growth remains. Should birth rates remain constant in a developing country, the number of annual births will obviously rise. Should mortality improvements continue, as is hoped, and death rates fall, population growth will accelerate.

The assumptions made regarding fertility by the UN, the U.S. Census Bureau, and many national statistical agencies in developing countries (where available)

Figure 49. Developing countries by age and sex, 2010

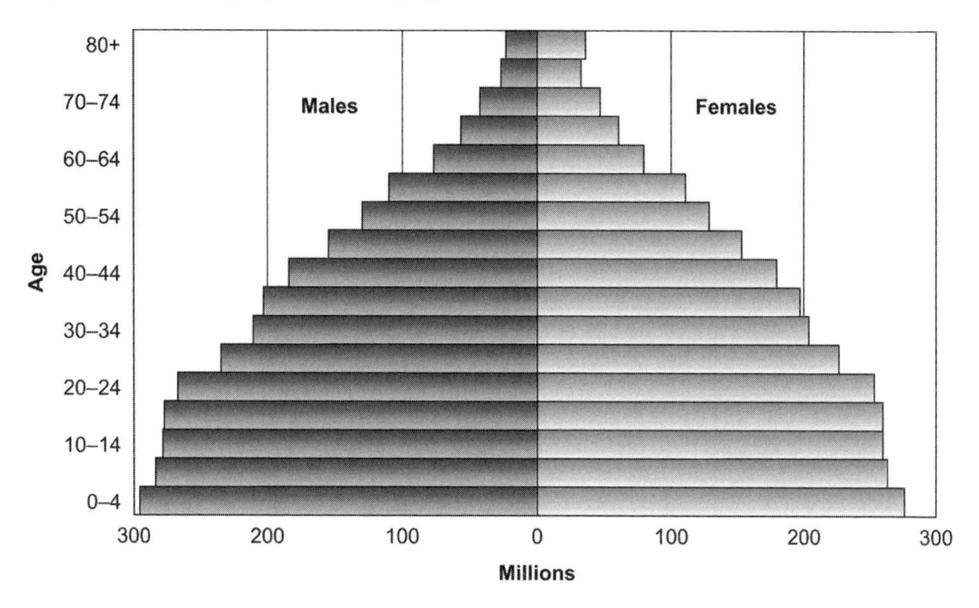

Source: Carl Haub.

make a rather uniform assumption for developing countries. It is assumed that fertility will decline from the present time in a smooth, uninterrupted manner to a value near the "two-child" family. In the past, the UN assumed that the decline would end at a TFR of 2.1—that is, with replacement-level fertility. More recently, the UN fixed the expected final value at a sub-replacement fertility rate, 1.85, simply because that is what happened in developed countries and even in some developing countries. This value is used in the medium variant scenario. The other principal variants, the low and high variants, assume that the final value of the TFR will be 1.35 and 2.35, respectively—that is, one-half child above and below the medium. Unfortunately, these other variants are often ignored, with the implicit assumption being that only the medium variant is expected. Another variant, the constant fertility variant, assumes no change in present levels of the TFR. It should be emphasized that these assumptions are applied uniformly to each country, including developed countries. No one expects individual countries to all conform to a single trend of fertility change. The value of the different variants comes very much into play at the country level.

The results of the UN projections (see Table 27) provide a potential range of world population size that can be expected for 2050. In fact, the range at the *world* level may be too great. The assumption that fertility will decline in developing

Table 27. Projected World Population, United Nations (millions)

	UN Medium 2010	TFR in 2010	UN Med. 2050	TFR in 2050	UN Low 2050	TFR in 2050	UN High 2050	TFR in 2050	UN Constant 2050
World	6,908	2.5	9,150	2.0	7,958	1.5	10,461	2.5	11,030
Developed countries	1,237	1.7	1,275	1.8	1,126	1.3	1,439	2.3	1,256
Developing countries	5,671	2.7	7,875	2.1	6,833	1.6	9,022	2.5	9,774
Africa	1,033	4.7	1,998	2.4	1,748	1.9	2,267	2.9	2,999
Sub-Saharan Africa	863	5.2	1,753	2.5	1,537	2.0	1,986	3.0	2,714
Asia	4,167	2.2	5,231	1.9	4,533	1.4	6,003	2.4	6,010
Latin America/ Caribbean	589	2.3	729	1.8	626	1.3	845	2.3	839
North America	352	2.0	448	1.9	397	1.4	505	2.4	468
Europe	733	1.5	691	1.8	609	1.3	782	2.3	657
Oceania	36	2.5	51	2.0	45	1.5	58	2.5	58

Note: TFRs in 2010 are from Population Reference Bureau, *2010 World Population Data Sheet*.

countries will almost certainly prove correct. As just mentioned, however, variation at the country level could be quite large. It may be that favoring one variant over another at the regional and country levels is a reasonable approach when analyzing the projections.

Table 27 gives 2010 and 2050 populations as projected by the UN for the three main variants (medium, low, and high), along with the constant fertility variant. The TFR is also shown for 2010 and for 2050 for the three main variants. For example, the medium variant projects a population of 1.753 billion for SSA in 2050. At that time, the TFR would have to have declined to 2.5. If the TFR had declined to 3.0, then the 2050 population of SSA would be 1.986 billion (and growing more quickly).

Examining Table 27, current prospects for Europe would seem to rule out an average TFR of 2.3 (high variant) in 2050 as being too high. Even a return to a TFR of 2.1 would surprise most observers given the relative stability of European TFRs today. For sub-Saharan Africa, the medium variant seems too optimistic to many, as the most recent surveys have indicated very little downward movement in the TFR. At the country level, the different variants are very instructive. In the case of Uganda, for example, the TFR is currently about 6.7 and has shown little sign of decline in consecutive surveys. Should Uganda's TFR decline at the rate assumed in the medium variant, its population would increase from 34 million in 2010 to 91 million in 2050. If Uganda's TFR were to remain constant, its population in 2050 would exceed 150 million *and* be growing at an annual rate of 4.0 percent or more—a rate that would double the population every 17 years. It is probably fair to say that some fertility decline can be expected in Uganda despite the current flat trend, but how much, and when? The UN Population Division is careful to point out that, in the case of developing countries, the assumption of fertility decline heavily depends on a variety of factors, the increase in family planning use among them.

In addition to fertility, the general assumption made regarding mortality is that it will improve gradually worldwide, as current trends in decreasing infant mortality and rising life expectancy seem to imply. For migration, it generally is assumed that current levels and directions will continue. Migration is often difficult to measure and even more difficult to project, however. It is also currently a very sensitive issue, particularly in Europe where loss or partial loss of cultural identity may mitigate against it to some degree.

Population projections are truly an indispensable tool for evaluating the future and for such uses as planning for potential food requirements and planning for health services delivery to improve global health and raise living standards. Nevertheless, population projections should be examined in some detail, rather than their results simply being taken at face value. They serve as a valuable source of discussion exercises and, in turn, must be continually evaluated to see just how well they track future population growth.

—*Carl Haub*

Further Reading

Haub, Carl. *2010 World Population Data Sheet.* Washington, DC: Population Reference Bureau. Accessed February 2011. www.prb.org/Publications/Datasheets/2010/2010wpds.aspx.

ICF Macro. "Demographic and Health Surveys." Accessed February 2011. www.measuredhs.com/.

United Nations Population Division. "International Migrant Stock, the 2008 Revision." Accessed February 2011. http://esa.un.org/migration/.

United Nations Population Division. "World Population Policies, the 2009 Revision." Accessed February 2011. www.un.org/esa/population/publications/wpp2009/WPP2009%20web/Countries/WPP2009%20Frame.htm.

United Nations Population Division. "World Population Projections, the 2008 Revision." 2009. Accessed February 2011. http://esa.un.org/unpp/index.asp.

United Nations Population Division. "World Urbanization Prospects, the 2009 Revision." 2010. Accessed February 2011. http://esa.un.org/wup2009/unup/index.asp?panel=1.

U.S. Bureau of the Census. "International Population Data Base." Accessed February 2011. www.census.gov/ipc/www/idb/.

Bibliography

Agriculture

Broek, Jan O., and John W. Webb. *A Geography of Mankind*. New York: McGraw-Hill, 1978, 264–280.

Brown, Lester R., Christopher Flavin, Sandra Postel, and Linda Starke. *State of the World: 1990*. New York: W. W. Norton, 1990, 59–78.

Dando, William A. "Food." In *Encyclopedia of Global Change*, edited by A. S. Goudie and D. J. Cuff. Oxford, UK: Oxford University Press, 2002, 455–461.

Dando, William A., and Vijay Lulla. "World Agriculture and Food Provisioning, A.D. 2100." *Climate Change and Variation: A Primer for Teachers*. Washington, DC: National Council for Geographic Education, 2007, 157–165.

Jones, Clarence F., and Gordan G. Darkenwald. *Economic Geography*. New York: Macmillan, 1954, 145–354, 287.

Phoenix, Laurel E. *Critical Food Issues: Problems and State-of-the-Art Solutions Worldwide*, Vol. 1. Santa Barbara, CA: ABC-CLIO (2009): xiv–xx.

Statistical Abstract of the United States: 2002. Washington, DC: U.S. Government Printing Office, 2002, 515–538.

World Almanac and Book of Facts 2001. Mahwah, NJ: World Almanac Books, 2001, 160, 860.

Agriculture, Economics, and Milk

Becker, Geoffrey S. *CRS Report for Congress—Livestock Feed Costs: Concerns and Options*. Washington, DC: Congressional Research Service, 2008.

Drake, Dawn M. "Connections between Mastitis and Climate: A Study of Holsteins of Pasture in Northampton, Pennsylvania." Master's Thesis, University of Delaware, 2008.

Economist Staff. "Cheap No More—Food Prices." *The Economist*, 8 Dec. 2007.

Giampietro, Mario, and Kozo Mayumi. *The Biofuel Delusion: The Fallacy of Large-Scale Agrobioful Production*. London: Earthscan, 2009.

Lowe, Marcy, and Gary Gereffi. *A Value Chain Analysis of the U.S. Pork Industry*. Durham: Center on Globalization Governance and Competitiveness, 2008.

Westhoff, Patrick. *The Economics of Food: How Feeding and Fueling the Plant Affects Food Prices*. Upper Saddle River: Peason Education, 2010.

Animal Husbandry

Dando, William A. *The Geography of Famine*. New York: John Wiley & Sons, 1980, 23–27.

Desilva, Udaya, and Jerry Fitch. *Breeds of Livestock*. Stillwater, OK: Oklahoma State University, 2010, 1–3.

FAO Production Yearbook. Rome: Food and Agriculture Organization of the United Nations, 1974 and 2008.

Fiala, Andrew. "Animal Welfare." In *Critical Food Issues: Problems and State-of-the-Art Solutions Worldwide. Vol. 2: Society, Culture, and Ethics*, edited by Lynn Walter. Santa Barbara, CA: ABC-CLIO, 2009, 228–238.

Grigg, D. B. *The Agricultural Systems of the World*. Cambridge, UK: Cambridge University Press, 1974, 241–255.

Harlan, J. "Agricultural Origins, Centers, and Noncenters." *Science* 174 (1971): 468–474.

Jacobson, N. L., and G. N. Jacobson. "Animals: Potentials and Limitations for Human Food." In *Dimensions of World Food Problems*, edited by E. R. Duncan. Ames, IA: Iowa State University Press, 1977, 136–149.

The New York Times 2010 Almanac. New York/London: Penguin Group, 2010, 316–318.

Sauer, Carl. *Agricultural Origins and Dispersals*. New York: American Geographical Society, 1952.

Animal Proteins

Diamond, Jared. *Guns, Germs, and Steel*. New York: W. W. Norton, 1997.

Insel, P., D. Ross, K. McMahon, and M. Bernstein. *Nutrition*, 4th ed. Sudbury, MA: Jones and Bartlett, 2011.

Pimental, D., and M. Pimental. "Sustainability of Meat-Based and Plant-Based Diets and the Environment." *American Journal of Clinical Nutrition* 78 (2003): 660S–663S.

Pond, W., D. Church, K. Pond, and P. Schoknecht. *Basic Animal Nutrition and Feeding*, 5th ed. Hoboken, NJ: Wiley, 2005.

Shlain, L. *Sex, Time and Power: How Women's Sexuality Shaped Human Evolution*. New York: Viking, 2003.

Whitney, E., and S. Rolfes. *Understanding Nutrition*, 12th ed. Belmont, CA: Wadsworth, 2001.

Arable Land

Baudet, M. B., and L. Cavreul. "The Growing Lust for Agricultural Lands." *Truth-out*, April 14, 2009, 1.

Dingding, X. "National Population and Planning Commission of China." *The China Daily*, 2007.

Edwards R. "Freezing Future." *New Science* 164, no. 6 (November 27, 1999).

Pimentel, D. "Pest Management in Agriculture." In *Techniques for Reducing Pesticide Use: Economic and Environmental Benefits*, edited by D. Pimentel. Chichester, UK: Wiley Press, 1997, 1–11.

Repetto, R. "The Second India Revisited: Population Growth, Poverty, and Environment Over Two Decades." Tata Energy Research Institute (TERI) and World Resources Institute (WRI). *Proceedings of the Conference on Population, Environment and Development*, Washington, DC, March 13–14, 1996, 2–31.

Smi, I. Vaclov. *Feeding the World: A Challenge for the 21st Century*: Cambridge, MA: MIT Press, 2000.

Southgate, D., and M. Basterrechea. "Population Growth, Public Policy and Resource Degradation: The Case of Guatemala." *Ambio* 21, no. 7 (November 1992): 460–464.

United Nations Food and Agriculture Organization. *Dimensions of Need: An Atlas of Food and Agriculture*. Rome: Author, 1995, 16–98.

Williamson, R. A. "Water and Space Systems." *Imaging Notes* 25, no. 1 (2010): 10–11.

Worldwide Institute. *State of the World 2011: Innovations that Nourish the Planet*. New York: W. W. Norton, 2011.

Biodiversity

Kauffmann, Robert, and Cutler Cleveland. *Environmental Science*. New York: McGraw-Hill, 2008.

Wilson, E. O., and Peter M. Frances, eds. *Biodiversity*. Washington, DC: National Academy Press, 1988.

Carbon and Conservation Tillage

Boerboom, C. *Facts About Glyphosate Resistant Weeds*. The Glyphosate, Weeds and Crops Series. West Lafayette, IN: Purdue University Agricultural Extension Service, December 2006.

Cherdantsev, G. N. "Arable Land in the USSR in 1954" [Map]. In *A Geography of the USSR: Background to a Planned Economy*, edited by J. C. Cole and F. C. German. London: Butterworths, 1961.

Derpsch, R., and T. Friedrich. *Development and Current Status of No-Till Adoption in the World*. Rome: UN–FAO, 2009.

Fagan, B. *The Great Warming: Climate Change and the Rise and Fall of Civilizations*. New York: Bloomsbury Press, 2009.

Faulkner, E. H. *Plowman's Folly*. Norman, OK: University of Oklahoma Press, 1943.

Gaston, G., T. Kolchugina, and T. S. Vinson. "Potential Effect of No-Till Management on Carbon in the Agricultural Soils of the Former Soviet Union." *Agriculture, Ecosystems and Environment* 45 (1993): 295–309.

Hobbs, J. A., and P. L. Brown. Effects of Cropping and Management on Nitrogen and Organic Carbon Contents of a Western Kansas Soil. *Kansas Agricultural Experiment Station Tech. Bulletin* 144 (1965).

Horowitz, J., R. Ebel, and K. Ueda. *"No-Till" Farming Is a Growing Practice*. Washington, DC: USDA Economic Research Service, Economic Information Bulletin 70, November 2010.

Houghton, R. A., J. E. Hobbie, J. M. Melillo, B. Moore, B. J. Peterson, G. R. Shaver, and G.M. Woodwell. "Changes in Carbon Content of Terrestrial Biota and Soils Between 1860 and 1980: A Net Release of CO2 to the Atmosphere." *Ecological Monographs* 53, no. 3 (1983): 235–262.

Huggins, D., and J. P. Reganold. "No-Till: The Quiet Revolution." *Scientific American* (2008): 71–77.

Jenkinson, D. S. "The Rothamsted Long-Term Experiments: Are They Still of Use?" *Agronomy Journal* 83 (1991): 2–10.

Kern, J. S., and M. G. Johnson. *Impact of Conservation Tillage Use on Soil and Atmospheric Carbon in the Contiguous United States*. EPA/600/3-91/056. Corvallis, OR: Environmental Research Laboratory, 1991.

Kobak, I. I. *Biotocial Compounds of the Carbon Cycle*. St. Petersburg: Hydrometeoizdat, 1988.

Kobak, I. I., and N. Kondrashova. *Distribution of Organic Carbon in Soils of the World*. Trudy, GGI 320 (1986): 61–76.

Kononova, M. M. *Soil Organic Matter*. Moscow: Nauka, 1963.

Mann, L. K. "Changes in Soil Carbon Storage after Cultivation." *Soil Science* 142, no. 5 (1986): 279–288.

Medvedev, Z. A. *Soviet Agriculture*. New York: Norton, 1987.

Phillips, R. E., R. L. Blevins, G. W. Thomas, W. W. Frye, and S. H. Thomas. "No-Tillage Agriculture." *Science* 208 (1980): 1108–1113.

Priputina, I. V. "Lowering of the Humus Content of Chernozem Soils of the Russian Plain as a Result of Human Action." *Soviet Geography* 30, no. 10 (1989): 759–762.

Ryabchikov, A. M., ed. *Geographic and Zonal Types of Landscapes of the World* [Map]. Moscow: Moscow State University School of Geography, 1988.

Suslov, S. P. *Physical Geography of Asiatic Russia*, translated by N. D. Gerchevsky. San Francisco: W. H. Freeman, 1961.

Symons, L. S. *Russian Agriculture: A Geographic Survey*. New York: Wiley, 1972.

Carrying Capacity

Leopold, A. *Game Management*. New York: Scribner, 1933.

Odum, E. P. *Fundamentals of Ecology*, 3rd ed. Philadelphia: Saunders, 1971.

Sayers, N. F. "The Genesis, History, and Limits of Carrying Capacity." *Annals of the Association of American Geographers* 98, no. 1 (2008): 120–134.

Cereal Foods

Anderson, Eugene Newton. *Everyone Eats: Understanding Food and Culture*. New York: University Press, 2005.

FAO Prospects and Food Situation Preview, No. 4. Rome: FAO, November, 2009.

Pearson, Debra. "The Effect of Agricultural Practices on Nutrient Profiles of Foods." In *Critical Food Issues: Environment, Agriculture, and Health Concerns*, edited by Laurel E. Phoenix. Santa Barbara, CA: ABC-CLIO, 2009, 177–193.

Vavilov, Nikolay Ivanovich. *Five Continents*, edited by L. E. Rodin and translated by Doris Love. Rome: IPGPRI; St. Petersburg: VIR, 1997.

Vavilov, Nikolay Ivanovich. "The Origin, Variation, Immunity, and Breeding of Cultivated Plants," translated and reprinted by K. Starr Chester. *Chronica Botanica* 13 (1949): 1–6.

Woolley, D. G. "Food Crops: Production, Limitations, and Potentials." In *Dimensions of World Food Problems*, edited by E. R. Duncan. Ames, IA: Iowa State University Press, 1977, 153–171.

Wortman, Sterling, and Ralph W. Cummings, Jr. *To Feed This World*. Baltimore, MD: Johns Hopkins University Press, 1978, 144–186.

Climate Change and World Food Production

Dando, William A., and Vijay Lulla. "World Agriculture and Food Provisioning." In *Climate Change and Variation*, edited by William A. Dando. Washington, DC: National Council for Geographic Education, 2007.

Hidore, John J. "Climate Change and the Living World." In *Climatology: An Atmospheric Science*, edited by John J. Hidore, John Oliver, Mary Snow, and Richard Snow. Upper Saddle River, NJ: Prentice-Hall, 2010, 219–230.

Hidore, John J. *Global Environmental Change*. Englewood Cliffs, NJ: Prentice-Hall, 1996.

Speer, James H. "Vegetation Response to Global Warming." In *Climate Change and Variation*, edited by W. A. Dando. Washington, DC: National Council for Geographic Education, 2007.

U.S. Department of Agriculture, Economic Research Service. *Household Food Security in the United States*. Washington, DC: U.S. Government Printing Office, 2009.

"Water: Our Thirsty World." *National Geographic Magazine* [Special issue]. Washington, DC: National Geographic Society, 2010.

Community-Supported Agriculture

Anderson-Wilk, Mark. "Does Community Supported Agriculture Support Conservation?" *Journal of Soil and Water Conservation* 62, no. 6 (2007): 126A.

Beauchesne, Audric, and Christopher Bryant. "Agriculture and Innovation in the Urban Fringe: The Case of Organic Farming in Quebec, Canada." *Tijdschrift Voor Economische En Sociale Geografie* 90, no. 3 (1999): 320.

Blanchard, Troy C., and Todd L. Matthews. "Retail Concentration, Food Deserts, and Food Disadvantaged Communities in Rural America." In *Remaking the North American Food System: Strategies for Sustainability*, edited by Clare C. Hinrichs and Thomas A. Lyson. Lincoln, NE/London: University of Nebraska Press, 2007, 210.

Brown, Cheryl. "The Impacts of Local Markets: A Review of Research on Farmer's Markets and Community Supported Agriculture (CSA)." *American Journal of Agricultural Economics* 90, no. 5 (2008): 1296.

Clancy, Kate, Janet Hammer, and Debra Lippoldt. "Food Policy Councils." In *Remaking the North American Food System: Strategies for Sustainability*, edited by Clare C. Hinrichs and Thomas A. Lyson. Lincoln, NE/London: University of Nebraska Press, 2007, 121.

Cone, Cynthia Abbott, and Andrea Myhre. "Community-Supported Agriculture: A Sustainable Alternative to Industrial Agriculture?" *Human Organization* 59, no. 2 (2000): 187.

DeLind, Laura B., and Anne E. Ferguson. "Is This a Women's Movement? The Relationship of Gender to Community Supported Agriculture in Michigan." *Human Organization* 58, no. 2 (1999): 190.

Girardet, Herbert. "Urban Agriculture and Sustainable Urban Development." In *CPULS Continuous Productive Urban Landscape: Designing Urban Agriculture for Sustainable Cities*, edited by Andre Viljoen. Oxford, UK: Elsevier, Architectural Press, 2005, 32.

Harrison, Charles H. *Tending the Garden State: Preserving Agriculture in New Jersey*. New Brunswick, NJ/London: Rivergate Books, 2007.

Heimlich, Ralph. "Agriculture Adapts to Urbanization." *Food Review* 14, no. 1 (January–March 1991): 1.

Henderson, Elizabeth, and Robyn Van En. *Sharing the Harvest: A Citizen's Guide to Community Supported Agriculture*, rev. ed. White River Junction, VT: Chelsea Green, 2007.

Lyson, Thomas A. *Civic Agriculture: Reconnecting Farm, Food and Community.* Lebanon, NH: Tufts University Press/University Press of New England, 2004.

Pfeiffer, Dale Allen. *Eating Fossil Fuels: Oil, Food, and the Coming Crisis in Agriculture.* Gabriola Island, BC, Canada: New Society Publishers, 2006.

Schnell, Steven M. "Food with a Farmer's Face: Community-Supported Agriculture in the United States." *Geographical Review* 97, no. 4 (2007): 550.

Trauger, Amy. "'Because They Can Do the Work': Women Farmers in Sustainable Agriculture in Pennsylvania, USA." *Gender, Place and Culture* 11, no. 2 (2004): 289.

Vandermeulen, Valerie, Ann Verspecht, Guido Van Huylenbroeck, Henk Meert, Ankatrien Boulanger, and Etienne Van Hecke. "The Importance of the Institutional Environment on Multifunctional Farming Systems in the Peri-urban Area of Brussels." *Land Use Policy* 23 (2006): 486.

Viljoen, Andre, and Joe Howe. "Cuba: Laboratory for Urban Agriculture." In *CPULs Continuous Productive Urban Landscapes: Designing Urban Agriculture for Sustainable Cities*, edited by Andre Viljoen. Oxford, UK: Elsevier, Architectural Press, 2005, 146.

Conservation and Sustainable Agriculture

Brown, L. *Plan B 4.0: Mobilizing to Save Civilization.* New York: W. W. Norton, 2009.

Diamond, J. "Evolution, Consequences, and the Future of Plant and Animal Domestication." *Nature* 418 (2002): 700–707.

Pollan, M. *The Botany of Desire: A Plant's-Eye View of the World.* New York: Random House, 2001.

Pretty, J. N. "Agricultural Sustainability: Concepts, Principles, and Evidence." *Philosophical Transactions of the Royal Society* 363 (2008): 447–465.

Pretty, J. N., A. D. Noble, D. Bossio, J. Dixon, R. E. Hine, F. W. T. Penning de Vries, and J. I. L. Morison. "Resource-Conserving Agriculture Increases Yields in Developing Countries." *Environmental Science and Technology* 40, no. 4 (2006): 1114–1119.

Shiva, V. *Stolen Harvest: The Hijacking of the Global Food Supply.* Cambridge, MA: South End Press, 2000.

Deficiency Diseases

Bollet, Alfred Jay. *Plagues and Poxes*, 2nd ed. Demos, NY: Medical Publishing, 2004.

Carpenter, Kenneth J. *Beriberi, White Rice, and Vitamin B.* Berkeley/Los Angeles: University of California Press, 2000.

Frankenburg, Frances Rachel. *Vitamin Discoveries and Disasters.* Santa Barbara, CA: Praeger/ABC-CLIO, 2009.

Holick, Michael F. *The Vitamin D Solution*. New York: Penguin Group, 2010.
Howson, Christopher P., Eileen T. Kennedy, and Abraham Horwitz, eds. *Prevention of Micronutrient Deficiencies*. Washington, DC: National Academy Press, 1998.
Jacobsen, Kathryn H. *Introduction to Global Health*. Sudbury, MA: Jones and Bartlett, 2008.

Desertification

Annan, K. *The Millennium Ecosystem Assessment*. Initiated by the United Nations Secretary General Kofi Annan in 2000, and supported by governments who are Parties to four multilateral environmental conventions.
Berry, L., J. Olson, and D. Campbell. "Assessing the Extent, Cost and Impact of Land Degradation at the National Level. Overview: Findings and Lessons Learned." *UNCCD Global Mechanism* (2003).
Bridges, E. M., I. D. Hannam, L. R. Oldeman, F. W. T. Penning de Vries, S. J. Scherr, and S. Sombatpanit, eds. *Response to Land Degradation*. Enfield, NH: Science Publishers, 2001.
Oasis Challenge Programme Proposal. ICARDA/ICRISAT (available on request).
Reynolds, J. F., et al. "Global Desertification: Building a Science for Dryland Development." *Science* 316, no. 11 (2007): 847–851.
Shah, J. *DNA: Daily News and Analysis*. Mumbai, December 25, 2009.
Thomas, R. J. "Addressing Land Degradation and Climate Change in Dryland Agroecosystems Through Sustainable Land Management." *Journal of Environmental Monitoring*, in press.

Drought

Burke, E. J., S. J. Brown, and N. Christidis. "Modelling the Recent Evolution of Global Drought and Projections for the 21st Century with the Hadley Centre Climate Model." *Journal of Hydrometeorology* 7 (2006): 1113–1125.
Cook, E. R., C. Woodhouse, M. Eakin, D. M. Meko, and D. W. Stahle. "Long-Term Aridity Changes in the Western United States." *Science* 306 (2004): 1015–1018.
Douville H., F. Chauvin, S. Planton, J. F. Royer, D. Salas-Melia, and D. Tyteca. "Sensitivity of the Hydrological Cycle to Increasing Amounts of Greenhouse Gases and Aerosols." *Climate Dynamics* 20 (2002): 45–68.
Parry, M. L., O. F. Canziani, J. P. Palutikof, P. J. van der Linden, and C. E. Hanson, eds. *Contribution of Working Group II to the Fourth Assessment Report of the Intergovernmental Panel on Climate Change*. New York: Cambridge University Press, 2007.
Piechota, T. C., H. Hidalgo, J. Timilsena, and G. A. Tootle. "The Western Drought: How Bad Is It?" *Eos* 85 (2004): 301–308.

Solomon, S., D. Qin, M. Manning, Z. Chen, M. Marquis, K. B. Averyt, M. Tignor, and H. L. Miller, eds. *Contribution of Working Group I to the Fourth Assessment Report of the Intergovernmental Panel on Climate Change*. New York: Cambridge University Press, 2007.

Evidence for and Predictions of Future Climate Change

Cline, W. R. *Global Warming and Agriculture: Impact Estimates by Country*. Washington, DC: Center for Global Development and Peterson Institute for International Economics, 2007.

Easterling, W., et al. "Food, Fiber, and Forest Products." In *Climate Change 2007: Impacts, Adaptation, and Vulnerability*, edited by M. L. Parry et al. Cambridge, UK: Cambridge University Press, 2007, 273–313.

Fischlin, A., et al. "Ecosystems, Their Properties, Goods and Services." In *Climate Change 2007: Impacts, Adaptation, and Vulnerability*, edited by M. L. Parry et al. Cambridge, UK: Cambridge University Press, 2007, 211–272.

Matthews, H. D., and K. Caldeira. "Stabilizing Climate Requires Near-Zero Emissions." *Geophysical Research Letters* 35(2008): L04705, 10.1029/2007GL032388.

Meehl, G. A., et al. "Global Climate Projections." In *Climate Change 2007: The Physical Science Basis. Contribution of Working Group I to the Fourth Assessment Report of the Intergovernmental Panel on Climate Change*, edited by S. Solomon et al. Cambridge, UK: Cambridge University Press, 2007, 747–846.

Overpeck, J. T., B. L. Otto-Bliesner, G. H. Miller, D. R. Muhs, R. B. Alley, and J. T. Kiehl. "Paleoclimatic Evidence for Future Ice-Sheet Instability and Rapid Sea-Level Rise." *Science* 311 (2006): 1747–1750.

Plattner, G. K., et al. "Long-Term Climate Commitments Projected with Climate Carbon Cycle Models." *Journal of Climate* 21 (2008): 2721–2751.

Schneider, S. H., et al. "Assessing Key Vulnerabilities and the Risk from Climate Change." In *Climate Change 2007: Impacts, Adaptation and Vulnerability. Contribution of Working Group II to the Fourth Assessment Report of the Intergovernmental Panel on Climate Change*, edited by M. L. Parry et al. Cambridge, UK: Cambridge University Press, 2007, 779–810.

Solomon, S., G. K. Plattner, R. Knutti, and P. Friedlingstein. "Irreversible Climate Change Due to Carbon Dioxide Emissions." *Proceedings of the National Academy of Sciences USA* 106 (2009): 1704–1709.

Weiss, J. L., J. T. O. Overpeck, and B. Strauss. "Implications of Recent Sea Level Rise Science for Low-Elevation Areas in Coastal Cities of the Conterminous U.S.A." *Climatic Change*. doi: 10.1007/s10584-011-0024-x.

Famine

Curry, B. "Is Famine a Discrete Event?" *Disasters* 16 (1992): 138–144.
Dando, William A. "Biblical Famines, 1850 B.C.—A.D. 46: Insights for Modern Mankind." *Ecology of Food and Nutrition* 13 (1983): 231–249.
Dando, William A. "Famine: Problems in Definition." *Bulletin of the Association of North Dakota Geographers* XXIX (1979): 41–48.
Dando, William A. *The Geography of Famine*. New York: John Wiley & Sons, 1980, 57–67.
Dando, William A., and Caroline Z. Dando. *A Reference Guide to World Hunger*. Hillside, NJ: Enslow, 1991, 45–53.
Field, John Osgood. *The Challenge of Famine*. West Hartford, CT: Kumarian Press, 1993.
Hyder, Massod. "Famine." In *The Oxford Illustrated Companion to Medicine*, edited by Stephen Lock, John M. Last, and George Dunea. Oxford, UK: Oxford University Press, 2001, 1301–1303.
Millman, S. "Hunger in the 1980s: Backdrop for Policy in the 1990s." *Food Policy* 15 (1990): 277–285.
Murton, Brian. "Famine." In *The Cambridge World History of Food*, edited by Kenneth K. Kiple and Kriemhild Conu Ornelas. Cambridge, UK: Cambridge University Press, 2000, 1411–1427.
Sen, Amartya. "Starvation and Exchange Entitlements: A General Approach and Its Application to the Great Bengal Famine," *Cambridge Journal of Economics* 1 (1977): 33–39.
West, K. P. "Famine." In *Encyclopedia of Human Nutrition*, 2nd ed., edited by Benjamin Caballero, Lindsay Allen, and Andrew Prentice. Oxford, UK: Elsevier Academic Press, 2005, 169–177.
Widdowson, E. M., and A. Ashworth. "Famine, Starvation, and Fasting." In *Encyclopedia of Food Science and Nutrition*, edited by Benjamin Caballero, Luiz C. Trugo, and Paul M. Finglas. Oxford, UK: Elsevier Science, 2003, 2243–2247.

Famine Early Warning Systems Network (FEWS NET)

Brown, M. E. *Famine Early Warning Systems and Remote Sensing Data*. New York: Springer Publishers, 2008, 313 pp.
Vanhaute, E. *From Famine to Food Crisis. What History Can Teach Us About Local and Global Subsistence Crises*. Munich, Germany: Munich Personal RePEc Archive Paper No. 17630, 2009.
Walker, P. *Famine Early Warning Systems: Victims and Destitution*. UK: Earthscan/Taylor and Francis Group, 2009.

Farm Adjustments to Climate Change

Glover, Jerry, et al. "Increased Food and Ecosystem Security via Perennial Grains." *Science* 328 (June 25, 2010): 1638–1639.

Mutel, Cornelia Fleischer, and John C. Emerick. *From Grassland to Glacier.* Boulder, CO: Johnson Printing, 1992.

Opie, John. *Ogallala: Water for a Dry Land*, 2nd ed. Lincoln, NE: University of Nebraska Press, 2000.

Peterson, Rorik. "Water Woes: Structure and Agency Influences on Adaptive Capacity in Southwest Kansas." *Geographical Bulletin* 47 (2006).

Pimenel, David. "Water Resources." *Agricultural and Environmental Issues* 54 (2004): 909–918.

"Planting Decisions Make Difference for Kansas Wheat Producer." *Hill City Times*, June 23, 2010, p. 18.

Postel, Sandra, L. "Entering an Era of Water Scarcity: The Challenges Ahead." *Ecological Applications* 10 (2000): 941–948.

Rosegrant, Mark W., Cai Ximing, and Sarah A. Cline. "Will the World Run Dry? Global Water and Food Security." *Environment* 45 (2003): 25–36.

"Teff Field Day on August 5." *Hill City Time*s, July 28, 2010, p. 4.

Tilman, David, et al. "Forecasting Agriculturally Driven Global Environmental Change." *Science* 292 (2001): 281–284.

Food Aid Policies in the United States: Contrasting Views

Allen, P. "The Disappearance of Hunger in America." *Gastronomica: The Journal of Food and Culture* 7, no. 2 (2007): 19–23.

Allen, P. "Reweaving the Food Security Safety Net: Mediating Entitlement and Entrepreneurship." *Agriculture and Human Values* 16 (1999): 117–129.

Anderson, S. A. "Core Indicators of Nutritional State for Difficult-to-Sample Populations." *Journal of Nutrition* 120, no. 11 (suppl, 1990): 1555–1600.

Brown, M. "Liberalism." In *The Dictionary of Human Geography*, 4th ed., edited by R. J. Johnston, M. Watts, D. Gregory, and G. Pratt. Malden, MA: Blackwell, 2000, 446–448.

Clark, B. *Political Economy: A Comparative Approach*, 2nd ed. Westport, CT: Praeger, 1998.

Cochran, C. E., L. C. Mayer, T. R. Carr, and N. J. Cayer. *American Public Policy: An Introduction*, 5th ed. New York: St. Martin's Press, 1996.

Collins, S. D. *Let Them Eat Ketchup! The Politics of Poverty and Inequality.* New York: Monthly Review Press, 1996.

Jansson, B. S. *The Reluctant Welfare State: A History of American Social Welfare Policies*. Belmont, CA: Wadsworth, 1988.

Kodras, J. E. “Breadlines.” In *Geographical Snapshots of North America*. New York: Guilford Press, 1992, 103–107.

Larner, W. “Neo-liberalism: Policy, Ideology, Governmentality.” *Studies in Political Economy* 63 (2000): 5–25.

Lobao, L., and G. Hooks. “Public Employment, Welfare Transfers, and Economic Well-Being Across Local Populations: Does a Lean and Mean Government Benefit the Masses?” *Social Forces* 82, no. 2 (2003): 519–556.

Mink, G., and R. Solinger. *Welfare: A Documentary History of U.S. Policy and Politics*. New York: New York University Press, 2003.

O’Connor, B. *A Political History of the American Welfare System: When Ideas Have Consequences*. Lanham, MD: Rowman & Littlefield, 2004.

O’Connor, A. *Poverty Knowledge: Social Science, Social Policy, and the Poor in Twentieth-Century US History*. Princeton, NJ: Princeton University Press, 2001.

Peck, J. *Workfare States*. New York: Guilford Press, 2001.

Peet, R. “Inequality and Poverty: A Marxist-Geographic Theory.” *Annals of the Association of American Geographers* 65, no. 4 (1975): 564–571.

Pinker, R. “The Conservative Tradition of Social Welfare.” In *The Student’s Companion to Social Policy*, 2nd ed., edited by P. Alcock, A. Erskine and M. May. Malden, MA: Blackwell, 2003, 78–84.

Piven, F. F., and R. A. Cloward. *Regulating the Poor: The Functions of Public Welfare*, updated ed. New York: Vintage Books, 1993.

Schram, S. F. *Words of Welfare: The Poverty of Social Science and the Social Science of Poverty*. Minneapolis, MN: University of Minnesota Press, 1995.

Starkey, L. J., K. Gray-Donald, and H. V. Kuhnlein. “Nutrient Intake of Food Bank Users Is Related to Frequency of Food Bank Use, Household Size, Smoking, Education and Country of Birth.” *Journal of Nutrition* 129, no. 4 (1999): 883–889.

Vaisse, J. *Neoconservatism: The Biography of a Movement*. Cambridge, MA: Belknap Press of Harvard University Press, 2010.

Food Assistance Landscapes in the United States

Brown, L. “Fat and Hungry: Will Political Leaders Ever Get It Right?” Keynote address to the Pennsylvania Hunger Action Center. Philadelphia, PA: Pennsylvania Nutrition Education Network, May 14, 2003.

Daponte, B. O., and S. L. Bade. *The Evolution, Cost, and Operation of the Private Food Assistance Network*. Discussion Paper No. 1211-00. Madison, WI: Institute for Research on Poverty, 2000.

Economic Research Service. *The Food Assistance Landscape: FY 2009 Annual Report*. Washington, DC: U.S. Department of Agriculture, 2010.

Economic Research Service. *Food Assistance and Nutrition Program Final Report: Fiscal 2009 Activities*. Washington, DC: U.S. Department of Agriculture, 2009.

Esping-Anderson, G. *The Three Worlds of Welfare Capitalism*. Cambridge, UK: Polity Press, 1990.

Hershkoff, H., and S. Loffredo. *The Rights of the Poor: The Authoritative ACLU Guide to Poor People's Rights*. Carbondale, IL: Southern Illinois University Press, 1997.

Kodras, J. E. "Economic Restructuring, Shifting Public Attitudes and Program Revision: The Politics Underlying Geographic Disparities in the Food Stamp Program." In *Geographic Dimensions of United States Social Policy*, edited by J. E. Kodras and J. P. Jones. London: Edward Arnold, 1990, 218–236.

Kotz, N. *Let Them Eat Promises: The Politics of Hunger in America*. Englewood Cliffs, NJ: Prentice-Hall, 1969.

Nord, M., A. Coleman-Jensen, M. Andrews, and S. Carlson. *Household Food Security in the United States, 2009*. Economic Research Report Number 108. Washington, DC: U.S. Department of Agriculture, 2010.

O'Brien, D., K. Prendergast, E. Thompson, M. Fruchter, and H. Torres Aldeen. *The Red Tape Divide: State-by-State Review of Food Stamp Applications*. Chicago, IL: America's Second Harvest, 1997.

Ohls, J., R. Cohen, F. Saleem-Ismail, and B. Cox. *The Emergency Food Assistance System: Findings from the Provider Survey. Volume II: Final Report*. Food and Nutrition Research Report No. 16–2. Washington, DC: Economic Research Service, U.S. Department of Agriculture, 2002.

Poppendieck, J. *Sweet Charity? Emergency Food and the End of Entitlement*. New York: Penguin Books, 1998.

Riches, G. "Advancing the Human Right to Food in Canada: Social Policy and the Politics of Hunger, Welfare, and Food Security." *Agriculture and Human Values* 16 (1999): 203–211.

Sen, A. "The Right Not to Be Hungry." *Contemporary Philosophy* 2 (1982): 343–360.

Theodore, N. "Welfare Reform, Work Requirements and the Geography of Unemployment." *Urban Geography* 22 (2001): 490–492.

Tiehen, L. "Use of Food Pantries by Households with Children Rose During the Late 1990s." In *Food Review*. Washington, DC: Economic Research Service, U.S. Department of Agriculture, 2002.

Food Diffusion

Barstow, C. *The Eco-foods Guide*. Gabriola Island, BC, Canada: New Society Publishers, 2002.

Broek, J. O. M., and J. W. Webb. *The Geography of Mankind*. New York: McGraw-Hill, 1968.

deLaubenfels, D. J. *A Geography of Plants and Animals*. Dubuque, IA: Wm. C. Brown, 1970.

FAO Production Yearbook, No. 23. Rome: United Nations, 1969.

Food: The Yearbook of Agriculture 1959. Washington, DC: U.S. Government Printing Office, 1959.

Grigg, D. B. *The Agricultural Systems of the World*. London: Cambridge University Press, 1974.

Nabhan, G. P. *Where Our Food Comes from*. London: Island Press/Shearwater Books, 2009.

Purnell, L. D., and B. J. Paulanka. *Transcultural Health Care*. Philadelphia: F. A. Davis, 2003.

Rogers, E. M. *Diffusion of Innovations*, 4th ed. New York: Free Press, 1995.

Food, Famine, and Popular Culture

Bremner, Robert H. *Giving: Charity and Philanthropy in History*. New Brunswick, NJ/London: Transaction Publishers, 1994.

Civitello, Linda. *Cuisine and Culture: A History of Food and People*. Hoboken, NJ: John Wiley & Sons, 2004.

Grivetti, Louis E., and Howard-Yana Shapiro, eds. *Chocolate: History, Culture and Heritage*. Hoboken, NJ: John Wiley & Sons, 2009.

King, Carla, ed. *Famine, Land and Culture in Ireland*. Dublin: University College Dublin Press, 2000.

Ó Gráda, Cormac. *Famine: A Short History*. Princeton NJ: Princeton University Press, 2009.

Rosenberg, Tina. "What Lara Croft Would Look Like If She Carried Rice Bags," *New York Times*, December 30, 2005, A24.

Schroeder, Fred E. H. *5000 Years of Popular Culture: Popular Culture Before Printing*. Bowling Green, OH: Bowling Green University Popular Press, 1980.

Food Poisoning

DeWaal, Caroline Smith, and Nadine Robert. *Global and Local: Food Safety Around the World*. Washington, DC: Centers for Science in the Public Interest, 2005.

McSwane, David, Nancy Rue, and Richard Linton. *Essentials of Food Safety and Sanitation*, 2nd ed. Englewood Cliffs, NJ: Prentice Hall, 2000.

Mokyr, Joel, and Cormac O'Grada. "What Do People Die of During Famines: The Great Irish Famine in Comparative Perspective." *European Review of Economic History* 6 (2002): 339–363.

Webb, Patrick, and Andrew Thorne-Lyman. *Entitlement Failure from a Food Quality Perspective: The Life and Death Role of Vitamins and Minerals in Humanitarian Crises*. Research Paper No. 20006/140. United Nations University, 2006.

Food Policy Debates: Global Issues of Acces

Bentley, A. *Eating for Victory: Food Rationing and the Politics of Domesticity.* Urbana, IL: University of Illinois Press, 1998.

Bluestone, B., and B. Harrison. *The Deindustrialization of America.* New York: Basic Books, 1982.

Cohen, L. *A Consumers' Republic: The Politics of Mass Consumption in Postwar America.* New York: Alfred A. Knopf, 2003.

Desjarlais, R., L. Eisenberg, B. Good, and A. Kleinman. *World Mental Health: Problems and Priorities in Low-Income Countries.* Oxford, UK: Oxford University Press, Inc., 1995.

Eisinger, P. K. *Toward an End to Hunger in America.* Washington, DC: Brookings Institution Press, 1998.

Glasmeier, A. K. *An Atlas of Poverty in America: One Nation, Pulling Apart, 1960–2003.* New York: Routledge, 2005.

Haynie, D. L., and B. K. Gorman. "Determinants of Poverty Across Urban and Rural Labor Markets." *Sociological Quarterly* 40, no. 2 (1999): 177–197.

Hobbs, K., W. MacEachern, A. McIvor, and S. Turner. "Waste of a Nation: Poor People Speak Out About Charity." *Canadian Review of Social Policy* 31 (1993): 94–104.

Iceland, J. *Poverty in America: A Handbook*, Vol. 2. Berkeley, CA: University of California Press, 2006.

Levenstein, H. *Paradox of Plenty: A Social History of Eating in Modern America.* Berkeley, CA: University of California Press, 2003.

Maney, A. L. *Still Hungry After All These Years: Food Assistance Policy from Kennedy to Reagan.* New York: Greenwood Press, 1989.

Maxwell, S. "Food Security: A Post-modern Perspective." *Food Policy* 21, no. 2 (1996): 155–170.

Miller, K., B. Weber, L. Jensen, J. Mosley, and M. Fisher. "A Critical Review of Rural Poverty Literature: Is There Truly a Rural Effect?" *International Regional Science Review* 28, no. 4 (2005): 381–414.

O'Connor, A. *Poverty Knowledge: Social Science, Social Policy, and the Poor in Twentieth-Century US History.* Princeton, NJ: Princeton University Press, 2001.

Poppendieck, J. *Breadlines Knee-Deep in Wheat: Food Assistance in the Great Depression.* New Brunswick, NJ: Rutgers University Press, 1986.

Sanchez, P. A. "Hunger in Africa: The Link Between Unhealthy People and Unhealthy Soils." *Lancet* 365 (2005): 442–444.

Schram, S. F. *Words of Welfare: The Poverty of Social Science and the Social Science of Poverty.* Minneapolis, MN: University of Minnesota Press, 1995.

Shaw, W. *The Geography of United States Poverty: Patterns of Deprivation, 1980–1990.* New York: Garland, 1996.

Trattner, W. I. *From Poor Law to Welfare State: A History of Social Welfare in America*, 6th ed. New York: Free Press, 1999.

White, M. "Food Access and Obesity." *Obesity Reviews* 8 (suppl 1, 2007): 99–107.

Wolch, J., and M. Dear. *Malign Neglect: Homelessness in an American City*. San Francisco: Jossey-Bass, 1993.

Yapa, L. "What Causes Poverty? A Postmodern View." *Annals of the Association of American Geographers* 86, no. 4 (1996): 707–728.

Food Safety

Arduser, Lora, and Douglas Robert Brown. *HACCP and Sanitation in Restaurants and Food Service Operations: A Practical Guide Based on the USDA Food Code*, 2005.

National Restaurant Association. *ServSafe Essentials*, 5th ed., 2010.

Food Sources

Anderson, J. *A Geography of Agriculture*. Dubuque, IA: William C. Brown, 1970, 40–41.

Barstow, Cynthia. *The Eco-foods Guide*. Gavriola Island, Canada: New Society Publishers, 2002, 13–20.

Corey, Helen. *Food from Biblical Lands*, 2nd ed. Terre Haute, IN: Helen Corey Press, 1990, vii–viii.

Eide, A. "The Human Right to Adequate Food and Freedom from Hunger." In *The Right to Food in Theory and Practice*. Rome: Food and Agricultural Organization of the United Nations, 1998, 1–5.

Lai, T., J. Ram, D. Perkins, and P. Cook, eds. *Hong Kong and China Gas Chinese Cookbook*. Hong Kong: Pat Printer Associates, 1978, 17–18.

McCall's Cookbook. New York: Random House, 1963, 6–14.

Pinstrup-Andersen, Per. "The Future World Food Situation and the Role of Plant Diseases." Paper presented at the joint meeting of the American Phytopathological Society and the Canadian Phytopathological Society, Montreal, Canada, August 8, 1999.

Genetically Modified Foods

Collier, Paul. "Can Biotech Food Cure World Hunger?" *The New York Times*, October 26, 2009.

The Environmental Food Crisis: The Environment's Role in Averting Future Food Crises. Nairobi, Kenya: United Nations Environmental Programme, 2009.

Estabrook, Barry. "Supreme Court on Modified Foods: Who Won?" *The Atlantic*, June 22, 2010.

Miller, G. Tyler, and Scott E. Spoolman. "Food, Soil, and Pest Management." In *Living in the Environment*. Belmont, CA: Brooks/Cole, 2009, 275–312.

Pickrell, John. "Introduction: GM Organisms." *New Scientist*, September 4, 2006.

U.S. Food and Drug Administration. *The FDA List of Completed Consultations on Bioengineered Foods*. Washington, DC: U.S. Government Printing Office, 2009.

Whitman, Deborah B. "Genetically Modified Foods: Harmful or Helpful?" *ProQuest*, April 2000.

Geotechniques (Remote Sensing and GIS): Tools for Monitoring Change

Abrams, M., J. Abrams, G. Asrar, et al. *Encyclopedia of Remote Sensing*. Encyclopedia of Earth Sciences Series. New York: Springer, June 2012.

Longley, P. A., and M. Goodchild. *Geographical Information Systems and Science*. New York: John Wiley and Sons, 2001.

Longley, P., and M. Batty. *Spatial Analysis: Modeling in a GIS Environment*. New York: John Wiley and Sons, 1997.

Mitchell, A. *The ESRI Guide to GIS Analysis, Volume 1: Geographic Patterns and Relationships*. Redlands, CA: ESRI Press, 2001.

Green Revolution

Bickel, L. *Facing Starvation: Norman Borlaug and the Fight Against Hunger*. Pleasantville, NY: Readers' Digest Press, 1974.

Brown, L. *Seeds of Change*. New York: Praeger, 1970.

Clever, H. "The Contradictions of the Green Revolution." *American Economic Review* 62, no. 2 (1972): 177–186.

Dovring, F. *Progress for Food or Food for Progress*. New York: Praeger, 1988.

French, C., J. Moore, C. Kraenzle, and K. Harling. *Survival Strategies for Agricultural Cooperatives*. Ames, IA: Iowa State University Press, 1980.

Ghai, Dharam, Azizur R. Khan, Eddy Lee, and Samir Radwan. *Agrarian Systems and Rural Development*. New York: Holmes & Meier, 1979, 22–112.

Grigg, D. *The World Food Problem: 1950–1980*. Oxford, UK: Basil Blackwell, 1985.

Nabhan, Gary Paul. *Where Our Food Comes from*. Washington, DC: Island Press/Shearwater Books, 2009.

Poleman, Thomas T., and Donald K. Freebairn. *Food, Population, and Employment: The Impact of the Green Revolution*. New York: Praeger, 1973.

Vallianatos, E. *This Land Is Their Land*. Monroe, ME: Common Courage Press, 2006.

Historiography of Food, Hunger, and Famine

Albala, K. *Eating Right in the Renaissance*. Berkeley/Los Angeles: University of California Press, 2002.

Albala, K. "History on the Plate: The Current State of Food History." *Historically Speaking* 10, no. 5 (November 2009): 6–8.

Bober, P. *Art, Culture, and Cuisine: Ancient and Medieval Gastronomy*. Chicago/London: University of Chicago Press, 1999.

Cronon, W. "A Place for Stories: Nature, History, and Narrative." *Journal of American History* 78, no. 4 (March 1992): 1347–1376.

Cullen, L. "The Politics of the Famine and of Famine Historiography." In *Comhdáil an Chraoíbhín 1996*. Roscommon, Ireland, 1997, 9–31.

Dennis, A. "From Apicius to Gastroporn: Form, Function and Ideology in the History of Cookery Books." *Studies in Popular Culture* 31, no. 1 (Fall 2008): 1–17.

Devereux, S. *Theories of Famine*. New York/London: Harvester Wheatsheaf, 1993.

Edkins, J. *Whose Hunger? Concepts of Famine, Practices of Aid*. Borderlines, Vol. 17. Minneapolis, MN/London: University of Minnesota Press, 2000.

Janmaat, J. "History and National Identity Construction: The Great Famine in Irish and Ukrainian History Textbooks." *History of Education* 35, no. 3 (May 2006): 345–368.

Marples, D. "Debate: Ethnic Issues in the Famine of 1932–1933 in Ukraine." *Europe-Asian Studies* 61, no. 3 (May 2009): 505–518.

Mintz, A., and C. Du Bois. "The Anthropology of Food and Eating." *Annual Review of Anthropology* 31 (2002): 99–119.

Torry, W. "Social Science Research on Famine: A Critical Evaluation." *Human Ecology* 12, no. 3 (1984): 227–252.

Hunger and Starvation

Barstow, C. *The Eco-Foods Guide: What's Good for the Earth Is Good for You*. Gabriola Island, Canada: New Society Publishers, 2002, 227–260.

Bray, G. A. "Hunger." In *Encyclopedia of Food and Culture*, edited by H. Katz and W. W. Weaver. New York: Charles Scribner's Sons, 2003, 219–222.

Dando, William A. *The Geography of Famine*. New York: John Wiley & Sons, 1980, 141–157.

Dolot, M. *Execution by Hunger: The Hidden Holocaust*. New York: W. W. Norton, 1985.

Grigg, D. *The World Food Problem: 1950–1980*. New York: Basil Blackwell, 1985, 2–53.

Halford, J. C. C., A. J. Hill, and J. E. Blundale. "Hunger." In *Encyclopedia of Human Nutrition*, 2nd ed., edited by Benjamin Caballario, Lindsay Allen, and Andrew Prentice. Oxford, UK: Elsevier Academic Press, 2005, 469–474.

Heldke, L. "Food Security: Three Conceptions of Access—Charity, Rights, and Co-responsibility." In *Critical Food Issues: Problems and State-of-the-Art Solutions Worldwide*, edited by Lynn Walter. Santa Barbara, CA: ABC-CLIO, 2009, 213–225.

Hughes, D. J. *Science and Starvation*. Oxford, UK: Pergamon Press, 1968, 1–157.

Keys, A., J. Brozek, A. Henschel, O. Mickelsen, and H. Taylor, eds. *The Biology of Human Starvation*. Minneapolis, MN: University of Minnesota Press, 1950.

Post, J. D. *The Last Great Subsistence Crisis in the Western World*. Baltimore, MD/London: Johns Hopkins University Press, 1977, 109–140.

Tucker, T. *The Great Starvation Experiment: The Heroic Men Who Starved So That Millions Could Live*. New York: Simon & Schuster, 2006.

Turner, B., ed. *The Statesman's Yearbook*. London: Macmillan, 2009, xv.

Malthus, Thomas Robert

Dando, William A. *The Geography of Famine*. New York: John Wiley & Sons, 1980, 193–194.

Fogel, R. *The Escape from Hunger and Premature Death*. Cambridge, UK: Cambridge University Press, 2004.

Harden, G. "Carrying Capacity as an Ethical Concept." In *Lifeboat Ethics*, edited by G. Lucas and T. Ogletree. New York: Harper and Row, 1976.

Harden, G. "Lifeboat Ethics: The Case Against Helping the Poor." *Psychology Today* 8, no. 4 (September 1974): 38–46, 123–124, 126.

Ingram, R. A. *Disquisitions on Population in Which the Principles of the Essay on Population, by the Rev. T. R. Malthus, Are Examined and Refuted*. London: J. Hatchard, 1808, 71.

Lappe, F., and J. Collins. "Food First." In *Global Perspective on Ecology*, edited by T. Emmel. Palo Alto, CA: Mayfield, 1977, 464.

Lappe, F., J. Collins, and P. Rosset. *World Hunger: Twelve Myths*. New York: Institute for Food and Development Policy and Grove/Atlantic, 1998.

Malthus, Rev. T. R. *The Essay on Population*, Vol. 2. London: J. M. Dent & Sons, 1914, 38–46.

Malthus, T. R. *First Essay on Population 1798*. London: Royal Economic Society and Macmillan & Co., 1926, 1–396.

Malthus, T. R. *Population: The First Essay*. Ann Arbor, MI: University of Michigan Press, 1959, 2–13.

Paddock, W., and P. Paddock. *Famine—1975: America's Decision: Who Will Survive*. Boston: Little, Brown, 1967.
Peterson, W. *Malthus*. Cambridge, MA: Harvard University Press, 1979, 1–20.
Smith, K. *The Malthusian Controversy*. London: Routledge & Kegan Paul, 1951, 3–7.
Turner, M. *Malthus and His Time*. New York: St. Martin's Press, 1986, 40–59.

Mapping the Geography of Hunger in the United States

Allen, P. "The Disappearance of Hunger in America." *Gastronomica* 7, no. 3 (2007): 19–23.
Amin, A. *Post-Fordism: A Reader*. New York: Wiley-Blackwell, 1995.
Barfield, J., and R. Dunifon. *State-Level Predictors of Food Insecurity and Hunger Among Households with Children*. Contractor and Cooperator Report No. 13. Washington, DC: Economic Research Service, U.S. Department of Agriculture, 2005.
Barfield, J., R. Dunifon, M. Nord, and S. Carlson. "What Factors Account for State-to-State Differences in Food Security?" *Economic Information Bulletin 20*. Washington, DC: Economic Research Service, U.S. Department of Agriculture, 2005.
Bernell, S. L., B. A. Weber, and M. E. Edwards. "Restricted Opportunities, Personal Choices, Ineffective Policies: What Explains Food Insecurity in Oregon?" *Journal of Agricultural and Resource Economics* 31, no. 2 (2006): 193–211.
Bickel, G., M. Nord, C. Price, W. Hamilton, and J. Cook. *Guide to Measuring Household Food Security: Measuring Food Security in the United States*. Washington, DC: Economic Research Service, U.S. Department of Agriculture, 2000.
Brown, L., and H. F. Pizer. *Living Hungry in America*. New York: Macmillan, 1987.
Citizens' Board of Inquiry into Hunger and Malnutrition in the United States. *Hunger, U.S.A*. Washington, DC: New Community Press, 1968.
Economic Research Service. *Food Security in the United States: Measuring Household Food Security*. Washington, DC: Economic Research Service, U.S. Department of Agriculture, 2010.
Edwards, M. E., B. Weber, and S. Bernell. "Identifying Factors That Influence State-Specific Hunger Rates in the U.S.: A Simple Analytic Method for Understanding a Persistent Problem." *Social Indicators Research* 81 (2007): 579–595.
McMurry, D. "Hunger in Rural America: Myth or Reality? A Reexamination of the Harvard Physician Task Force Report on Hunger in America Using Statistical Data and Field Observations." *Sociological Spectrum* 11 (1991): 1–18.

Nord, M. "Does It Cost Less to Live in Rural Areas? Evidence from New Data on Food Security and Hunger." *Rural Sociology* 65, no. 1 (2000): 104–125.

Nord, M., K. Jemison, and G. Bickel. *Prevalence of Food Insecurity and Hunger, by State, 1996–1998*. Washington, DC: Economic Research Service, U.S. Department of Agriculture, 1999.

Physician Task Force on Hunger in America. *Hunger in America: The Growing Epidemic*. Middletown, CT: Wesleyan University Press, 1985.

Ryu, H. K., and D. J. Slottje. "Analyzing Perceived Hunger Across States in the U.S." *Empirical Economics* 24 (1999): 323–329.

Taponga, J., A. Suter, M. Nord, and M. Leachman. "Explaining Variations in State Hunger Rates." *Family Economics and Nutrition Review* 16, no. 2 (2004): 12–2.

U.S. Congress. *Hunger and Malnutrition in the United States*. Senate Subcommittee on Employment, Manpower, and Poverty, 90th Congress, 2nd Session. Washington, DC: U. S. Government Printing Office, May 23 and 28, 1968/ June 14, 1968.

Watts, M. "Entitlements or Empowerment? Famine and Starvation in Africa." *Review of African Political Economy* 51 (1991): 9–26.

Watts, M. "The Great Tablecloth: Bread and Butter Politics, and the Political Economy of Food and Poverty." In *The Oxford Handbook of Economic Geography*, edited by G. Clark, M. Gertler, and M. Feldman. Oxford, UK: Oxford University Press, 2000, 195–214.

Watts, M., and H. Bohle. "The Space of Vulnerability: The Causal Structure of Hunger and Famine." *Progress in Human Geography* 17 (1993): 43–67.

Williamson, E. "Some Americans Lack Food, But USDA Won't Call Them Hungry." *The Washington Post*, November 16, 2006, A01.

Yardley, J. "Democrats Criticize Bush as Being Out of Touch on Poverty." *The New York Times*, December 24, 1999.

New Energy Sources

Bagotskiĭ, Vladimir S. *Fuel Cells: Problems and Solutions*. Hoboken, NJ: John Wiley and Sons, 2009.

Hazeltine, Barrett, and Christopher Bull. *Appropriate Technology: Tools, Choices, and Implications*. San Diego: Academic Press, 1999.

Hirsch, Robert L., Roger H. Bezdek, and Robert M. Wendling. *Peaking of World Oil Production: Impacts, Mitigation, and Risk Management*. New York: Novinka Books, 2007.

International Atomic Energy Agency. *Energy, Electricity, and Nuclear Power Estimates for the Period up to 2030*. Vienna: Author, 2008.

Leone, Daniel A., ed. *Is the World Heading Towards an Energy Crisis?* Farmington Hills, MI: Greenhaven Press, 2006.

Lior, Noam. "Energy Resources and Use: The Present Situation and Possible Paths to the Future." *Energy* 33 (2008): 842–857.

Muller-Kraenner, Sascha. *Energy Security: Re-measuring the World.* London: Earth Scan, 2008.

Murray, Raymond LeRoy. *Nuclear Energy: An Introduction to the Concepts, Systems, and Applications of Nuclear Energy*, 6th ed. Burlington, VT: Butterworth-Heinenmann/Elsevier, 2009.

Smil, Vaclav. E*nergy in Nature and Society: General Energetics of Complex Systems*. Cambridge, MA: MIT Press, 2008.

Sternberg, R. "Hydropower: Dimensions of Social and Environmental Coexistence." *Renewable and Sustainable Energy Reviews* 12, no. 6 (2008): 1588–1621.

Tertzakian, Peter, and Keith Hollihan. *The End of Energy Obesity: Breaking Today's Energy Addiction for a Prosperous and Secure Tomorrow.* Hoboken, NJ: John Wiley and Sons, 2009.

North American Agroclimatic Regions, 2000–2100 CE

Chang, J. "Agroclimatology." In. *The Encyclopedia of Climatology*, edited by J. Oliver and R. Fairbridge. New York: Van Nostrand Reinhold, 1987, 16–22.

Corcoran, W. T., and E. Johnson. "Climate of North America." In *The Encyclopedia of World Climatology*, edited by J. Oliver. Dordrecht, the Netherlands: Springer, 2004, 525–534.

Dando, William A., and Bharath Ganesh Babu. *Mapping the Shifting Agroclimatic Parameters and Agricultural Regions of North America: A.D. 2100.* Paper presented at the Annual Meeting of the National Council for Geographic Education, Detroit, MI, October 9, 2008.

Dando, William A., and Bharath Ganesh Babu. "Climates of the Future: Case Studies of the Middle East and North America." In *Climate Change and Variation: A Primer for Teachers*, edited by William A. Dando. Washington, DC: National Council for Geographic Education, 2007, 31–48.

Nordhaus, W. D. "Climate Change's Global Warming Economics." *Science* 294, no. 5545 (2001): 1283–1284.

Nuttonson, M. "Ecological Crop Geography of the Ukraine and Agroclimatic Analogues in North America." *Institute of Crop Ecology* 1 (1947): 1–24.

Watson, R. T., et al. *The Regional Impacts of Climate Change: An Assessment in Vulnerability.* IPCC. Cambridge, UK: Cambridge University Press, 1998.

Nutrition

Brown, Judith, Janet Isaacs, Bea Krinke, and Ellen Lechtenberg, *Nutrition Through the Life Cycle*, 4th ed. Belmont, CA: Wadsworth, 2010.

Duyff, Roberta. *ADA Complete Food and Nutrition Guide*, 3rd ed. Chicago: American Dietetic Association, 2006.

Heijmans B. T., E. W. Tobi, A. D. Stein, H. Putter, G. J. Blauw, E. S. Susser, P. E. Slagboom, and L. H. Lumey. "Persistent Epigenetic Differences Associated with Prenatal Exposure to Famine in Humans." *Proceedings of the National Academy of Science of the USA* 105 (2008): 17046–17049.

Popkin, B. M., K. E. D'Anci, and I. H. Rosenberg. "Water, Hydration, and Health." *Nutrition Reviews* 68 (2010): 439–458.

Roseboom, T., and R. Painter. "The Dutch Famine and Its Long-Term Consequences for Adult Health." *Early Human Development* 82 (2006): 485–491.

Organic Agriculture

Ammann, Klaus. "Integrated Farming: Why Organic Farmers Should Use Transgenic Crops." *New Biotechnology* 25 (2008): 101–107.

Avery, Alex. " 'Organic Abundance's Report: Fatally Flawed." *Renewable Agriculture and Food Systems* 22 (2007): 321–323.

Badgley, Catherine, Jeremy Moghtader, Eileen Quintero, Emily Zakem, M. Jahi Chappell, Katia Avilés-Vázquez, Andrea Samulon, and Ivette Perfecto. "Organic Agriculture and the Global Food Supply." *Renewable Agriculture and Food Systems* 22 (2007): 86–108.

Kirchmann, Holger, Lars Bergström, Thomas Kätterer, Olof Andrén, and Rune Andersson. "Can Organic Crop Production Feed the World?" In *Organic Crop Production: Ambitions and Limitations*, edited by Holger Kirchmann and Lars Bergström. Netherlands: Springer, 2008, 39–72.

Lu, Yanhui, Kongming Wu, Yujing Jian, Bing Xia, Ping Li, Hongqiang Feng, Kris A. G. Wyckhuys, and Yuyuan Guo. "Mirid Bug Outbreaks in Multiple Crops Correlated with Wide-Scale Adoption of Bt Cotton in China." *Science* 328 (2010): 1151–1154.

Niggli, Urs. "Organic Agriculture: A Productive Means of Low-Carbon and High Biodiversity Food Production." In *Trade and Environment Review 2009/2010*, edited by Ulrich Hoffmann and Darlan F. Marti. UNCTAD, 2010, 112–118.

Paarlberg, Robert. *Starved for Science: How Biotechnology Is Being Kept Out of Africa*. Cambridge, MA: Harvard University Press, 2008.

Pretty, J. N., A. D. Noble, D. Bossio, J. Dixon, R. E. Hine, F. W. T. Penning De Vries, and J. I. L. Morison. "Resource-Conserving Agriculture Increases Yields in Developing Countries." *Environmental Science and Technology* 40 (2006):1114–1119.

Ronald, Pamela C., and Raoul W. Adamchak. *Tomorrow's Table: Organic Farming, Genetics and the Future of Food*. New York: Oxford University Press, 2008.

Willer, Helga, and Lars Kilcher, eds. *The World of Organic Agriculture: Statistics and Emerging Trends 2009*. Bonn: IFOAM; Frick: FiBL; Geneva: ITC, 2009.

Protein and Protein Deficiency

Bassett, Thomas J., and Alex Winter-Nelson. *The Atlas of World Hunger*. Chicago: University of Chicago Press, 2010.

Gardner, Gary T. *Underfed and Overfed: The Global Epidemic of Malnutrition*. Washington, DC: Worldwatch Institute, 2000.

Haerens, Margaret. *Malnutrition*. Farmington Hills, MI: Greenhaven Press, 2009.

Jacobsen, Kathryn H. *Introduction to Global Health*. Sudbury, MA: Jones and Bartlett, 2008.

Remote Sensing of Food Production

Copenhaver, Ken. *Development of a Monitoring System for the Proactive Detection of Insect Pest Resistance to Transgenic Pesticidal Crops Using Geospatial Technologies*. EPA/NASA Technical Report. Chicago: Energy Resources Center, University of Illinois at Chicago, 2010.

Jensen, John R. *Remote Sensing of the Environment: An Earth Resource Perspective*. Englewood Cliffs, NJ: Prentice-Hall, 2006.

Lu, D., M. Batistella, P. Mausel, and E. Moran. "Mapping and Monitoring Land Degradation Risks in the Western Brazilian Amazon Using Multitemporal Landsat TM/ETM+ Images." *Land Degradation and Development* 18 (2007): 41–54.

Lu, Dengsheng, Paul Mausel, Eduardo Brondizio, and Emilio Moran. "Change Detection Techniques." *International Journal of Remote Sensing* 25, no. 12 (2007): 2365–2407.

Mausel, P., D. Lu, and N. Dias. "Remote Sensing of Amazon Deforestation and Vegetation Regrowth (Succession): Inputs to Climate Change Research." In *The Global Climate System: Pattern, Process and Teleconnections*, edited by Howard Bridgman and John Oliver. Cambridge, UK: Cambridge University Press, 2006, 79–95.

Mausel, P., Y. Wu, Y. Li, E. Moran, and E. Brondizio, "Spectral Identification of Successional Stages Following Deforestation in the Amazon." *Geocarto International* 8 (1993): 1–11.

Wardlow, B. D., and S. L. Egbert. "Large-Area Crop Mapping Using Time-Series MODIS 250 M NDVI Data: An Assessment for the U.S. Central Great Plains." *Remote Sensing of Environment* 112, no. 3 (2008): 1096–1116.

Soil

Challenges for Agricultural Research. Co-operative Research Programme on Biological Resource Management for Sustainable Agricultural Systems, together with the Czech Republic's Ministry of Agriculture. Washington, DC/Paris, France: OECD Publishing, 2011.

Kotkin, J. *The Next Hundred Million: America in 2050*. New York: Penguin Press, 2010.

Soil Survey Staff. "Soil Survey Manual: Chapter One." In *Soil Taxonomy: A Basic Systems of Soils Classification for Making and Interpreting Soil Surveys*. Lincoln, NE: Natural Resources Conservation Service, U.S. Department of Agriculture, 1999.

Soil Survey Staff. "The Twelve Orders of Soil Taxonomy." In *Soil Taxonomy: A Basic Systems of Soils Classification for Making and Interpreting Soil Surveys*. Lincoln, NE: Natural Resources Conservation Service, U.S. Department of Agriculture, 1999.

Verheye, W. H. "Land Cover, Land Use and Global Change." In *Encyclopedia of Life Support Systems*, Vol. I, edited by W. H. Verheye. Washington, DC: National Science Foundation, 2010.

Sustainable Food Production

Bein, F. L. "Response to Drought in the Sahel." *Journal of Soil and Water Conservation* 35, no. 3 (1980): 121–123.

Bein, F. L., and Christopher Hill. "Four Storey Agriculture in the District of Massinga, Province of Inhambane, Along the Mozambican Coast." *Focus on Geography* 53, no. 4 (2009): 62–65.

Bein, F. L., and Gilbert Nduru. *Traditional Agriculture: Indigenous Crops, Production and Utilization in Kenya*. Paper presented at the annual meeting of the Association of American Geographers, Las Vegas, NV, March 2009.

Bonar, James, ed. *Malthus and His Work*. New York: Augustus M. Kelley, 1966.

Dando, William A. *The Geography of Famine*. London: Arnold, 1980.

Diamond, Jared M. *Collapse: How Societies Choose to Fail or Succeed*. New York: Viking Press, 2005.

Dover, M. J., and L. M. Talbot. *To Feed the Earth: Agro-ecology for Sustainable Development*. Washington, DC: World Resources Institute, 2006.

Edwards, Clive A., Rattan Lai, Patrick Madden, Robert H. Miller, and Gar House, eds. *Sustainable Agricultural Systems*. Soil and Water Conservation Society, 1989.

Powel, Keven S. "Pest and Natural Enemy Survey of Kamiali Garden Area." In *Biodiversity Inventory of the Kamiali Wildlife Management Area*, edited by F. L. Bein. Lei, Papua New Guinea: Papua New Guinea University of Technology, 1997.

Teaching Concepts

Overview: Understanding, Measuring, Overcoming Poverty. Washington, DC: World Bank, 2010.

The State of Food Insecurity in the World: Economic Crisis—Impacts and Lessons Learned. Rome: Food and Agriculture Organization of the United Nations, 2009.

Statistics by Area/Child Survival and Health. Rome: UNICEF, November, 2009.

Teaching Data Sources

Bassett, Thomas J., and Alex Winter-Nelson. *The Atlas of World Hunger.* Chicago, IL: University of Chicago Press, 2010.

Beckmann, David, et al. *Hunger Report 2011 Our Common Interest: Ending Hunger and Malnutrition.* Washington, DC: Bread for the World Institute, 2011.

Dando, William, Mary Snow, Richard Snow, Melissa Martin, and Kathleen Kozenski, eds. *Climate Change and Variation: Lesson Plans, Data Sets, and Activities, Vol. II.* Washington, DC: Pathways in Geography Series No. 37, 2008.

Hubbard, Jim. *Lives Turned Upside Down: Homeless Children in Their Own Words and Photographs.* New York: Aladdin Paperbacks/Simon & Schuster for Young Readers, 2007.

Jensen, Eric. *Teaching with Poverty in Mind: What Being Poor Does to Kid's Brains and What Schools Can Do about It.* Alexandria, VA: Association for Supervision & Curriculum Development, 2009.

Kogan, Felix, Alfred Powell, and Oleg Fedorov. *Use of Satellite and In-Situ Data to Improve Sustainability.* New York: Springer, 2010.

National Geographic Student Atlas of the World. Washington, DC: National Geographic Children's Books, 2009.

Parkinson, Claire L. *Earth from Above: Using Color-Coded Satellite Images to Examine the Global Environment.* Herndon, VA: University Science Books, 1997.

Shaw, D. John. *The World's Largest Humanitarian Agency: The Transformation of the UN World Food Programme and of Food Aid.* New York: Palgrave Macmillan, 2011.

Warner, Carl. *Carl Warner's Food Landscapes.* New York: Abrams Image, 2010.

Teaching Definitions

Coyle, L. Patrick. *The World Encyclopedia of Food.* New York: Facts on File, 1982.

Dando, William A., and Caroline Z. Dando. *A Reference Guide to World Hunger.* Hillside, NJ: Enslow, 1991.

DiSalvo-Ryan, Dyanne. *Uncle Willie and the Soup Kitchen.* New York: Morrow Junior Books, 1997.

Kaye, Cathryn Berger. *A Kid's Guide to Hunger and Homelessness: How to Take Action!* Minneapolis, MN: Free Spirit Publishing, 2007.

Lappe, Frances Moore, Joseph Collins, Peter Rosset, and Luis Esparza, eds. *World Hunger: 12 Myths*, 2nd ed. New York: Grove Press, 1998.

Markandaya, Kamala. *Nectar in a Sieve*. New York: Penguin Group, 1982.

Merriam-Webster's Dictionary and Thesaurus. Chicago, IL: Encyclopaedia Britannica Company, 2007.

Park, Linda Sue. *A Long Walk to Water: Based on a True Story*. New York: Clarion Books/Houghton Mifflin Harcourt, 2010.

Pfeffer, Susan. *Life as We Knew It*. New York: Scholastic, 2006.

Williams, Karen Lynn, Khadra Mohammad, and Doug Chayka. *Four Feet Two Sandals*. Grand Rapids, MI: Eerdmans Books for Young Readers, 2007.

Unbounded Food Trade: A Deterrent to Hunger and Famine

Bennett, M. K. *The World's Food*. New York: Harper & Brothers, 1954.

Connor, J. M., and W. A. Schiek. *Food Processing: An Industrial Powerhouse in Transition*. New York: John Wiley & Sons, 1997.

Enzer, S., R. Drobnick, and S. Alter. *Neither Feast nor Famine: Food Conditions to the Year 2000*. Toronto: Lexington Books, 1978.

Global Food Markets. Washington, DC: Economic Research Service, U.S. Department of Agriculture, 2009.

Hebebrand, C., and K. Wedding. *The Role of Markets and Trade in Food Security*. Washington, DC: Center for Strategic and International Studies, 2010.

Hopkins, R. F., and D. J. Puchala. *Global Food Interdependence: Challenge to American Foreign Policy*. New York: Columbia University Press, 1980.

Johnson, D. G. *Trade and Agriculture*. New York: John Wiley & Sons, 1950.

U.S. Agricultural Trade. Washington, DC: Economic Research Service, U.S. Department of Agriculture, 2009.

World Agricultural Supply and Demand Estimates. Washington, DC: Economic Research Service, Foreign Agricultural Service, 2011.

U.S. Farm Machinery Industry in a Global Market

Gerstner, John, ed. *Genuine Value: The John Deere Journey*. Moline: Deere & Company, 2000.

Gibbard, Stuart. *The Ford Tractor Story, Part Two: Basildon to New Holland 1964 to 1999*. East Yorks/Ipswich: Japonica Press/Old Pond Publishing, 1999.

Grooms, Lynn. "Got Food? Can Crop Production Keep Pace with the Growing Global Population?" *Farm Industry News* 43, no. 5 (2010): 34–38.

IBISWorld. *IBISWorld Industry Report: Tractors & Agricultural Machinery Manufacturing in the US*. Santa Monica, CA: IBIS World. 2009.

Marsh, Barbara. *A Corporate Tragedy: The Agony of International Harvester Company*. Garden City: Doubleday & Company, 1985.

McMahon, Karen. "AGCO Shifts Gears." *Farm Industry News* 43, no. 9 (2010): 22, 24.

Petersen, Walter F. *An Industrial Heritage: Allis Chalmers Corporation*. Milwaukee, WI: Milwaukee County Historical Society, 1978.

Stonehouse, Tom, and Eldon Brumbaugh. *JI Case Agricultural and Construction Equipment 1956–1994*, Vol. 2. St. Joseph, MO: American Society of Agricultural Engineers, 1996.

World Agricultural Systems

Amelia, M. "The (Not So) Mysterious Resilience of Russia's Agricultural Collectives." *World Bank's Transition Newsletter*, November/December 1991.

Anderson, J. *A Geography of Agriculture*. Dubuque, IA: Wm. C. Brown, 1970.

Cone, C. A., and Andrea Myhre. "Community-Supported Agriculture: A Sustainable Alternative to Industrial Agriculture?" *Human Organization* 59, no. 2 (2000): 187.

Dando, W. A., and C. Z. Dando. *A Reference Guide to World Hunger*. Aldershot, UK: Enslow, 1991.

Dando, W. A., and J. D. Schlichting. *Soviet Agriculture Today: Insights, Analysis, and Commentary*. Washington, DC: National Council for Soviet and East European Research, 1987.

Fourier, C. *Theory of Social Organization*. New York: C. P. Somerby, 1876.

Grigg, D. B. *The Agricultural Systems of the World: An Evolutionary Approach*. Cambridge, UK: Cambridge University Press, 1974.

Higbee, E. *American Agriculture*. New York: John Wiley & Sons, 1958.

Lockwood, G. B. *The New Harmony Movement*. New York: D. Appleton and Company, 1905.

World Food Day

"Achieving Food Security in Times of Crisis." *World Food Day/TeleFood 2009 Note*. Rome: FAO, 2009, 1–2.

Boustany, Nora. "Warrior for the Hungry Masses." Washington Post Foreign Service, *Washington Post*, October 25, 2008, A12.

Conversations with Patricia Young, U.S. Coordinator of World Food Day, beginning in 1984 and continuing until 2010.
Eide, A. "The Human Right to Adequate Food and Freedom from Hunger." *The Right to Food in Theory and Practice*. Rome: FAO, 1998, 1–5.
Emergencies. Rome: Food and Agricultural Organization of the United Nations, 2008, 1–2.

World Population and Demographic Projections to 2050

Haub, Carl. *2010 World Population Data Sheet*. Washington, DC: Population Reference Bureau, 2010.
ICF Macro. *Demographic and Health Surveys*. ICF Macro. Various issues.
United Nations Population Division. *International Migrant Stock, the 2008 Revision*. Rome: United Nations, 2009.
United Nations Population Division. *World Population Policies, the 2009 Revision*. Rome: United Nations, 2010.
United Nations Population Division. *World Urbanization Prospects, the 2009 Revision*. Rome: United Nations, 2010.

About the Editor and Contributors

Editor

Editor **William A. Dando** is Professor Emeritus, Department of Geography, Geology, and Anthropology, Indiana State University. He has taught courses on the geography of food and famine, as well as on climatology. His interests lie in agriculture, hunger/famine in Russia, climate and food, and application of geotechniques to food, famine, and agricultural problems. He is a member of the Association of American Geographers and the National Council for Geographic Education.

Contributors

Stephen Aldrich is Assistant Professor of Geography, Department of Earth and Environmental Systems, Indiana State University. Dr. Aldrich studies human–environment interaction, focusing on how land conflict affects deforestation in the Brazilian Amazon. He employs geospatial techniques such as remote sensing and geographic information systems in research and teaching, and is particularly interested in GIS for environmental science.

Frederick L. Bein is Professor of Geography at Indiana University–Purdue University, Indianapolis. He has written extensively on traditional agriculture systems in the tropics of Brazil, Africa, and Papua New Guinea. In 2009, he received the campus Chancellor's Faculty Award for Excellence in Civic Engagement, and in 2004–2005, he was a Fulbright Scholar in Mozambique, Faculdade de Agronomia e Florestal, Universidade de Eduardo Mondlane.

Sara A. Blackburn is a Clinical Associate Professor of Nutrition and Dietetics at Indiana University–Purdue University, Indianapolis. She is a Registered Dietitian. Her area of interest is malnutrition in seniors and the critically ill.

Joyce V. Cadwallader is a Professor of Biology at Saint Mary-of-the-Woods College in Indiana. Her interests lie in women in science, history of science, and global health issues.

Christina E. Dando is Associate Professor of Geography and Graduate Studies Coordinator at the University of Nebraska at Omaha. She received her master's degree and PhD in Geography from the University of Wisconsin–Madison. Dr. Dando previously worked in the Alliance Program at the National Geographic Society and as the Education Director for the Association of American Geographers. Her research explores the intersection of landscape, gender, and the media. Dando has published on a range of topics including media depictions of the American West and women and cartography.

Christine Drake is Professor of Geography Emerita at Old Dominion University in Norfolk, Virginia. She has authored several books and published many articles in geographical and other scholarly journals.

Dawn M. Drake has just accepted a teaching position in the Department of Geography at Winona State University in Winona, Minnesota. She received her PhD from the Department of Geography at the University of Tennessee. Her research focuses on developing a new theory in business and management that can better account for the role of place in firm performance, operations, and competitive advantage. Dr. Drake is currently engaged in a project using case studies of the three largest U.S. farm machinery firms to empirically test Porter's theory of competitive advantage.

Bharath Ganesh-Babu is an Assistant Professor of Geography at Valparaiso University in northwest Indiana. His research interest lies in the use of remote sensing and geographic information systems to study the interactions between humans and the environment. He teaches classes in biogeography, environmental conservation, globalization, remote sensing, and GIS.

Gregory Gaston is a physical geographer. He is currently an Associate Professor in the Department of Geography at the University of North Alabama. His dissertation research at Oregon State University focused on using GIS and remote sensing tools to understand and quantify the terrestrial carbon cycle in the former Soviet Union.

Deborah Greenwood is a PhD candidate in the Geography Program at Rutgers University. She received a B.S. in human ecology–public policy and a master's in geography from Rutgers. Ms. Greenwood is a teaching assistant and part-time lecturer in the Human Ecology Department. Her dissertation research focuses on the urbanization and feminization of agriculture—trends that together may give rise to sustainable alternative food production systems in the United States.

Carl Haub is Senior Demographer at the nonprofit Population Reference Bureau in Washington, D.C., where he has been employed since 1979. He has produced the Bureau's annual *World Population Data Sheet* since joining the organization and has worked on demographic projects in Africa, Asia, Latin America, and Europe. He holds a B.A. degree in political science and a M.A. degree in demography from Georgetown University.

John J. Hidore is Emeritus Professor of Geography at the University of North Carolina–Greensboro. He has authored books on physical geography, global environmental change, and climatology; the most recently published is *Climatology: An Atmospheric Science (2010).* Dr. Hidore has served as a visiting faculty member and has participated in research in universities in Israel, Saudi Arabia, Sudan, and Nigeria. In addition to authoring books and contributing to a number of books, he has published articles in 11 different scientific journals.

Ann Myatt James is a PhD candidate in geography at The Pennsylvania State University. As an economic geographer, Ms. James is generally curious about healthy bodies. To guide her inquiries, she draws on a diverse body of scholarship from the social, health, law, and policy fields.

Michael W. Kerwin is Associate Professor of Geography and Director of the Environmental Science and Geology Programs at the University of Denver. Dr. Kerwin's MS, PhD, and postdoctoral research involved reconstructing past temperature and hydrologic variability in northeastern North America and the western United States over the past 12,000 years, and validating predictive climate models that are now being used to predict future changes on Earth.

Kathleen Lamb Kozenski is the Executive Director for the Geography Educators' Network of Indiana, Inc. (GENI). She has provided leadership in writing state geography academic standards, geography/social studies assessment tools, educator licensure requirements, preservice education requirements, and diverse curriculum. Ms. Kozenski's educational background is in agricultural meteorology from Purdue University and in secondary science/math education from Indiana University. Her areas of teaching interest involve water, soils, climate, and interdisciplinary possibilities.

Samuel T. Ledermann is a PhD candidate in geography at Rutgers University. His doctoral research analyzes the socioeconomic impact of organic agriculture in Tanzania. His master's thesis, prepared at the University of Florida, investigated the relationship between agricultural export production and income inequalities in Africa. His other interests include the use of econometric techniques to advance understanding of contemporary development issues.

Robin Leichenko is Associate Professor in Geography and Director of the Initiative on Climate and Society at Rutgers University. Dr. Leichenko's current research focuses on economic and social vulnerabilities to climate change in U.S. cities and regions. Her book *Environmental Change and Globalization: Double Exposures* (Oxford University Press) received the 2009 Meridian Book Award for Outstanding Scholarly Contribution in Geography from the Association of American Geographers.

Vijay Lulla graduated with a doctorate degree in geography from Indiana State University in August 2010. He presently is a postdoctoral research associate at Indiana University–Purdue University, Indianapolis. Dr. Lulla's interests include remote sensing, geographic information systems, statistics, spatial databases, computer programming, and epistemology. Currently he is working with the U.S. Centers for Disease Control and Prevention on modeling vulnerability in urban areas, specifically related to extreme heat events that affect human health.

Melissa Martin, a graduate of Purdue University, is a former seventh-grade (middle school) social studies teacher. She is an active member of the Geography Educators' Network of Indiana (GENI), where she continues to write lesson plans and conduct teacher workshops. Ms. Martin also has completed all requirements and has been awarded the title of National Geographic Teacher Consultant. She is a frequent contributor to GENI's publications, and recently she published in two National Council for Geographic Education books, *Climate Change and Variation: A Primer for Teachers,* Volumes I and II.

Paul Mausel is Professor Emeritus of Geography, Department of Earth and Environmental Systems, Indiana State University (ISU). Dr. Mausel founded and directed the Remote Sensing and GIS Lab at ISU (1974–1988). He has published more than 160 articles and book chapters specializing in remote sensing/ GIS applications and has received approximately $2 million in basic research and science education grants from the National Aeronautics and Space Administration, National Science Foundation, Environmental Protection Agency, and other agencies.

Kacey Mayes, a native of Alabama, is a biologist, ecologist, and geographer who graduated with a BS from the University of North Alabama in December 2010. She currently is looking for a professional position that will allow her to combine her training as a biologist and geographer.

Kausar F. Siddiqi is a Registered Dietitian at Larue D. Carter Memorial Hospital in Indianapolis, Indiana. Her area of expertise is nutrition in mental health. She has been a contributor to the *Indiana Diet Manual*.

Dennis Skelton is President/CEO of SEROS, a Science Education and Research Opportunities for Students nonprofit organization. Dr. Skelton has taught science and computer education for 38 years in the Vigo County School Corporation in Indiana. He is co-author of several nationally distributed NASA CDs and Web-based science materials and serves as co-principal investigator and education consultant on selected science education projects at Indiana State University, including those funded with grants from NASA.

Richard Snow is Associate Professor of Meteorology and **Mary Snow** is Professor of Meteorology in the Department of Applied Aviation Sciences at Embry-Riddle Aeronautical University. Mary is their first female full professor in the college. They have published numerous articles on the environment and climate change and are co-authors of the third edition of *Climatology: An Atmospheric Science* (2010).

Stephen J. Stadler received his PhD from Indiana State University, specializing in applied climatology. Professor of Geography at Oklahoma State University, he is also co-director of the Oklahoma Wind Power Initiative, has served as the President of the Oklahoma Renewable Energy Council, and is the State Geographer of Oklahoma.

Danny M. Vaughn is a national board-certified mapping scientist (Geographic Information Systems/Land Information Systems, R132GS), registered with the American Society for Photogrammetry and Remote Sensing. His research has been in fluvial geomorphology, catastrophic events, and the mapping sciences. He is a former Professor of Geosciences, Director of Geomatics: Applied Mapping Sciences, and Geospatial Analysis Programs at Weber State University. Dr. Vaughn currently works as an independent scholar with Indiana State University, and is a board member of the SEROS (Science Education and Research Opportunities for Students) group, a nonprofit corporation advocating for the advancement of science education.

Andrew Walter is an economic-urban geographer with interests in the ways in which social forces shape human landscapes and how that interaction relates to the well-being of different people in particular places. Specifically, he has studied the relationship of hunger in the United States to geographical political economic change, the everyday geographies of homelessness, the spatiality of farmworker struggles for improved wages and rights, and the processes and effects of faith-motivated neighborhood change in poor urban areas. As a member of the faculty at the University of West Georgia, he teaches courses in human geography, including those focusing on Atlanta's geographies, the geographies of global financialization, and globalization.

Kay Weller is Associate Professor of Geography and Co-Director of the Geographic Alliance of Iowa at the University of Northern Iowa. She has authored many articles in geographical and other scholarly journals. Dr. Weller's entry is based on her personal experiences as a farm owner, growing crops and raising cattle in Kansas.

Donald J. Zeigler is Professor of Geography at Old Dominion University–Virginia Beach. His early publications centered on natural and technological hazards. More recently, he has focused on urban geography and the Middle East. He is co-editor of *Cities of the World* (2008). In 2009, he received the Gilbert Grosvenor Honors Award from the Association of American Geographers.

Index